JAPANESE NAVAL
FIGHTER ACES

0 11557 01167 8

The Stackpole Military History Series

THE AMERICAN CIVIL WAR
Cavalry Raids of the Civil War
Ghost, Thunderbolt, and Wizard
In the Lion's Mouth
Pickett's Charge
Witness to Gettysburg

WORLD WAR I
Doughboy War

WORLD WAR II
After D-Day
Airborne Combat
Armor Battles of the Waffen-SS,
   1943–45
Armoured Guardsmen
Army of the West
Arnhem 1944
Australian Commandos
The B-24 in China
Backwater War
The Battle of France
The Battle of Sicily
Battle of the Bulge, Vol. 1
Battle of the Bulge, Vol. 2
Beyond the Beachhead
Beyond Stalingrad
The Black Bull
Blitzkrieg Unleashed
Blossoming Silk against the Rising Sun
Bodenplatte
The Brandenburger Commandos
The Brigade
Bringing the Thunder
The Canadian Army and the Normandy
   Campaign
Coast Watching in World War II
Colossal Cracks
Condor
A Dangerous Assignment
D-Day Bombers
D-Day Deception
D-Day to Berlin
Decision in the Ukraine
Destination Normandy
Dive Bomber!
A Drop Too Many
Eagles of the Third Reich
The Early Battles of Eighth Army
Eastern Front Combat
Europe in Flames
Exit Rommel
The Face of Courage
Fist from the Sky
Flying American Combat Aircraft of
   World War II
For Europe
Forging the Thunderbolt
For the Homeland
Fortress France

The German Defeat in the East,
   1944–45
German Order of Battle, Vol. 1
German Order of Battle, Vol. 2
German Order of Battle, Vol. 3
The Germans in Normandy
Germany's Panzer Arm in World War II
GI Ingenuity
Goodwood
The Great Ships
Grenadiers
Guns against the Reich
Hitler's Nemesis
Hold the Westwall
Infantry Aces
In the Fire of the Eastern Front
Iron Arm
Iron Knights
Japanese Army Fighter Aces
Japanese Naval Fighter Aces
JG 26 Luftwaffe Fighter Wing War Diary,
   Vol. 1
JG 26 Luftwaffe Fighter Wing War Diary,
   Vol. 2
Kampfgruppe Peiper at the Battle of
   the Bulge
The Key to the Bulge
Knight's Cross Panzers
Kursk
Luftwaffe Aces
Luftwaffe Fighter Ace
Luftwaffe Fighter-Bombers over Britain
Luftwaffe Fighters and Bombers
Massacre at Tobruk
Mechanized Juggernaut or Military
   Anachronism?
Messerschmitts over Sicily
Michael Wittmann, Vol. 1
Michael Wittmann, Vol. 2
Mission 85
Mission 376
Mountain Warriors
The Nazi Rocketeers
Night Flyer / Mosquito Pathfinder
No Holding Back
On the Canal
Operation Mercury
Packs On!
Panzer Aces
Panzer Aces II
Panzer Aces III
Panzer Commanders of the
   Western Front
Panzergrenadier Aces
Panzer Gunner
The Panzer Legions
Panzers in Normandy
Panzers in Winter
Panzer Wedge, Vol. 1
Panzer Wedge, Vol. 2

The Path to Blitzkrieg
Penalty Strike
Poland Betrayed
Red Road from Stalingrad
Red Star under the Baltic
Retreat to the Reich
Rommel's Desert Commanders
Rommel's Desert War
Rommel's Lieutenants
The Savage Sky
Ship-Busters
The Siege of Küstrin
The Siegfried Line
A Soldier in the Cockpit
Soviet Blitzkrieg
Stalin's Keys to Victory
Surviving Bataan and Beyond
T-34 in Action
Tank Tactics
Tigers in the Mud
Triumphant Fox
The 12th SS, Vol. 1
The 12th SS, Vol. 2
Twilight of the Gods
Typhoon Attack
The War against Rommel's Supply Lines
War in the Aegean
War of the White Death
Winter Storm
Wolfpack Warriors
Zhukov at the Oder

THE COLD WAR / VIETNAM
Cyclops in the Jungle
Expendable Warriors
Fighting in Vietnam
Flying American Combat Aircraft:
   The Cold War
Here There Are Tigers
Land with No Sun
MiGs over North Vietnam
Phantom Reflections
Street without Joy
Through the Valley
Two One Pony

**WARS OF AFRICA AND THE
MIDDLE EAST**
Never-Ending Conflict
The Rhodesian War

GENERAL MILITARY HISTORY
Carriers in Combat
Cavalry from Hoof to Track
Desert Battles
Doughboy War
Guerrilla Warfare
Ranger Dawn
Sieges
The Spartan Army

JAPANESE NAVAL FIGHTER ACES

1932–45

**Ikuhiko Hata, Yashuho Izawa,
and Christopher Shores**

STACKPOLE
BOOKS

Published in paperback in 2013 by
STACKPOLE BOOKS
5067 Ritter Road
Mechanicsburg, PA 17055
www.stackpolebooks.com

Cover design by Tracy Patterson

Printed in the United States of America

10 9 8 7 6 5 4 3 2 1

Library of Congress Cataloging-in-Publication Data

Hata, Ikuhiko, 1932–
  [Japanese Army Air Force fighter units and their aces, 1931–1945]
  Japanese naval fighter aces 1932–45 / Ikuhiko Hata, Yasuho Izawa, and Christopher Shores.
     pages cm. — (Stackpole military history series)
  Revision of: Japanese Army Air Force fighter units and their aces, 1931–1945. Grub Street, [2011].
  Includes bibliographical references.
  Translated from the Japanese.
  ISBN 978-0-8117-1167-8
  1. World War, 1939–1945—Aerial operations, Japanese. 2. Fighter pilots—Japan—Biography. 3. Japan. Kaigun. Kokutai—Biography. 4. Naval aviation—Japan—History—20th century. I. Izawa, Yasuho, 1943– II. Shores, Christopher F. III. Hata, Ikuhiko, 1932– Japanese Army Air Force fighter units and their aces, 1931–1945. IV. Title.
  D792.J3H297 2001
  954.54'5952—dc23
                                    2012045636

Contents

Introductory Notes

In turning the text of this book from approximate English into a somewhat more readable narrative for the Anglo-Saxon type of reader, a number of matters became immediately obvious to me. The first thing which became abundantly clear was the manner in which the Imperial Japanese Navy recorded the claims submitted on conclusion of a mission without any particular attempt apparently being made to check and seek to verify or confirm what was reported. In many cases this led to a very high level of over-claiming. This was in no way very different than happened with the initial claims made by many air forces, but does require to be borne in mind.

Although Appendix D endeavours to set out the claims and losses of each side during the bigger engagements in order to allow a more balanced view to be obtained, this listing is frequently incomplete and the sources of the claims and losses listed for the opposing forces are not always clear. I have assiduously inserted the words 'claim', 'claims' or 'claimed' to indicate that the figures should not simply be accepted as indicating precisely what had really happened. In agreement with my Japanese colleagues, I have on an occasional basis indicated what actual losses are now known to be, in order to show the level of overclaiming which had taken place. In a number of cases I have also indicated what Allied claims were in comparison with actual IJN losses, to show that the Japanese pilots were in no way alone in this matter and that their opponents were frequently little less optimistic and culpable in what they believed at the time they had accomplished.

Consideration was given to providing comparable figures in each case (additional to those in Appendix D referred to above), but this was felt to require a degree of re-writing and additional research more than was justifiable, since in many cases these alternative figures can be found in other already-published sources. It should be noted, however, that it has not been possible to provide to the same degree, such 'actual' figures in relation to claims made against the Chinese Nationalist air force during the 1937-1942 period due to a lack of available source documentation. It may reasonably be assumed that the claims made by the IJN units are probably inflated to a similar degree to those made against US and Allied aircraft during the Pacific War of December 1941-August 1945.

The initial translation resulted in all the proper aircrew names ending in the letter 'o', being written as 'oh' in order to reproduce the 'breathy' ending – 'ohhh' – rather than the perhaps flatter and harder Anglo-Saxon 'o'. However, since the majority of books in this language referring to Japanese aircrew have employed the straight 'o' ending, I have reverted them to this basis in order that they may more easily be found in the indices, etc, of existing works.

Aircraft

In referring to IJN aircraft I have employed the Japanese Type number for early types, and the letters/numbers designations for all types, together where appropriate, with their popular Japanese name. I have not employed the Allied codenames applied to them during the war, although I list these here so that there may be no confusion regarding the aircraft identified by the Allied forces.

a) Pre Pacific War types as employed over China only:

Fighters

Nakajima Type 3   A1N
Nakajima Type 90   A2N
Nakajima Type 95   A4N

Dive-Bombers

Aichi Type 94   D1A1
Aichi Type 96   D1A2

Torpedo-Bombers (generally referred to by the IJN as Attack Aircraft)
Mitsubishi Type 13   B1M
Mitsubishi Type 89   B2M
Yokosuka Type 92   B3Y

b) Aircraft Identified by Allied Code Names during the Pacific War

Fighters

Mitsubishi Type 96	A5M		'Claude'
Mitsubishi Type 0	A6M2	Zero-Sen (Reisen)	'Zeke'
Mitsubishi Type 0	A6M3	Zero-Sen 32 (Reisen 32)	'Hamp'
Mitsubishi Type 0	A6M5	Zero-Sen 52 (Reisen 52)	'Zeke 52'

Interceptors (land based)

Mitsubishi	J2M	Raiden (Thunderbolt)	'Jack'
Kawanishi	N1K1-J	Shiden (Lightning Flash)	'George'
Kawanishi	N1K2-J	Shiden-kai (improved Shiden)	'George'
Mitsubishi	J8M	Shusui (Shining Blade)	-

Night Fighters

Nakajima	J1N1-S	Gekko (Moonlight)	'Irving' *
Yokosuka	D4Y2-S	Suisei (Comet)	'Judy' *
Yokosuka	P1Y1	Ginga (Milky Way)	'Frances' *
Nakajima	C6N	Saiun (Irridescent Cloud)	'Myrt' *

*These were all bomber or reconnaissance types converted for night fighting, as also were a fairly considerable number of A6M Zero-Sens.

Dive-Bombers

Aichi Type 99	D3A		'Val'
Yokosuka	D4Y	Suisei (Comet)	'Judy'

Torpedo-Bombers

Yokosuka Type 96	B4Y		'Jean'
Nakajima Type 97	B5N		'Kate'
Nakajima	B6N	Tenzan (Heavenly Mountain)	'Jill'
Aichi	B7A	Ryusei (Shooting Star)	'Grace'

Medium Bombers (land based)

(These were referred to generically as 'Rikkos', and this term was applied to all aircraft of this nature. Thus a Rikko unit might well have been equipped with two types – e.g. G3Ms and G4Ms, or G4Ms and P1Ys, but all would be referred to as Rikkos.)

Mitsubishi Type 96	G3M		'Nell'
Mitsubishi Type 1	G4M		'Betty'
Yokosuka	P1Y	Ginga (Milky Way)	'Frances'

Reconnaissance Aircraft (carrier based)

Yokosuka Type 2	D4Y		'Judy'
Nakajima	C6N	Saiun (Irridescent Cloud)	'Myrt'

Reconnaissance Aircraft (land based)

Mitsubishi Type 98	C5M		'Babs'
Nakajima Type 2	J1N		'Irving'

Float Fighters

Nakajima Type 2	A6M2N		'Rufe'
Kawanishi	N1K1	Kyofu (Mighty Wind)	'Rex'

Floatplanes (reconnaissance/bombing)

Nakajima Type 95	E8N1		'Dave'
Aichi Type 0	E13A1		'Jake'
Kawasaki	E15K1	Shiun (Violet Cloud)	'Norm'
Mitsubishi Type 0	F1M2	Zero-kan	'Pete'
Aichi	E16A1	Zui-un (Auspicious Cloud)	'Paul'

Special Attack Aircaft

	Oka		'Baka'

c) Units:

Administrative Name	Operational Name
i. Carrier Based	
Carrier Division (Koko-Sentai)	Striking Force
Air Group composed of:	
Hikotai (leader – Hikotaicho)	
One each for:	
Carrier Fighter	
Carrier Dive-Bomber	
Carrier Torpedo-Bomber (Attack)	
Organised into:	
Hikobuntai (leader – Buntaicho – Division Officer)	
Divided among each of the carriers of the	
Koku-sentai to form:	Carrier Air Group (Hikotai)
	With:
	Carrier Fighter Unit
	Carrier Bomber Unit
	Carrier Attack Unit
ii. Land Based	
Air Fleet	Base Air Force
Air Flotilla (Koku-sentai)	Air Attack Force
Air Group (Kokutai)	
Hikotai (Hikotaicho – Group Leader) *	
Hikobuntai (Buntaicho – Division Officer)	

Flight Formations	Typical Numbers pre-1944	Typical Numbers 1944-1945
Daitai (squadron)	18-27	16
Chutai (flight)	9	8
Shotai (section)	3	2
Kutai	-	4

The term Buntai was also used; essentially, this means simply a unit of men and was usually equivalent to a Chutai. A Lieutenant could thus be referred to both as a Chutai-cho and a Buntai-cho.

* Until 1944 there was only one hikotai for each main aircraft type in the group. From early in 1944 air groups were re-organised with two or three numbered hikotais, which usually would all operate the same type of aircraft – i.e. fighters, dive-bombers, etc.

d) Ranks

Officers

Japanese	English Translation and Abbreviation		Equivalent US Navy	Equivalent Royal Navy
Taisho	Admiral	Adm	Admiral	Admiral
Chujo	Vice-Admiral	V.Adm	Vice-Admiral	Vice-Admiral
Shosho	Rear Admiral	R.Adm	Rear-Admiral	Rear-Admiral
Taisa	Captain	Capt	Captain	Captain
Chusa	Commander	Cdr	Commander	Commander
Shosa	Lieutenant Commander	Lt Cdr	Lieutenant Commander	Lieutenant-Commander
Dai-i	Lieutenant	Lt	Lieutenant	Lieutenant
Chu-i	Lieutenant (junior grade)	Lt(jg)	Lieutenant (junior grade)	Sub-Lieutenant
Sho-i	Ensign	Ens	Ensign	Midshipman

Non-Commissioned Ranks

Prior to June 1941

		Abbreviations	
English Pronounciation	Japanese Pronunciation	English	Japanese
Flying Warrant Officer	(Koku Heisocho)	Wt Off	Kusocho
Flying 1st Petty Officer	(Itto Koku Heiso)	PO1c	Ikkuso
Flying 2nd Petty Officer	(Nito Koku Heiso)	PO2c	Nikuso
Flying 3rd Petty Officer	(Santo Koku Heiso)	PO3c	Sankuso
Flying Seaman, 1st Class	(Itto Koku Hei)	Sea1c	Ikku
Flying Seaman, 2nd Class	(Nito Koku Hei)	Sea 2c	Niku
Flying Seaman, 3rd Class	(Santo Koku Hei)	Sea 3c	Sanku

Post June 1941

Flight Warrant Officer	(Hiko Heisocho)	Wt Off	Hisoucho
Flight Chief Petty Officer	(Joto Hiko Heiso) *	CPO	Johiso
Flight 1st Petty Officer	(Itto Hiko Heiso)	PO1c	Ippiso
Flight 2nd Petty Officer	(Nito Hiko Heiso)	PO2c	Nihiso
Flight 3rd Petty Officer	(Santo Hiko Heiso) +	PO3c	Sanpiso
Flight Leading Seaman	(Hiko Heicho) *	LdgSea	Hiheicho
Flight Superior Seaman	(Joto Hiko Hei) *	SupSea	Johi
Flight Seaman, 1st Class	(Itto Hiko Hei) *	Sea1c	Ippi
Flight Seaman, 2nd Class	(Nito Hiko Hei) +	Sea 2c	Nihi
Flight Seaman, 3rd Class	(Santo Hiko Hei) +	Sea 3c	Sanpi

* Rank established in November 1942
+ Rank abolished in November 1942

SECTION 1

Japanese Naval Aviation Fighter Operations

THE EARLY PERIOD AND THE SHANGHAI INCIDENT

The Initial Period

Although the history of Japanese Naval Air Force fighter units is deemed to encompass the period 1932-1945, the history of Japanese naval aviation in fact commenced in 1909. During that year the IJN began its involvement in aviation in company with the Army and various other official bureaux with the formation of the Rinji Gunyo Kikyu Kenkyukai (Temporary Military Balloon Research Association). Three years later the Navy parted company from this organisation in order to undertake its own investigations, acquiring two float-fitted aircraft from abroad. The first of these, a Farman, undertook its first flight on 6 November 1912 in the hands of Lt Yozo Keneko.

During the summer of 1914 war broke out in Europe, and at an early date Japan declared her intention to participate on the Allied side. Her early intention was to try and seize the German-leased territory at Qingdao(Tsingtao) on the Shandong Peninsula of mainland China. In support of the warships despatched for this purpose was the *Wakamiya-Maru*, a converted floatplane tender. The four aircraft carried by this vessel undertook their first flights over Qingdao on 5 September 1914, and by the time that the annexation had been successfully completed on 7 November, they had made 50 sorties. Although the Germans had an aeroplane at Qingdao, no aerial engagements occurred, the Japanese floatplanes being involved purely on reconnaissance and bombing duties. This proved to be the only contribution Naval aircraft were to make to Japanese operations during the rest of the war.

In April 1916 the first land-based flying unit, the Yokosuka Kokutai*, was set up at Yokosuka, a large port close to Tokyo. Initially floatplanes were employed, operating from the shore, but an airfield was soon constructed within the base. The main task of this new unit was the education and training of aircrew and the testing of new aircraft. From this beginning, Naval kokutais gradually increased both in number and in scale, so that my 1930 four more had come into being, and plans for organising 17 tais** (units) were underway.

During World War I European aviation had developed greatly and that of Japan had been left very much in the state which it had reached by 1914. In 1921 a British Mission

---

* 'Kokutai'; may be translated as 'Air Group', but it differs in many ways from that of UK or US air groups; no attempt at a translation for this text is made. The abbreviation of 'Kokutai' is 'Ku', which is generally understood by most Western historians and is used here.
** 'tai' (unit) is a numerical standard of air unit and is different from time to time, and from one type of aircraft to another.

1

reached Japan and commenced training the Navy on the various categories of aircraft which by now existed – including fighters. In consequence 50 Gloster Sparrowhawks were purchased from the United Kingdom. Following this Mitsubishi produced the first domestically-constructed carrier fighter, the Type 10, which was designed for the Japanese manufacturer by Herbert Smith, one of the Sopwith Company's senior designers. 128 were to be built, becoming the first carrier-borne fighter of the IJN.

In 1922 *Hosho*, the first Japanese aircraft carrier, was completed and training began with a complement of Type 10 carrier fighters. Two more carriers, *Akagi* and *Kaga*, followed during 1927 and 1928 respectively. These were both large vessels, converted from a battlecruiser and a battleship which had been under construction prior to their conversion and completion for this new role. In 1925 the Special Duty Ship *Notoro* was converted to become a floatplane tender, replacing the now ageing *Wakamiya*. With a speed of only 12 knots/hour, this vessel had become too slow to accompany other warships. At the same time battleships and cruisers were also being fitted to carry floatplanes for reconnaissance and gunnery spotting.

The next fighter to appear was Nakajima's Type 3 carrier fighter (A1N)*** which was based upon the British Gloster Gambit. This aircraft won a competition against machines designed by Mitsubishi and Aichi, and some 150 were to be produced. By the time the new aircraft had been introduced to carrier service, conflict between Japan and China had commenced.

THE SHANGHAI INCIDENT (1932)

On 29 January 1932 the Shanghai Incident broke out and the Japanese 3rd Fleet was despatched to that city, including the 1st Air Flotilla comprising the carriers *Hosho* and *Kaga*. Aboard *Hosho* were ten Type 3 Carrier Fighters and nine torpedo aircraft, while *Kaga* carried 16 Type 3s and 32 torpedo-bombers. Also present was *Notoro* with eight floatplanes, together with one more such aircraft carried by the light cruiser *Yura*.

On 5 February three Type 3 Fighters led by Lt Mohachiro Tokoro and two torpedo aircraft were launched from *Hosho*, being engaged by nine Chinese fighters whilst flying over Zhenru, Shanghai. No claims were made during this first combat, but in fact one Chinese pilot had been seriously wounded although he managed to land safely. Another pilot then took off in his aircraft, but due to the damage it had suffered during the fighting, it crashed, the pilot losing his life.

Two days later the flying units of the 1st Air Flotilla flew ashore to Kunda in Shanghai, and from here on 22 February Lt Nokiji Ikuta, PO3c Toshio Kuroiwa and Sea1c Kazuo Takeo took off in three *Kaga* Type 3 fighters to escort three of the carrier's torpedo-bombers. The formation was attacked by a Boeing 218 fighter (an export version

---

*** Confusingly, the IJN identified its aircraft in two or three different ways. The first would read 95-shiki (Type 95) carrier fighter, and the number represented the year of adoption into service. In the case of the Type 10 carrier fighter and Type 3 carrier fighter, the first came from Year of Tenno (Type 10 derives from Taisho 10 = 1921; Type 3 from Showa 3 = 1928). From Type 89 onwards the number represents the Japanese year (which was 660 years longer than the Western A.D., so that Type 95 denotes the Japanese Year 2595 = 1935 A.D.

The second form of designation employs an alphabetic letter to denote the type of aircraft, one or two numbers denoting the order in which the type entered service, and a further alphabetic letter denoting the company of manufacture. Thus, the A6M indicates a carrier fighter (Type A), the sixth of its type to enter service, and that it is manufactured by Mitsubishi (M). A third basis, adopted circa late 1944, arose from naming the aircraft in, for instance, the British manner. For example, the J2M 'Raiden' (Thunderbolt).

Top: A Nakajima Type 3 (A1N) taking off from *Hosho* during the Shanghai Incident of January 1932.
Bottom: The victors of the first official aerial victory of the IJN. From left to right are Lt Nokiji Ikuta, PO3c Toshio Kuroiwa and Sea1c Kazuo Takeo. Their victim was an export demonstrator Boeing 83.

of the USAAC's P-12), flown by US test pilot Robert Short, who had come to China to demonstrate the aircraft to the Chinese. His fire killed the navigator in the lead torpedo-bomber, but he was then shot down and killed by the three escorting Type 3 pilots who thereby achieved the first IJN aerial victory.

On 26 February six fighters from *Hosho* were led by Lt Tokoro to escort nine *Kaga* bombers to Hangchow. Following the bombing, the Japanese formation was attacked by five Chinese fighters as it headed for its base, the Type 3 pilots claiming three of the interceptors shot down – one each by Tokoro, PO3c Saito and by the second element of three, led by Lt Atsumi. A cease-fire on 3 March brought further such action to a close, and no further actions occurred.

Meanwhile a "17-unit Scheme" had been completed during the autumn of 1931, although only two of these were to be fighter-equipped. To counter the US Navy's expansion programme which was perceived to be underway at this time, the IJN

Top: In the foreground is the carrier *Hosho*, whilst beyond may be seen *Kaga*. This photograph was taken during July 1937.

Bottom: Nakajima Type 95 carrier fighter (A4N) in service with 12 Ku.

introduced the First (in 1931) and Second (in 1933) Expansion Programmes under which 39 units were to be set up by the end of 1937. Just prior to the outbreak of the China Incident, the IJN had formed 35 land-based units including five and a half fighter units. Apart from training units, all flying units had twice the standard numbers of aircrew on establishment. By the close of 1936 the service had to hand 419 land-based and 261 carrier-based aircraft; to operate these there were 1,600 aircrew available for the former and 701 for the latter.

A replacement for the Type 3 had been deemed to be desirable, and for this purpose Nakajima produced the first purely Japanese-designed fighter, a radial-engined biplane which was adopted initially as the Type 60 carrier fighter (A2N1) in April 1932. Following some redesign the aircraft re-appeared as the Type 95 (A4N1) in January 1936, and 221 of the type were produced.

Meanwhile, however, during February 1934 prototype examples were ordered from Mitsubishi and Nakajima of an all-metal monoplane fighter of much more modern design. Following extensive testing, the Mitsubishi aircraft, a radial engined semi-gull-wing monoplane with a fixed, spatted undercarriage, was declared the winner, and was

An example of the first production batch of the Mitsubishi Type 96 carrier fighter (A5M1).

adopted as the Type 96 carrier fighter (A5M). Production commenced early in 1937, Mitsubishi turning out 782 examples whilst other companies added 200 more, built under licence. Although with its open cockpit and fixed undercarriage, the aircraft looked somewhat dated when compared to new fighters appearing in other nations, it was in performance generally their equal, while it was also supremely manoeuvrable. Indeed, even its successor, the redoubtable Zero, was considered to be its inferior in dog fighting. It would subsequently receive the Allied codename 'Claude'.

THE CHINA INCIDENT

The Early Period in China (1937)

On 7 July 1937 a military clash between Japanese Army and Chinese Army units broke out near Beijing (Peking), while early in August a Japanese Marine officer was killed in Shanghai. These incidents were employed as 'Cause célèbres' to allow what rapidly became a full-scale war to develop between the two countries which would last until 1945. Nevertheless, it remained known to the Japanese as 'The China Incident' throughout its course.

Available as the 'incident' broke out, a number of new Kokutais had been recently formed within existing Kokutais. At Saeki Kokutai in Kyushu the 12th Kokutai, equipped with 12 Type 95 fighters, 12 dive-bombers and 12 torpedo aircraft was formed, while at Ohmura Kokutai – also located in Kyushu – the 13th Kokutai also came into existence with 12 new Type 96 Fighters, six dive-bombers and torpedo aircraft.

Thus at the start of the incident, the IJN had available to take part some 84 carrier aircraft, 118 land-based aircraft and 62 floatplanes (including those on ships of the 3rd Fleet). This represented some 65% of the service's 408 first-line machines at the time.

An inter-air force agreement was reached with the Japanese Army on 11 July 1937, whereby the IJN would be responsible for aerial operations over central China. In mid August the carriers *Kaga*, *Hosho* and *Ryujo* were despatched to Shanghai. *Kaga* carried 16 Type 90 fighters, 14 dive-bombers and 22 torpedo-bombers; *Hosho* and *Ryujo* each had aboard 21 Type 90 fighters, 12 dive-bombers and nine torpedo-bombers.

Nakajima Type 95 (A4N1) of 12th Ku over China in the hands of PO1c Yukiharu Ozeki.

As the IJN prepared for an initial strike on Chinese airfields and other targets on 14 August, a typhoon struck the ships offshore, preventing take-off. This allowed the Chinese to strike first, a force of about 40 fighters and bombers attacking the vessels and the Japanese Marine Headquarters in Shanghai. In the conditions prevailing only the single floatplanes from the cruiser *Izumo* and the light cruiser *Sendai* managed to get off, claiming victories over a Curtiss Hawk biplane fighter and a Northrop single-engined bomber. Fortuitously for the Japanese, aerial combat had become one of the items taught to two-seat floatplane crews since 1932, and such aircraft would frequently be launched to intercept incoming raiders.

Two days later on 16 August, another crew from *Izumo* claimed a Hawk destroyed, while on 21st six floatplane crews claimed six, suffering only damage to one of their number – although this was of a fairly serious nature. On 2 September three floatplanes engaged nine Curtiss Hawks, claiming three shot down and one forced down for a loss of a single floatplane. However, as these aircraft were also engaged in bombing and reconnaissance, their rate of attrition was rather high, 25% of the aircraft available and 7% of the crews being lost during a period of just one month. No floatplane aces were produced during the China Incident, but aircraft of this type had gained a good number of early victories. As a result, the idea began to be adopted that float fighters were an inexpensive and speedy expedient which allowed the creation of a floatplane base, rather than having to construct a front line airfield.

With an improvement in the weather, operations by the carrier-based air groups soon got underway, and on 16 August six of *Kaga*'s Type 90 fighters led by Lt Chikamasa Igarashi encountered four Chinese fighters, claiming three of them shot down over Jiangwan, Shanghai. Next day four more such aircraft led by Wt Off Mitsuo Toyoda claimed two further successes. At the end of the month the first of the new Type 96 monoplanes arrived by carrier from the homeland, and on 4 September Lt Tadashi Nakajima led two of these to their first victories, claiming three Curtiss Hawks shot down. On 7 September Lt Igarashi, now flying one of the new aircraft, led his flight of three to claim five victories over Taihu, three of which were credited to Igarashi personally.

Meanwhile the other carrier air groups had also been piling up successes. Lt Tadashi Kaneko leading four Type 90 fighters from *Ryujo* spotted 18 Curtiss Hawks over Baoshan on 22 August and 'bounced' these, he and his pilots claiming six without any of their aircaft suffering a single hit in return. Next day four more *Ryujo* fighters led by Lt(jg) Minoru Suzuki took on 27 fighters in the same area, and despite their numerical inferiority, claimed nine for no loss, three of them by Suzuki. The *Hosho* fighters were less fortunate in finding opponents. Their only success during this initial period amounted to a single twin-engined Martin monoplane bomber, shot down on 25 August by a trio of Type 90s led by Lt(jg) Harutoshi Okamoto.

Mitsubishi Type 96 (A5M) of 12 Ku. The fuselage carries two red bands, which means that it is the aircraft of the buntai leader. The number '3' on the tail indicates the unit with which it was serving.

Early in September Kunda airfield at Shanghai became available for use by the IJN, and at once the 12th Ku (12 Type 95 fighters, 12 Type 94 dive-bombers and 12 Type 92 torpedo-bombers) and the 13th Ku (12 Type 96 fighters, six Type 96 dive-bombers and six Type 96 torpedo-bombers) flew in. They were joined by six Type 90 fighters, six Type 96 fighters, 18 dive-bombers and 18 torpedo-bombers from *Kaga*.

In an effort to neutralise the Chinese fighters defending the Nanking area, Type 96 Fighters from Zhenru, Shanghai undertook 11 sweeps over Nanking between 19-25 September, accompanied by dive-bombers, torpedo-bombers and floatplanes. On 19th, the initial day of the attacks, 19 Type 96 fighters of the 13th Ku led by Lt Shichiro Yamashita, escorted 17 dive-bombers to the area, accompanied by 16 floatplanes. More than 20 intercepting fighters were encountered, the Type 96 pilots claiming 15 and three probables without loss while the floatplane crews claimed 12 for a single loss. By 25 September the IJN units had claimed a total of 42 destroyed and six probables. Following these raids, *Kaga* returned to Japan with its air group embarked.

Aircraft of the Chinese Air Force were now beginning to appear over southern China, and fearing that these might interfere with the Japanese naval blockade, or even launch attacks on Japanese bases in Formosa, attacks by Rikkos (medium bombers) were launched, while both *Ryujo* and *Hosho* took part in an attack on Canton on 21 September. Lt Cdr Yasuna Kozono, the Hikotai leader on *Ryujo*, led six fighters from each carrier to escort torpedo-bombers and dive-bombers to attack two airfields at Canton. Here the escorts engaged ten or more Curtiss Hawks, each unit claiming six shot down. However, during the return flight five *Hosho* fighters force-landed in the sea when they ran out of fuel; all the pilots were rescued safely by Japanese destroyers. A second attack followed, during which nine *Ryujo* fighters escorted the bombers, claiming five more shot down and one probable. Carrier aircraft were to launch further attacks on two following days, but encountered no opposition.

On land around Shanghai the situation became steadily more favourable to the Japanese despite overwhelming numerical odds. During November three Japanese infantry divisions landed in Hangchow Bay, south-west of Shanghai. The Japanese command now believed that if Nanking, the capital of China at that time, could be taken by the end of November, the 'incident' could be concluded satisfactorily. During this period the A5Ms (Type 96s) of the 13th Ku were regularly in action over the Nanking area. The commanding officer, Lt Yamashita, had been obliged to force-land in hostile territory and had become a prisoner of war, leadership of the unit being taken over by Lt Mochifumi Nango. On 12 October the unit's pilots claimed five victories, while on 2 December six pilots fought more than 30 interceptors, claiming 13 shot down.

The 12th Ku, saw no action with its older A4Ns(Type 95s) at this time, becoming engaged in defensive patrols and ground support sorties as the unit began gradually to convert to the superior A5Ms. On 13 December Nanking fell to the Japanese, but far from giving up, the Chinese government fled to Hankow, determined to continue the war from there.

The Later Period in China

Following the fall of Nanking, the 12th and 13th Ku moved forward to Daixiaochang airfield, close by the city, and from there launched attacks on Nanchang and Hankow, both major bases of the Chinese Air Force. Here the Chinese fighter pilots concentrated their intercepting attacks on the leading aircraft of the Japanese formations with some telling effects. On 22 December, although the Japanese pilots claimed 17 victories whilst over Nanchang, they lost Lt Norito Ohbayashi, who had succeeded Lt Nango as buntai leader in the 13th Ku. His replacement, Lt Shigeo Takuma, was lost on 24 February 1938. In the 12th Ku, meanwhile, buntai leaders Lt Ryohei Ushioda was lost on 7 January and Lt Takashi Kaneko on 18 February.

At the end of March 1938 the Rikko units of the 1st Combined Kokutai returned to Japan, and in their stead the 13th Ku became a Rikko (medium bomber) unit. Consequently, the 12th Ku took on all the remaining fighters, increasing in strength to 30 A5Ms and more than 50 pilots.

Action over southern China continued during this period, *Kaga*'s air group making ten attacks between October and the end of December 1937. Only on one occasion was any opposition encountered in the air, when four Chinese interceptors were claimed shot down. By contrast the IJN's floatplanes saw considerable action here at this time, while

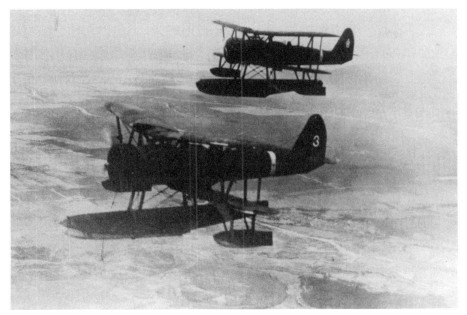

A pair of Nakajima Type 95 reconnaissance floatplanes (E8N1) from the tender *Kamui* in flight over China.

making a similar number of raids on airfields. On 8 January 1938 two floatplanes, one each from the heavy cruiser *Myoko* and the floatplane tender *Kagu Maru*, attacked Nanning for the first time; they were engaged by more than ten fighters, returning to claim seven of these shot down between them! On 24 February aircraft from the tenders *Notoro* and *Kinugasa Maru*, attacked Nanxiong, just after three-seat floatplanes had bombed the target, 15-16 Chinese fighters attacked, but 12 E8N two-seaters intercepted these, claiming eight shot down for the loss of two floatplanes. Lt Kunihiro Iwaki was credited with three of these, although his observer was killed. On return his aircraft was found to contain 138 bullet holes; it was sent home and exhibited in the Togo Shrine in Tokyo.

The tail number '9' indicates that this is an A5M4 of 14 Ku. It was referred to as a '4-go' aircraft, meaning that it came from the largest production batch of these fighters.

During April 1938 the 14th Kokutai with one fighter unit of 12 aircraft, half a dive-bomber unit of six aircraft and one and a half torpedo-bomber units (18 aircraft) was formed in southern China, moving in May to Sanzao Island. April also saw further attacks on southern China targets by carrier aircraft from *Kaga*, which continued until the end of September. On 13 April three A5Ms and three A4Ns fought at least 20 Gloster Gladiators over Canton, their pilots claiming six for the loss of one A5M and two A4Ns. Six carrier fighters engaged 21 Chinese fighters over Nanxiong on 30 August, claiming nine destroyed and two probables for the loss of two aircraft which included that flown by the Japanese formation leader, Lt Hideo Teshima.

In central China fighters from the 12th Ku raided Hankow and Nanchang from Wuhu and Anking, fighting a number of very heavy battles. On 29 April 27 A5Ms escorted Rikkos (Mitsubishi G3Ms) to Hankow where the fighters claimed 28 destroyed and seven probables for the loss of two A5Ms. In bad weather on 31 May 30 A5Ms swept over Hankow, but due to the poor visibility only nine pilots spotted a force of 50 Chinese fighters, 20 of which were claimed shot down without loss. Over Nanchang on 4 July 23 A5Ms which were again escorting G3Ms, claimed 36 destroyed and nine probables shared with the gunners in the Rikkos; only two A5Ms and none of the bombers were lost in return. Thus from the end of April until 19 July 1938 Japanese fighter pilots claimed 100 destroyed and 12 probables in the Hankow area for the loss of just five aircraft.

In the meantime on 25 April 1938 nine fighters, 18 dive-bombers and nine torpedo-bombers from the new carrier *Soryu*, which had been completed in August 1937, moved to Nanking, then to Wuhu, and finally to Anking. The fighters intercepted SB 2 bombers on 25 June, one of which Wt Off Sakae Kato shot down. However, his aircraft then stalled and he crashed to his death. It was thought that he had been weakened by an insufficient supply of drinking water at Anking. On 10 July the *Soryu* units returned to the carrier, but a fighter buntai led by Lt Nango was left behind to form the core of a new 15th Kokutai which had been formed a few days earlier. At this time A5Ms were becoming in short supply, and the new unit was formed with both these aircraft and A4Ns.

A5M4 of 15 Ku with Type 95 (A4N) fighters of the same unit in the background, seen here at Anking, China.

On 18 July Lt Nango led six of the new unit's A5Ms to escort bombers to Nanchang, engaging 11 fighters of which the Japanese fighter pilots claimed seven shot down and two probables. However, the now-renowned ace Lt Mochifumi Nango collided with one of the Chinese aircraft which he had just shot down, and fell to his death. On the same date Lt(jg) Shoichi Ogawa and his four dive-bombers force-landed on an enemy airfield – apparently due to shortage of fuel; the crews successfully burned their aircraft and made good their escape.

The advance of the Japanese Army had resulted in the capture of Hankow on 26 October 1938, but again the Chinese government had gone, this time to Chungking, where the capital was again established. IJN flying units then moved to Hankow, but the capture of Chungking was not planned due to the distances involved. The radius of action of the A5M was 400 km, but the distance from Hankow to this city was 780 km. As a result, the A5Ms were restricted to defensive patrols over Hankow while the Navy's Rikkos and Army heavy bombers raided the new capital and other targets in Sichuan Sheng without fighter escort until summer 1940. With little left to do, the fighter element of the 13th Ku was disbanded on 15 November 1938, while that of the 15th Ku followed on 1 December.

In southern China the occupation of Canton was completed during the autumn of 1938 which meant that there were no chances of aerial combat for the fighters of the 24th Ku. However, in conjunction with fighters from *Akagi* and the 12th Ku, a series of attacks were made on Kweilin (Guilin) and Liuchow. On 30 December 1939 Lt Aioi led 14 fighters from the 14th Ku and others of 12th Ku to Liuchow, engaging Polikarpov I-15 and I-16 fighters, 14 of which were claimed shot down for a single loss. 10 January 1940 saw Lt Igarashi lead 14 of the 14th Ku and 12 of the 12th Ku to Kweilin (Guilin) where 16 were claimed shot down and nine more destroyed on the ground – all without loss.

Debut of the Zero Fighter

During May 1937 an initial request was issued to Mitsubishi and Nakajima for a 12-shi carrier fighter to succeed the manoeuvrable but short-ranging A5M. In October the requirement was formalized, indicating that the new aircraft should have a maximum speed in excess of 500 km/hr, and be able to climb to a height of 3,000 m within three and a half minutes. Perhaps the most difficult requirement was that it should be able to fly for 1.2 to 1.5 hours at full engine power, or for six-eight hours at cruising speed. The armament was to be radically increased from the usual Japanese requirement for two rifle-calibre machine guns, to include not only two such guns of 7.7mm calibre, but also a pair of 20mm shell-firing cannon. Finally, it was to be as agile as the A5M!

In the face of such a challenging brief, Nakajima declined to participate, but at Mitsubishi Jiro Horikoshi, chief designer of the A5M, once more filled this post for the new project. He created a radial air-cooled engine low wing monoplane with a retractable undercarriage, a variable-pitch propeller and a fully-enclosed cockpit, constructed in the main of ultra-duralumin. The first prototype was completed in March 1939, making its maiden flight on 1 April. Initially two early test aircraft suffered mid-air disintegration accidents due to severe vibration, but these faults were remedied with remarkable swiftness, and by the end of July 1940 the aircraft had been adopted as Rei-shiki (Type 0) carrier fighter (A6M). Ultimately, the Zero fighter would have many sub-types from the initial 11, 21 (the first main production sub-type), 32 with shortened wings, 22 and 52 (the most numerous sub-type to be built). Mitsubishi would construct 3,880 A6Ms while Nakajima produced 6,545 more. Though the Zero proved inferior to the A5M in horizontal turning, it was superior in the vertical plane, and in virtually all other respects.

Formation of 12 Ku's initial batch of Mitsubishi Type 0 (A6M) Zero-Sen Model 11 fighters over China.
Inset: On 7 October 1940 the Zero-Sens of 14 Ku returned from Kunming to Hanoi. The A6M in the foreground was flown by Lt Motonari Suho.

In summer 1940 the long-awaited Reisen (Zero fighter) arrived in Hankow where the first six A6Ms arrived on 21 July, led by Lt Tamotsu Yokoyama. They flew in from the Yokosuka Kokutai where they had been service-tested, and would quickly be followed by nine more.

The first mission was undertaken by the new aircraft on 19 August which were able to reach Chungking – unlike their predecessor. Here, however, the Chinese Air Force chose to avoid combat. Finally, on 13 September during the Zero's fourth sortie, 13 pilots led by Lt Saburo Shindo caught more than 30 opposing fighters in the air. The Japanese fighters had passed over the city and then reversed course to catch the Chinese in the air after they had 'scrambled' to intercept. An incredible 27 for no loss was claimed. They were to return regularly, and by the following summer of 1941 the A6Ms of the 12th Ku (joined by elements of the 14th Ku) had claimed 103 victories in the air and 163 on the ground during strafing attacks. Not a single A6M was shot down by hostile fighters during this period, although three fell victim to anti-aircraft fire.

The 14th Kokutai had received nine A6Ms during September 1940, and on 7 October seven of these had accompanied 27 G3M bombers from Hanoi in French Indochina to attack Kunming in the far south-west of China. Led by Lt Mitsugu Kofukuda, these fighters engaged I-15s, I-16s and Curtiss Hawks, claiming 13 shot down in 15 minutes of frenzied combat, while four more fighters were claimed destroyed on the ground; all the A6Ms returned safely. After this operation the unit was not to see any further opposition in the air, but on 12 December seven A6M pilots, guided by a reconnaissance aircraft, flew 630 km to Xiangyun, Kunming airfield, where 22 Chinese aircraft were claimed burned on the ground by strafing.

On 15 September 1941 the 12th Ku in central China and the 14th Ku in southern China were disbanded, the air over China being left entirely to the Imperial Japanese Army Air Force.

THE PACIFIC WAR BEGINS

The Expansion of Navy Aviation

In September 1939 the Second World War commenced in Europe, although as yet Japan and the USA were not participants. However, further expansion of the Imperial Japanese Navy was underway, the number of aviation units now targetted to increase by 75, so that a total of 128 might be reached by 1944.

The keels for the aircraft carriers *Zuikaku* and *Shokaku* had also been laid down during 1938, the two vessels being completed during September and August 1941 respectively. Just before the outbreak of war in the Pacific the conversion of the *Kasuga-Maru* to become the aircraft carrier *Taiyo*, and the construction of the carrier *Zuiho* were commissioned.

Meanwhile, during October 1941 the Tainan Kokutai was formed from within the Chitose Kokutai, equipped initially with both fighters and Rikkos (G3M medium bombers), while the 3rd Kokutai was converted from Rikkos to fighters. The 3rd Kokutai would initially have a strength of 54 fighters (plus 18 spares) together with nine land-based reconnaissance aircraft. Aircraft and pilots arrived at Kanoya (Kyushu), following which the greater part of the unit moved to Takao airfield, Formosa, during October 1941. Here, under the guidance of hikotai leader* Lt Tamotsu Yokoyama, the fighter pilots

---

* The chain of command in a Kokutai was Shirei (Commanding Officer) – Fukuchou (Executive Officer) – Hikocho (Section Leader) – hikotaicho (hikotai leader) – buntaicho (buntai leader).

commenced training in long-distance flights involving minimum fuel consumption. The initial target was to allow a flight to be made to Manila and back. The distance to Luzon in the Philippines where this city was based, was 500 sea miles. The unit subsequently managed to stretch its flying to 1,200 sea miles with a spare 20 minutes to allow for combat over the target. Consequently, Capt Yoshio Kamei, the commanding officer of the kokutai, recommended that the proposed attack on Manila should be made by units from Formosa rather than from carriers. On its formation the 3rd Ku welcomed to its ranks many ex-members of the 12th Ku who had considerable combat experience over China. Indeed, it was recorded that even of those amongst the newcomers who had not already been in action, all had at least 1,000 hours of flying time in their logbooks.

The Tainan Kokutai also received an initial complement of 54 fighters plus 18 spares, backed up by six land-based reconnaissance aircraft. This unit also engaged in extensive training, particularly in regard to long-distance flights under the guidance of its hikotai leader, Lt Hideki Shingo.

At the start of December 1941 the IJN had the following first line units available:

1st Air Fleet aboard six aircraft carriers; this force incorporated 108 Mitsubishi A6M fighters, 126 Aichi D3A dive-bombers and 144 Nakajima B5N torpedo-bombers.

4th Fleet in the central Pacific with the Chitose Ku with fighters and Rikkos; the Yokohama Ku with flyingboats; four floatplane kokutais and one floatplane tender. This force included 36 A5Ms, 38 Rikkos, 24 flyingboats and 54 floatplanes.

11th Air Fleet in Formosa with the 21st and 23rd Air Flotillas, the 3rd and Tainan Kokutais, two and a half Rikko kokutais and a flyingboat kokutai. Also available were three floatplane tenders and the aircraft carrier *Ryujo* for operations over the Philippines. This force totalled 90 A6Ms, 33 A5Ms, 122 Rikkos, 25 transport aircraft, 12 C5M reconnaissance aircraft, 24 flyingboats and 38 floatplanes, while aboard *Ryujo* were a further 12 A5Ms and 12 B5Ns.

22nd Air Flotilla in Indochina which incorporated an attached fighter unit and a further two and a half Rikko kokutais, a flyingboat kokutai at Palau and three floatplane tenders. Available for operations over Malaya, therefore, were 25 A6Ms, 12 A5Ms, 99 Rikkos, six C5Ms and 31 floatplanes.

Additionally, in the homeland and northern Pacific there were 27 fighters, 20 torpedo-bombers and eight floatplanes, while aboard the battleships and cruisers of the various fleets were a further 90 floatplanes.

The Attack on Hawaii
At the end of November 1941 the warships of the 1st Air Fleet gathered at Hitokappu Bay, Etorofu Island, before departing eastwards into the north Pacific. At 0750 on 7 December (Hawaiian time) a force of 43 A6Ms, 89 D3As and 51 B5Ns reached Pearl Harbor, Oahu Island, where Cdr Mitsuo Fuchida, the overall commander of the force, radioed back to the fleet the message "Tora, Tora, Tora" ("I have succeeded in achieving a surprise attack").

The escorting fighter pilots found various aircraft in the air, PO1c Akira Yamamoto from *Kaga* claiming one shot down, while a second was claimed by PO1c Hirano and

Top: One of the Imperial Japanese Navy's first aircraft carriers, *Akagi* is seen here shortly after completion of her original construction. Note that at this stage the vessel has three flight decks, similar to those on some of the early Royal Navy carriers.

Bottom: Kaneohe airfield, Oahu, under attack on 7 December 1941.

US Navy battleships near Ford Island when Oahu, in the Hawaiian Islands, was under attack on 7 December 1941. Note the great water spout caused by a torpedo strike.

PO1c Iwama from *Akagi*. These were both identified as 'training aircraft' but were in fact civilian machines. More importantly, Sea1c Isao Doigawa and PO3c Shin-ichi Suzuki (both from *Soryu*) intercepted 18 Douglas SBD Dauntless dive-bombers from the aircraft carrier USS *Enterprise* (which fortuitously for the Americans, was not in harbour); the Japanese pair claimed three and two shot down respectively for a total of five SBDs. Having covered the bombers in their attacks, the A6Ms then undertook strafing attacks during which one *Akagi* fighter and two from *Kaga* were lost.

Circa 0900 the second wave reached the target with 167 aircraft. The escort comprised 35 A6Ms led by Lt Saburo Shindo, a buntai leader from *Akagi*. Having ascertained that there were no hostile aircraft in the air, each section proceeded to chosen targets and strafed. The section from *Soryu*, led by Lt Fusata Iida, attacked the Kaneohe flyingboat base, but here, as they formed up following their attack, they were 'bounced' by six Curtiss P-36 fighters. Lt(jg) Iyozo Fujita and PO2c Jiro Tanaka each claimed one shot down, but two A6Ms were lost. One of these was the aircraft flown by Lt Iida, which had been hit during his strafing attack. Meanwhile, *Hiryu*'s fighters, led by Lt Sumio Nono, also became engaged in combat whilst strafing, PO1c Tsuguo Matsuyama claiming two US fighters shot down but PO1c Shigenori Saikaichi's A6M was hit and he force-landed on Niihau. Finding himself surrounded by locals, he committed suicide. From this second wave attack two more fighters failed to return – both aircraft from *Kaga*'s section.

The Invasion of the Philippines

Although the Imperial Japanese Army Air Force (IJAAF) had 189 aircraft, including 72 fighters, based on Formosa, these did not have the range capability to attack the southern parts of Luzon where the major US bases were located. Consequently, this area was initially solely the responsibility of the IJN.

On 8 December (actually the same day as the attack on Pearl Harbor, but located on the other side of the International Date Line) therefore, 44 A6Ms of the Tainan Ku led by Lt Hideki Shingo, the hikotai leader, escorted Rikkos to Iba and Clark airfields on Luzon Island, claiming nine aircraft shot down before undertaking a strafing attack which the pilots reported left some 60 aircraft burning on the ground. One A6M was shot down and four more failed to return.

Pilots of 3 Ku at Takao, Formosa, being briefed for the first attack on the Philippines on 8 December 1941.

Iba and Clark were also the targets for 53 3rd Ku A6Ms led by Lt Tamotsu Yokoyama, the hikotai leader. This unit also escorted Rikkos to the area, its pilots claiming seven destroyed and three probables against the intercepting fighters which were encountered. 22 more were also claimed destroyed on the ground, all for the loss of two aircraft and their pilots. USAAF losses during these attacks included nine Curtiss P-40s shot down and others crashed due to fuel shortage, whilst about 49 aircraft of a variety of types were destroyed on the ground.

These two attacks had both been delayed by thick fog over their bases, but the IJAAF bases in the southern part of the island had not been similarly affected, allowing the Army units to strike their targets in northern Luzon as planned. This had worked to the advantage of the IJN, since these earlier attacks had caused US fighters in the south to take off in anticipation of an attack which did not then occur as anticipated. As a result, the American fighters were just about to land, low on fuel, as the Navy units finally arrived. This allowed the majority to be caught on the ground by the Rikkos and strafing fighters.

On 10 December 34 A6Ms from the 3rd Ku were led back to the Manila area by Lt Yokoyama. On this occasion claims were submitted for 42 shot down and four probables in the air, plus a further 42 destroyed on the ground; two of the Zero fighters were lost. The 3rd Ku returned again to southern Luzon on 12 and 13 December. On the first of these sorties eight more aerial victories were claimed, while next day 14 were claimed

A6M Zero-Sens of 3 Ku just prior to take-off on the first mission to Luzon. The aircraft with two fuselage bands in the foreground is believed to be that flown by the formation leader, Lt Tamotsu Yokoyama.

burnt on the ground by strafing, with seven more damaged. On subsequent incursions no further US aircraft were seen. The Tainan Ku also returned to the area on 10th, 12th and 13th, but this unit was able only to claim three destroyed and two probables on 10th , following which only strafing attacks were made.

During 10 December the Japanese Army landed in northern Luzon, securing Vigan and Aparri airfields, while on 12th *Ryujo*, accompanied by some of the floatplane tenders, covered landings at Legaspi on the south-eastern tip of Luzon. This area too was soon secured by the Army. While patrolling over Vigan, Tainan Ku pilots made the first claim for a Boeing B-17 four-engined heavy bomber from a small raiding force of these aircraft – the first claim for such an aircraft to be made by IJN fighter pilots. On 14 December one chutai from this kokutai moved to Legaspi, led by Lt Masuzo Seto.

At Davao on Mindanao, the southern-most of the Philippine islands, 13 B5Ns from *Ryujo* were escorted to attack this base by nine A5Ms led by Lt Takahide Aioi on 8 December, but no aerial resistance was met. During a second wave attack, however, one A5M was shot down by anti-aircraft fire. Units of the Japanese Army landed near Davao on 19th, and within two days the area had fallen into Japanese hands. On 25 December landings were made on Jolo Island, located between Mindanao and Borneo, and this too was quickly secured.

The 3rd Ku began moving to Davao on 23 December, most of the unit having become established there by 29th. From there seven A6Ms were led by Lt Ichiro Mukai and by a C5M as navigator to attack Tarakan in Borneo. Here six intercepting Dutch Brewster Buffalos were claimed shot down.

Starting on 26 December, the Tainan Ku moved to Jolo, 41 of the unit's fighters being based there by 7 January 1942. On 11 January Japanese forces landed at Tarakan, allowing A6Ms to start moving there at once. From there attacks were launched on Balikpapan on 18th and Bandjermasin on 20th. Meanwhile, on 11 January IJN paratroops dropped on Menado in the northern Celebes. As soon as this location was secure, 3rd Ku fighters began moving in next day, and from this latest base were able to launch an attack on Ambon on 19th. These most recent landings were covered by floatplane tenders which made a considerable contribution. On one occasion Mitsubishi F1M biplanes from *Mizuho* were able to claim three Curtiss P-40s shot down over Lamon Bay, eastern Luzon.

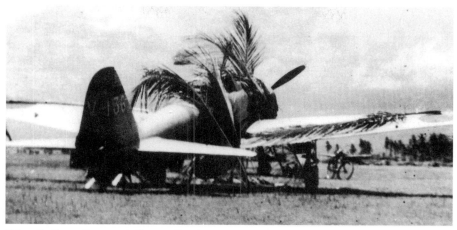

Tainan Ku Zero at Denpasar, Bali, camouflaged with palm fronds. Note 'V-136' on the tail.

The Advance through the Malayan Peninsula

Two capital ships of the Royal Navy had recently arrived at Singapore (the battleship *Repulse* and the battlecruiser *Prince of Wales*), posing a threat to operations planned here which the IJN sought to challenge by air attack. Two and a half kokutais of Rikkos (Mitsubishi G3M and G4M medium bombers) were to be despatched from Indochina for this purpose. To provide escort to these aircraft, one chutai each from the 3rd Ku and the Tainan Ku were detached to the area where they were amalgamated into the 22nd Air Flotilla, becoming "the fighter unit attached to the 22nd Air Flotilla". This unit, which comprised 25 A6Ms, 12 A5Ms and six C5Ms, was to be commanded by Cdr Yutaka Yamada, with Lt Cdr Mohachiro Tokoro as hikotai leader; it would frequently be referred to more simply as 'the Yamada Unit'.

From 6 December the Zero fighters of this unit began patrolling over ships heading for the northern Malay peninsula, while the C5Ms undertook reconnaissance sorties over Singapore. On 8 December patrols were flown over Singora, site of the initial landings, and here Lt Tadatsune Tokaji's element claimed a single Blenheim shot down at around 0930. However, Tokaji's A6M was hit during this engagement and he force-landed in the sea, suffering injuries. Three more A6Ms also came down in the sea, but all four pilots were picked up safely by Japanese vessels.

On 10 December the Rikkos of the 22nd Air Flotilla made their famous attack on the two British capital ships, resulting in the sinking of both vessels. The supporting fighter unit took no part in this operation, however, continuing to patrol over the transport ships further north. On 22 December Lt Kikuichi Inano led nine A6Ms to Miri in northern Borneo to take over the air defence of the Kuching area from floatplanes. Here on 28th the pilots claimed to have shot down all of three Blenheims attacking their airfield. Subsequently, they escorted Rikkos to Ledo airfield on 25th and to Tarakan on 28th.

On 26 December 19 A6Ms and five C5Ms moved from Indochina to Kota Bharu in Malaya where they were shortly joined by the Inano chutai from Miri. From this base the unit became involved in air raids on Singapore from 12 January 1942 onwards. Bad weather during the first three raids resulted in no opposing aircraft being seen. However, on 16 January, as 12 A6Ms escorted G3Ms of the Genzan Ku to the island, Buffalo fighters, estimated as more than 20 strong, were encountered, claims being submitted for nine shot down plus one probable. There is no record of any Buffalos being lost in combat on this date. In return only the navigating C5M which had accompanied the unit, was lost. Thereafter the two chutais took it in turns to escort formations of Rikkos to Singapore on a daily basis. When these raids ceased with the fall of the island, the unit had flown 150 sorties, claiming 40 victories for the loss of two A6Ms, one C5M, and two of the Rikkos in their care.

Air Battle over the Dutch East Indies

On 24 January 1942 Japanese forces landed on Balikpapan, Borneo – famous for its oilfields. Tainan Ku A6Ms patrolled over these landings, and as soon as the area was secure, the unit moved there itself, joined by the Headquarters of the 23rd Air Flotilla. Quickly, the greater part of the 3rd Ku also reached this new base, initially coming under the command of the Tainan Ku.

On 3 February Lt Yokoyama of the 3rd Ku led 27 A6Ms and two C5Ms to Surabaya in Java. Here a sizeable force of Dutch and US fighters intercepted, but suffered heavy losses. The IJN pilots claimed 37 destroyed and five probables in the air, then adding claims for 13 more burned and six badly damaged on the ground; one C5M and three A6Ms failed to return.

During the previous day 17 Tainan Ku pilots had operated over Maospati, where two aircraft were claimed shot down, but on 3rd, while the 3rd Ku was attacking Surabaya, 14 Tainan Ku A6Ms and a single C5M were led by Lt Shingo to Malang. Here two more aircraft – one of them a B-17 – were claimed in the air, plus seven more aircraft, five of which were B-17s, on the ground; one A6M was lost. Thus during these two major raids on 3 February the IJN had achieved almost complete air superiority over southern Java; Allied losses actually amounted to some 16 fighters, but for the restricted forces available to them, that was quite bad enough.

Further north 21 A6Ms and four C5Ms of the 22nd Air Flotilla attached unit moved to Kuching on 5 February. Meanwhile, nine of this unit's A5Ms were transferred to the Bihoro Ku, moving first to Kuantan and then to Kahang. On 9 February 15 A6Ms and a C5M from the main body of the unit attacked Batavia in northern Java, and capital city of that island. 12 intercepting fighters were claimed shot down and eight more burned in strafing attacks before the unit returned to Kuching unscathed.

This unit's A6Ms patrolled over transport vessels heading for Palembang, Sumatra, on 13th and 14 February, claiming ten raiding Blenheims shot down during these two days. On 23 February a move was made to newly-occupied Muntok. From here two days later the unit's fighters escorted Genzan Ku G3Ms on a raid during which four victories were claimed against interceptors for the loss of one A6M and a C5M.

At the other end of Java the Tainan and 3rd Ku fighters were active on a daily basis. On 5 February 3rd Ku pilots claimed nine destroyed and two probables over Bali Island, while on 19th the Tainan Ku claimed 14 shot down and three more probably so over Surabaya for the loss of Lt Asai. On 20 February the Tainan Ku moved to Denpasar on Bali, followed by the 3rd. The latter unit took part in the sinking of the US aircraft carrier *Langley* on 27 February. Early in March the Allied forces in Java surrendered.

During February the 3rd Ku also supported the advance of the Japanese Army through the eastern Dutch East Indies. This took them from Kendari in the southern Celebes to Koepang, Timor and Ambon, undertaking patrol, reconnaissance and ground attack sorties. With the close of the Java campaign, the unit was stationed at Kendari and Koepang, ready for attacks on Allied air units in Australia.

The Yamada Unit with the 22nd Air Flotilla moved from Muntok to Saigon during mid March, the A5Ms and A6Ms then patrolling over the Rikko bases and sea lanes, but seeing no action. 20 aircraft were then moved to Tavoy in southern Thailand to undertake patrols over the Andaman Islands in the Indian Ocean from 29 March-3 April. Again, no opposition was encountered, and finally the unit moved to Bangkok where it became the fighter group of the Kanoya Ku.

Aircraft Carrier Attacks on Australia, the Dutch East Indies and Ceylon

Following the Pearl Harbor attack, the 1st Air Fleet returned to Japan. The 1st and 5th Air Flotillas then co-operated in the occupation of Rabaul, New Britain, and Lae, New Guinea. The 1st then rendezvoused with the 2nd Air Flotilla to undertake an attack on Port Darwin in north-west Australia on 19 February 1942. Here nine *Hiryu* A6Ms provided direct close escort to the attacking bombers, while nine more fighter pilots from Akagi claimed four aircraft shot down as did nine from *Kaga*. A similar number from *Soryu* claimed only one in the air but then strafed three airfields, claiming 26 more aircraft destroyed on the ground, all for the loss of just one A6M.

Thereafter from the end of February the carrier-borne aircraft attacked Allied destroyers and transports off southern Java, also making one major attack on the port of Tjilatjap on the southern coast of the island during which no aerial combat occurred. At that stage *Kaga*, which had suffered a minor scratch to her bottom when she touched a reef prior to the Darwin raid, departed on 15 March for repairs in Japan.

The main carrier force than attacked Ceylon, the main base of the British Eastern Fleet following the fall of Singapore. For this operation the 1st Air Fleet was joined by the two carriers of the 5th Air Flotilla which had been held in Japan in case of attack by US carriers. These ships set off for Celebes on 17 March, where they rendezvoused with the other three vessels and their covering warships, departing on 26 March for their new target. Colombo was raided on 5 April, but found the defences alerted already. 36 A6Ms were heavily involved, claiming 40 fighters, a Defiant and ten torpedo-carrying Swordfish shot down for the loss of a single A6M from *Soryu*. This carrier's eight surviving pilots claimed 11 and three probables, *Akagi*'s nine claimed nine and seven probables, while *Hiryu*'s nine claimed 19 and two probables. On this same date Japanese dive-bombers also sank the two heavy cruisers, *Cornwall* and *Dorsetshire*.

The first aircraft carrier to be sunk by air attack alone was the Royal Navy's HMS *Hermes*, seen here under the fatal attack off Ceylon (Sri Lanka) on 9 April 1942.

Trincomalee was then attacked on 9 April, where ten *Shokaku* A6M pilots claimed 23 aircraft shot down for one loss, ten more from *Zuikaku* claiming 20 for the loss of two A6Ms. While the attack was underway, nine Blenheims attacked *Akagi* and the heavy cruiser *Tone*. Combat Air Patrol (CAP) A6Ms managed to claim five of these shot down, the two IJN vessels suffering no damage. Meanwhile, the British aircraft carrier *Hermes* had been spotted and was sunk by D3A dive-bombers. Returning from escorting this attack, one A6M pilot managed to shoot down one more fighter.

Following these operations the 1st Air Fleet returned to Japan.

AIR BATTLES OVER THE EASTERN AND NORTH-EASTERN PACIFIC

Wake Island

While the front line in the Pacific Ocean ran between the Marshall Islands and Wake Island, the 24th Flotilla in the Marshalls had only 36 A5Ms of the Chitose Ku as its fighter force. Rikkos from this kokutai undertook 'softening-up' raids on Wake, but the A5Ms lacked the range to accompany them. A handful of Grumman F4Fs and the coastal batteries on Wake repulsed the first attempt by the Japanese to land on the island. However, the 2nd Air Flotilla aboard *Soryu* and *Hiryu* then lent a hand, wiping out the surviving defending US Marine fighters during 21-23 December 1941 (two F4Fs were shot down, the rest of VMF-211's aircraft being written off on the ground). The island was then invaded and occupied, as were Guam and Tarawa Atolls in the Gilbert Islands, the latter without any aerial combat occurring.

Taken by the observer aboard a dive-bomber from the carrier *Hosho*, this photograph shows *Hiryu* sinking during the Midway battle on 6 June 1942.

US carriers launched a series of attacks on Marshall Island targets on 1 February 1942, closing in sufficiently for accompanying cruisers to undertake a bombardment. Over Roi (Kwajalein Atoll), 12 A5Ms intercepted an attacking force of USN aircraft, claiming five shot down, while at Maloelap 22 A5M pilots claimed nine destroyed and three probables, although one Japanese fighter was shot down; the pilot baled out safely. USN losses amounted to four Douglas SBD Dauntless dive-bombers over Roi and one over Maleolap, while claims for three Japanese fighters were made. 13 A5Ms then bombed the cruiser USS *Chester*, claiming two bomb hits. Only a few G3M Rikkos were available still and these counter-attacked. The bomber flown by Lt Kazuo Nakai struck the edge of the deck of the carrier USS *Enterprise* when falling after being shot down.

Midway
The IJN planned to take both Midway Island and the Aleutian Islands at the same time. To achieve this the available aircraft carriers were divided into three forces. That to assault Midway, Nagumo Force, departed Japan on 27 May, launching the opening attack during the morning of 5 June (Japanese time). Nine A6Ms from *Akagi* were led by Lt Shirane to engage defending interceptors, 11 of these being claimed shot down, the pilots also reporting strafing and destroying a B-17 on the ground; one A6M fell to AA fire. Escorting the attacking bombers were nine A6Ms from *Kaga* and a similar number from *Hiryu*. The *Kaga* pilots claimed six more interceptors shot down, while Lt Shigematsu's *Hiryu* force claimed an astonishing 18 more.The defending US Marine Corps fighters based on the island actually lost 15 F4Fs and F2As shot down and seven more damaged during these attacks.

However, the Americans had broken the Japanese Navy codes and had been able to decipher these to an extent whereby they were well aware of Japanese intentions. Patrol aircraft rapidly found Nagumo Force, and attacks then soon followed each other in rapid succession. Initially, the superior IJN fighters appeared on the point of inflicting another defeat upon their opponents. Intercepting these raids, *Akagi*'s fighter unit claimed 51 victories during a total of 12 different raids, albeit 30 of them shared with the pilots of other carrier groups. *Kaga* fighters claimed 32 in 32 sorties, while *Soryu*'s men claimed another 32 in only 18 sorties. Raiders included USN Grumman TBF Avengers and USAAF Martin B-26 Marauders from Midway, five TBFs and two B-26s being lost, and three damaged. A second attack by 16 USMC SBDs resulted in eight lost and six damaged, while 11 Vought SB2Us which followed suffered four losses. Attacks by the three US Navy carrier air groups resulted in the losses of 37 Douglas TBD Devastators, 16 SBDs and 14 F4Fs – although two of the SBDs and ten F4Fs were actually lost when they ditched out of fuel after failing to find the Japanese carriers. Thus some 74 American aircraft had been lost to a combination of fighters and AA fire compared with something over 100 claims by the defending Zero pilots.

But while intercepting the incoming TBD torpedo-bombers at low level, the defenders failed to stop a strike by SBD dive-bombers which approached at higher altitude and which gained hits on *Kaga*, *Akagi* and *Soryu*, all of which began to burn fiercely. In that moment the issue of the battle was determined.

Only *Hiryu* was in a position to fight on, and she sent out two waves of bombers which badly damaged the carrier USS *Yorktown*. While escorting the dive-bombers on these attacks, four A6M pilots claimed seven US aircraft over their own fleet, but lost three of their number, only Lt Shigematsu, the formation leader, returning. Six more A6Ms, including two from the stricken *Kaga*, escorted the torpedo-bombers. Finding themselves up against an estimated 30 defenders, the pilots claimed 11 shot down but suffered two losses, including that of their leader, Lt Mori.

Other *Hiryu* fighters continued to fly sorties in defence of the remains of Nagumo Force, claiming 43 more victories in 33 sorties for the loss of five pilots. However, their carrier too was hit during the afternoon and by next day all four carriers had sunk – a staggering defeat for the previously triumphant Japanese. At the close of the battle total US aircraft losses to all causes amounted to 150, 109 of them carrier-based types.

Battles over the Aleutians

As already mentioned, the IJN attempted to occupy the western Aleutian Islands while also engaged in the Midway operation. Air groups from the carriers *Ryujo* and *Junyo* attacked Dutch Harbour during the early morning* of 4 June, A6M pilots from *Junyo* claiming victories over two Consolidated PBY Catalina flyingboats. These attacks proved insufficient, and were repeated next day when five A6Ms, again from *Junyo* and led by Lt Yoshio Shiga, engaged two formations of P-40s, claiming seven shot down for the loss of two D3A dive-bombers. *Ryujo* A6M pilots claimed one PBY, but one of their number was hit by AA and force-landed. This aircraft was salvaged by the Americans, and was later repaired, becoming the first flyable Zero available to them for testing and analysis. Following these two raids, the two carriers returned to Japan.

Despite the defeat at Midway, the occupation of Attu and Kiska Islands in the Aleutians was undertaken on 8 June and a floatplane base was prepared at Kiska on the same day. US raids commenced from 12 June, and early on a destroyer was damaged and a transport sunk by these. On 5 July a float fighter unit belonging to the Toko Ku (one of the flyingboat units) moved to Kiska on 5 July with six A6M2Ns. These were Nakajima-built A6M Zero fighters fitted with floats. These aircraft were soon to be faced by a USAAF force which grew to three bomber squadrons, four squadrons of fighters and a squadron of flyingboats. The small Japanese floatplane unit was to be the sole opposition for this force since the Japanese had not intended to build an airfield in the Aleutians.

The first interception by the A6M2Ns occurred on 8 July, when their first victory was claimed over a Consolidated B-24 Liberator four-engined bomber. On 5 August the unit became the 5th Kokutai and had attached to it a half-unit of three-seat floatplanes, ten E13As reaching Kiska during the month. Air battles continued in low cloud and foul weather; during August one victory was to be claimed as a result of five engagements. On 8 August Kiska was bombarded by a US force of cruisers and destroyers during which one floatplane was destroyed.

From September onwards US forces attacked Kiska from low altitude and aerial fighting became quite intense. On 14th the float fighters claimed a Lockheed P-38 Lightning probably destroyed, while next day four A6M2Ns intercepted 12 heavy bombers and 28 fighters, claiming five destroyed and a probable, but lost three aircraft and two pilots, leaving only one of the unit's aircraft available next day. During this particular engagement PO2c Gi-ichi Sasaki had claimed four of the victories.

On 25 September *Kimikawa Maru* reached Kiska to deliver six new A6M2Ns and two E13As. During the next ten days the pilots of these claimed three victories, two probables and four damaged during nine fights, but lost four pilots and all the A6M2Ns.

The 5th Ku was renamed 452nd Kokutai on 1 November 1942, and during this period half of all the A6M2Ns produced were sent to the Aleutians. This was not, however, the

---

* Dates and times used here relate to Japanese Standard Time which was in use by the Japanese forces at the time. The Aleutians are to the east of the International Date Line, so the date on site is minus 1 from the description.

A6M2Ns of 5 Ku on the shoreline at Kiska, Aleutians, being battered by high waves.

level of reinforcement which might be imagined, since at this time only 12 per month were coming out of the factory!

Kimikawa Maru returned on 6 November, but docked at Attu since the danger of attack by US ships and aircraft at Kiska was too great until the float fighter cover had been replaced. However all the new aircraft delivered were almost immediately destroyed during a storm next day and a strafing attack on 10th. A further replenishment trip by *Kimikawa Maru* reached Attu on 25 December, carrying seven A6M2Ns and replacement aircrews, six of the float fighters moving over to Kiska next day.

On 31 December 1942 and 1 January 1943 the new aircraft undertook interceptions, the pilots claiming two victories on each occasion, and this time without loss. Nonetheless, appalling weather on 4 January reduced the number of serviceable aircraft to nil again. Two A6M2Ns were rendered available by repair, and on 24 January the pilots of these spotted US vessels at Amchitka Island. If this island should fall into the enemy's hands and an airfield be constructed there, it was immediately realised, this could be very serious as it was only 160 km from Kiska and could render its continued occupation unviable. With this threat before them, the two float fighters were each loaded with a pair of 60 kg bombs, and so armed made attacks on three occasions during the closing days of January.

On the last day of the month six new A6M2Ns and one E13A were delivered to Attu by the ubiquitous *Kimikawa Maru*, all flying to Kiska next day. However, two were shot down by AA on 2 February whilst attempting to attack US shipping at Amchitka. The survivors were involved in interceptions during February but achieved only one victory. On 19th the attack on Amchitka was repeated, but two pilots were lost, one of them being the recently-promoted PO1c Gi-ichi Sasaki, who had claimed four individual and five shared victories, plus one shared probable.

From February onwards, with few chances of resupply, Lt Yamada restricted activities only to those circumstances which appeared to offer favourable results. On 17 March seven A6M2Ns were engaged in a fight with ten P-38s during which the Japanese pilots claimed two shared victories for the last successes over Kiska. In May 1943 a US force landed on Attu and from Kiska, now isolated, the remaining pilots of the 452nd Ku were evacuated by submarine, returning to Yokosuka to reform. The float fighter pilots had claimed 15 victories and five probables for the loss of 12 aircraft and ten pilots over the Aleutians. In total all 35 A6M2Ns that had been sent to Kiska and Attu had been lost, a large part of these losses being due to the very adverse weather conditions in this demanding area.

AIR WAR OVER NEW GUINEA AND THE SOLOMON ISLANDS

The Capture of Rabaul
With Truk Atoll as the major Japanese naval base, the IJN planned to take Rabaul and Kavieng in the Bismarck Islands, which could otherwise become a threat to that base. Fighters and Rikkos of the Chitose Kokutai were therefore moved to Truk from where Rikkos began raiding Rabaul from 4 January 1942. The potential for Rabaul to become a threat was highlighted when RAAF bombers from there bombed Truk during the night of 16 January as a riposte to the IJN's raids.

Rabaul and Lakunai airfields, New Britain.

To support the occupation of Rabaul and Kavieng, the carriers *Akagi*, *Kaga*, *Zuikaku* and *Shokaku* were despatched, raiding Rabaul on 20 and 22 January, and Kavieng on 21st. A6M pilots shot down four out of five intercepting Australian fighters (which were in fact only Wirraways – an armed version of the AT-6 Harvard trainer) during the first attack. *Zuikaku* and *Shokaku* attacked Lae and Salamaua in eastern New Guinea on 21 January, while units of the Japanese Army landed at Rabaul and Kavieng on 22nd, securing both next day.

To defend Rabaul 16 A5Ms of the Chitose Ku had been carried aboard *Zuikaku* and *Shokaku*, and these took off on 26 January, although bad weather caused them to turn back and land back aboard. Off again next day, their pilots found Rabaul hidden by dense cloud, so landed at Kavieng, moving to their original destination on 31st when the weather improved.

Thirteen more A5Ms led by Lt Shiro Kawai had been withdrawn from the Tainan and 3rd Ku just before the outbreak of war and had been despatched to Peleliu for the air defence of the Palau Islands. These too were now sent down to Rabaul.

Rabaul was raided during the night of 24 January and again by night on 3 February. Against these raids some A5Ms were scrambled, PO1c Hiroyoshi Nishizawa claiming one shot down for his first success, while Lt Shiro Kawai claimed a probable.*

On 10 February the 4th Kokutai with fighters and Rikkos was formed at Rabaul, and the fighter units of the Chitose Ku and Kawai Unit became the fighter element of this kokutai. Next day four A5Ms that had moved to Surumi (Gasmata) in New Britain which had just been occupied, intercepted three bombers, the pilots claiming all shot down for the 4th Ku's first victories; on 13th they claimed a Hudson over Surumi.

Shoho delivered seven A6Ms to Rabaul on 15 February, bringing in another 19 on 9 March. The 4th Ku gradually converted to these aircraft, PO2c Motosuna Yoshida claiming the first A6M victory here by shooting down a B-17 on 23 February. After undertaking escort duty over forces heading for Port Moresby on 13th (see below), 4th Ku A6Ms moved to Lae on 8 March, claiming a Hudson shot down there on 11th. Five A6Ms attacked Port Moresby on 13th, 12 A6Ms escorting Rikkos to Horn Island next day. Here six of the fighters were led by Lt Kawai to strafe while the other six, including one flown by PO1c Nishizawa, engaged at least ten P-40s, claiming six destroyed and two probables for the loss of two Zeros.

4th Ku fighters were to claim 11 more destroyed and two probables by the end of the month, including six and a probable over Port Moresby. On 1 April all the unit's fighters were transferred to the Tainan Ku, the 4th Ku then becoming an all-Rikko unit.

Air Battles over Eastern New Guinea

On 1 April 1942 the only fighter unit at Rabaul was the Tainan Ku (45 fighters and a few land-based reconnaissance aircraft). On paper there was also supposed to be the Yokohama Ku, a flyingboat unit, which had a float fighter buntai with nine A6M2Ns. It was not until 3 June, however, following the formation of the Yokosuka Ku, that these floatplanes finally arrived at Rabaul.

The main bodies of the Tainan Ku were still on Bali Island and in the Philippines at this time, and a number of veteran pilots including Lt Hideki Shingo, the hikotai leader, were transferred to other units at this time, including training units at home in Japan. The remaining members of the unit moved to Rabaul by transport ship during the following two weeks where they joined the ex-4th Ku aircrews. During this period the ex-4th Ku pilots claimed 20 victories and a probable for the loss of two pilots during two attacks on Port Moresby and four interceptions over Lae.

The main body of Tainan Ku pilots took part in attacks on Port Moresby from 17 April, and by 25th of that month they had to hand six A5Ms and eight A6Ms at Rabaul plus 24 A6Ms at Lae. The unit would not be brought up to full strength until August, and for the time being could only count on being able to put up 20 aircraft at a time. While the majority of pilots thus served at Lae, they were despatched alternatively to Rabaul for

---

* This was in fact a Catalina which was badly damaged and landed back at base. See *Bloody Shambles II*.

a rest from operations.

Between April and the end of July Tainan Ku pilots were involved in attacks on Port Moresby 51 times, flying 602 sorties over the target area, claiming 201 shot down and 45 probables. They also made claims over Lae, Rabaul and a convoy to Buna, the unit's total score reaching around 300 victories during this period.

During the latter part of May 15 A6Ms from the 1st Ku and the Chitose Ku, led by Lt Joji Yamashita, came under the temporary command of the Tainan Ku in New Guinea until 27th of the month. At that date some of the pilots of these units were transferred to the Tainan Ku, the rest returning to their original units in the Marshall Islands. Those joining the Tainan Ku at this time would remain with it until its return to Japan during November.

Battle of The Coral Sea

The Japanese command now planned to take Port Moresby by an invasion from the sea, which it was hoped would also bring out the US aircraft carriers and allow them to be destroyed. In preparation for these actions, during the early morning of 3 May three fighters and three torpedo-bombers from the small carrier *Shoho* supported landings at Tulagi in the southern part of the Solomon Islands. The carrier then rendezvoused with transport vessels carrying the invasion forces, her fighters then patrolling overhead. Just after 0900 on 7 May 93 USN carrier aircraft attacked the transports, *Shoho* herself receiving seven torpedos and 13 bombs which caused her to sink within 20 minutes. Six of her fighters on CAP claimed four destroyed and a probable (USN admitted three losses) but lost three A6Ms while the three others landed on a nearby island from which the pilots were later rescued. Despite this loss, *Shokaku* and *Zuikaku* were unable to launch counter-strikes due to bad weather.

On the next day (8 May) the first naval battle occurred between carriers on both sides during which there were no sightings of opposing ships other than from the air. It was named the Battle of the Coral Sea almost at once. As IJN bombers and torpedo aircraft headed to attack the US vessels, nine escorting A6Ms from *Shokaku* claimed 30 American aircraft shot down. Three more fighters from *Zuikaku* escorting torpedo-bombers from that carrier, plus six more giving cover to other torpedo-bombers from *Shokaku*, claimed a further 26 destroyed and three probables.

Over the Japanese carriers ten *Zuikaku* A6Ms intercepted attacking US carrier aircraft, claiming 13 fighters, six torpedo-bombers and five dive-bombers, two of these 24 claims being classified as 'probables'. In all the day's operations, only one A6M was lost, this aircraft force-landing in the sea. During the USN attacks a further 21 claims were submitted by the pilots of nine defending *Shokaku* fighter pilots to bring total IJN fighter claims for the day to 99 plus five probables; US losses in fact amounted to 33 aircraft in combat, although a further 36 went down with the *Lexington*. Amongst the *Shokaku* victories during the attack on their own ship, PO2c Takeo Miyazawa was seen to shoot down one torpedo-bomber and then ram another which was just about to release its torpedo. Despite this, *Shokaku* was hit by bombs and set on fire. This was put out by the damage control parties, allowing Wt Off Yukio Hanzawa to land aboard while black smoke was still pouring from the vessel and the arrester wires could not be deployed.

During this first carrier-versus-carrier battle, the USS *Lexington* was sunk, more than matching the serious damage inflicted on *Shokaku*, but the real victory lay in the fact that the IJN plans to invade Port Moresby had to be abandoned. Following this reversal, Japanese Army units landed instead at Buna on the eastern shore of New Guinea on 21

July, advancing on foot over the Owen Stanley mountain range which contained many peaks of 2,000 metres in an effort to reach Port Moresby by this route. Late in August Japanese marines landed at Milne Bay on the south-easternmost tip of New Guinea where they tried to capture the RAAF airfield located there. Resistance proved too strong, however, and on 6 September the survivors were withdrawn.

To provide cover for these various actions, on 11 August Tainan Ku A6Ms took part in attacks on Rabi (Gurney and Turnbull airfield), six pilots of the Sasai chutai claiming seven destroyed and two probables for one force-landed Zero. The fighter unit of the 2nd Ku which had just reached the South-East Area, moved forward to Buna on 22 August with part of the Tainan Ku to operate jointly. On 24th during a fight with Bell P-39 Airacobras over Rabi, the 2nd Ku pilots claimed seven and two probables, while Tainan Ku fighters claimed five and two probables. However, on 26 August a group of 2nd Ku fighters were surprised as they were taking off from Buna and lost two of their number, while next day all of five Tainan Ku Zeros failed to return from an attack on Rabi.

Guadalcanal

At Tulagi, almost at the south-eastern tip of the Solomon Islands, the main party of the flyingboat-equipped Yokohama Ku arrived during May, commencing patrol sorties from there. The float fighter buntai of this kokutai, it will be recalled, were at Rabaul where they were engaged on interception duties. However, between 10-29 June they were able to attack raiders on only one occasion, and no positive results were obtained. At the start of July these aircraft were flown to Tulagi to join the main body of the kokutai.

Nakajima-built A6M2N float fighter of 802 Ku near Jaluit, Marshall Islands.

At this stage the IJN had commenced construction of an airfield on Guadalcanal Island, located next door to Tulagi, but each day B-17s or B-24s passed overhead, reconnoitring what was happening. From 5 July, as soon as they had become established at Tulagi, float fighter A6M2Ns commenced intercepting these aircraft, the first victory being claimed over a B-24 on 10 July. Further interceptions took place on 23 July and on

1st and 2 August. On the latter of these dates claims for one destroyed and one probable were submitted. Records for this period are incomplete, but on 4 August one B-17 was brought down by ramming by a float fighter, while loss records indicate the death in action of Sea1c Shigeto Kobayashi on this same date, leading to the belief that he was responsible for the ramming attack.

The Occupation of Guadalcanal

During the morning of 7 August 1942 a telegram was received at IJN High Command from the HQ of the Yokohama Ku at Tulagi advising that an enemy fleet had appeared and that enemy troops had been landed. Thereafter no further word was received from this unit. In fact the US 1st Marine Division had landed on Guadalcanal to occupy the new airfield which had been constructed there, while a smaller party had come ashore at Tulagi and had completely destroyed the Japanese units based there.

At once 27 Rikkos of the 4th Ku departed Rabaul, escorted by 17 A6Ms led by Lt Cdr Tadashi Nakajima. Three hours later, after a flight of 1,040 km, they arrived over Tulagi where they encountered a reported 60 US aircraft, the Zero pilots claiming 36 shot down and seven probables for the loss of two of their own fighters. During this engagement the later well-known pilot PO1c Saburo Sakai was hit by the rear gunner of an SBD, but managed to fly all the way back to base in a desperately wounded condition. Flying as a wingman on this occasion, PO1c Nishizawa claimed six of the victories. At Tulagi, all the A6M2Ns had already been destroyed on the water during a strafing attack by USN Grumman F4F Wildcat fighters.

The Japanese command decided on 13 August that Guadalcanal must be retaken at whatever cost. Thus began the fighting over the Solomon Islands which was to last for one year and seven months. Steadily now, the fighters of the US Navy, Marine Corps and the USAAF operating both from Guadalcanal and from Port Moresby in New Guinea would begin to pick off the veteran pilots and aces of the Tainan Ku. From August until early November 1942, this kokutai would lose 32 of its pilots, despite submitting claims for 164 aircraft destroyed and 37 more probables.

On 6 August 15 A6Ms and 16 D3As of the 2nd Kokutai arrived at Rabaul. However this unit's fighters were A6M3s (Type 32) which had a shorter range than the previous model and were therefore unable to reach Guadalcanal, thereby providing little relief for the hard-pressed Tainan Ku. The latter unit escorted Rikkos to attack the US shipping off Tulagi on four consecutive days, but on the final day found no vessels remaining present – they had withdrawn due to the persistent air attacks. However, on 20 August 19 F4Fs and 12 SBDs of the US Marine Corps flew in to Guadalcanal. This was of vital importance, for from this day onwards aerial superiority over Guadalcanal and the immediate surrounding area passed more and more to the Americans.

When a detachment of troops from the Ichiki Unit landed on Guadalcanal on 18 August, they came ashore without meeting initial resistance. However, when the unit's second detachment sought to land on 25 August, followed by a detachment of the Kawaguchi Unit on 28th, both were turned back by the air raids inflicted on them by the USMC aircraft.

Following the Midway battle, the surviving IJN carriers had regrouped as the 3rd Fleet and these departed Japan on 16 August to challenge the US Navy around Guadalcanal and to attack American bases on the island and on Tulagi. This led to the Battle of the Eastern Solomons on 24 August. Initially, 15 A6Ms from *Ryujo* were led by Lt Notomi to escort six torpedo-bombers to attack targets on Guadalcanal. Here the escort fought intercepting F4Fs, claiming 15 of the defenders shot down. In the meantime *Ryujo* was attacked by US carrier aircraft, and was sunk despite ten A6M pilots on CAP over her claiming 11 of the attackers shot down. Six more Zeros from *Zuikaku* then escorted an attack on the American ships, their pilots claiming six victories for three losses. Four *Shokaku* fighters led by Lt Shigematsu claimed four for one loss. The attack force managed only to obtain a single hit on USS *Enterprise*, damaging, but not sinking her. Following this engagement, the remaining fighter pilots from *Ryujo* were transferred to the land-based Tainan Ku.

Following several days of action in eastern New Guinea, on 21 August 13 A6Ms had flown over Guadalcanal, engaging 13 USMC F4Fs for the first time, four of these being claimed shot down plus three probables without loss. During 25th and 26 August Zeros escorted Rikkos over the island, their pilots claiming nine destroyed and one probable on the latter date. However, the three losses suffered in return included Lt(jg) Jun-ichi Sasai, who had by that time become the highest-scoring officer pilot in the area.

Following the carrier battle related, the IJN carriers departed the area for the homeland, leaving the fighters from *Shokaku* and *Zuikaku* on detachment at Buka Island; Buka is located more than 100 sea miles closer to Guadalcanal than is Rabaul. During 28 August and 4 September 15 *Shokaku* fighters undertook three sorties to Guadalcanal, claiming 15 victories for the loss of six pilots. Additionally, Lt Shingo, the hikotai leader and Lt Ibusuki were each obliged to force-land on one occasion. The 15 *Zuikaku* Zeros led by Lt Hidata also took part in these attacks, but details of their claims and losses failed to survive the war.

The Struggle for Guadalcanal

Fighters from the land-based units attacked Guadalcanal five times during August, totalling some 57 sorties despite being turned back by bad weather on several occasions. This did not allow them to undertake more than one patrol over destroyers and other shipping carrying reinforcements to the army units on the island. Towards the end of August, therefore, the floatplanes of the 11th Air Flotilla (floatplane tenders *Kamikawa*

F1Ms and A6M2Ns of *Kamikawa Maru* at Shortland, August 1942.

Maru, Chitose, Sanyo Maru and *Sanuki Maru*) all arrived at a new floatplane base at Shortland Island. From here Mitsubishi F1Ms began patrols over the convoys at morning and evening, earning the gratitude of the destroyer crews. When *Kamikawa Maru* reached Shortland on 4 September, she brought with her not only two F1Ms, but also 11 A6M2Ns. These float fighters also commenced convoy patrols, but also operated from Rekata Bay on the north shore of Santa Isabel Island, just 150 miles from Guadalcanal, where a floatplane base had also been prepared. Other elements of the float force would subsequently move here also.

In an effort to obtain reinforcements, the 11th Air Fleet requested that the 6th Kokutai be despatched to the South-East Area. This unit was at Kisarazu in Japan, where it had been engaged in training new fighter pilots following the losses at Midway. Consequently, it still contained many 'green' pilots, and the decision was taken only to send veteran pilots to the Solomons. Lt Kofukuda, the unit's new hikotai leader, led 18 A6Ms via Iwo Jima, Saipan and Truk to Rabaul, guided by a few Rikkos for navigation purposes. These fighters all arrived safely at Rabaul on 21 August, completing an unprecedentedly long flight for single-seat aircraft, but one which would soon be followed by other reinforcements. The new unit commenced its first operation which was to be an attack on Rabi, Milne Bay, but the Zeros were turned back by the weather. They then participated in attacks on Guadalcanal where the first two claims were made on 12 September.

Unit of aces; Tainan Ku pilots on 4 August 1942. Front row, 2nd from left is Takeichi Kokuba; 4th from right Kazushi Uto; 2nd row, 2nd from left Toraichi Takatsuka; 4th from left Jun-ichi Sasai; 5th Shiro Kawai; 6th Yasuna Kozono; 7th Masahisa Saito; 8th Tadashi Nakajima; 12th Takeyoshi Ohno; 3rd row, 6th from left Mototsuna Yoshida; 7th Saburo Sakai; 8th Hiroyoshi Nishizawa; 10th Yoshio Ohki; 13th Toshio Ohta; 14th Masuaki Endo; 4th row, 3rd from left Sadao Uehara; 4th Keisaku Yoshimura; 9th Ichirobe Yamazaki.

Meanwhile, from the Marshall Islands ten A6Ms from the 24th Air Flotilla (1st and Chitose Ku) flew down to Rabaul on 2 September for attachment to the Tainan Ku. On 16 September the fighter unit of the Kanoya Ku from South-East Asia moved to Kavieng with their Rikko unit, while next day 21 A6Ms and four C5Ms of the 3rd Ku also reached Rabaul, led by Lt Cdr Kiyoji Sakakibara, the Hikocho – although this unit too came under the command of the Tainan Ku's commanding officer. This Rabaul detachment of the 3rd Ku was to operate over Guadalcanal and to undertake convoy patrols and interceptions until early November 1942, claiming 48 destroyed and 20 probables for the loss of eight pilots.

Despite this influx of fighter units, the numbers of A6Ms available told a different story. By late September there were eight Tainan Ku aircraft, 25 of the 6th Ku, 20 of the 3rd Ku, eight of the Kanoya Ku and 16 of the 2nd Ku – a total of 57 aircraft, of which 29 were A6M3s lacking the range for round trips to Guadalcanal from Rabaul.

On Guadalcanal itself, some 900 men of the Ichiki Unit assaulted the US lines during the morning of 21 September, but were annihilated. Following this failure, the IJN sought to despatch some 5,000 men of the Kawaguchi Unit and the remaining elements of the Ichiki Unit to the island in 34 destroyers during 11 sorties to the island's north-east coast during the nights of 29 August and 7 September*. These troops launched a number of all-out attacks during 12th and 14 September, but failed to reach Henderson Field, as the island's main airfield was now known.

Tainan Ku pilots about to commence a mission led by Lt Cdr Nakajima (extreme left).

In the meantime Zeros (including those of the 2nd and 6th Ku) escorted Rikkos to Guadalcanal during the period 5-14 September. Indeed, attacks were launched every day between 9-14, during which the fighter escorts claimed 45 destroyed and 12 probables for the loss of seven pilots. Included amongst the latter were aces such as Wt Off Toraichi Takatsuka and PO3c Kazushi Utoh.

* The Japanese termed these destroyer-borne reinforcement runs 'Rat Transport', while the Allies titled them 'Tokyo Express'.

On 13 September *Kamikawa Maru* sent two A6M2N float fighters on a reconnaissance over Guadalcanal, the pilots of these claiming one F4F shot down for the unit's first victory. Next morning three more float fighters undertook a similar mission, but all failed to return. That same day in the afternoon two more of these aircraft were intercepted by two F4Fs, claiming one shot down for one loss. During the evening of 14th eight F1Ms from *Chitose* undertook a bombing raid on Henderson Field but were intercepted by seven F4Fs. The floatplane crews claimed two shot down against the loss of one F1M destroyed and a second badly damaged.

Supplying Guadalcanal

In spite of the failure of the infantry attacks during September, the Japanese General Staff still did not abandon the idea of a reconquest of Guadalcanal. It was by now clear, however, that artillery support would be necessary, and this required the delivery of heavy items which could not be carried aboard destroyers; vessels with an appreciably greater loading capacity would be needed, but by their nature these would also be slower. Initially the floatplane tender *Nisshin* was used, undertaking sorties on 3rd and 8 October, while *Chitose* was also used on 11 October.

A6M2N and F1M floatplanes at Shortland. 'Y11' on the tails indicate that these aircraft belonged to the tender *Kamikawa Maru*.

To provide patrol cover over these convoys while also flying from Rabaul imposed many problems and restrictions, and between 11-21 September only four such missions, totalling 28 sorties, could be accomplished. In an effort to ease the situation, on 28 September 21 A6M2s of 2nd Ku were moved to Buka Island, and after a new airstrip at Buin on Bougainville Island had been completed, the 6th Ku moved in there. From here patrols could be flown both over the convoys, and over the anchorage at Shortland. More float fighters also arrived in the shape of the unit of these aircraft forming part of the 14th

3 Ku A6M at Rabaul.

Kokutai, which reached Shortland on 12 October aboard *Kiyokawa Maru*. Next day they intercepted a B-17, claiming it probably shot down for the loss of one of their number.

Attacks on the shipping bringing supplies and reinforcements were made in the main by US dive-bombers and were generally carried out at dawn or dusk when patrolling Zeros tended to be absent. Dusk on 11 October proved to be disastrous for the fighters. 6th Ku aircraft undertook 21 sorties, but during the second patrol in very bad weather all three A6Ms taking part failed to return. A 4th Ku patrol by six Zeros remained until after sundown as a result of which they could not fly back to their base and were forced to ditch in the sea near the convoy; three of the pilots, including ace and veteran PO1c Juzo Okamoto, were lost. The medical officer on board the destroyer *Natsugumo* later painted the scene, showing the sky behind New Georgia Island to appear dull red in colour while the A6M flown by young Lt(jg) Kazuto Kubo was trying to land on the black sea, covered by his two wingmen.

Between 28 September and 12 October it was mainly fighters from the 2nd and 6th Ku which patrolled over the convoys and Shortland, undertaking some 220 sorties during this period but seeing action on only three occasions, claiming three dive-bombers shot down and two probables. By contrast the floatplanes of the 11th Air Flotilla were engaged on ten days between 24 September and 12 October, claiming 20 destroyed and three probables. Amongst this total were three B-17s claimed by F1M crews, and one destroyed by Sea1c Kiyomi Katsuki, who rammed it with his F1M. Destroyer crews rated the floatplanes more highly than the Zeros at this period, but attrition was high. Eight F1Ms were lost by 10 October, while at one point none were serviceable at all.

A new convoy comprised of six high-speed transports departed Shortland during the night of 13 October, arriving off Guadalcanal next evening where they were successful in off-loading 80% of their cargo, then making good their escape. To support this effort, two battleships bombarded Henderson Field during the night of 13th, two heavy cruisers repeating this effort next night. 22 A6Ms from the Tainan and 6th Ku patrolled over the departing convoy on 14th, but were not challenged. That evening, however, Shortland was attacked by F4Fs and SBDs. Eight F1Ms rose to intercept, their crews claiming the destruction of two fighters and one dive-bomber.

Early on 15th flights from Henderson Field were seen approaching, these aircraft attacking shipping off-shore. Destroyers, float fighters and other floatplanes from Rekata Bay sought desperately to defend the vessels, but three were set on fire and destroyed. A6Ms from Buka Island finally reached the area; the second patrol launched by the Tainan Ku (nine fighters including three of the 3rd Ku's aircraft) claimed 12 destroyed and six probables, PO1c Takeo Okumura claiming four of the former and one of the latter, while PO2c Yoshiro Hashiguchi of the 3rd Ku claimed three and a probable. In total 41 sorties were flown over the remnants of the convoy, and during the third patrol nine 3rd Ku A6M pilots claimed a further eight destroyed for one loss. However, during their return flight to Buka Island, four more had to force-land in the sea when their aircraft ran out of fuel; all were rescued. The three remaining ships of the convoy reached Shortland during 16th.

During this same period, 19 September-15 October, apart from four days of impossible weather, 325 sorties were flown by Zeros escorting formations of Rikkos to attack Guadalcanal. By this time the number of serviceable A6Ms had become so small that only about 18 could be put into the air on each occasion, usually drawn from two, or even three of the available units.

Claims during this period reached 64 destroyed and 28 probables, the IJN pilots feeling that they had acquired a degree of air superiority over the island, although eight pilots were lost in combat and four more due to fuel shortage or technical problems during the return flights. It was felt that when the Zeros appeared, intercepting aircraft tended to evade them. On 9 October no opposition at all was encountered, while on 14th US aircraft did not get close enough to engage either the Zeros or the Rikkos.

Lt Kenjiro Notomi, hikotai leader of *Zuikaku* (back to camera) briefs his pilots before a mission on 7 April 1943.

The Battle of Santa Cruz

To assist the second major attack which the Army was about to launch on Guadalcanal, three aircraft carriers of the 3rd Fleet and two of the 2nd Fleet were again sent southwards to the Solomons. On 14 October fighters from *Zuikaku* shot down a shadowing flyingboat while another such was despatched by three *Zuiho* Zero pilots next day. Patrolling over a friendly convoy on this same day (15th), eight *Hiyo* A6M pilots shot down a dive-bomber and a third shadower.

On 17 October nine A6Ms from *Junyo* took part in an attack on Guadalcanal, claiming six victories without loss – but eight dive-bombers which they were escorting were all shot down by the defences. During the day contact was also established with units of the US Fleet. While escorting the first attack launched against the American vessels, nine A6Ms led by Lt Tadashi Kaneko encountered 15 F4Fs, the Japanese pilots claiming to have shot down seven of these opponents.

At this stage *Hiyo* suffered engine trouble during 20 October, and was obliged to withdraw to Truk. Part of her air party went aboard *Junyo*, but the greater part flew off to Rabaul. From here on 25 October Lt Iyozo Fujita led 15 of the *Hiyo* fighters to escort Rikkos to Guadalcanal. Here about ten US fighters were met, three of which were claimed destroyed plus one probable, for the loss of one Zero.

Having withdrawn north, the remaining carriers then again headed south to support the second big offensive on Guadalcanal by destroying the US Fleet. This led to another major carrier-versus-carrier battle which commenced on 26 October. It would prove to be the last such engagement until the Battle of the Philippine Sea during 1944.

The first sighting was made by an IJN scout plane which sent "a carrier found" message at 0450 hours*. It then took time to estimate the precise location of the American vessels, and consequently it was 0525 before 21 A6Ms from the three carriers took off to escort the first attack force. The Americans had been quicker off the mark, and at 0452, just after the first sighting report had been received, US dive-bombers arrived overhead. Although defending Zero pilots were able to claim two shot down, two more dived on *Zuiho*, one bomb hitting her aft flying deck, causing a large hole in this; as a result, *Zuiho* was forced to withdraw at once.

The first main attack forces passed each other, the US force comprising six fighters and eight dive-bombers from USS *Enterprise*. Nine *Zuiho* A6Ms led by Lt Hidaka attacked these, claiming all of them destroyed for the loss of two Zeros; a third was heavily damaged and returned to the Japanese Fleet at once, while two more subsequently failed to return. The US formation actually lost three F4Fs and three TBFs with one more of each damaged.

The other aircraft of the first attack force found the US carrier *Hornet* at 0655, claiming hits with six bombs and two torpedos, while four A6M pilots from *Shokaku* claimed three destroyed and a probable during two engagements, but lost Wt Off Hanzawa. Eight *Zuikaku* fighters led by Lt(jg) Ayao Shirane took on the bulk of the defending aircraft, engaging a reported 30 plus, claiming 14 shot down for the loss of two Zeros.

Meanwhile five A6Ms and 19 D3As took off from *Shokaku* at 0610, followed by four A6Ms and 16 B5Ns from *Zuikaku* at 0645 and then 12 A6Ms and 17 D3As from *Junyo* at 0714. The *Shokaku* D3As bombed the second US carrier, *Enterprise*, instead of the stationary and still burning *Hornet*. The five escorting Zero pilots were able to claim only a flyingboat which was observing events, as probably shot down. The B5Ns attacked the

* Japanese Standard Time, which was Greenwich Time plus nine hours.

A6Ms and D3A Type 99 Dive-Bombers on *Shokaku* in October 1942.

same carrier at 0900 while their escorting *Zuikaku* A6Ms fought with some ten intercepting F4Fs, their pilots claiming six of these shot down and one probably so; two A6Ms force-landed in the sea during their return flight. At 0921 the *Junyo* D3As also attacked *Enterprise* while their 12 escorting Zeros were led by Lt Yoshio Shiga to attack the defending aircraft, claiming nine destroyed and five probables.

For the first time since the war had begun, the Japanese carriers had been fitted with radar sets, which at 0640 reported the approach of hostile aircraft, USN aircraft then attacking the Japanese vessels in three waves. 24 A6Ms were launched from *Shokaku*, but only four made contact with the raiders, although PO1c Shigetaka Ohmori reportedly shot down five before ramming an aircraft which was about to bomb, both aircraft falling burning into the sea. The other three pilots claimed three probables. From *Zuikaku* 27 fighters took off, eight pilots claiming eight victories for the loss of three A6Ms. *Zuiho* launched 14 Zeros, eight of which made interceptions, their pilots claiming six destroyed for the loss of three A6Ms. Despite these interceptions, however, at 0730 *Shokaku* was hit by four bombs, and while her crew were successful in extinguishing the fires caused, she too was obliged to withdraw.

Now from *Junyo* a second attack force of seven B5Ns and eight A6Ms took off at 1106, attacking the stricken *Hornet* at 1310. Behind them came two D3As, six B5Ns and five A6Ms from *Zuikaku* which also attacked *Hornet* and other vessels at 1325. The escorting Zero pilots of both formations encountered no US aircraft during this attack, but the *Junyo* fighters lost PO1c Kiyonobu Suzuki and another pilot, both of whom failed to return. Three more A6Ms force-landed, but all the pilots of these were picked up safely.

Finally, at 1333 *Junyo* launched six more A6Ms and four D3As which attacked the drifting *Hornet* at 1510, again meeting no opposition in the air. So the Battle of Santa Cruz ended, somewhat to the Japanese advantage. *Shokaku*, *Zuiho* and the heavy cruiser *Chikuma* had all suffered damage, but the US Navy lost *Hornet*, which was subsequently sunk by her own escorts, and the destroyer USS *Porter*. Additionally, the battleship USS *South Dakota*, cruiser USS *San Juan*, and destroyer USS *Smith* were all damaged. The battle cost the IJN 69 aircraft and the USN 74 (of which only 20 were actually shot down, the other 54 being lost to other operational causes), so honours appeared about even. However, from the Japanese point of view many veteran crews, including a considerable number of leaders, were lost, and these were irreplaceable.

CRITICAL FIGHTING OVER THE SOLOMON ISLANDS AND NEW GUINEA

Redesignation of Flying Units

On 1 November 1942 the designation of fighting air units was changed. All units serving abroad were now to be identified by a three-figure number. Fighter units were to be designated with a 2 or 3 as the first figure in the number, while units of dive-bombers or torpedo-bombers were to have their number commence with a 5. Rikko (medium bomber) units were to begin their number with a 7. For example, the Tainan Kokutai became the 251st Kokutai (abbreviated to 251 Ku), the 2nd Kokutai became the 582nd Kokutai (582 Ku), and the Kanoya Kokutai became the 751st Kokutai (751 Ku).

Consequently, the following changes took place amongst the IJN fighter force:-

Chitose Kokutai became the 201st Kokutai (201 Ku);
3rd Kokutai became the 202nd Kokutai (202 Ku);
6th Kokutai became the 204th Kokutai (204 Ku);
The Fighter Unit of the 1st Kokutai was amalgamated into 201 Ku;
Tainan Kokutai became 251st Kokutai (251 Ku);
The Fighter Unit of the Genzan Kokutai became the 252nd Kokutai (252 Ku);
The Fighter Unit of the Kanoya Kokutai became part of the 751st Kokutai (751 Ku), but shortly thereafter became an independent unit and received the designation 253rd Kokutai (253 Ku);
However, the Fighter Unit of the 2nd Kokutai (2 Ku) remained as part of what became the 582nd Kokutai (582 Ku) – (things can never be entirely simple !)

The headquarters tent of 204 Ku at Buin. A signboard on a tree reads 'Headquarters of Morita Unit'; Morita was the name of the unit's commanding officer.

From the South-Eastern Front (Solomons and New Guinea), 251 Ku at the start of November returned to the homeland for recuperation. At much the same time 202 Ku finished its operations in this area and returned to the South-Western Front (Dutch East Indies). To take their places, 252 Ku moved to Rabaul on 9 November aboard the carrier *Taiyo*. Thus, the fighter units remaining in the area became 204 Ku, 252 Ku, 253 Ku and 582 Ku. The fighter unit from *Hiyo* continued to operate over the area until mid December.

Fighting over Buna

In March 1943 the Allies went onto the offensive in New Guinea as well as in the Solomons. By the previous September, the US Army Air Force had 250 fighters and 175 bombers in this latter area, and these were constantly added to at Milne Bay (Gurney Field) and Port Moresby. These were being employed to attack Japanese Army units in the Owen Stanley Range, and to strike at the bases at Lae and Buna.

Leading ace Kan-ichi Kashimura (centre) was assigned to 582 Ku in December 1942 when this photograph was taken at Buna airfield. He is surrounded by young NCOs, including future aces Mitsuo Hori (back row, 4th from right) and Kiyoshi Sekiya (back row, 2nd from left).

By this time IJN fighter units were almost totally committed to the fighting over Guadalcanal, and consequently only took part in two attacks on Port Moresby, and in some patrols over a convoy to Buna. On 5 October 1942 five A6Ms of the 3rd Ku intercepted five North American B-25 Mitchell bombers over this convoy, claiming one shot down and one probable. 15 of *Hiyo*'s fighters escorted a formation of Rikkos undertaking a supply drop to Japanese units at Kokoda.

On 1 November, 11 251 Ku pilots made claims for one victory in return for one loss; during this same day six A6Ms from 202 Ku were engaged over Lae, where Lt(jg) Usaburo Suzuki claimed one P-40 shot down while LdgSea Kiyoshi Ito claimed a P-39.

About 3,000 men of the Japanese Army and Navy were stationed at Gona and Buna. During the opening weeks of October the Allies completed construction of an airstrip at Wanigela, only 80 km south-east of Buna, following which the USAAF airlifted a regiment of infantry there. Following this, on 16 November Japanese reconnaissance indicated that Allied troops had landed on the coast east of Buna, while at the same time they discovered the Wanigela airstrip at last. 1,500 Marines were at once ordered by the Japanese high command to be carried in destroyers to Basabua, west of Buna, during 17-18 November. Next day, however, attacks on the Japanese-held areas commenced from three different directions, all launched at regimental strength. On 21 November a further airstrip at Dobodura, only 30 km from Buna, was completed by Allied construction units.

By this time the IJN flying units were suffering severe exhaustion and it was also now very difficult to reinforce them properly. Air Fleet commanders ordered that training of new fighter pilots to operational level should be put in hand at Kavieng until 15 December. Meanwhile, on 17 November the fighters and dive-bombers of 582 Ku and the fighters of 252 Ku moved forward to Lae.

Meanwhile, on 16 November 253 Ku aircraft had joined 582 Ku to escort the dive-bombers from Rabaul to Buna. Here next day four 253 Ku A6Ms were engaged over Buna, claiming one victory, but then returned to Rabaul. The unit's personnel were then involved in training new 'green' crews as required by the Air Fleet's order, until the middle of December.

Over Lae one aircraft of 582 Ku and two of 252 Ku were lost on 18 November. On 22nd, however, CPO Bunkichi Nakajima and one other 252 Ku pilot claimed between them to have shot down four Martin B-26 Marauder bombers and a 'Lockheed' (presumably a Hudson patrol-bomber). Over Buna on this date eight more 252 Ku A6Ms escorted dive-bombers, becoming engaged with 24 Allied aircraft. The fighter pilots claimed one of these shot down and seven probables in what was obviously a very confused fight, and two of their number failed to return. Two pilots from 582 Ku, the fighters of which had been providing the close escort to the six dive-bombers, managed to claim one destroyed and two probables in return.

On the ground the Japanese garrisons fought desperately against an attacking force ten times their number, but to no avail. Gona fell to the Allies on 9 December and Buna on 14th. During this period the two fighter units continued to take part in support missions for the ground forces at Buna, patrol sorties over supply convoys, and interceptions over Lae. On 24 November PO2c Kiyoshi Sekiya twice undertook single-handed combats with formations of P-39s, claiming one shot down on each occasion. On the same day six pilots of 252 Ku took on seven P-39s, claiming one shot down and one probable. Two days later five from 582 Ku and six from 252 Ku intercepted two transport aircraft and seven escorting P-39s, claiming both transports and one Airacobra for the loss of one Zero.

582 Ku returned to Rabaul on 28 November, but two days later sent 12 A6Ms and six D3As back to attack hostile forces around Buna. Here the fighters attacked ten P-40s and six P-39s, claiming 11 shot down and two probables. Three of these were claimed by PO2c Ki-ichi Nagano, but the unit suffered two losses, including the formation leader, Lt(jg) Tomoyuki Sakai.

At the start of December 252 Ku claimed a B-25 shot down over Buna on 1st, while next day the unit's pilots intercepted three B-17s, claiming one shot down; the unit then withdrew to Rabaul. From here 582 and 252 Ku continued to operate over Buna, joined by 253 Ku when it completed its training function at Kavieng, action here continuing until the fall of the area to the Allies.

Thus during 7 December 582 Ku pilots claimed two P-39s and a B-17 over Buna for the loss of two D3As and an A6M. 252 Ku meanwhile submitted claims for one destroyed and one probable while escorting Rikkos. Next day, whilst over a convoy evacuating troops from Buna, 582 Ku fighters intercepted B-17s again, claiming one

A6Ms of 582 Ku at Buin. The Zero marked with double chevrons was that flown by Lt Cdr Saburo Shindo.

destroyed and two probables, while 252 Ku pilots also engaged these big bombers, claiming one shot down and one probably so. On 14 December both units attacked a formation of 24 Consolidated B-24 Liberators, each unit claiming one shot down.

Failure of the Resupply of Forces on Guadalcanal

The transport of supplies and reinforcements to Guadalcanal by destroyers remained in constant need, and was therefore restarted on 29 October. Next day when 25 A6Ms of the 6th Ku undertook patrols over such a convoy, 11 of them, led by Lt Miyano, encountered five F4Fs. This time the Japanese pilots came away empty handed, less one Zero shot down and three more damaged, the pilot of one of which was wounded.

Early on the morning of 30 October six F4Fs and two SBDs attacked Rekata Bay. Four of the resident A6M2Ns intercepted, but while PO3c Hisao Jito claimed two of the attackers shot down, one float Zero was shot down and a second was damaged beyond repair. Jito had claimed his first victory on 17 October.

Even after the Battle of Santa Cruz and the failure of the Army attack during late October, the Japanese General Staff remained convinced that the recovery of Guadalcanal was still a possibility. To achieve this reinforcement of men, heavy guns, ammunition and food to the island remained vitally necessary. The first run was made by one light cruiser and 16 destroyers, which reached Guadalcanal during the night of 1 November. Thereafter, the 'Tokyo Express' continued to 'run' during the nights of 4th, 6th, 7th and 9 November.

The US forces responded vigorously, and two destroyers were badly damaged by aircraft on 7th. Patrols by float fighters from Shortland and Rekata Bay, and by A6Ms from Buin, continued. However, on 7 November six A6M2Ns of 802 Ku (ex-14th Ku since 1 November) led by Lt Hidero Goto were bounced by eight F4Fs of VMF 121 led by Capt Joe Foss, and were all shot down. This meant that the float fighter unit from *Kamikawa Maru* had been virtually annihilated, 802 Ku being left only with two A6M2Ns, which were then retained only for patrolling over Shortland for the time being. Next day, however, ten F1Ms which attempted to cover the destroyers, suffered seven losses. While their crews had been able to claim a few victories in return, there was no

Zero-Sens at Lakunai, Rabaul. 'T2' on the tail of the fourth aircraft indicates that it was from 204 Ku.

doubt that the floatplanes were now in great difficulties when faced with F4Fs. As an example, while F1M crews from *Chitose* were able to claim ten victories between 9 September-30 December, this had cost them the loss of 14 aircraft.

As already mentioned, the warships of the 'Tokyo Express' could not carry heavy weapons and other bulky supplies, so now a convoy of 11 fast transports was organised for this purpose. Before this could set sail, it was extremely desirable that Guadalcanal's air striking power should be reduced as much as possible. Increased raids by the Rikkos proved nearly impossible due to adverse weather conditions, and these bombers were only able to attack under fighter escort on 5th, 11th and 12 November.

On 5th no intercepting fighters were encountered, but on 11th it was a different story. 11 A6Ms of 252 Ku on their first mission to Guadalcanal, engaged three F4Fs, claiming one shot down. Nine 253 Ku Zeros patrolling in the area met nothing, but six out of 15 582 Ku pilots claimed one victory; however, four Rikkos failed to return.

Also during 11 November, nine D3As attacked US shipping off Lunga, but lost five of their number. Lt Tadashi Kaneko had led 12 of *Hiyo*'s A6Ms and six from 204 Ku to escort the dive-bombers, and these fighters encountered some 30 American aircraft. In the fight which followed Lt Kaneko personally claimed two shot down and a probable, while Lt Iyozo Fujita also claimed two and Wt Off Mitsugu Mori claimed one, other of the unit's pilots claiming eight more while the 204 Ku contingent added a further six destroyed and four probables; two *Hiyo* aircraft failed to return.

Wt Off Yoshio Wajima led 12 562 Ku A6Ms to the island on 12 November, claiming five destroyed and four probables from a force of 20-plus interceptors for one loss; Wajima himself claimed two and one probable. On the same day 12 A6Ms of 252 Ku and six of 253 Ku fought F4Fs, the former unit claiming seven shot down and a probable without loss, while the 253 Ku pilots added two and a probable, also without loss. Despite this successful conclusion for the fighters, they failed to prevent the loss of 14 out of 19 Rikkos which were attempting a torpedo attack on a US convoy making for the island.

That night the 11 fast transports set course for Guadalcanal, while two battleships and 15 destroyers went in to bombard the island's airfields. This force was challenged by two Allied heavy cruisers, three light cruisers and eight destroyers as a result of which the incoming Japanese convoy was held back. During the night of 13/14th four IJN heavy cruisers again closed in to bombard Henderson Field, following which the transports sailed from Shortland next morning.

Meanwhile, following the naval engagement of the night of 12/13th, next day 13 A6Ms were ordered to the area to provide cover for the battleship *Hiei*, which had been severely damaged during the firefight. However, in poor visibility six 252 Ku Zeros could not get through to the area and all force-landed in the water of Rekata Bay, while six more from 204 Ku also proved unable to find *Hiei*. Eight fighters from *Junyo* – still land-based – reached Savo Island, led by Lt Shigematsu. Here they spotted five US torpedo aircraft, but before they could attack, they were 'bounced' by about 15 F4Fs, losing three A6Ms while only able to claim one and one probable in return.

As the convoy approached on 14 November, six 204 Ku fighters patrolling overhead encountered a lone B-17 at 0530, then claiming two dive-bombers shot down. Six 252 Ku A6M pilots then attacked 12 dive-bombers, 12 torpedo-bombers and nine F4Fs at 1050. They claimed two shot down and two probably so,

Lt Masaji Suganami, hikotai leader of 252 Ku, who failed to return from a sortie on 14 November 1942.

but lost three of their own fighters. Between 1240-1315 six pilots of 252 Ku engaged eight B-17s and ten or more F4Fs of which they claimed 14 destroyed for the loss of their leader, Lt Masaji Suganami.

On the right of this group is Lt Cdr Tadashi Kaneko, hikotai leader from the carrier *Hiyo*, who was also lost on 14 November 1942.

On the next patrol *Hiyo*'s six fighters engaged 18 aircraft, the pilots claiming eight of these shot down, including four by CPO Tanaka alone; however, in this fight the unit lost its redoubtable leader, Lt Tadashi Kaneko. On the final patrol of the day six A6Ms from 204 Ku intercepted about 15 dive-bombers between 1520-1540, claiming one shot down and one probable for the loss of two Zeros. Throughout the hours of daylight 14 F1M floatplanes also engaged in patrols, but suffered heavy losses.

Despite these efforts, seven of the vital transports had been sunk prior to reaching their destination. It was obvious that it would not be possible for the four survivors to anchor, unload, and then withdraw safely, so instead they were all run aground at 0200 that night. 2,000 men and four days' supply of rice for the troops on the island were unloaded, and this together with 260 cases of ammunition for the artillery already on the island, was all that reached the Army's hands. The four ships burned until dawn, while eight F1Ms patrolled overhead.

The effect of the convoy, while providing food for 30,000 soldiers until 20 November, also increased the number of mouths to be fed without providing sufficient reinforcement or supplies to allow a further serious attack to be launched. On 15 November four A6Ms

CPO Hideo Watanabe took over the leadership of 204 Ku in the air with great distinction on several occasions after the unit's young officer pilots had been wiped out during aerial combats.

from *Hiyo*'s unit and three of 204 Ku escorted seven dive-bombers to attack US shipping off Guadalcanal. Nine F4Fs intercepted, one of which was claimed shot down in return for the loss of one Zero.

Resupply Efforts Continue

Following the failure of the major convoy, transportation of supplies to Guadalcanal remained critical, and therefore continued. However, growing US strength on the island rendered the dangers greater and losses continued to mount, while the effectiveness of what was undertaken, fell away. From 18 November until the end of December, 17 destroyers and one light cruiser were damaged, 12 of them by air attack. At the same time, as has been recorded, the imminent danger of the loss of Buna in New Guinea, led to the despatch of 252, 253 and 582 Kokutais to this area. The Solomons front was thus left only with the fighters of 204 Ku and the *Hiyo*. During the latter part of November these units were involved in interceptions over Buin, but gained no successes and lost one aircraft and its pilot. On the morning of 1 December, however, 204 Ku fighters intercepted B-17s, LdgSea Shoichi Sugita ramming the right wing of one of these bombers with his tail and right wing, which brought it down.

Now the 'Tokyo Express' was frequently attacked by torpedo boats as its ships neared the Guadalcanal coast, so a method was devised to float ashore sealed drums of supplies. In this area on 3 December four 204 Ku fighters met no opposition, but 12 F1Ms encountered a reported 40 aircraft which raided their base between 1450-1640. The floatplane crews claimed four shot down and five probables for the loss of five F1Ms. Four days later Zeros again failed to encounter any hostile aircraft, while eight F1Ms fought a reported 16 at 1640, sharing with the AA gunners on the destroyers they were escorting in claiming three shot down for a single loss. However, the delivery of supplies was thwarted by US torpedo boats on this occasion.

On 10 December a new opponent was encountered when fighters of *Hiyo* and 204 Ku intercepted 11 B-17s which were being escorted by five Lockheed P-38 Lightning fighters. Two Zeros were lost during a long pursuit, including that flown by CPO Jiro Tanaka of *Hiyo*. Two A6M2N pilots of 802 Ku claimed one P-38 shot down and a second probable over Shortland on this date.

A Zero of 204 Ku taxies for take-off at Buin. The aircraft carries the unit tail marking 'T2-135'.

Towards the end of November the Japanese commenced the construction of an airfield at Munda, New Georgia, 315 km from Guadalcanal; from 6 December US attacks on this airfield commenced. Against such raids 204 Ku began patrols from 19th. On 14th the runway was completed, and on 23rd 15 A6Ms of 252 Ku made the four-hour flight from Rabaul to this new base. Even as they were landing, six F4Fs, four P-38s, five B-17s and nine SBDs attacked. Even in such unfavourable circumstances, the Japanese pilots were able to claim eight of the attackers shot down and six probables for the loss of one Zero, with four more suffering damage. Next day five more A6Ms flew in.

Next morning 27 more US aircraft attacked, nine A6Ms taking off to join three which were already on patrol, and two victories were claimed. In return two Zeros were shot down, two were unable to take off and a pilot was injured, while another pilot was badly wounded during the aerial fighting and 11 A6Ms were damaged on the ground. Two days later on 26 December five more were strafed and burnt.

Zero 32s and 22s of 204 Ku over Buin.

During the afternoon of 27 December four A6Ms which were in the process of landing, were 'bounced' by 18 US aircraft. One failed to return, while the pilot of a second bailed out; the two surviving pilots claimed two P-39s shot down, but Wt Off Toshiyuki Sueda was wounded and force-landed. The higher command then decided that to station fighters at Munda was impossible, the 252 Ku personnel being flown back to Rabaul in Rikkos on 29 December and 3 January 1943.

Withdrawal from Guadalcanal
The deliveries of sealed drums by destroyers commenced on 29 November, and in December was undertaken on three occasions, two of which actually succeeded. In practice the Army on the island recovered only 430 drums (about 210 tons) from 2,700 which had been dropped. This allowed just one rice ball per three days for each soldier (over 20,000 men). While this allowed men to live (just) it was certainly not sufficient to fight on, and at this stage within the Japanese General Staff arose a discussion of the need to withdraw from Guadalcanal. After some six weeks of agonising, it was finally agreed in council in the Imperial presence on 31 December 1942 that it had to be done.

The time agreed to exercise this difficult manoeuvre was set for the period 30 January-7 February, when the moon was on the wane. The sea lane running roughly north-south between the two lines of islands forming the Solomons, had become known as 'The Slot'. While patrolling here over the continuing sorties by the 'Tokyo Express' on 11th, 15th, 19th and 23 January, pilots of 204 Ku lost six of their number, but claimed 22 destroyed and seven probables during this period, while 252 Ku on a similar mission on 15th claimed three victories for one loss. In the same period 204 Ku claimed three more destroyed and two probables over Buin, while on 24th during an interception over Rabaul 252 and 582 Ku made one claim.

On 25 January in an effort to gain air superiority over Guadalcanal during the withdrawal period, 22 A6Ms of 204 Ku, 18 of 252 Ku and 18 from 253 Ku undertook a sweep. In foul weather conditions 252 Ku lost seven fighters (though only one pilot) while the other two units claimed two destroyed and three probables for two losses. 21 fighters of 582 Ku escorted dive-bombers to attack ships off Santa Isabel Island on 1 February, encountering at least ten F4Fs. Six Zero pilots of the 2nd chutai, led by Lt(jg) Suzuki, claimed five of these shot down and a probable for the loss of one failed to return and a second force-landed. The 3rd chutai, led by Wt Off Tsunoda, suffered one loss but made no claims.

The withdrawal was made during 1st, 4th and 7 February, under frequent air attack. On the first day 128 patrolling 252 Ku fighters claimed nine destroyed and a probable for one loss, while on the second day 12 pilots of 204 Ku claimed four. On the final day 15 582 Ku aircraft claimed two shot down and a probable for one loss. Under this cover, the withdrawal was successfully concluded.

Meanwhile, on 17 January *Zuikaku* had sailed south to Truk from where her 36 fighters, led by Lt Notomi, flew down to Rabaul on 29th, and then on to Buin. From here these aircraft also flew patrols over the withdrawing convoy before returning to Truk to rejoin the carrier on 17 February. Whilst at Buin 19 of this unit's pilots claimed 11 F4Fs shot down and two

CPO Susumu Ishihara saw considerable service over the Solomons with 204 Ku, subsequently transferring to 202 Ku.

probables on 1 February, all for the loss of a single A6M. On 4th nine of these fighters claimed 16 victories, while six more of its 2nd chutai aircraft claimed two for one loss.

Thus it was that the land, sea and air battle for Guadalcanal that had lasted for seven months, finally came to a close. During this time the tide of war had turned greatly in the Allies' favour.

Resupply of New Guinea

In an effort to save the worsening situation in New Guinea, during December 1942 units of the Japanese Army landed at Wewak on the island's northern coast. The flying units from the carrier *Junyo* flew in here on 17 January 1943, flying patrols over the supply convoys heading there. In the period 19-24 January six interceptions were made of B-24s, and six of these were claimed shot down, including one by a ramming attack by CPO Ryosuke Sato on 23rd. One was claimed on 24th by PO1c Shizuo Ishi-I, but on this date the ace, Wt Off Saburo Kitahata, was lost.

In early January transport of a division to Lae was undertaken, but on 5th, just before the convoy was due to depart, US aircraft attacked Rabaul. Pilots of 582 Ku and 253 Ku each claimed one B-24 shot down. The convoy, comprising five transports and five destroyers, began at noon on 5 January with a voyage ahead of them of 800 km (430 miles). In an effort to prevent interference by Allied aircraft, an attack on Port Moresby was attempted on 6th, but bad weather prevented this. Fighters instead visited bases on New Guinea's east coast, three units each claiming one B-24 destroyed.

Over the convoy during 6th, Nakajima Ki 43 fighters (Allied codename 'Oscar') engaged raiders, but during the early morning of 7th one transport sank as a result of night attack. With daylight, seven A6Ms of 582 and 252 Ku undertook patrols together with the Ki 43s, the former unit engaging 18 attacking aircraft, claiming three destroyed including the formation leader. Two claims were made by Wt Off Kashimura, but one pilot was killed and PO2c Mitsuo Hori had to bale out, although he was rescued. The second patrol of nine 252 Ku aircraft also lost one killed and one force-landed in the sea, but the other pilots were able to claim a further five destroyed and one probable.

The convoy reached Lae during the afternoon of 7th, and here above the anchorage 253 Ku's nine fighters intercepted Douglas A-20 Havocs and P-38s at 1230, claiming three destroyed and two probables. However, 252 Ku was able only to claim damage to two B-17s, while one 582 Ku A6M force-landed in the sea. Despite the concentrated attacks, the convoy had suffered the loss of only the one vessel, and no further such would be inflicted prior to its return sailing on 10th.

On 17 January 1943 A6Ms of 252 Ku (21 aircraft), 253 Ku (nine aircraft) and 682 Ku (18 aircraft) escorted Rikkos on a raid on Rabi (Gurney). Here 253 Ku pilots claimed one of three P-40s shot down, while 582 Ku made claims for one P-39 and a second as a probable. However, Ens (Reserve) Mitsuoki Asano's aircraft suffered an oil leak which caused its engine to fail, and he was killed when the Zero crashed.

The Tragedy of the Lae Supply Convoy
Following the transport operation of early January, the activities of the IJN fighter units subsided. Only on 19 February did 253 Ku undertake two interceptions over Surumi, claiming two victories for one loss. The Headquarters of the 18th Army which was to command the forces in New Guinea, was still at Rabaul, and it was necessary that both this headquarters and reinforcement troops should be transported to Lae as soon as possible. Thus at midnight on 28 February eight transport ships and escorting destroyers sailed for this destination. As the ships passed to the north of New Britain, 204 Ku and 253 Ku Zeros, JAAF fighters, and F1Ms undertook patrols overhead. During the day fighters from *Zuiho* reached Kavieng to be available for operations in the area.

There were no attacks on 1 March, and next morning found the ships off the western point of New Britain. Nine A6Ms from 253 Ku arrived over the convoy at 0730, becoming engaged with some 40 US aircraft. Claims were made for a B-17, a B-24 and a P-38, plus one more of the latter as a probable, for the loss of one A6M. Nine more of the unit's Zeros arrived at 0800, intercepting B-17s without result. At 0940 14 fighters from 204 Ku arrived, but saw nothing. Meanwhile, three B-17s bombed from 3,000m, and were successful in sinking one transport.

Next morning, 3 March, the convoy was off Finschhafen at the eastern end of New Guinea, the IJN being briefed to provide patrols at this time. First, therefore, 14 A6Ms of 253 Ku took off from Surumi, arriving overhead at 0600. They were followed by 12 204 Ku fighters at 0700, while 18 of the *Zuiho* aircraft from Kavieng arrived at 0805, at which time 204 Ku and 253 Ku were patrolling at 6,000m.

Attacks at 0515 and 0723 were not intercepted, but no damage was suffered in these raids. At 0755 20 B-17s approached at the middle level altitude that had been anticipated. Unexpectedly, however, 30 A-20s and B-25s came in at low level, escorted by P-38s and P-40s. 12 B-25s skip-bombed from various directions, and between 0800-0806 17 bombs of 500 lbs hit amidship on seven transports, all of which sank by the end of the day (Japanese estimates at the time were that 29 bombs had hit these vessels). Three of the escorting destroyers were also hit and these all sank between 0809 and 0810; during the afternoon a further destroyer would also be sunk during a repeat attack.

The Zeros dived to the attack, but were unable to save the ships. 253 Ku's pilots claimed three A-20s and P-38s by 0830, also claiming probable successes against an A-20, a B-17 and a P-38. 204 Ku claimed seven destroyed and two probables, but lost two pilots. The *Zuiho* pilots engaged mainly the B-17s, claiming two destroyed including one rammed by LdgSea Masanao Maki. They also claimed one P-38 and a third B-17 as a probable. At 1250 eight 252 Ku A6Ms arrived over a stationary transport, providing cover when some 40 or more US aircraft attacked again at 1315. Two A-20s and three P-38s were claimed shot down, but for this IJN unit the only loss was a single aircraft which force-landed on the way back to base due to fuel shortage.

Zuiho's fighters did not withdraw at once following this disastrous day. On 8 March 15 of this unit's Zeros escorted Rikkos to Oro Bay, claiming one destroyed and two probables when P-38s intercepted. Accompanying them, 252 Ku pilots met no opposition. 11 of the carrier's A6Ms escorted JAAF heavy bombers to Buna on 11th, claiming four destroyed (one by Wt Off Akira Yamamoto) and one probable, but losing two of their number. On this date nine other A6Ms of 252 Ku enjoyed a 'field day', claiming seven P-38s and five P-39s shot down, with another five P-38s and a single P-39 as probables – and all without loss. The pilots of 253 Ku were almost equally successful on this occasion, 18 of them, led by Lt Iizuka claiming five P-38s and three P-39s, with four more P-38s and one P-39 as probables, again all achieved without loss.

It would appear that the destruction of the convoy had been fairly thoroughly avenged.

The *Zuiho* fighter group returned to Truk on 13 March, at which time 252 Ku also withdrew, going to the Marshall Islands. Over the Solomons and New Guinea, this latter unit had claimed in total 109 aircraft destroyed plus 36 probables for the loss of 16 pilots.

Japan Turns to the Defensive

Over the Solomons, a new float fighter unit within 802 Ku which had been formed at Yokosuka, commenced operations from 14 January 1943. Four days later nine A6M2Ns encountered six P-39s, claiming two shot down for the loss of two of their own aircraft and one pilot. They subsequently engaged in interceptions on 20th (one victory) and 2 February (one loss).

Prior to the Lae convoy operations, nine 204 Ku Zeros and 20 from 252 Ku operated over Buin and Ballale on 13 February. Here pilots of 204 Ku claimed three B-24s, four P-38s and two P-40s for one loss, while the 252 Ku men claimed three P-24s, three P-38s and four P-39s, plus one P-38 probable, also suffering but one loss. 11 float fighters were also involved in interceptions, claiming one P-38 and one B-24, with one P-39 and two B-24s as probables.

Next day 253 Ku intercepted a raid on Buin, 24 A6Ms taking part, their pilots claiming a B-24, nine P-38s, a P-39 and two new Vought F4U Corsairs, for one loss. They also shared a further B-24, a P-38, two P-39s and an F4U with the 204 Ku. This latter unit also claimed an additional two B-24s, two F4Us and four P-38s, while ten A6M2N pilots of 802 Ku claimed two more B-24s and probably a P-40, without loss.

Following these two days of heavy combat, no further raids were experienced for the time being. In March 802 Ku moved to the Marshall Islands and the activities of float fighters over the Solomons came to an end.

During the Lae transport operations detailed in the preceding sections of this chapter, 582 Ku remained in the Solomons. On 6 March 18 of this unit's A6Ms escorted dive-bombers to Russell Island, where about ten fighters intercepted. The efforts of the Zero pilots kept their charges' losses to only one, but in doing so, veteran ace Wt Off Kan-ichi Kashimura was killed. On 10th 129 A6Ms again escorted the unit's dive-bombers to the same target where six US fighters were engaged. Wt Off Tsunoda claimed one F4U shot down, but again one dive-bomber was lost, as were three A6Ms and two of their pilots. Thereafter, the rest of March remained quiet.

"RABAUL KOKUTAI"

Operation 'I-go'

By the end of March 1943 the Japanese High Command no longer had any way to turn the tide of events in the Solomons and New Guinea. Nonetheless, the IJN now planned to seize air superiority over the area at least, even if only for a short period. The method

to be adopted involved the transfer to the area of most of the remaining carrier-based fighter groups. Units already established and operating in the South-West Area at this time remained 204 Ku, 253 Ku and 582 Ku. Reinforcements of Rikkos and Zeros from Truk on 26 March raised these units almost to their established strength; for instance 204 Ku now had 45 aircraft, 253 Ku had 36, and the fighter element of 582 Ku had 27 A6Ms. At the same time the mobilization for land-based duty of the flying units from the carriers *Zuikaku*, *Zuiho*, *Junyo* and *Hiyo* added a further 103 A6Ms and 81 other types. The operations to be undertaken by this enhanced force were given the codename Operation 'I-go'.

Top: Admiral Isoroku Yamamoto, IJN C-in-C of the Combined Fleet, watches the departure of part of the 'I-go' operation force in April 1943. Note the hastily-painted camouflage on the carrier-based Zero in the background.

Bottom: Zeros of 204 Ku lined up, including examples of both the 22 and 32 sub-types.

On 28 March, therefore, 12 A6Ms from 204 Ku and 27 from 253 Ku escorted 18 D3As to Oro Bay, New Guinea, where 204 Ku pilots claimed six Allied fighters shot down while those of 253 Ku also claimed six, but added seven probables, losing two of their own. 11th Air Fleet A6Ms intercepted a raid on Munda on 1 April, pursuing the departing raiders as far as the Russell Islands. During this engagement with an estimated 110 hostile fighters, 12 204 Ku pilots claimed 11 destroyed for two losses, 26 from 253 Ku claimed 23 and five probables, but lost six pilots, while 20 582 Ku pilots claimed 19 destroyed and two probables for one loss.

Having arrived at Rabaul from Truk, the 1st Air Flotilla (*Zuikaku* and *Zuiho*) flew on to Ballale and Buin on 6 April, where they were joined next morning by the 2nd Air Flotilla (*Junyo* and *Hiyo*). At noon on that day (7th) a big raid was launched against Guadalcanal. First over the target were 21 A6Ms from 253 Ku, the pilots of which claimed six intercepting fighters for the loss of two Zeros. Immediately behind them were 26 more fighters from 204 Ku, this unit claiming five more victories for one loss. The assault was then taken over by 27 *Zuikaku* and three *Zuiho* fighters which escorted dive-bombers to the island; the carrier pilots claimed two and a probable, but three *Zuikaku* D3As were lost.

The next wave comprised more dive-bombers, this time from the 582 Ku, escorted by 22 A6Ms, including three more *Zuiho* aircraft. Dense clouds prevented the bombers of this wave being able to attack suitable targets, although their escorts did claim a further two victories for one loss. A third wave of dive-bombers drawn this time from the *Hiyo* air group, and escorted by 24 *Hiyo* and six *Zuiho* A6Ms, attacked transport shipping in Sealark Channel. The fighter pilots of this wave claimed three destroyed and five probables, but lost one Zero; three D3As also failed to return. The final wave incorporated more D3As from *Junyo*, escorted by 23 A6Ms from that carrier group, plus three *Zuiho* aircraft. These fighters became involved in a tough engagement, losing six of their number while claiming five destroyed and four probables. One dive-bomber was also lost and a second crashed on landing.

This day of heavy combat allows some comparison between the IJN losses and the claims of the defending US fighter pilots. 12 'Val' (D3A) dive-bombers were claimed plus one probable by US Marine Corps pilots; seven of these and the probable were credited to one man, 1st Lt James E. Swett of VMF 221, in a single engagement. Marine pilots also claimed 17 Zeros shot down with two more as probables and one damaged. In the same combat, USAAF pilots claimed a further ten Zeros plus one probable, raising the total number of claims for Japanese fighters to 27 destroyed, three probables and one damaged. Actual losses, as have been described, amounted to seven D3As, plus one crashed on return (presumably due to damage suffered) and 12 A6Ms*.

It is important to note, however, that IJN fighter claims amounted to 23 destroyed and nine probables against an actual loss of six or seven F4Fs (one of which – that flown by Swett – reportedly ditched some time after being hit by fire from the rear gunner in one of the dive-bombers), one P-38 and one P-39 and its pilot. The latter was the only US pilot fatality during this action. US records indicate that the D3As did enjoy some success, sinking a 14,500 ton tanker, the destroyer USS *Aaron*, and a New Zealand corvette, HMNZS *Moa*. Given the weather conditions prevailing and the circumstances

---

* However, in his book *The Wildcat in WWII*, Barrett Tillman states on page 161 "…..Actual losses were twenty-nine. including a dozen Vals, in exchange for seven Wildcats." This differs from the information provided by the Japanese historians in the preparation of this book by ten aircraft. Possibly these figures referred both to aircraft lost, and to those damaged and requiring repair before being useable again.

relating to each side, this level of over-claiming and general over-assessment of results was absolutely typical of the Pacific war and of the Solomons fighting in particular. It should certainly not be taken as an adverse reaction to the courage, fortitude and honesty of the pilots of either side.

The second attack in the Operation 'I-go' series occurred on 11 April and involved an attack on Oro Bay on New Guinea's eastern coast; this attack was undertaken by the carrier group aircraft only. 15 *Zuiho* A6Ms arrived over the area first, their pilots claiming nine destroyed and two probables. Two of these claims were made by Wt Off Akira Yamamoto and one by Wt Off Tsutomu Iwai; no losses were suffered. Next on the scene were 27 *Zuikaku* fighters and the dive-bombers they were escorting. PO1c Yoshio Oh-ishi claimed one destroyed and one probable, other pilots adding five more destroyed and a second probable; however, two A6Ms and three D3As were lost. Finally 21 *Hiyo* and nine *Junyo* fighters arrived as cover for *Hiyo* D3As. Only Wt Off Mitsugu Mori was able to submit a claim for one interceptor shot down, but one more dive-bomber was lost.

Next day Port Moresby was the target. 23 *Zuikaku* A6Ms provided top cover with 15 *Junyo* and 17 *Hiyo* fighters, while 18 253 Ku aircraft and 14 from *Zuiho* escorted a first wave of Rikko bombers. Behind them came 24 204 Ku and 20 582 Ku Zeros with the second wave of Rikkos. Against intercepting Allied fighters the *Zuikaku* pilots claimed one shot down, those from *Junyo* claimed four and one probable,

CPO Shizuo Ishii from the carrier *Junyo*, was involved in much action over the Solomons, but was killed after being transferred to 204 Ku.

while the *Hiyo* men claimed nine and five probables, including three victories for Wt Off Hideo Morinio. These top cover units suffered no losses. From the first bombing wave, however, six Rikkos were shot down while their escorts from *Zuiho* were able only to make one claim while attempting to protect them. During the second wave attack pilots from 204 Ku made four claims without loss, but several of the Rikkos which they were escorting were damaged.

Milne Bay was the target on 14 April, where Rikkos escorted by 21 204 Ku, 17 253 Ku and 18 582 Ku A6Ms were joined by D3As covered by 75 Zeros from the carrier groups. Three Rikkos and three D3As were lost during this attack. Seven victories were claimed by 204 Ku, including one and one probable by CPO Ozeki and one destroyed by LdgSea Shoichi Sugita. 253 Ku pilots claimed two and one probable, while those from 582 Ku claimed six and a probable; all three units returned without loss. Meanwhile, 20 of the *Zuikaku* pilots on top cover claimed one victory for one loss, while from the dive-bomber escorts, 20 *Hiyo* pilots claimed three and a probable and *Junyo* claimed for its 18 pilots, seven destroyed and one probable; one of this unit's A6Ms force-landed in the sea off Gasmata during the return flight.

This raid concluded Operation 'I-go', and between 14th-18th, the carrier units returned to Truk. On this latter date Admiral Yamamoto, Commander-in-Chief of the Japanese Combined Fleet, was killed when the aircraft in which he was flying to Buin on a morale-raising visit, was shot down by USAAF P-38s. American code-breakers had identified the time and place of this visit, allowing an aerial ambush to be put in place. Six A6Ms from 204 Ku were flying escort to the two bomber aircraft carrying the Admiral and his staff, and these claimed four destroyed and two probables against the attackers, two of these claims being submitted by LdgSea Shoichi Sugita. The American pilots claimed two Zeros shot down as well as the two G4M Rikkos, but suffered the loss of only one P-38.

The Appearance of Japanese Night Fighters
Following Operation 'I-go', on 25 April 12 582 Ku fighters escorted Rikkos to attack Gatukai Island. Six F4Us were reported to have intercepted and all were claimed shot down. (In fact a flight of four F4Us from a newly-arrived US Marine unit were returning from a bomber escort mission of their own when they came upon the large IJN formation by chance, but attacked at once. Two Corsairs were shot down and a third landed, hit by some 80 bullets; the returning American pilots claimed that they had shot down six of the Zeros!)

Nakajima J1N Gekko night fighter of 251 Ku at Rabaul, carrying the unit identity code 'UI' on the tail.

On 7 May 253 Ku claimed to have shot down a B-17 over Kavieng, but one of the unit's pilots was killed during this encounter. The unit claimed the destruction of two F4Us over Buin on 13 May, again for the loss of one pilot, but following this combat, the unit then withdrew to Saipan. As one of the transport aircraft involved in this move was taking off, it crashed; some 253 Ku pilots were killed in this accident. During some eight months in the Solomons, the unit's pilots had claimed about 110 victories for the loss of 20 of their number.

Elsewhere on 13 May, 22 204 Ku A6Ms and 18 from 582 Ku undertook a sweep over the Russell Islands, becoming engaged with intercepting US fighters. While the 582 Ku pilots claimed four and two probables for one loss, those of 204 Ku claimed 18 and seven probables for two losses. Once more, in a whirling and confused dogfight, neither side's claims bore much relation to the actuality of events. 15 Corsairs from two units had taken part, the Marine pilots claiming 15 or 16 Zeros plus four probables for the loss of three F4Us, with at least one more slightly damaged.

By this time 251 Ku had recuperated at Toyohashi in central Honshu, and now returned to the Solomons on 10 May, when 58 of the unit's A6Ms landed at Lakunai airfield on New Britain. (The kokutai's fixed establishment was 60 fighters, eight reconnaissance aircraft and four transports). By 17 May seven Type 2 land-based reconnaissance aircraft and two 13-shi converted night fighters also arrived there. These were basically models of the same aircraft, but to avoid confusion, a little detail on their development is probably desirable at this point. During the summer of 1938 the IJN Aviation HQ had requested Mitsubishi and Nakajima to produce designs for a twin-engined fighter. At the time Mitsubishi was deeply involved in production of the A6M fighter, G3M and G4M bombers and the F1M floatplane, so declined the invitation. Nakajima took up the challenge, but the assets demanded of the aircraft proved so demanding that the required performance could not be met. Armed with one 20mm cannon and two 7.7mm machine guns firing forward, the aircraft also had four more machine guns mounted in two turrets to fire to the rear.

The first prototype flew on 2 May 1941, becoming the 13-shi two-engine fighter (J1N1), featuring a crew of three; nine of these initial models were built. Following test flights, the aircraft was adopted as the Type 2 land reconnaissance aircraft (J1N1-C – J1N1-R) and was ordered into production in July 1942 as a replacement for the Mitsubishi C5N. However, in this production version the four rear-firing guns and their turrets were deleted, although the crew remained as three; 26 were to be constructed by the end of 1942, and indeed the reconnaissance buntai of the Tainan Ku employed the aircraft in this role. Other fighter units such as 202 Ku and 204 Ku used Mitsubishi Ki 46s obtained from the Army in preference, so the J1N1-R cannot be said to have been an overwhelming success.

At the instigation of Cdr Kozono, two of the prototype J1N1s were converted to night fighters, fitted with two 20mm cannon firing upwards at an angle of 30 degrees, and two more firing at a similar angle downwards, all these guns being mounted in the rear fuselage in what had been the third crew member's compartment. Thus it was that two aircraft so modified were sent to Rabaul in May 1943, together with seven reconnaissance-type J1N1-Rs, as mentioned above.

These aircraft commenced patrol flights as soon as they arrived, and during the early morning hours of 21 May the first engagement with night-flying US heavy bombers took place. CPO Shigetoshi Kudo engaged two B-17s above Rabaul, stalking these from below and firing on them with the pair of obliquely-mounted 20mm cannon with which his aircraft was fitted. Finally, both B-17s were shot down, their demise being witnessed by many of the troops on the ground below.

During the night of 10 June Wt Off Satoru Ono attacked B-24s, claiming one shot down and one probable (following evidence received from a spy, the latter was subsequently confirmed as destroyed). More successes followed, for during the early hours of 11 June CPO Kudo claimed another B-17, followed by his fourth such success early on 13th. Two nights later Wt Off Ono was again credited with shooting down two B-24s. CPO Kudo claimed two more B-17s early on 26th, and then yet one more on 30th.

Thus during a period of 40 nights, Kudo had claimed seven victories and Ono four. Early in July the little night fighter unit moved south to Ballale with two aircraft and five crews.

As a result of these successes, telegrams were despatched from Rabaul reporting favourably, and the aircraft was adopted in August as the Gekko 11 (J1N1-S). Further production resulted in the Gekko 21 (J1N2), which had the fuselage windows originally included for the crew member deleted, and the Gekko 23 (J1N3) with the pair of downward-firing cannon also deleted. By December 1944 477 J1N aircraft would have been built.

From Ballale night fighting over Buin commenced during the night of 7 July when Kudo claimed a Hudson shot down. In the early hours of 17 July another of the unit's pilots claimed a B-24 while one of his fellows claimed a second probably brought down. During the night of 18/19th the crew of a J1N was reported to have brought down a B-17 in three attacks with the heavy expenditure of 260 rounds of 20mm ammunition. This indicated that the pilot had also employed the downward-firing obliquely-mounted cannon with which this aircraft was fitted. Next night a further crew accounted for both a B-17 and a B-24, but on this occasion the observer was killed by return fire from the bombers; both engines were also hit and the pilot had to force-land on the sea where the aircraft at once sank. Finally, on 27 July one crew claimed two B-24s shot down, bringing the total claimed by the Ballale-based night fighters to seven and one probable.

Japanese Fighter Pilots Try to Stem the Tide

The newly-arrived 251 Ku recommenced operations on 14 May, despatching 33 A6Ms to escort 18 Rikkos to Oro Bay. Here six of the bombers, including that flown by the formation leader, were shot down. The Zero pilots responded fiercely, claiming eight of the interceptors shot down plus five probables. CPO Nishizawa claimed two and two probables while also protecting Lt(jg) Takashi Oshibuchi, a young officer on his first operational sortie; Oshibuchi would later become a noted fighter leader. Other new pilots involved in this mission included Lt(jg) Yoshishige Hayashi, who would become the leader of the famous 343 Ku; all returned safely.

Next day, however, 12 251 Ku fighters escorting Rikkos on a search mission for a force-landed aircraft, encountered five B-25s near Surumi. These were attacked, one being claimed shot down and one probably so, but three A6Ms and two pilots were lost in this engagement.

251 Ku then undertook attacks on Wau on 18th, 21st and 22 May, while 204 Ku aircraft strafed ships here on 19th. During the mission on 21st, 21 A6Ms led by Lt(jg) Ohno fought 12 P-38s for 40 minutes, two being claimed destroyed plus one probable (one by Ohno) without loss. The formation then met three B-24s, claiming one of these as a probable also.

In order to try and delay the new Allied offensive which was clearly in the making, the IJN planned a series of attacks to be launched from 7 June onwards. Before this could begin, on 5 June about 60 fighters and bombers attacked Buin. Interceptions were undertaken by 582 Ku pilots who flew a total of 28 sorties during two scrambles. 14 of the attacking aircraft were claimed shot down plus four probables for the loss of three pilots.

Next day 251 Ku moved up to Buka Island while 204 Ku moved to Buin. On 7 June Lt Cdr Shindo of 582 Ku led 81 fighters southwards to attack a variety of targets. From 24 A6Ms launched by 204 Ku, eight (now carrying bombs) dropped these weapons on an airfield in the Russell Islands. These pilots also then engaged about 50 US fighters, claiming nine destroyed and two probables; three of these Zeros failed to return, while three more force-landed on Munda and Kolombangara. From 21 A6Ms of 582 Ku taking part in the operation, only six of the 3rd chutai became engaged, the pilots of these claiming four F4Fs shot down. Four of the 32 251 Ku aircraft turned back, but the other 28 fought for 20 minutes against a reported 100 enemy aircraft, claiming 18 destroyed and five probables. Included amongst these claims were two destroyed by Nishizawa, plus one and one probable by PO1c Yamazaki. Six pilots were lost, including PO1c Masu-aki Endo, and three others force-landed.

Lt Miyano led 24 A6Ms of 204 Ku, 32 of 251 Ku and 21 of 582 Ku to the Russell Islands on 12 June, where again the Japanese fighters became involved in dogfights with defending aircraft. 204 Ku claimed eight victories, some of which were shared with the other units. Claims included one by Miyano himself, one and one shared each by PO2c Sugita and PO2c Kanda. The cost was just one Zero damaged. The 251 Ku pilots claimed ten shot down and one probable, but lost four aircraft and three pilots. 582 Ku also lost three pilots, claiming in return eight destroyed and five probables.

On 16 June 24 D3As of 582 Ku, newly brought up to strength with these aircraft, attacked shipping off Lunga where one transport was claimed sunk and other ships damaged. However, no less than 13 of the dive-bombers were lost in this attack. Of 70 escorting fighters, 16 from 582 Ku which had arrived over the target first, claimed two victories but lost four A6Ms. Having noted in previous attacks that dive-bombers seemed frequently to be shot down after they had released their bombs, Lt Miyano had proposed that the fighters should provide cover at this point, which required them to fly at low altitude for this purpose. Miyano now, therefore, led 24 A6Ms of 204 Ku in this manner. As the D3As were completing their attacks, the Zero pilots were indeed able to claim 14 victories (including six shared and one probable), but the cost was high. Four pilots and their aircraft failed to return, including Miyano himself and PO2c Saji Kanda, who had recently been a fairly prolific scorer, while three more force-landed. 251 Ku's 30 pilots claimed nine destroyed and one probable, but lost six of its own fighters, two of which were seen to be hit by AA fire; one more force-landed in the sea on the way back to base.

Fighting around Munda, New Georgia

These recent Japanese raids had achieved nothing in delaying US plans, and minor landings were commenced on New Georgia and its surrounding smaller islands. Then on 30 June a substantial Allied force came ashore on the northern coast at Rendova, on the opposite side of the island to Munda, in the eastern part of New Georgia where an airfield had been completed.

In response the Japanese immediately sent to the island 15 A6Ms from 582 Ku and 12 from 204 Ku; the latter unit had no officer pilots remaining by now, and was led by CPO Hideo Watanabe. 14 of the Zeros carried bombs, and the formation reached the target area after only 50 minutes in the air. However, patrolling US fighters made it impossible to drop the bombs, though the 582 Ku pilots claimed ten destroyed and two probables on return, those from 204 Ku claiming 14 and four probables – all without loss.

During the early afternoon 24 251 Ku A6Ns escorted 26 torpedo-laden Rikkos to attack shipping in Branche Channel. Defending fighters shot down 20 of the bombers, 18 of the crews being lost, while the escorting Zero pilots claimed eight US fighters shot down and one probable. In doing so they lost eight of their own number, including the hikotai leader, Lt Ichiro Mukai – a young officer ace – and Lt(jg) Takeyoshi Ohno.

The final mission of the day was undertaken by ten dive-bombers escorted by 24 A6Ms from the 582 and 204 Ku, led by Lt(jg) Usaburo Suzuki. The dive-bombers carried out their attack, but whilst the Zeros became involved in a combat with opposing fighters, their pilots made no claims, the whole Japanese formation returning to base without loss.

204 Ku at Buin, September 1943. Notable pilots present included: 2nd row, 5th from left, Kiyokuma Okajima; 6th, Cdr Asaichi Tamai; 7th, Akimasa Igarashi; 9th, Sumio Fukuda; 11th, Matsuo Hagiri; 12th, Susumu Ishihara; 3rd row, extreme left, Kiyoshi Sekiya; 4th row, 5th from left, Takao Banno; 10th, Tomita Atake; 11th, Wataru Nakamichi; 5th row, 5th from left, Yoshio Nakamura; 9th, Isamu Ishii; 11th, Toshihisa Shirakawa.

Next day, 1 July, 32 Zeros escorted six D3As to Rendova, 12 pilots of 582 Ku claiming eight victories without loss, while 12 from 204 Ku, led by CPO Watanabe, claimed nine and one probable at a cost of one of their own. The remaining ten fighters, all from 251 Ku, claimed four but lost an equal number. Next day 24 more Zeros, drawn equally from the 582 and 204 Ku, returned to Rendova, each unit submitting claims for four victories to which the 204 Ku pilots added one more as a probable. Again, no losses were suffered. On 3 July each of the three IJN units provided 16 A6Ms for a further sweep over Rendova where on this occasion claims amounted to three each by 582 and 204 Ku, and two by 251 Ku – without loss.

To take part in the defence of the central Solomons, 11 fighters and 13 dive-bombers from *Ryuho* arrived at Rabaul on 2 July while other units of 11th Air Fleet received replacements. JAAF heavy bombers now also arrived to take part in the attack on the new US bases. By 5th the remaining aircraft from *Ryuho* and a detachment from 253 Ku (a single chutai led by Lt(jg) Saburo Saito) also reached Rabaul.

Thus reinforced, on 4 July 51 A6Ms from four fighter units, accompanied by JAAF Kawasaki Ki 61 ('Tony') fighters, escorted 17 Mitsubishi Ki 21 heavy bombers of the JAAF 14th Sentai to attack Rendova. Here 251 Ku pilots claimed 14 destroyed (including three probably so) for one loss, but eight of the bombers were shot down by AA. Following these heavy losses, JAAF units were never again to take part in any operations over the Solomon Islands.

A further Allied landing at Rice Bay on the eastern side of New Georgia took place on 5 July, seven D3As escorted by 48 Zeros being sent to attack the shipping. The fighter pilots claimed 13 destroyed and three probables without loss. Two days later 41 A6Ms provided escort to six Rikkos to Rendova, located between Rendova and Munda. On this occasion 16 *Ryuho* pilots claimed ten for one loss while those from 204 and 251 Ku claimed nine and two probables for one loss, while a third loss was suffered by 253 Ku.

On 9 July 38 Zeros (including eight carrying bombs) attacked Rendova again, claiming five victories for one loss. Two days later 55 A6Ms provided cover for eight Rikkos to Enogai Inlet, south of Rice Bay, where their pilots claimed 19 victories, four of them probables. Three Zeros failed to return, one each from 582 Ku, 253 Ku and the *Ryuho* detachment. 12 D3As were escorted by 40 Zeros to attack seven Allied destroyers but failed to find their target. Instead, various targets ashore were attacked while the fighters claimed three destroyed and two probables against defending fighters. On this occasion, however, 582 and 251 Ku each lost two Zeros.

Following these operations the fighter unit of 582 Ku was disbanded (the formal date for this actually being 1 August), and 13 of this unit's pilots, including Lt(jg) Usaburo Suzuki, were transferred to 204 Ku to continue operations. The remainder returned to Japan where they were transferred to other units. To compensate for the loss of this unit, 201 Ku which had been recuperating in the homeland, arrived at Rabaul between 12th and 15 July, with 52 A6Ms. Cdr Chujiro Nakano, the commanding officer, recorded the abilities of his pilots at this time as including eight who were skilled, 20 who were competent and 24 who were inexperienced. During this same period the flying unit from *Junyo* also arrived at Rabaul from Truk.

'W1' and '2' on the tails of these Zeros show them to be aircraft of 201 Ku at Buin in summer 1943.

201 Ku commenced undertaking convoy patrols on 15 July, but on that date 12 A6Ms of 204 Ku, flown in the main by pilots transferred from 582 Ku (including Lt(jg) Suzuki), joined 16 from 251 Ku, four of 253 Ku and 12 from *Ryuho* to escort nine Rikkos to Rubiana. The 204 Ku pilots claimed eight destroyed and three probables, while 251 Ku claimed 15 and one probable for a single loss, and those from *Ryuho* claimed four, also for one loss. However, 253 Ku lost three aircraft including that flown by the unit's leader, Lt(jg) Saburo Saito, while five of the Rikkos failed to return and a sixth force-landed. Thereafter daylight missions by the Rikkos were suspended and the 253 Ku detachment, which had suffered heavy losses, was withdrawn to Saipan.

Whilst the Japanese units were concentrating their attacks on the landing areas, on 17 July an Allied force comprising 78 bombers and 114 fighters attacked Buin. Amongst the intercepting IJN fighters, 18 pilots of 204 Ku claimed ten of the raiders destroyed and shared four more with other units, losing two Zeros in doing so. Eight *Ryuho* pilots claimed two victories for three losses. At this time 19 *Junyo* Zeros were on their way to Buin from Rabaul, joining the fray on their arrival. As the lead element attempted to attack US dive-bombers, they were bounced and all four pilots were shot down. Two out of three aircraft forming the third element were also lost, but the unit's surviving pilots claimed six destroyed and one probable; three 251 Ku aircraft also took part in the interception, but one of these was also lost. (On this occasion US aircrews claimed a total of 52 Japanese aircraft shot down against an actual loss of 12.)

On 18 July 18 204 Ku A6M pilots claimed 12 destroyed and two shared for two losses, while eight *Ryuho* pilots claimed four and 11 from *Junyo* added six more for one loss.

Thereafter operations continued unabated. On 21 July 15 fighters from 201 Ku took part in offensive missions for the first time since the unit's arrival, suffering one loss. Next day over *Nisshin* eight of this unit's Zeros led by CPO Masami Shiga, claimed seven destroyed and one probable, but were unable to prevent the vessel being sunk. A force of 54 Zeros attacked Rendova again on 25th, claiming 24 victories in return for five losses. During the next ten days IJN units attacked four times, but on 5 August Munda airfield was captured by Allied ground forces, and nine days later two US Marine fighter squadrons flew in to this new base from where they at once became operational.

During August Japanese attacks were made on Rendova on 1st, 4th (25 victories claimed for five losses), 7th and 10th (four destroyed and two probables claimed for three losses).

Battle for Vella Lavella

The strength of the Japanese Army on Kolombangara had been augmented in anticipation of further US advances, but on 15 August a landing was made by the latter at Barakoma Beach on Vella Lavella. On learning of this news, 47 Zeros were despatched from Buin at 0500 to escort six D3As to attack the landing ships. Here at a cost of five A6Ms, the Japanese pilots were able to claim seven destroyed and eight probables. Between 0910-0930, 45 more Zeros were led by Lt Kawai to escort 11 D3As for a further attack, on this occasion six destroyed and four probables being claimed. During this fighting Wt Off Masaichi Kondo from *Junyo* was seriously wounded. That afternoon 24 more Zeros escorted eight dive-bombers to the area, claiming eight more victories for the loss of three fighters.

Following these initial attacks, Zeros escorted dive-bombers to Vella Lavella on 18th, 21st, 24th, 25th, 30th and 31 August, a total of 30 Allied aircraft being claimed shot down plus one probable for one loss.

In return Japanese bases such as Buin were raided frequently. On 12 August 26 aircraft from 204 Ku, 15 from 201 Ku, 13 from *Junyo* and six from 251 Ku rose to intercept. 201 Ku, led by CPO Kazuo Umeki, made 117 claims for a single loss while the other units claimed a further 12 and four probables without loss. Raids were intercepted on 18th, 23rd and 25th, while on 26th Zeros were twice in the air, 251 Ku claiming ten for one loss. However, CPO Hideo Watanabe, who had led elements of 204 Ku, was hit in the face and PO2c Shoichi Sugita who had made a number of multiple victory claims, was also wounded, both these pilots being evacuated. On 30 August during the final interception of the month, 201, 204 and 251 Ku claimed 16 victories (including four probables during two engagements), losing four of their pilots in doing so.

During September 251 Ku was reorganised as a night-fighter unit with 24 night fighters and two transport aircraft. At the same time the day fighter element of this Kokutai was disbanded and at least ten of the pilots, including Lt(jg) Yoshio Ohba and

Ens Chitoshi Isozaki, were transferred to 201 Ku. At least eight of the other remaining pilots, including CPO Hiroyoshi Nishizawa and Wt Off Kosaku Toyoda, were transferred to 253 Ku; all these pilots continued to serve in the Solomons following these transfers.

At this stage the units from the 2nd Air Flotilla were transferred into those of the 11th Air Fleet. Specifically, the fighter elements from *Junyo* (12 pilots including Lt(jg) Tetsushi Ueno, Lt(jg) Sumio Fukuda and CPO Shizuo Ishi-i) and *Ryuho* (nine pilots including Lt(jg) Sueichi Ohshima and CPO Takehiko Baba) were all transferred to the 204 Ku. Meanwhile, 253 Ku which had been recuperating on Saipan, moved back to Rabaul to re-join operations from 4 September.

On 4 September Allied shipping appeared off Lae, New Guinea, which indicated to the Japanese forces based there, a landing. Three fighter units escorted the attack force which was at once deployed from Rabaul to attack this new threat, and here eight 201 Ku pilots claimed 17 victories for a single loss, while 14 253 Ku pilots claimed only one and one probable in exchange for one Zero. Allied forces did indeed land at Hopoi, to the east of Lae, while next day paratroops were dropped at Nadzab, to the west. Consequently, the Japanese forces at Lae faced attacks from three sides. Rikkos undertook several attacks, A6Ms undertaking escort missions on 6th and 7th. A new threat developed on 22 September when a further Allied landing took place at Finschhafen, north of Cape Cretin on the eastern tip of New Guinea. 38 Zeros were despatched to cover eight Rikkos attacking these landings. Heavy combats again erupted, 23 pilots from 253 Ku claiming four destroyed and four probables in return for two losses, while a dozen 201 Ku men claimed two and a probable but lost four of their number, including their leading ace, CPO Takeo Okumura. Only three 204 Ku aircraft were present, their pilots claiming two and losing one. However, the Rikkos were hard hit, losing six of their number.

During this same period attacks on Vella Lavella continued, 201 and 204 Ku participating in raids here on 9th, 13th, 18th, 23rd and 25th. During the 23 September mission ten victories (including some probables and shares) were claimed by the 27 204 Ku pilots taking part, all for just the one loss. Five of the claims were submitted by CPO Ishi-i, but Wt Off Matsuo Hagiri, who had been claiming steadily as a flight leader, was hit and severely wounded.

Fighting over Buin

While these actions over New Guinea and Vella Lavella had been taking place, the main scene of action during September had shifted to Buin, where Allied air attacks became intense. Zeros were constantly in the air, intercepting these intrusions, but they were having their effect. Late in the month a second airfield on Bougainville was completed at Kara as a substitute for Buin which was frequently being rendered inoperable by the bombing.

201 Ku aircraft intercepted three raids during 2nd and 3rd of the month, claiming six destroyed and eight probables without loss. On 6th, however, one pilot was killed and one badly wounded against claims for 13 hostile aircraft shot down.

On 14 September 201 and 204 Ku undertook five interceptions over Buin. At 0600 20 201 Ku and 23 204 Ku fighters engaged 11 B-24s and 30 escorting fighters; on the second sortie 25 204 Ku A6Ms were again engaged, while on the third occasion both units fought a similar number of intruders. Buin was rendered unusable during these initial attacks, causing the defenders to land at Ballale instead. Just after 1000 both units were off again, this time encountering an estimated 50 dive-bombers and 50 fighters.

Finally, at 1115 16 A6Ms arrived from Rabaul, led by Lt(jg) Fukuda, and these joined the resident units in intercepting what became the biggest fight of the day. During this attack 201 Ku lost two aircraft and 304 Ku lost four. Just before noon ten or more B-24s again attacked, but on this occasion the IJN aircraft were unable to catch them. When stock was taken at the end of the day, 201 Ku claimed 30 destroyed and three probables to which 204 Ku added 18 destroyed and three probables. The outstanding pilot of the day had been CPO Takeo Okumura who had claimed a total of ten, including seven F4U Corsairs, two P-39 Airacobras and a single TBF Avenger.

Next day 16 Zero pilots from 201 Ku made eight claims while 24 from 204 Ku added nine more, plus three probables and two shared; two A6Ms were badly damaged and one force-landed. One 201 Ku pilot was killed on 16th, and one other Zero badly damaged against claims for eight and three probables by this unit, while 26 204 Ku pilots claimed three destroyed, four probables and seven shared, but lost five pilots.

During the morning of 23 September over Buin 13 more victories and three probables were claimed by 201 Ku, but no further successes were gained on 26th, 27th or 29th. On 30th eight P-38s were claimed shot down and two more probably so by 23 pilots from 204 Ku (including one and one probable by Lt(jg) Fukuda), while 35 more pilots from 201 Ku claimed a further five aircraft shot down and a sixth probably.

On 30 September 32 A6Ms from 204 Ku took part in an attack on Munda, claiming a single F4U destroyed here, but losing two pilots killed and one badly wounded. Next morning, 1 October, 31 aircraft from 201 and 204 Ku attacked Vella Lavella, claiming nine victories for three losses.

Japanese forces now began pulling out of Kolombangara between 28 September and 2 October, while on 6 October ground troops on Vella Lavella were also withdrawn. On 7th, 18 204 Ku Zeros attacked Vella Lavella, where PO2c Takao Banno was reported to have shot down two P-38s before falling himself as he pursued a third. 16 pilots from 201 Ku who were also taking part in this raid claimed two more victories.

From Buin on 4 October both defending units put up 54 A6Ms between them. Three destroyed, three probables and three damaged were claimed for one loss by 201 Ku. 8 October, however, saw the withdrawal of 204 Ku from Buin (Kahiri) to Rabaul, leaving only 201 Ku to continue interceptions here. On the morning of 10th, therefore, 25 of this unit's Zeros claimed five victories, including two B-24s, for one loss, while during the afternoon 223 pilots were up again, this time claiming four destroyed and a probable. However, on 11th 201 Ku suffered the loss of five aircraft and four pilots, while on 18th another five went down, although one pilot managed to bale out successfully. Four days later on 22 October five F4Us were claimed shot down for the loss of two pilots, but following this combat, 201 Ku was also forced to retreat to Rabaul.

Night Fighting Activities Develop

251 Ku had become a night fighter unit with an establishment of 24 aircraft (including eight spares); by 1 September 1943 the unit had nine aircraft to hand of which only five were serviceable – and this was considered exceptional. By early September the unit rarely had more than five or six aircraft on hand, of which perhaps three or four were serviceable. On 18 September, the commanding officer, Cdr Kozono, was posted to Yokosuka, his place being taken by Cdr Ikuto Kusumoto who arrived on 25 September.

As the situation in the Solomons worsened for the Japanese, two or three night fighters were maintained at Rabaul, and one at Ballale. During the night of 15 September the crew of one J1N pursued a pair of B-25s, one of these being claimed shot down. In the hours of daylight on 20th CPO Kudo gave chase to a B-24 between Rabaul and Ballale, claiming heavy damage; this was later judged to have been fatal, and was classed as an aircraft destroyed. Ballale came under attack from 16 September, and in mid October the

Detachment here was withdrawn to Rabaul. During a period of some 100 days 251 Ku's crews had claimed five B-24s and one probable, two B-17s, one B-25 and a Hudson, plus one torpedo-boat sunk.

Over Rabaul the raiding bomber crews were obviously very conscious of the presence of night fighters, and the Gekko crews were never able to manoeuvre into position to make an attack during this period. During December the aircraft were employed for ground attack missions, all of which were accomplished without loss. Goodenough and Kiriwina Islands were attacked on 13 December, while early morning sorties were flown to Torokina airfield, Bougainville, on 24 December. In the early hours of 28th Finschhafen airfield was attacked, whilst that night a similar attack was made on Cape Gloucester.

During December CPO Saburo Sue arrived on posting to the unit, and during the night of 3 January 1944 he was able to claim a PBY flyingboat destroyed and a second probably so. In March this pilot claimed a B-24 during the night of 5th and a second of these bombers early on 11th. Another early morning interception by Wt Off Kunishige Hasegawa on 26 March brought claims from him for damage to both a B-24 and a B-25. Already on 1 February the main body of 251 Ku had moved back to Truk. The remains of the unit, now led by Hasegawa, withdrew to this base during 25th and 26 April. By this time the unit had claimed 23 destroyed, four probables and five damaged.

Heavy USAAF Raids on Rabaul
There was to be no respite for the Japanese fighter units which had withdrawn from Bougainville Island to Rabaul, for now a number of heavy raids were made by the US 5th Air Force from New Guinea, over 100 aircraft taking part on each occasion. On 12 October 63 B-24s, 107 B-25s and 106 escorting P-38s raided the New Britain base. At 0825 18 A6Ms of 253 Ku took off, their pilots claiming three B-25s and one P-38 for one loss. Ten minutes later 204 Ku put up 22 more Zeros which claimed two and one shared. Finally, at 1915, 16 253 Ku pilots were able to claim only one B-24, losing one Zero in return.

In response to this attack, a raid by D3A dive-bombers was made on Oro Bay on 15 October, 40 A6Ms escorting the bombers. Here 204 Ku's 24 pilots claimed eight destroyed and seven shared for three losses, whilst the balance of the escorts from 253 Ku claimed three P-38s but lost three Zeros. A repeat attack was made on Oro Bay two days later, 28 pilots of 253 Ku claiming four P-38s and a single P-40, plus one more P-38 as a probable, but on this occasion the unit lost five of its own. During this mission 32 more Zeros from 204 Ku were present, their pilots claiming nine victories including three by CPO Shizuo Ishi-i and two by CPO Susumu Ishihara.

Between 23-25 October US aircraft attacked Rabaul almost constantly. On 23rd 204 Ku intercepted 32 aircraft, claiming nine destroyed and one probable for the loss of three aircraft and two pilots. On the same day 15 201 Ku pilots claimed four P-38s at a cost of just one Zero badly damaged, while 22 more pilots from 253 Ku claimed 14 victories (including four probables) for one loss.

On 24th CPO Nishizawa led 13 253 Ku Zeros into the attack, the unit claiming three B-25s and 13 P-38s shot down (one of the latter being classed as a probable) for the loss of two aircraft, the pilot of one being killed and the other seriously wounded. A further six P-38s were claimed by 26 pilots of 201 Ku, while in a very heavy fight 204 Ku's 26 pilots claimed 11 victories but lost seven Zeros and six of their pilots, including the successful CPO Shizuo Ishi-i.

On the third day of this series of attacks ten Zeros of 201 Ku and 19 of 204 Ku attacked B-25s, but achieved no successes, one of the latter unit's fighters being lost. 16

more A6Ms of 253 Ku were credited with one P-38 shot down and one probable without loss. US 5th Air Force claimed 70 victories during these three days of attacks, 50 of them by fighters. Once more actual losses were considerably less, 15 Zeros having been destroyed, two badly damaged and one force-landed; US losses had, of course, also been very much lighter than had been claimed.

Air Battles over Bougainville Island

During late October the headquarters of the Combined Fleet anticipated that from recent US shipping movements further invasions were imminently likely. The Combined Fleet commander therefore issued orders on 28 October for the flying units of the 1st Air Flotilla to move to Rabaul under the codename Operation 'Ro'; these units arrived on 1 November.

Indeed, one of the reasons for the series of heavy raids launched by US 5th Air Force had been in support of further landings in the Solomon Islands, and on 27 October troops went ashore on the Treasury Islands, just 40 km from Shortland. At once 32 Zeros were sent off to escort D3As to attack naval vessels off the Treasuries, where the fighter pilots claimed one aircraft shot down and one probable for one loss.

Following this move, on 1 November further landings were made at Cape Torokina in the central part of Bougainville Island which is only 350 km from Rabaul. 11th Air Fleet aircraft attacked three times during the day. The first raid proved costly; 44 Zeros escorted dive-bombers, but while the pilots of these reported two victories, no less than 12 Zeros failed to return and two more force-landed. During the second raid 16 A6Ms from 204 Ku attacked alone, claiming one destroyed and one probable, but returning with six Zeros damaged. Finally, on the third mission a dozen 201 Ku aircraft again escorted dive-bombers, but whilst claiming three victories, this unit also suffered the loss of four Zeros.

After the escorting US fleet had been spotted by reconnaissance on 2 November, 89 fighters were despatched to attack. This force included 65 from the 1st Air Flotilla (*Zuiho*, *Zuikaku* and *Shokaku*), and 18 D3As. Over the target area the Japanese fighter pilots claimed six US fighters shot down. Before a further attack could be organised, 100 B-25s and 100 P-38s raided Rabaul. The biggest force yet was launched to intercept, incorporating 38 1st Air Flotilla Zeros and 74 from the resident 11th Air Fleet. Claims were quite incredible, amounting to 121, of which 26 were probables. In return 15 Zeros were lost and four more sustained heavy damage.

Next day 35 1st Air Fleet fighters escorted dive-bombers to Torokina, claiming five of ten intercepting fighters shot down for the loss of two Zeros. Two days later on 5 November more than 200 US aircraft again raided Rabaul, the 71 Zeros that were launched to intercept bringing claims for 29 destroyed and 20 probables. One pilot each from *Zuikaku* and 291 Ku were killed and one aircraft of 204 Ku was also lost. On 7 November 38 1st Air Flotilla fighters and 50 from 11th Air Fleet intercepted 20 B-24s and 45 P-38s, claiming three of the bombers and 13 of the escorting fighters shot down without loss.

On 8 November 71 A6Ms escorted 26 D3As to attack transports off Torokina, but here ten of the dive-bombers were lost. The escorting Zero pilots engaged some 60 defending fighters, claiming 12 and four probables for the loss of five, including Lt Kenjiro Notomi, a veteran fighter leader who was bounced by F4Us from above, and was killed. This is perhaps another useful opportunity to compare losses with US claims. All three US services were involved, US Navy carrier pilots claiming four dive-bombers and eight fighters (four of them unidentified or identified as IJAAF Ki 61 'Tonies'); the Marines claimed three dive-bombers and three more as probables, while the USAAF pilots claimed seven 'Val' dive-bombers and a 'new-type' dive-bomber, plus seven Zeros. This was almost exactly twice the actual losses, and demonstrates the oft-noted phenomenon that claims against bombers are usually considerably more accurate than those against fighter aircraft.

During the morning of 11 November 30 B-24s, 20 P-38s, 60 dive-bombers and 70 carrier fighters attacked Rabaul. 35 1st Air Flotilla Zeros, 24 204 Ku aircraft and 19 from 201 Ku intercepted, claiming 35 destroyed and 11 probables for the loss of nine pilots killed.

Realising that the dive-bombers had clearly come from aircraft carriers, an attack was launched against these by 67 Zeros, 27 dive-bombers and 14 torpedo-bombers. By now the two aircraft carriers which had provided cover for the Torokina landings and had launched the initial carrier aircraft strike on Rabaul on 5th, had been joined by Task Force 50.3 which included three more modern carriers. These had made the morning attack on Rabaul where the Navy pilots had claimed 38 Japanese fighters shot down and nine probables. Now the fighters, joined by land-based Navy F4Us, rose to protect their ships which the attackers discovered close to Mono Island in the Treasuries. The Japanese crews pressed home their attacks fiercely, believing that some ships were sunk; they were not, although several near misses were achieved. However, all 14 torpedo aircraft and 17 of the dive-bombers were lost, together with two Zeros when 33 1st Air Flotilla fighters arrived over the US vessels, claiming five destroyed and one probable. 32 more Zeros from 253 Ku turned back without finding their target. Heavy as these losses were, they paled into near-insignificance compared with the claims of the still relatively 'green' Navy pilots, who claimed 33 torpedo-bombers, 50 dive-bombers and nine Zeros !

Following this debacle, the 1st Air Flotilla units were pulled out of combat next day, returning to Truk. In their few days at Rabaul the three carrier groups had lost 42 of 82 fighters, 38 of 45 dive-bombers and 34 of 40 torpedo-bombers. The loss of experienced fighter leaders had also been severe; it would take a long time to rebuild these precious and highly-experienced carrier units.

Between 5-16 November, Japanese aircraft attacked the US fleet five times, but except for the assault on 11th, these were all made at night and without any notable success. However, the IJN reported its apparent successes without making any reconnaissance check as to their veracity.

Rabaul Becomes Isolated

Following the departure of the 1st Air Flotilla units, further reinforcement of the units remaining at Rabaul took place. 16 Zeros forming two chutais from 281 Ku at Chishima were led to Rabaul's Lakunai airfield by Lt(jg) Torajiro Haruta on 14 November. Eight pilots, including Haruta and Wt Off Iwamoto were transferred with their aircraft to 201 Ku, while the other eight, including Ens Suzuo Ito and CPO Nobuo Ogiya, joined 204 Ku.

After 12 November the heavy raids on Rabaul ceased, allowing 11th Air Fleet to plan a dive-bomber attack on a convoy of transports off Torokina. 56 Zeros undertook the escort of this mission on 17th, but while claims were made for two defending fighters shot down, each kokutai lost three Zeros, and four out of ten dive-bombers were also lost. 48 Zeros from the two units returned to Torokina on 22 November, fighting F4Us against which claims for two destroyed and two probables were lodged, but two Zeros failed to return.

Another US landing on 15 December saw Allied troops ashore at Arawe in the western part of New Britain Island, which was getting very close to home. 54 Zeros at once provided escort to eight dive-bombers, 15 of these also carrying out strafing attacks. Led by Wt Off Tadao Yamanaka, who had recently joined 204 Ku from 202 Ku, all the strafing fighters were hit by AA fire, and while two pilots had to force-land, all crews ultimately returned safely to Rabaul.

Next day, 16th, 28 of 204 Ku's fighters (eight of which, led by CPO Igiya, were to strafe), together with 11 from 201 Ku and 17 from 253 Ku, covered seven dive-bombers to Arawa. One flight led by Wt Off Hideo Maeda claimed three victories and a probable for the loss of one pilot who baled out. In the afternoon 54 Zeros returned to the area, 15 to strafe on this occasion. Four victories were claimed (including two by CPO Ogiya) and one probable, but two Zeros failed to return.

A new raid on Rabaul occurred on 17 December, 72 A6Ms rising to intercept. Their pilots claimed 15 destroyed and one probable but lost three aircraft and two pilots. Now seven fighters from each of the carrier groups of 1st Air Flotilla returned to Rabaul to reinforce the defenders, these 21 pilots all joining 253 Ku at Dobera airfield; from 19th Lt Kenji Nakagawa now led this unit in the air. On the day of their arrival a further Allied raid hit Rabaul and this time 94 Zeros were scrambled, but lost five aircraft. An even bigger force of 98 A6Ms intercepted a raid on 23 December, this time claiming 14 of the attackers shot down and four probables. Notable amongst the defenders on this occasion was Wt Off Tatsuzo Iwamoto who had recently joined 204 Ku, and who made two claims. However, six pilots were lost including Ens Kagemitsu Matsuo of 253 Ku, one of the rare college graduate pilots of the IJN.

Two attacks were made on the Arawe area on 21 December, 64 Zeros escorting dive-bombers on the first such raid, but four of the fighters were lost. During the second mission eight fighters strafed, claiming to have sunk two landing craft and set fire to about 30 small vessels.

Yet another landing took place on 26 December, this time at Cape Gloucester on the north-western end of New Britain. As usual, this was met by a dive-bombing attack which was covered by 51 Zeros.The pilots of these claimed 15 destroyed and three probables, CPO Kanamaru claiming four on this occasion. Five Zeros and three pilots were lost. Next day 76 Zeros from three units returned to the area, 23 of the pilots being briefed to strafe on this occasion, led by Wt Off Masao Taniguchi. Over Cape Gloucester 14 defending US aircraft were claimed shot down and three probables for the loss of four Zeros and three pilots. An attack was made on Arawe on the last day of the year during which the 24 A6M pilots taking part claimed two destroyed and one probable in return for the loss of four aircraft and three pilots.

Meanwhile on 27 December 75 fighters scrambled to intercept raiders over Rabaul, claiming 18 aircraft shot down and seven probables, but losing six Zeros and five pilots. Next day 71 Zeros intercepted, claiming 19 and one shared for the loss of four aircraft and three pilots. The final raid of the year on 30th was met by 82 Zeros, 253 Ku's pilots claiming three without loss, while 201 and 204 Ku claimed six destroyed, one shared and one probable for the loss of a single Zero.

At the start of 1944 the main focus of the war had already moved to the central Pacific with the resurrection of the US Navy's aircraft carrier force. However, on the south-eastern front severe air battles over Rabaul continued. On New Year's Day 1944 some 100 US carrier aircraft attacked Kavieng. 12 A6Ms drawn from 201 and 204 Ku had been moved there for convoy patrol duties, and here they had been joined on 27 December by 36 more fighters from the carriers *Hiyo* and *Ryuho*. The pilots of these latter aircraft claimed two destroyed and ten probables for the loss of four Zeros, while the land-based fighters claimed eight – two by CPO Ogiya – for two losses.

On this same date B-24s and F4Us raided Rabaul, being intercepted by 55 A6Ms from the three resident units. 253 Ku claimed four victories while 201 and 304 Ku pilots claimed eight B-24s shared between them plus two F4Us, the defenders suffering no losses. Next day 78 Zeros were off again, this time claiming seven and three probables, although on this occasion three fighters failed to return and two more were badly damaged.

On 3 January 1944 201 Ku was ordered to Saipan, transferring some of its pilots to 204 Ku before it departed. Since July 1943 the unit had claimed some 450 victories.

Having suffered the loss of the major part of the striking force available, together with Bougainville and the western part of New Britain, the fighters remained the only hope of the whole Japanese force at Rabaul. Now the pilots had regularly to face attacks by 100 aircraft, and from mid January this would rise to 200 on occasion. At this time 253 Ku was usually led by Lt Nakagawa and 204 Ku by Lt Sadao Yamaguchi, who had flown with 202 Ku and who had been posted to his current unit only fairly recently.

On 3rd, 4th and 6 January a total of 159 A6M sorties brought claims for 28 destroyed and 11 probables for the loss of six pilots. 7 January saw 41 253 Ku Zero pilots claim 18 destroyed and eight probables without loss, while 33 204 Ku pilots claimed 16 destroyed and three probables for the loss of only two pilots. 41 pilots of 253 Ku claimed another 18 and eight probables on 9 January, but lost six aircraft and two pilots. 204 Ku also claimed eight and probably five; four more of the unit's aircraft dropped 3-go bombs on attacking aircraft, Wt Off Yamanaka and SupSea Shibagaki claiming six destroyed and six probables with these weapons – the highest total ever to be claimed by this method.

Claims multiplied during this period. On 14 January 253 Ku claimed 28 destroyed and nine probables for the loss of three fighters and two pilots from the 44 engaged. On this same date 204 Ku claimed 25 (including eight probables), Wt Off Tetsutaro Kumagaya claiming two and two probables, while Lt Yamaguchi and Wt Off Maeda claimed two each.

Three days later on 17th 36 pilots from 253 Ku claimed ten without loss, but this was 204 Ku's day, the unit claiming 47 destroyed and 20 probables, also without loss. Wt Offs Iwamoto and Maeda claimed five apiece and CPO Ogiya claimed four. Next day only 253 Ku aircraft intercepted, claiming five but losing four.

20 January saw 38 253 Ku pilots claim 23 destroyed and four probables, and 204 Ku adding 11, including four again by Ogiya. This latter unit lost two aircraft on this occasion, but both pilots managed to bale out.

Still the claims rose. During 22 January 28 pilots of 204 Ku claimed 35 destroyed (including eight B-25s) and five probables for the loss of three aircraft and two pilots, while two more force-landed. On this occasion notable performances were recorded by Ldg Sea Isamu Hashimoto (three B-25s and an F4U), Wt Off Iwamoto (four destroyed and two probables) and Ldg Sea Matao Ichioka (three destroyed and two probables). 25 253 Ku pilots claimed a further seven (including two probables) but lost one pilot.

Next day, 23rd, on the first mission of the day 33 pilots from 253 Ku claimed nine destroyed and six probables for the loss of two, while on a second mission 30 of the unit's pilots claimed a single success. However, on this mission four pilots were killed, one more baled out and one force-landed. During the first mission 34 more pilots of 204 Ku claimed 25 destroyed and nine probables for the loss of three pilots, but during the second 28 pilots managed to claim only two and two probables for the loss of three pilots with a fourth force-landed.

Finally, on 24 January 18 253 Ku pilots claimed one and one probable without loss while 26 from 204 Ku led by Ens Ito claimed ten, three of them by CPO Ogiya, plus five probables. One Zero was badly damaged and a second force-landed.

At this point 204 Ku was ordered to withdraw to recuperate, and on 25 January eight A6Ms led by Lt Yamaguchi departed for Truk while Wt Off Iwamoto, CPO Ogiya and others were transferred to 253 Ku. In the event 204 Ku was to suffer severe losses at Truk in mid February and would in consequence be disbanded on 4 March 1944. It was reputed to have claimed some 1,000 victories since its days as 6th Ku.

To replace 204 Ku, 69 fighters of the 2nd Air Flotilla (*Junyo*, *Hiyo* and *Ryuho*) now moved to Lakunai airfield, accompanied into the 'mincing machine' of the Solomons by 18 dive-bombers and 18 torpedo-bombers.

The Last of the "Rabaul Kokutai"

The newly reorganised fighters at Rabaul were in action on 26 January. 54 of the 2nd Air Flotilla fighters claimed 19 destroyed and ten probables for the loss of four, while 38 from 253 Ku, including the newly-transferred pilots from 204 Ku, claimed ten destroyed and five probables. They lost three aircraft including that flown by PO1C Kiyoshi Shimizu, who had been flying with Wt Off Iwamoto since their 281 Ku days.

In the first of two interceptions next day while 253 Ku claimed nine destroyed and four probables without loss, 31 2nd Air Flotilla fighter pilots added 15 and six probables but lost six pilots, including Lt(jg) Fujikazu Koizumi – a high-scoring veteran. A second interception brought no successes, but no losses either.

28 January brought more heavy claiming. 28 2nd Air Flotilla pilots claimed 16 destroyed and eight probables for the loss of four pilots, while 37 from 253 Ku claimed an astonishing 63 for the loss of three aircraft and two pilots.

Next day the carrier Zero pilots claimed another six destroyed and seven probables for one killed and one baled out. At the same time 34 253 Ku pilots claimed 14 destroyed and a probable, this time for the loss of four pilots. During an initial interception on 30 January, 29 2nd Air Flotilla pilots claimed one and four probables for one loss, while 28 253 Ku Zero pilots claimed 11 without loss. Later that day 26 carrier Zeros produced claims for 23 victories (including six probables); four pilots were killed and one baled out. During the same mission 18 253 Ku pilots led by Wt Off Iwamoto claimed ten F4Us shot down and one probable. One Japanese pilot baled out, seriously wounded. On the last day of January 20 2nd Air Flotilla pilots claimed three destroyed and two probables without loss, while 31 253 Ku men claimed 16 destroyed and four probables for two losses.

At this time the remaining pilots from the 1st Air Flotilla who had remained at Rabaul with 253 Ku, left the unit, some of them being returned to Japan during February. Meanwhile, on 1 February US forces landed on Kwajalein Atoll in the Marshall Islands. In support, 2,891 carrier aircraft sorties were flown against Rabaul between 3rd and 15 February. During the interception of the initial raid on 3 February, the 253 Ku put up 35 Zeros, the pilots of which claimed nine destroyed and four probables, while 2nd Air Flotilla pilots led by Lt Hohei Kobayashi added ten more, five of them probables. On a later interception the two units again suffered no losses, but were able only to add one F4U shot down.

During four consecutive days of US raids, 4-7 February, the 2nd Air Flotilla claimed 29 destroyed and 17 probables while 253 Ku's claims totalled 46 destroyed and ten probables. This unit lost Wt Off Teruo Sugiyama while two more pilots baled out; three carrier Zero pilots were also lost.

Between 9th and 15 February IJN fighters made eight interceptions. During these days the 2nd Air Flotilla pilots undertook 151 sorties, claiming 24 destroyed and 17 probable, but lost eight pilots, while 253 Ku pilots made 197 sorties, claiming 116 victories, of which 26 were probables. This unit also lost eight pilots, including CPO Nobuo Ogiya, CPO Kaoru Takaiwa and Lt(jg) Susuo Ito.

From 6-19th 253 Ku undertook six interceptions totalling 146 sorties, claiming 21 destroyed and seven probables. 12 Zeros and eight pilots were lost. 2nd Air Flotilla was engaged on 17th and 19th, undertaking 47 sorties to claim seven destroyed and six probables at a cost of four pilots.

While these engagements were occurring, on 17th and 18 February US carrier aircraft raided Truk, the largest Japanese Naval base. Although initially in some trepidation, great opposition being anticipated, in practice the Americans practically annihilated the Japanese forces there. With the greater part of the Marshall Islands in US hands, Rabaul became a solitary outpost surrounded by enemy-held territory. It was no longer of any particular use strategically or tactically, the US commanders having decided to allow it to 'wither on the vine'. It was now virtually impossible to re-supply, but would continue to be attacked much as a training exercise, or introduction to operations for new Allied aircrews.

From Rabaul 14 A6Ms of the 2nd Air Flotilla and 25 of 253 Ku flew to Truk on 20 February where from 22nd the latter unit commenced patrols over Truk from Moen Island. The rest of the remaining pilots at Rabaul were moved back to Truk gradually. However, 400 ground crews of the 2nd Air Flotilla and other flying units were evacuated by ship, the vessels involved making for Palau. However, all were intercepted and sunk by air raids and all these personnel were lost.

At the end of February 20 unserviceable Zeros and four Gekkos remained at Rabaul. Under the leadership of Wt Off Shigeo Fukumoto these undertook a few interceptions – seven sorties on 3 March, nine on 10th and seven on 12th. They also undertook some night attacks and reconnaissance flights, but without supplies the number available gradually decreased and on 5 May 1945 the 25th Air Flotilla was formally disbanded, what remained at Rabaul becoming the 105th Air Base Unit on 15 June, which continued to function until the end of the war.

ACTIONS OVER AUSTRALIA AND THE CENTRAL PACIFIC

Attacks on Australia during 1942
By 1 April 1942 the fighter units of the Kanoya Ku and the Genzan Ku were located at Sabang and in Thailand respectively, ready to operate over the Indian Ocean. The Kanoya aircraft were ex 22nd Air Flotilla, and included 27 combat-ready A6Ms and nine spares, while the Genzan unit was similarly-equipped. In the Dutch East Indies, positioned for action against Australia, was the 3rd Ku, which had to hand 45 A6Ms (plus 15 spares) and six C5Ms with two more as spares.

On 3 March nine Zeros from 3rd Ku had been led by Lt Zenjiro Mitano to attack Broome harbour in Western Australia, which was full of flyingboats, evacuating people from the Indies; several more aircraft were located on the nearby airfield. The Japanese pilots claimed 24 aircraft burned on the water or the ground, and three of four shot down in the air. (In fact 21 aircraft were destroyed by the strafing attack and two were shot down, a third falling to the Zeros during their return flight.) One pilot, Wt Off Osamu Kudo, was shot down and killed by fire from the ground. Later that month Lt Tamotsu Yokoyama, the hikotai leader of the unit was posted to the newly-formed 6th Ku, his place being taken by Lt Takahide Aioi.

Port Darwin then became the main target for the 3rd Ku, reconnaissance and strafing missions being flown on 4th and 22nd March, while on 30th and 31st Rikkos were escorted to bomb the area. On the first of these dates the Zero pilots claimed six defending fighters shot down and three probables, while next day they claimed three more and two probables. Their opponents were P-40s of the 9th Fighter Squadron from the US 49th

Two A6Ms of Kanoya Ku at Penang, Malaya. The code 'K-125' can be seen on the tail of the aircraft on the left.

Fighter Group, which had just arrived in the North-Western Territories, gaining their first success over a 3rd Ku C5M on 22 March. They were soon followed to the area by the Group's other two squadrons. In fact there is no evidence of any US fighter losses on 30 March, while only two P-40s were brought down on 31st.

Lt Takahide Aioi led 15 Zeros to cover 24 G4M Rikkos of the Takao Ku to Darwin on 25 April, encountering all three squadrons from the 49th, which they assessed to be 25-strong. The Zero pilots claimed seven destroyed and two probables for the loss of one fighter and four of the Rikkos. The American pilots actually had three P-40s hit which force-landed, one ending up on its nose; they claimed 12 of the raiders shot down.

Two days later 21 3rd Ku A6Ms led by Lt Takeo Kurosawa escorted 16 Rikkos to Darwin, again reporting interception by 25 P-40s. 13 of these were claimed destroyed and six probably so for the loss of one bomber. The Americans claimed three bombers plus five Zeros and a sixth probable. Three P-40s were lost in which two of the pilots were killed, while two more were slightly damaged and the pilots slightly wounded. On leaving the target area the IJN formation flew back to Koepang, Timor, but on 29 April they moved back to Kendari, Celebes, to avoid strafing attacks by Australian aircraft. The levels of claiming in these early raids are indicative of many of the engagements to follow.

At Koepang RAAF Hudsons attacked several times between 22 March and 20 May, 3rd Ku fighters intercepting on five occasions, claiming five raiders shot down and two probables for the loss of one Zero. On 22 May nine twin-engined aircraft attacked Ambon where six Zeros engaged them, their pilots claiming four destroyed.

In June 1942 the 3rd Ku attacked Darwin on four consecutive days. On the first occasion (on 13th), Lt Aioi led 45 A6Ms to escort Rikkos, claiming ten of 20 plus P-40s, and two more as probables for the loss of two Zeros. Next day 27 fighters undertook a sweep without bombers, meeting ten plus P-40s and claiming five and three probables without loss. G4Ms were again escorted on 15th, by 21 Zeros which claimed ten and three probables, again without loss. Finally, on 16th 27 Zeros again escorted Rikkos, once more meeting a reported 20 plus P-40s against which eight were claimed shot down and six more probably so – once more without loss.

Through June and July 3rd Ku aircraft made five interceptions over Koepang – mostly against B-17s; two of these big bombers were claimed shot down, but two Japanese pilots were lost to the gunners in these. On 29 July two Zeros intercepted Hudsons over Ambon, claiming one shot down, while next day 26 of the unit's fighters escorted Rikkos to Darwin, claiming 13 destroyed and three probables for the loss of one Zero.

Top: Fighter leaders at Lakunai airfield, Rabaul, in September 1942. First row, extreme left, Lt Toshitaka Ito (Kanoya Ku), Lt Takahide Aioi (3 Ku hikotai leader); Lt Cdr Sakakibara (3 Ku, detachment commander); Capt Masahisa Saito (Tainan Ku commanding officer); unknown; Lt Cdr Tadashi Nakajima (Tainan Ku hikotai leader). Third row, left to right, Lt(jg) Sadao Yamaguchi (3 Ku); Lt(jg) Baba Naoyoshi (Kanoya Ku); Lt(jg) Takeyoshi Ohno (Tainan Ku).

Bottom: Zero 21s of 3 Ku are seen flying over Celebes on their way to raid Port Darwin in Australia's Northern Territories. The leading aircraft is flown by Lt Takahide Aioi, while on his wing is that flown by Yoshiro Hashiguchi.

During August interceptions were twice carried out over Ambon and twice over Timor. On 15 August one hostile aircraft was claimed over Ambon while on 21st one more and one probable were claimed, this time for the loss of one A6M. Two days later the only attack of the month on Darwin was made by 27 G4Ms escorted by 27 A6Ms. Once again 20 plus defending fighters were reported, the 3rd Ku breaking up into chutais to fight these. 22 victories and two probables were claimed, but the lead chutai of three Zeros all failed to return, Lt Tadatsune Tokaji being one of the pilots lost. A fourth A6M was lost by the 2nd chutai, as was one of the Rikkos. On this occasion the US pilots claimed 15 raiders, their only loss actually being a single P-40 which force-landed after running out of fuel. It was, however, the 49th Fighter Group's last action over Darwin for it then moved to New Guinea, handing over the defence of Darwin to an Australian Kittyhawk unit.

However, following the raid on 23 August, the Rikkos were restricted to night sorties for a time, while on 1 September a part of the 23rd Air Flotilla was ordered to Rabaul. 21 3rd Ku A6Ms and four of its C5Ms were led to this area by the Hikocho, Lt Cdr Sakakibara. After engaging in some actions here, some of the unit's pilots were transferred to 582 Ku (see preceding section), while the rest returned to Kendari on 15 November.

Meanwhile, on 1 November 3rd Ku had been renumbered 202 Ku, and during the absence of the greater part of its pilots and aircraft, those remaining in the Indies flew defensive sorties. Interceptions were undertaken three times during September, twice in November and only once during December. These were all against intruding twin-engined aircraft over Ambon, six of which were claimed shot down during this period. On 1 December the establishment of 202 Ku was reduced to 36 A6Ms plus 12 spares.

Zeros Dominate the Spitfires

During January 1943 B-24s began to attack Kendari, but every time 202 Ku pilots attempted to intercept, they failed to shoot any of these down. These bombers also raided Ambon during January and February, and here PO1C Yoshiro Hashiguchi and others managed to claim three destroyed and six probables thanks to the availability of radar here. At Koepang the kokutai claimed one B-24 probably destroyed on 23 January, while a Beaufighter was claimed shot down by Wt Off Shigeo Sugi-o and other pilots on 28th.

During this period three squadrons of Spitfires VCs arrived in Australia from England to form the 1st Fighter Wing under the leadership of Wing Commander C.R. 'Killer' Caldwell. The Wing was to be stationed around Port Darwin.

Spitfires were first encountered over Darwin on 2 March when Lt Cdr Aioi led 21 A6Ms to escort Rikkos to the area, also strafing Bachelor airfield. During a 20 minute combat the Japanese pilots claimed five destroyed and a probable without loss. No Spitfires were actually lost on this occasion, however, their pilots claiming two Zeros shot down.

Immediately after this engagement Aioi left the unit, being replaced by Lt Minoru Kobayashi. By mid March the unit had only added a single claim for a B-24 over Ambon, but on 15th the new commanding officer led 27 Zeros over Darwin where ten Spitfires were claimed shot down and five probables for a single loss; (four Spitfires were actually shot down, while the pilots of the others claimed seven Zeros shot down and four probables in return). Within days Kobayashi had been taken ill and in April he too was replaced, Lt Cdr Minoru Suzuki, a long-serving and battle-tested leader, taking over the unit, training his pilots in over-sea navigation.

During late March and early April 202 Ku made three interceptions over Ambon, two over Babo in western New Guinea and once over Kendari. One B-24 probable was claimed over Ambonia on 18 March, one B-24 and one probable from five attacked over Ambon on 29 April, one B-25 over Babo on 30 March, when one Zero failed to return from the engagement, and one more B-24 in this area next day.

Another raid on Darwin took place on 2 May when 26 Zeros provided escort for Rikkos. Led by Lt Cdr Suzuki, the fighters became involved in an exchange with Spitfires following the bombing, their pilots claiming 17 destroyed and four probables. (On this occasion five Spitfires were actually shot down, but the rest remained in the air too long, eight crashing or force-landing due to fuel starvation or engine failure.) Seven A6Ms were hit and damaged, but all returned to base. On their arrival, Koepang was strafed by three Beaufighters which destroyed two Zeros on the ground, these attackers then evading pursuit by five of the Japanese fighters.

Nine Zeros were led by Ens Morio Miyaguchi from Langgur, Kai Islands, to attack Miringimbi (Stewart) airfield on 10 May, this base being located well to the east of Darwin. The pilots of the three leading A6Ms, Miyaguchi himself, PO1c Kuraichi Goto and PO2c Kiyoshi Ito, claimed six aircraft shot down while the other six strafed, destroying a Hudson on the ground. One Zero failed to return.

On 28 May seven A6Ms again attacked Miringimbi from Langgur, engaging Spitfires and claiming one and one probable without loss. A month later on 28 June Lt Cdr Suzuki led 27 Zeros from Lautem, Timor, to escort G4M Rikkos to Darwin. Following the bombing 12 Zeros of the 3rd and 4th chutais claimed one and two probables. Three Zeros of the element at the tail end of the formation were hit and one pilot was badly wounded, but all managed to return home after a flight of four hours and 45 minutes.

Two days later 27 Zeros were again led by Suzuki to escort Rikkos, this time the target being Brooks Creek airfield which was believed to be the base of the B-24s. On this occasion defending Spitfires attacked before the bombing. The Japanese pilots claimed 12 destroyed and three probables, gunners aboard the Rikkos claiming one more shot down (five Spitfires were actually lost). One bomber was lost but all the rest of the formation returned after a flight which this time took five and a half hours. 27 Zeros were again involved in escorting Rikkos to Brooks Creek on 6 July, this time led by Lt(jg) Shiozuru, but on this occasion three G4Ms were lost and two Zeros were damaged.

For a time thereafter further attacks were suspended, but the bases in the Kai Islands (Langgur and Tual) came under attack by Allied aircraft during May and August. One raider was claimed shot down here on 15 May and three probables were added on 17th. Over Koepang A6Ms succeeded in intercepting five times, claiming three B-24s and a probable on 6 June, and two twin-engined aircraft and one probable subsequently. Other bases such as Kendari, Makassar and Mimika received one attack each, but only on 15 August were 202 Ku fighters successful, claiming a B-25 over Mimika (western New Guinea) on this date.

202 Ku pilots newly returned from an attack on Port Darwin. In the centre with his right hand extended, is Lt Cdr Minoru Suzuki, the leader of the mission.

On 7 September 36 A6Ms escorted a JAAF reconnaissance aircraft over northern Australia where 15 Spitfires were claimed destroyed plus three probably so for the loss of one Zero from the leading element. The successes claimed included four in 20 minutes by PO1c Kuraichi Goto. Claims remained as optimistic as ever, three Spitfires actually being lost. On this occasion the Australian pilots were more modest in their own claims which amounted to one Zero, one probable and three damaged. Two days later 27 A6Ms attacked Merauke in south-western New Guinea where the unit lost the recently successful Goto. Following this mission the kokutai undertook no further raids or interceptions prior to the end of October.

Float Fighters versus Beaufighters

The 36th Kokutai was formed with floatplanes in June 1942 at Balikpapan, and in November was redesignated 934 Ku. At the end of February 1943 it was decided that a float fighter element should be attached, and this was formed at Yokosuka, arriving in the Dutch East Indies in mid March. By then a floatplane base had been set up at Maikoor in the Aru Islands, located between northern Australia and western New Guinea. It was to here that 934 Ku was sent late in April, and was joined by the fighter element.

Float fighter pilots of 934 Ku at Maikoor, Aru Islands, in mid 1943. Front row, 2nd from right is WO Takeru Kawaguchi (leader of the detachment), while Hidenori Matsunaga is 2nd from right in the back row.

Maikoor was quite near the airfields in north-western Australia, but was beyond the range of single-engined fighters from there, so it was that all attacking aircraft were twin-engined types, or larger, but particularly the Beaufighters of 31 Squadron, RAAF. Two A6M2N float fighters first engaged Beaufighters here on 26 April, claiming their first victory over one of these aircraft, However, Beaufighters usually came in very low to make their attacks and on 26 May they achieved complete surprise, destroying four float fighters and three other float aircraft on the water. Next day two A6M2N pilots were able to claim two Hudsons shot down from a formation of five, nonetheless.

On 12 June six Beaufighters struck again, destroying four A6M2Ns and a single E13A on the ground. Three other A6M2Ns and two F1Ms intercepted, claiming one shot down for the loss of one of the F1Ms. The unit's E13A floatplanes had made attacks on shipping in the Arafura Sea, where one had been shot down on 11 May, and were now to be escorted by float fighters. On 18 June one more E13A was lost, but on 20th a float fighter pilot reportedly shot down a twin-engined aircraft with a single shot.

Four A6M2Ns and two F1Ms intercepted three B-24s on 22 June, claiming one of these bombers shot down. Two A6M2Ns and an F1M then engaged nine Beaufighters, claiming one destroyed but losing the F1M. Two Beaufighters were claimed shot down on 24 July, but during a mission on 10 August an E13A and an A6M2N strayed too close to the Australian coast where they were intercepted by a pair of Spitfires. The E13A was shot down, but Lt Toshiharu Ikeda in the float fighter, despite his aircraft being hit and damaged, was able to claim one of the Spitfires shot down.

Two interceptions on 17 August brought claims for one and two probables at a cost of one pilot killed. Three float fighters intercepted six Beaufighters on 21 August, four of the latter being claimed for one loss. During two further interceptions float fighter pilots claimed one victory on each occasion. Replacement crews then arrived for the kokutai, some of whom took part in a fight between three float fighters and six Beaufighters on 17 September, claiming five shot down and the sixth force-landed, all without loss.

Around this time a number of new airfields were being constructed in western New Guinea to support the operations of the IJAAF which were based along the northern coast of that island. Float fighters were sent to Manokwari for the air defence of one of these new airfields while it was still being built. Here on 21 November two A6M2N pilots claimed to have shot down a B-24 from a formation of seven, and to have badly damaged two others. On the same day over Maikoor two more float fighters fought three Beaufighters, claiming one shot down for one loss. An attempt was made to intercept another formation of seven B-24s, but on this occasion this resulted only in the loss of one of the A6M2Ns. On 10 December 934 Ku was evacuated from Maikoor.

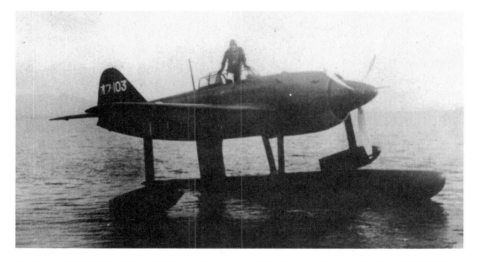

The Kawanishi N1K1 Kyofu float fighter was designed as such, but became the basis of the later N1K1-J Shiden and Shiden-kai fighters. This aircraft belonged to the Otsu Ku (on the tail is painted 'Otsu-103'), and the pilot was Lt Yasushi Kuwajima, who saw action over the Solomons area as hikotai leader with *Sanyo Maru*. 934 Ku took this new aircraft into action for the first time, several claims being made by pilots flying it.

The A6M2N was constructed from a standard A6M fighter by the addition of floats. However, the Kawanishi Kyofu was designed as a float aircraft from the start. Following the remedy of certain initial teething troubles, a number of these aircraft were transported to Singapore on a small aircraft carrier. New replacement pilots, including PO1c Kiyomi Katsuki, who had rammed and brought down a B-17 in the Solomon Islands during 1942, flew these aircraft from Singapore to Ambon via Java. Although rated highly for its high climbing speed, it proved inferior to the A6M2N in terms of manoeuvrability and range, so was not well-favoured by pilots. The first victory for the type was claimed by PO1C Katsuki over a B-24 on 16th or 19 January 1944.

During this period float fighters undertook interceptions over Ambon and Manokwari. On 16 January seven pilots of such aircraft claimed one of six B-24s shot down over Amboina, while three days later seven more intercepted 20 B-24s, claiming two destroyed and one probable for one loss over Ambon. On 1 March the float fighter element of 934 Ku was removed and the aircraft were sent back to Surabaya in Java. Some of the pilots returned to Japan, but seven or eight were transferred to 381 Ku to fly land-based Zeros until the end of the war. The records of 934 Ku are known not to be fully complete, but the float fighters gained at least 29 victories and seven probables for the loss of five pilots.

Quiet in the Marshall Islands

In the spring of 1942 there were based in the Marshall Islands the Chitose Ku (42 fighters and 29 land-based torpedo aircraft), the 1st Ku (17 fighters and 30 Rikkos) and the 14th Ku (16 flyingboats), plus a few floatplanes. In May 15 fighter pilots led by a chutai leader of the 1st Ku were sent to the South-East area, finally returning to the central Pacific area in November 1942 following involvement in severe fighting.

The Chitose Ku was redesignated 703 Ku in November, while 1st Ku became 752nd Ku. The fighter unit of the new 703rd Ku was then separated to become the 201st Kokutai (45 A6Ms and 15 spares). At the same time the fighter unit of the 752nd Ku was also removed and was amalgamated into the new 201st Ku to bring it to its full planned strength.

On Nauru Island which is situated to the north-north-east of the Solomons Islands, raids by US heavy bombers were occasionally experienced. On 7 February 1943 two A6Ms probably shot down a B-17, while on 9th of that month five more pilots led by Wt Off Sahei Yamashita claimed one of these bombers shot down, though Yamashita himself was also lost during this combat.

At the end of February 252 Ku (ex-Genzan Ku) was withdrawn from the heavy fighting over the South-East area, returning to the central Pacific to take the place of 201 Ku, which returned to the homeland. The centre of activity here now fixed on Nauru, due to its being closest to the Solomons, and on 18 March the float fighter element of 802 Ku joined the main body of that kokutai at Emidg in Jaluit Atoll. The float fighter unit then moved to Makin Island in the Gilbert Islands between late August and the start of October, but saw no action. Consequently, on 15 October the unit was deleted from 802 Ku, the seven A6M2Ns involved and their maintenance crews moving to Truk to join 902 Ku.

Meanwhile on 21 April 1943, 24 B-24s had attacked Nauru but were intercepted by four A6Ms led by CPO Ryoichi Hanabusa, the pilots of these claiming two bombers destroyed, one probable and one damaged. During May B-24s also attacked Wake Island and on 18th of that month seven Zeros were sent to Wake from Roi. Here they were to claim one destroyed and three probables.

An A6M2N of 802 Ku at Emidg, Jaluit Island, in the Marshalls group.

By 9 June 252 Ku had 20 A6Ms based on Wake, 16 on Roi, nine on Taroa and 14 on Nauru, while at this time 802 Ku still had its 13 A6M2Ns on Emidg. Three Zero pilots claimed one B-24 probably destroyed on 29 June, but based on intelligence information more fighters were sent from Roi to Wake. Here on 25 June they were engaged, claiming two shot down but losing four of their number. Two days later 18 A6Ms intercepted eight B-24s, reporting that four had been badly damaged; one Zero failed to return.

The US Carriers are Coming!

By now the US Navy had rebuilt its aircraft carrier force with new *Essex*-class carriers and Grumman F6F Hellcat fighters. On 1 September 1943 these new forces undertook their first major operation, attacking Marcus Island. However, on 25 September it was a force of 17 B-24s which attacked Tarawa. A dozen Zeros rose to challenge the bombers, one destroyed and seven damaged being claimed; one Zero pilot was seriously wounded by return fire.

On 6 October US carrier aircraft attacked Wake Island. The first incoming raid was detected by radar at 0242, 23 Zeros led by Lt Motonari Suho being scrambled at 0305 to intercept. These engaged some 100 incoming attackers just before dawn, but suffered a major defeat, 16 A6Ms failing to return including that flown by Wt Off Toshiyuki Sueta. Those pilots who did return claimed ten victories and four probables, but three of them had been wounded. Three Zeros were all that were available to take off for a second mission; these were able only to engage a flyingboat which was not claimed, but which resulted in another Zero being lost. Ashore, the strafing and bombing US aircraft destroyed 16 G4M bombers and damaged seven more.

From Maloelap Lt Yuzo Tsukamoto led seven A6Ms as escorts to seven Rikkos heading for Wake. Before reaching the island, F6Fs were met and three Zeros were shot down, including CPO Bunkichi Nakajima's aircraft. The surviving fighter pilots claimed four destroyed and two probables, but while these four Zeros continued to Wake, the Rikkos turned back to Maloelap. Next day there were further raids on Wake and all available Zeros at Nauru, Maloelap, Roi and Kwajalein were readied to oppose them. However, after this date US attacks on the Marshalls ended for the time being.

252 Ku at Roi, Kwajalein, in the Marshall Islands in July 1943. 2nd row, extreme left is Toshiyuki Sueda; 5th from left, Yuzo Tsukamoto; 6th, Yoshitane Yanagimura (CO); 4th row, 6th from left is Bunkichi Nakajima; 7th, Isamu Miyazaki.

On 15 November 252 Ku A6Ms were located as 16 on Roi, 18 on Taroa and 12 on Nauru. Occasional raids by heavy bombers were being made on Nauru in mid November, and on this date six Zeros intercepted, one being claimed shot down. Three days later on 18th Lt Suho led 15 fighters to attack eight bombers, one being claimed destroyed and three probably.

US carrier aircraft returned to the Marshalls on 19 November, but only over Nauru did six Zeros manage to engage them, intercepting about 20 F6Fs escorting a like number of SBD dive-bombers. The pilots claimed two of the attackers shot down and two probables, one Zero force-landing. Following this raid US Marines landed on Makin Island and on Tarawa in the Gilbert Islands on 21 November, and on Apamama (Abemama) on 27th.

20 A6Ms led by Lt(jg) Sumio Fukuda took off from Taroa on 23 November flying to Mille where 13 four-engined bombers were encountered. Two B-17s were claimed shot down and a B-24 damaged, but two Zeros were hit and suffered serious damage when landing as a result. Next day Lt Suho led 15 Zeros on an offensive reconnaissance mission to Makin, but as they reached a point 85 miles north-north-east of that island they met 30 F6Fs. Although claims for five of these shot down and two probables were made by the survivors, nine of the Japanese aircraft were lost with their pilots. On 25th Lt(jg) Fukuda led 24 A6Ms to the same area, but after a flight of one hour 40 minutes these too encountered a strong force of F6Fs – identified as 52 in number. This time seven Zeros failed to return and two more were damaged, the survivors returning to claim seven Hellcats shot down and four probables. Before long, however, the Gilbert Islands were in US hands and units of the US 7th Air Force were moving to the airfields here.

In an effort to strengthen the Marshall Islands, 30 A6Ms from the three aircraft carriers of the 1st Air Flotilla reached Roi and Kwajalein on 25 November. They were followed by more of these aircraft from 281 Ku which had been stationed at Chishima in the Kurile Islands. The first 21 aircraft reached Roi on 3 December, followed by 17 more between 6th and 8th.

Meanwhile, on 5 December the US carrier task force again attacked the Marshalls. From Roi 26 1st Air Flotilla Zeros and 27 from 281 Ku rose to intercept, claiming between them seven and two probables from amongst the attackers. However, the cost was high again, 281 Ku losing ten pilots including a chutai leader, while 1st Air Flotilla lost six. 12 more Zeros were destroyed on the ground. Immediately after this attack the 1st Air Flotilla was ordered to return to Truk.

281 Ku was then left with 25 fighters at Roi, Kwajalein, while 252 Ku had 30 more at Taroa, Maloelap. During the period 1-21 December 252 Ku Zeros made about 14 interceptions of heavy bombers (full details of these operations have not survived). Claims totalled 13 B-24s and three B-17s destroyed with five more B-24s and a B-17 claimed as probables. Two of the A6Ms involved were lost with one of the pilots being killed.

On 14 December, 13 281 Ku Zeros led by Lt Hasuo were ordered to Mille. From here on 19th they attacked 14 B-24s, claiming one and one damaged, but one Zero pilot was killed and one badly wounded during this engagement. Next day the detachment intercepted eight P-39s, claiming three shot down for the loss of one A6M. Further combats with P-39s and A-20s took place on 23rd and 24th, during which at least two aircraft were claimed to have been shot down plus three probables, for the loss of two Zeros. (The Airacobras encountered in each case were P-39Qs of the 72nd Fighter Squadron of the 21st Fighter Group, US 7th Air Force.) The detachment then returned to Roi.

During late November IJN bombers made several attacks on US carriers, claiming numbers of these sunk. In practice none were even hit.

Loss of the Marshall Islands and the Great Attack on Truk

From 22 December 20-24 A6Ms of 252 Ku undertook interceptions over Maloelap, led usually by Lt(jg) Fukuda. From 22nd to 3 January 1944 B-24s raided in formations of 9-15 aircraft, six interceptions bringing forth claims for ten destroyed, eight probables and seven damaged; no Zeros were lost and only a few suffered some damage from return fire. Typically, on 27 December over Wotje six pilots attacked 16 B-24s, claiming one destroyed and two probables.

From January the US bombers turned to night bombing and against this new challenge four Zero pilots tried to intercept on seven consecutive nights from 7-13 January. It was to no avail, as without radar or other proper equipment they were not able to make any successful attacks.

From 5 January onwards B-25Gs of the 41st Bomb Group, operating from the Gilberts, executed very low altitude attacks. This model of the B-25 carried a nose-mounted 75mm gun which proved very effective in airfield strafing, and against these aircraft the Zero pilots at first found it very hard to achieve worthwhile results. During the first three such attacks 252 Ku pilots were able only to claim some damage inflicted but during the next 12 attacks they were more successful, pursuing the bombers to the limit of their range and claiming 18 destroyed, two probables and 30 damaged for only one loss. However, on 27 January the Japanese pilots found themselves ambushed by a dozen P-40Ns fitted with drop tanks, which had been stationed along the B-25s' return route. In the fight which followed four of the US fighters were claimed shot down, but at a cost of four Zeros. (This combat was recorded by the USAAF as occurring on 26 January, due to the proximity of the International Date Line. The unit involved was the 45th Fighter Squadron, the pilots of which claimed ten Zeros and one probable on this occasion.)

Moen Island, Truk Atoll; many aircraft can be seen on and around the airfield.

At Roi, Kwajalein, Lt(jg) Isamu Mochizuki led 281 Ku's A6Ms in a number of interceptions of bombers, single B-24s being claimed on 29 December and 5 January. Further such sorties brought no more victories but did result in the loss of two pilots.

On 30 January (29th in US time) carrier aircraft again attacked the Marshalls. Over Maloelap 11 Zero pilots led by Lt(jg) Fukuda claimed five F6Fs for the loss of one Zero and three force-landed. Over Roi the Zeros of 281 Ku may have appeared, but no records survive as all personnel of the unit were killed when US forces landed on the island. Next day the US Navy aircraft were back, while on 1 February American troops landed on Roi and Kwajalein Islands in the Kwajalein Atoll which was the main Marshalls base. Here every Japanese on the islands had been killed by 6 February. During the night of 5 February those aircrews who remained on Wotje, Taroa and Eniwetok were picked up by flyingboats or Rikkos and flown out to Truk. From here members of 252 Ku were despatched home to Japan to recuperate. US forces also landed on Mejuro, so that apart from Eniwetok, the western-most of the group, the greater part of the Marshalls was now in American hands.

With the sea to the east of Truk (nicknamed the Gibraltar of the Pacific) controlled by the US Task Force by early February, the greater part of the Japanese Fleet based there left on 10th. Large numbers of aircraft which it had been intended to send to Rabaul were also located there at this time, including a number of fighter units. 501 Ku, which had been formed in July 1943 and which had been operating over the Solomons, had been withdrawn to Truk towards the end of January. In February a Zero fighter-bomber unit was to be added to this kokutai, which had begun training on Eten Island.

204 Ku had also been withdrawn from Rabaul at the end of January and had commenced recuperation with a number of new pilots being posted in. Similarly, 201 Ku had arrived at Saipan from Rabaul earlier in January. It received new Zeros on arrival back in Japan, then returning to Saipan. From here eight Zeros were led to Truk by Wt Off Masao Taniguchi in mid February. The first night fighter unit, 251 Ku, moved to Eten Island in the Truk Atoll on 1 February, going on to Param Island on 7th.

In January 1944 F1Ms and A6M2Ns of 902 Ku moved to Mortlock Island, where several of the former aircraft may be seen.

Finally, there was the float fighter unit of 902 Ku at Dublon Island. This unit had moved to Greenwich Island during November and then to Mortlock Island in January. Both these islands were located between Truk and Rabaul. From here patrol sorties had been flown but there had been no interceptions, although aircraft had been lost on the water to a low-flying B-24 at Greenwich, after which the unit had returned to Truk.

The Japanese Navy received an alarm on 15 February, indicating an attack by US carrier forces on Truk. This intelligence had been obtained by listening services, tuned in to US transmissions. However, search aircraft found nothing, and next day the alert condition at Truk was cancelled. At 0430 that morning Truk radar stations indicated approaching hostile aircraft, but someone on the spot incorrectly identified these as heavy bombers approaching at high altitude. As a result no immediate alert was issued to the fighter units, and it was not until the first F6Fs appeared overhead at low level that the Zero kokutais started to take off. By then, it was too late.

Eight 201 Ku A6Ms led by Wt Off Taniguchi got off first, engaging in six interceptions during which seven F6Fs and two TBF Avengers were claimed destroyed and four F6Fs probably so. Three of these victories were claiming by PO2C Kantaro Yamaguchi, but two of the unit's pilots were killed, one badly wounded and a fourth less seriously so. 501 Ku took part in combats throughout the day between 0510 and 1530, claiming ten F6Fs and three TBFs, but losing 11 Zeros including that flown by the unit's leader. The 31 204 Ku pilots to take off claimed 23 F6Fs and five bombers shot down plus two F6Fs as probables, but 18 of them were killed. Eight of 902 Ku's A6M2Ns were also airborne, PO1C Takeji Ito and LdgSea Mitsuo Suzuki each claiming two F6Fs following which the former pilot baled out and the latter force-landed on the sea. PO1C Sotojiro Demura claimed one F6F shot down, but was then shot down himself, while four of the other float fighter pilots were killed. Only one A6M2N survived to make a second sortie, but this was then also hit and force-landed on the sea. 251 Ku had nine Gekkos on Param Island, but only two of these managed to take off, the other seven being destroyed on the ground. Those two sought to take refuge from the marauding F6Fs, but one failed to return and the other was eventually destroyed after landing from such a flight.

Next day the carrier aircraft attacked again, meeting no opposition in the air. 24 A6Ms which flew down from Saipan to Truk failed to encounter any raiders. During the two days of operations, despite the claims made against them, only eight F6Fs were actually lost, several of them to AA fire rather than to opposing aircraft. Following these raids, US forces landed on Eniwetok Atoll on 19 February, securing an anchorage for the carriers to prepare for their next venture. It had been a stunning American victory.

THE INDIAN OCEAN, CHINA AND THE MARIANAS

The China Front

Following the outbreak of the Pacific War in December 1941, the air over China was left to the IJAAF. However, the sea route along the Chinese coast became dangerous due both to US submarines and to the attacks of US aircraft from Chinese bases. The IJAAF was not able to deal with these threats. On 4 May 1943, B-24s and fighters attacked Hainan Tao where some of the IJN's training bases and some important harbours were located.

In consequence, nine fighters of the Sasebo Ku moved to Sanya on Hainan Tao on 12 May, while on 1 October 1943 the 254 Kokutai was formed from the Sasebo detachment, commanded by Lt(jg) Hiroshi Maeda. The unit was initially equipped with four torpedo-bombers and 24 fighters. On 26 October six of this unit's A6Ms intercepted four B-25s, PO1C Takeda ramming one B-25, but losing his life in doing so. The unit would claim one further victory, but a second pilot was killed.

In mid January 1944 a detachment from 254 Ku was sent to Hong Kong where on 23rd 15 P-40s and nine B-25s attacked. Six Zeros intercepted, the pilots claiming three destroyed and one probable. The latter and one of the former were claimed by LdgSea Kunimichi Kato during his first combat. He would soon become the unit's top-scoring pilot, his skill in downing B-25s becoming a text for other young pilots.

Two pilots, believed to be Lt Yamazaki and Ens Toyoda, in the command post at Longhua airfield, Shanghai.

Hong Kong was attacked by 12 B-25s and 15 fighters on 11 February, six 254 Ku Zeros again intercepting. Kato claimed one of each, while two other pilots were credited with a victory apiece, although both were killed during the fight. Three days later Lt(jg) Yoshi-ichi Minami led nine A6Ms of a detachment sent by 254 Ku to Haikou, to the Nanning area. Here three P-40s were encountered, one being claimed shot down and one probably so in return for the loss of one Zero.

For the defence of the central China coastline the 256 Ku was formed on 1 February 1944, with 24 fighters and eight torpedo-bombers. Hikotai and buntai leader was Lt Hiroshi Mori-i. The unit undertook convoy patrol sorties and defensive flights over the base while training pilots, but three accidents were suffered during the first month alone.

On 5 April three chutais of fighters attacked Nanning. Two of the chutais were formed from instructors from the Sanya Ku and Haikou Ku, all equipped with bomb-carrying Zeros, while nine standard 254 Ku aircraft provided escort. The target area was found to be shrouded in cloud, and while searching for their target beneath this overcast, the two fighter-bomber chutais were bounced by US fighters, and nine were shot down including a veteran fighter pilot, Lt(jg) Nakahara. Flying above the clouds, the 254 Ku escorts failed to see anything, but dived down to strafe an airfield, then returning to base unscathed. The fighter-bombers had all fallen to P-40s of the 51st Fighter Group, the pilots of which thought they were engaging Japanese Army aircraft and claimed an extaordinarily accurate nine shot down, three probables and three damaged.

Immediately following this disastrous encounter, 254 Ku intercepted bombers over Sanya on 6th, 8th, 9th and 12 April, claiming four B-25s and one B-24; two of the B-25s were accounted for by LdgSea Kato. However, the unit's leader, Lt Maeda, and a second pilot were killed, Maeda on 8th. On 18 April the Hong Kong detachment also claimed one B-24 but in mid May this detachment turned its interception duties over to 256 Ku and returned to Hainan Tao. Sanya was attacked by 25 B-25s, six B-24s and ten P-38s on 29 July. 18 254 Ku Zeros intercepted, three P-38s being claimed shot down plus two probables and three B-24s damaged, but four A6Ms were lost.

In October the Philippines Campaign commenced and almost all the fighters of 254 and 256 Ku were sent to Tainan, Formosa, and then on to Luzon to take part in the fighting here. Those that survived returned to China in mid November.

Meanwhile, B-29 Superfortresses had arrived in China and began bombing northern Kyushu, western parts of the Japanese homeland, and Manchuria. In mid October three new Mitsubishi J2M Raiden interceptors were sent to Shanghai where flying with 256 Ku, led by Lt(jg) Chiyoyuki Shibata, they tried without success to catch bombers returning from their raiding sorties. On 11th and 12 November B-29s attacked Shanghai, and here too 256 Ku Zeros and Raidens sought to intercept the bombers, but again without success.

256 Ku was disbanded on 15 December 1944, its aircraft being amalgamated into 951 Ku and returned to Japan prior to participating in the Okinawa operations. 254 Ku was also disbanded on 1 January 1945, being amalgamated into a 901 Ku detachment. 11 of this unit's A6Ms intercepted P-38s and P-51s on 5 January, claiming four destroyed and one probable for the loss of three Zeros, while on 16 January 10-14 Zeros fought eight US Navy dive-bombers and 11 F6Fs, claiming five victories but losing three of their own. In April the unit returned to Shanghai where little action was seen until the war ended.

Indian Ocean
Over the Indian Ocean, Andaman and Nicobar Islands, Army fighter units had been operational for some time by 1943, but these were gradually withdrawn one at a time to be cast into the fierce battles raging in New Guinea. On 1 July 1943 therefore, 331 Ku was

formed at Saeki with 24 fighters and 24 torpedo-bombers under the veteran hikotai leader and fighter pilot, Lt Cdr Hideki Shingo. The unit went aboard the carrier *Junyo* at the end of August for transport to the area, having increased the unit's establishment to 36 fighters, but with only 18 torpedo aircraft. These all took off to fly to Sabang, a small island situated at the northern tip of Sumatra. On 1 September the torpedo-bombers were detached from the unit to form a new 551 Ku.

Allied aircraft occasionally flew over the Andamans and Nicobars on reconnaissance sorties, or to drop a few bombs, and in consequence 331 Ku sent detachments to Port Blair, Mergui, Waingapu and to Sumba, west of Timor. On 22 September Lt Cdr Shingo led nine Zeros on their first interception in the area, intercepting a Liberator of the RAF's 160 Squadron over Nicobar and this was shot down, although one A6M was also lost to the aircraft's gunners. A further Liberator – again an aircraft of 160 Squadron – was intercepted and shot down in the same area on 26 October by nine Zeros led by Ens Hiroshi Suzuki.

On the Burma Front a major bombing attack on Calcutta, India, was planned to incorporate both JAAF and IJN bombers and escorts, the only time during the war when such a joint assault would be made. The raid occurred on 5 December 1943 and was intercepted only by RAF Hurricanes, Spitfire units being located too far away. The results proved costly for the defenders, and five fighters of 176 Squadron which attempted to attack the nine IJN G4Ms taking part,

Lt Cdr Hideki Shingo, hikotai leader, addresses his pilots at Sabang, an island just off Sumatra.

found themselves disadvantageously placed beneath the 27 escorting A6Ms provided by 331 Ku. Three Hurricanes were shot down and a fourth was badly damaged; Lt Cdr Shingo, Wt Off Sadaaki Akamatsu, Wt Off Masao Taniguchi and PO1C Moriji Sako each claimed one, and two more were claimed as probables.

On 23 January 1945 four A6Ms led by Ens Hiroshi Suzuki reported intercepting 13 B-24s over Mergui, Thailand, claiming two shot down while a third force-landed with engine failure. The intruders were actually 11 RAF Liberators of 356 Squadron on a special duties mission to Indo-China; three were indeed lost on this occasion.

Meanwhile on 1 October 1943 the 381 Ku had been formed at Miho, central Honshu with a strength of 36 fighters. Later some J2M Raidens were added to the unit's Zeros, and early in March 1944 a move was made to Balikpapan in Borneo to defend the oil refinery there. The kokutai was led by Lt Takeo Kurosawa who had served with 3rd Ku during 1941-42 as a buntai leader. Here on 1 April the unit was considerably expanded, with Sento 602nd Hikotai (abbreviated as S602) with 48 fighter-bombers, led by Lt Kurosawa, S902 with 24 night fighters led by Lt Hideo Matsumura who had previously served with 934 Ku, and S311 with 48 fighters, led by Lt Kunio Kanzaki and formed with some of the aircraft and pilots of 202 Ku. While based at Balikpapan, 381 Ku despatched detachments to Kendari and Surabaya.

At Iwakuni, western Honshu, a new 331 Ku with 24 fighters was formed in March 1944. One buntai led by Lt Akira Tanaka moved to Sabang in mid-May, then moving to

Penang, Malaya, due to the serious situation which had arisen in the Mariana Islands. 12 of the unit's aircraft then moved to Peleliu on 17 June, patrolling over this island until the end of the month. Nothing having been seen, the detachment returned to Penang.

Sento 311 was transferred from 381 Ku to 153 Ku on 5 May, taking part in fighting over Biak, 5-16 June during which period 32 victories were claimed for the loss of 15 fighters in the air and most of the rest of the unit's aircraft on the ground.

During October 1944, 331 Ku became a mixed unit with Sento 309 (48 fighters led by Lt Kisuke Hasegawa) and Kogeki 253rd Hikotai (48 torpedo-bombers), moving from Penang to Balikpapan. On 1 September, meanwhile, 381 Ku moved 40 A6Ms of S602 and eight Gekkos of S902 to Balikpapan, 32 A6Ms, nine Raidens and two Gekkos to Kendari, and two Gekkos to Surabaya.

From September US B-24s began raiding Menado on the north-eastern tip of Celebes. To counter this a detachment from S602 led by Wt Off Keishu Kamihara undertook interceptions using 3-go bombs (air-bursting phosphorus), claiming several of the attackers destroyed. Heavy bombers then raided Balikpapan four times, starting on 30 September. 331 and 381 Ku fighter pilots claimed more than 80 victories during these attacks (US sources admitting the loss of 19 B-24s and six escorting fighters). On both the first raid and during one on 3 October Gekkos of S902 and F1M floatplanes from the 11th Special Base Unit took part in these interceptions, Lt(jg) Masaji Nioshiwaki of the latter unit claiming to have brought down three B-24s with one 3-go bomb. This pilot had earlier claimed two P-40s shot down over the Philippines during the early stages of the war when flying from the floatplane carrier *Mizuho*. He later served on the staff of the University of Tokyo where he became famous as the 'Whale Professor'.

During raids on 10th and 14 October, US fighters escorted the B-24s all the way to their targets. This prevented the Gekkos and F1Ms taking any further part in the fighting, while losses amongst the defending Zeros increased.

Detachments from both 331 Ku and 381 Ku were sent to the Philippines during October, while in February 1945 S309 with the greater part of the remaining fighters withdrew to Singapore. From here convoy patrols were flown along the coast of Indo-China. In mid March S902 moved to Bali Island to cover the withdrawal from Timor, following which early in April the greater parts of S602 and S902 were sent home to Japan. In May S309 was ordered to follow, but on arrival at Ohmura was disbanded on 15 May. Following the disbandment of 11th Ku, 12th Ku and 13th Ku, the aircraft of these units were amalgamated into 381 Ku which included 24 fighters, 12 torpedo-bombers, 12 Rikkos and 24 advanced training aircraft. As this force prepared to move to Sumatra, the war ended.

Before the Battle of the Philippine Sea
i) The Organisation of the 1st Air Fleet and the February 1944 Marianas Raid

On 1 February 1944 just before the end of the battle for the Marshall Islands, the 1st Air Fleet composed of land-based air units was formed. It would become the mainstay of the air battles in the Marianas and Caroline Islands. The 1st Air Fleet comprised the 61st Air Flotilla with ten newly-formed kokutais and the 62nd Air Flotilla with 11 kokutais, also newly formed.

Amongst these new kokutais were ten equipped with fighters. The first of these, 261 Ku, had an establishment of 54 Zeros as at 1 February, but to this and all the other units were added an additional one third of the established number as spares. The other units were:-

61st Air Flotilla
263 Ku (54 Zeros)
321 Ku (54 Gekkos)
341 Ku (54 Shidens)
343 Ku (54 Shidens)

62nd Air Flotilla
265 Ku (54 Zeros)
221 Ku (54 Zeros)
345 Ku (54 Shidens)
332 Ku (18 Gekkos – formed on 15 March, the unit becoming S804)
361 Ku (36 Shidens – became S407 on 15 March)

n.b. The units of the 61st Air Flotilla were formed before those of the 62nd Air Flotilla.

The full establishment of 732 aircraft required for the ten kokutais forming the 61st Air Flotilla could not be provided quickly, and all units were understrength from the start. For instance amongst the better provided for, 261 Ku had 76% of its establishment and 263 Ku had 80%. It had been intended that the new fighter, the Shiden, would be the main element of the fighter force, but teething troubles prevented it making its major debut during the Marianas fighting. Indeed, 343 Ku was obliged to enter action with A6Ms instead, as did 265 Ku, while 341 Ku had no aircraft at all at this time. The Shiden was the Kawanishi N1K1-J, which later received the Allied codename 'George'. It was a land-based derivative of the Kyofu float fighter. Additionally, the greater part of the fighter pilots in these new kokutais were hastily trained and newly qualified, and had received inadequate and insufficient training for their critical role.

Time for further training had run out, however, and when US carrier aircraft attacked Truk on 17 February, all units of the 61st Air Flotilla were ordered to the Mariana Islands. Amongst these 263 Ku arrived on Tinian with 18 Zeros on 21 February, joined by 12 Gekkos of 321 Ku. Next day 68 Zeros of 261 Ku flew from Katori to Iwo Jima. Even as they arrived, reconnaissance aircraft spotted the American task force for the second time; it had previously been seen south of Eniwetok on 20th. On receipt of this second sighting, the commander of the 1st Air Fleet ordered the Rikkos of 761 Ku to mount a night attack. 29 bombers went out in three waves to undertake this order, but 15 were lost and no hits were gained.

Before dawn on 23 February 321 Ku launched five Gekkos to search for the US Fleet, but these were swiftly intercepted by US fighters and four were shot down. Around the same time nine Rikkos from a formation of 11 which had taken off just before the Gekkos, were also shot down. Four dive-bombers of 523 Ku, two of which had taken off from Saipan and two from Tinian, attempted to attack the task force. Two were thought to have achieved hits on carriers, but the other two failed to return. Carrier aircraft then raided Saipan and Tinian in three waves. Here on the two islands at this time were 18 Zeros of 263 Ku, four of 201 Ku, and 12 Gekkos of 321 Ku. Details of the fighting which followed are not known, but 201 Ku lost all its aircraft, 321 Ku lost ten Gekkos and 263 Ku suffered the loss of 17 aircraft, either in the air or on the ground.

ii) The Reorganisation of IJNAF Air Units and the March 1944 Raid by US Carrier Aircraft

On 4 March Navy aviation separated the air echelons of units from their ground organisation, instituting instead a rather complicated and confusing system of numbering. The flying elements became independent from the kokutais, being given instead a three digit number preceded by the class of aircraft; the hikotai was inserted after the number. Thus a flying unit with 36 fighters and 12 spares became Sento 301st Hikotai (abbreviated to S301) but remained under the command of 202 Ku. The flying unit of the old 331 Ku became S603, and also came under the command of 202 Ku.

S601 was now the flying unit of 301 Ku, with a newly-formed S316 (36 Zero fighter-bombers plus 12 spares), but remained under the command of 301 Ku. 201 Ku's flying unit became S305, while 204 Ku was disbanded, the pilots of this unit forming the core of S306 (36 Zeros and 12 spares); S305 and S306 came under the command of 201 Ku. 552 Ku (a dive-bomber unit) was disbanded, the pilots becoming the core of the newly-formed S351 (36 Zero fighter-bombers and 12 spares), the hikotai coming under the command of 501 Ku. The flying unit of 253 Ku which had 72 Zeros and 24 spares was separated and at the same time sub-divided to form S309 and S310, each with 36 and 12 spares, both remaining under the command of 253 Ku. 251 Ku's flying unit became S901 with 18 Gekkos and six spares.

Having taken the Marshall Islands, US forces then occupied the Admiralty Islands during February and March, B-24s being brought forward to airfields here. From 15 February, therefore, these bombers raided Ponape, then also beginning attacks on Truk in the Carolines group. Before dawn on 29 March a 251 Ku Gekko attempted to attack a B-24, but failed to do so. At 1030 that morning, however, 39 Zeros of 202 Ku and 19 of 253 Ku managed to intercept 16 B-24s, claiming one shot down and two probables. However four 202 Ku pilots, including the veteran Wt Off Mitsu-omi Noda, were lost. Next day Wt Off Tetsuzo Iwamoto led 34 202 Ku aircraft and 16

WO Hisamitsu Yamazaki and a Gekko night fighter of 202 Ku.

from 263 Ku after ten B-24s, two destroyed and two probables being claimed. During the night of 31 March two Gekkos of 351 Ku intercepted B-24s from Kwajalein, inflicting damage on one of them. However, they did not prevent the bombers from hitting the Japanese airfields which were their targets, destroying 20 Zeros, four Rikkos, two reconnaissance aircraft and two transports.

On 20 March ten Zeros of S351 had moved to Peleliu, 26 more from S305 going from Saipan to Peleliu over the next three days. During the morning of 28 March a search aircraft from Meleyon Island spotted aircraft carriers to the south. Combined Fleet Headquarters decided that the carrier group was planning merely a raid on Palau as there appeared to be no landing force accompanying the task force.

As anticipated, a force of 456 US carrier aircraft raided Palau Island throughout 30 March, attacking in 11 waves. 20 Zeros of 201 Ku took off at 0530, engaged 50 F6Fs and claimed 15 of these shot down (including three probables) plus two TBFs. Four victories

Isley airfield, Tinian Island.

were claimed by PO1C Shinsaku Tanaka, but the unit lost nine Zero pilots including veteran ace Ens Koshiro Yamashita. Four further Zeros were badly damaged and another two force-landed, while all other aircraft in the Japanese formation were hit. 12 S351 Zeros also got off, claiming two and two probables, but five were shot down and the fighters which survived were all destroyed on the ground by later waves of attackers. 52 more Zeros of 261 and 263 Ku flew in after the raid, but eight of these were damaged on landing.

During 31st there were raids all morning. 261 Ku intercepted, this unit's pilots claiming 15 destroyed and three probables but 20 Zeros were shot down, four more were badly damaged and four were destroyed on the ground, leaving the kokutai with no aircraft. 263 Ku claimed five victories but lost 15 of its own aircraft shot down, two destroyed on the ground and one badly damaged. During these raids, therefore, the 26th Air Flotilla lost about 50 aircraft while 1st Air Fleet lost around 90, the number remaining available to the latter command being reduced to 118 of which only 92 were serviceable. Moreover, the headquarters staff of the Combined Fleet were flown out to Davao, Philippines, in two flyingboats. That carrying the commander-in-chief simply disappeared while the second aircraft landed on the sea, those aboard, which included the chief of general staff, were captured by guerilla forces although they were later handed back. This was a notorious incident known as "Kaigun Otsu Jiken" (Navy Number 2 Incident); the whereabouts of the aircraft carrying the commander-in-chief was never discovered, while how and why the 'high brass' in the second aircraft came to be returned by guerillas was never explained by the survivors. Certainly, none in that latter group were accorded any blame by their superiors.

iii) Preparation for the 'A-go' Operation

Following the raids of March, the land-based air force in the Marianas and Carolines was strengthened. By 15 April strength of the fighter units had increased to:-

261 Ku 25 Zeros on Saipan; 16 Zeros on Meleyon
263 Ku 22 Zeros on Guam
343 Ku 12 Zeros on Tinian
321 Ku 5 Gekkos on Guam
202 Ku 27 Zeros at Moen, Truk
253 Ku 32 Zeros at Eten
251 Ku 6 Gekkos at Eten
381 Ku 24 Zeros and 2 Gekkos at Balikpapan
301 Ku in Japan, receiving aircraft
501 Ku under training at Davao, Philippines

Through April US heavy bombers continued to attack Truk, and on 2nd 33 Zeros of 253 Ku intercepted a formation of 26 B-24s, claiming five destroyed, three probables and three damaged for the loss of one Zero. Next day five 202 Ku aircraft made a further interception, one shot down and two probables being claimed. During the night of 4 April six Zeros from 202 Ku managed to find the bombers, claiming two shot down. A pair of Gekkos also took off but were not able to achieve any success.

At 1930 on 6 April three 202 Ku aircraft led by Lt Cdr Suzuki, the hikotai leader, again achieved a night interception, claiming two more B-24s. Success for the Gekko crews finally came on 11th when aircraft of S901 and a detachment from 321 Ku which had reached Eten earlier in the month, between them put up three of the night fighters, the pilots of which claimed two B-24s. Next night six more Gekkos made a further interception resulting in claims for one more shot down and one probable. By day on 18th eight Zeros of 202 Ku intercepted 25 bombers, claiming one. The effects of the bombing which had in the main been aimed at airfields, had resulted in decreasing the strength of 202 Ku to ten aircraft, 251 Ku to five and 253 Ku to 25 by the end of the month.

253 Ku at Truk, 29 April 1944. Front row, extreme left, Tetsuzo Iwamoto; 3rd from left, Harutoshi Okamoto; 2nd row, 6th from left, Ken-ichi Takahashi.

Then came the carrier aircraft. Radar on Truk picked up the first echoes at 0430 on 30 April, allowing 28 Zeros of 253 Ku to make the first interception at a height of 3,000 m over Polle. Returning pilots claimed four F6Fs and a single SBD, but 20 of their comrades failed to get back. Four more of this kokutai's Zeros took off at 0635, but two of the pilots were killed and a third seriously wounded. Seven 202 Ku fighters led by Lt Shiro Kawakubo then escorted attack aircraft to strike at the carriers, the Zero pilots claiming 16 defending fighters shot down and two probables for the loss of four Zeros.

The 'A-go' operation was planned to be the first decisive carrier-versus-carrier battle since that at Santa Cruz in October 1942. Separation of the carrier flying units from their ground crews had also taken place as with the land-based units. The flying units of the 1st Air Flotilla from the carriers *Taiho*, *Shokaku* and *Zuikaku* had joined to become 601 Ku; those of 2nd Air Flotilla from *Junyo*, *Hiyo* and *Ryujo* had become 652 Ku; those of 3rd Air Flotilla from *Chiyoda*, *Chitose* and *Zuiho* had become 653 Ku, while those of 4th Air Flotilla from the battleship carriers *Ise* and *Hyuga* became 634 Ku.

The new aircraft carrier *Taiho*.

When the old 1st Air Flotilla units had become 601 Ku on 15 February 1944, they had already lost 71% of their aircraft and 45% of the pilots mainly during the air battles over Bougainville and Rabaul during the previous November. As a result the main body of the unit had become new graduates from the flying schools, the core of the unit containing just ten experienced CPOs. The new pilots generally had only 100-150 flying hours in their logbooks when they joined the unit. They were then posted to Lingga, south of Singapore, for further training for here at least, there was an abundant supply of fuel.

The 2nd Air Flotilla flying units had fought over the Solomons until late in February 1944, losing many aircraft and men. After returning to Japan for recuperation, they became 652 Ku on 10 March at Iwakuni. At this time fighter-bombers were introduced even to the carrier flying units, so that the pilots of this kokutai received training not only in aerial fighting, but in bombing as well. The new Kugisho D4Y Suisei dive-bomber (Allied codename 'Judy') was built in large numbers at this time but suffered production difficulties with its inline engine which delayed its supply to the front line units. As a result time was not available to allow dive-bombing training to be undertaken to a level where a good rate of hits on the target could be anticipated.

653 Ku had formed on 10 February with *Zuikaku*'s flying unit as the core. However, this unit too had been in action until the end of January 1944 and had suffered many losses. In consequence, formation was delayed and the main attacking strength of the new 653 Ku would be fighter-bombers. The pilots were therefore trained mainly for this role, scarcely receiving any instruction in aerial combat techniques at all.

By early April the pilots of 601 Ku could manage to take off from an aircraft carrier but could only land at an airfield ashore. Those of 653 Ku were not carrier qualified at all and could be counted on only to take off from a carrier for the first strike. Men and machines of 601 Ku were aboard their carriers on 7 May in the Singapore area, reaching the anchorage off Tawi Tawi Island at the north-eastern tip of Borneo. The carriers of the 2nd and 3rd Air Flotillas left Saeki, Kyushu, on 11 May, reaching Tawi Tawi on 16th. Unfortunately, there were no airfields ashore here, whilst US submarines were abundant around the anchorage. Every time ships left the anchorage to undertake training, they were attacked by these predators, so training of the aircrews progressed no further and their skills, if anything, deteriorated.

The Battle of the Philippine Sea
i) Aerial Fighting Before the Battle

In New Guinea US forces landed at Hollandia on 22 April, Japanese Army and air force units in the eastern part of northern New Guinea then being cut off from the main force on the island and isolated. USAAF units swiftly moved into Hollandia and B-24 raids on Guam commenced from there. Here on 7 May Zeros intercepted the first of these bombers, claiming damage to two but losing three of their number. A dozen Zero pilots claimed one B-24 and three damaged on 29 May, while during a raid on Saipan on that same date 41 intercepting fighters were only able to add claims for one more shot down and three damaged.

Meanwhile raids on Truk continued and here a little more success was achieved on 10 May when Lt(jg) Ko and PO1C Teruo Eguchi between them claimed three B-24s shot down and two damaged. On 21 May six 202 Ku Zeros led by Lt Cdr Suzuki attacked 42 B-24s, claims this time totalling one shot down and eight damaged; one Zero suffered damage in return. Following that raid 202 Ku moved to the Vogelkopf area of western New Guinea, leaving only 253 Ku to defend Truk. This kokutai intercepted B-24s on 2nd, 3rd, 7th and 9 June, employing between 17 and 26 aircraft on each occasion. Results were:-

on 2nd, one claimed destroyed for two losses;
on 3rd three victories for one loss;
on 7th PO1C Eizo Narita rammed the tail of a B-24, being posthumously credited with this as a probable victory;
on 9th three claimed destroyed and one probable for the loss of Lt Ryutaro Masuda, the formation leader.

During the period 11-17 June 253 Ku intercepted every day:-
11th three probables;
12th two pilots lost;
13th one probable;
16th one probable, one pilot lost;
17th one pilot lost.

During this period US forces landed on Wakde Island on 17 May and on Biak on 27th. The defence of Biak proved more effective than anticipated and the Japanese command decided to support the defenders. Ships and 153 Ku moved from Kendari to Babo, four of the latter unit's Zeros strafing and bombing the landing area; all failed to return. 46 Zeros from 202 Ku then moved to Sorong between 31 May and 3 June. From here on 2 June, 42 of the unit's aircraft escorted Suisei dive-bombers and JAAF Ki 43 fighters to attack the US forces on the island; three Zeros failed to return. Next day 22 Zeros of 202 Ku and nine of 503 Ku, together with Ki 43s, claimed six aircraft shot down but lost five pilots and seven aircraft. On this same date 18 Zero pilots were engaged over their base at Babo, claiming four shot down but at a cost of six of their own aircraft plus one more badly damaged,

On 4 June 17 A6Ms from S603 led by Lt Toshiharu Ikeda, escorted Suiseis and Ki 43s to attack Biak where two F6Fs were claimed at a cost of one Zero shot down and a second badly damaged. However, 153 Ku had seen its strength diminished by every day's raids and on 7 June OC 23rd Air Flotilla ordered the unit to withdraw. To replace the unit, 25 Zeros of 261 Ku and 31 from 265 Ku, together with P1Y1 Gingas (the latest Rikko type) and dive-bombers moved to Halmahera. As will be recorded below, on 11 June US carrier aircraft struck the Mariana Islands again, and on 13th American battleships bombarded Saipan. At this point the Japanese command ordered all forces to prepare for Operation 'A-go', and the fighting over Biak Island was suspended.

ii) US Forces land on Saipan

A reconnaissance aircraft from Guam had discovered the main US carrier task force at 1150 on 11 June, following which aircraft from the task force attacked every one of the Mariana Islands. Eight Zeros from 263 Ku took off from Guam to meet the raiders, their pilots claiming four shot down and three damaged for the loss of four A6Ms. Eight 261 Ku aircraft then engaged some 50 F6Fs over Guam, claiming five but losing five. A dozen 343 Ku and 301 Ku aircraft also intercepted during the day. Next morning 14 fighters again took off from Guam but all but one were lost. Following this, there were virtually no fighters left to the defenders.

US battleships bombarded Saipan and Tinian on 13th while destroyers swept mines around these islands, finally persuading the Japanese General Staff that this was not just another raid, but a full-scale amphibious operation. At that point they ordered Operation 'A-go' to be put in hand. From Wasule, Halmahera, 261 and 265 Ku together with Gingas and Suiseis of the 2nd Attack Group moved back to Yap on 14th.

On 15 June American forces landed on Saipan. At once 11 Zeros escorted Gingas and Suiseis from Yap to attack shipping around the island. Zero pilots claimed two aircraft shot down but the majority of the Gingas were lost. During the day the carrier aircraft also raided Iwo Jima where 18 Zeros of 301 Ku had remained due to bad weather. These took off, led by Lt Shigeo Juji, to intercept, but only PO1C Mitsugu Yamazaki returned. S401 was a Shiden unit, but Lt Motoi Kaneko and 12 other pilots from this unit had been picked to try and fly Zeros out to Saipan. 11 of them took off from Iwo Jima but these too met F6Fs. PO1C Masao Ono returned to claim one victory, but the other ten, Lt Kaneko included, failed to return. Over the island the Hellcat pilots of US Navy Fighter Squadrons VF-1, 2 and 15 claimed 40 Zeros shot down plus six probables.

31 Zeros now moved back from Wasile to Peleliu, while the units gathered for the Biak operations withdrew from Halmahera to Palau, Yap and Guam.

iii) The Battle is Joined

In preparation for Operation 'A-go', the 1st Mobile Fleet departed Tawi Tawi on 13 June, stopped off at Guimaras Island between Panay and Negros, to take on fuel and other supplies before leaving again on 15th. Rendezvous was made with shipping around Biak in the afternoon of 16th, and after completing taking on further supplies, this force set out during the early evening of 17th. As of 15 June the 1st Air Flotilla (601 Ku) had 214 aircraft including 91 fighters; 2nd Air Flotilla (652 Ku) had 135 including 53 Zero fighters and 27 Zero fighter-bombers; 3rd Air Flotilla (653 Ku) had 90 aircraft including 18 Zero fighters and 45 fighter-bombers. With reconnaissance aircraft, etc, the total aboard amounted to 450 aircraft.

At 1540 on 18 June a search aircraft spotted three groups of US carriers 300 miles to the east. As an attack at this range would have to return during the hours of darkness, an immediate raid was cancelled. The carriers were found again next morning at 0634 and at 0725 the first wave of aircraft from 3rd Air Flotilla were launched. This comprised 43 fighter-bombers, seven Nakajima B6N Tenzan torpedo-bombers, and 14 Zero fighters. The formation was intercepted by F6Fs at 0935, losing 31 fighter-bombers, two Tenzans and eight Zeros. The survivors reported that hits had been obtained on one carrier and a cruiser.

Close behind the first wave came the aircraft of 1st Air Flotilla, which had taken off at 0745. This wave was composed of 29 Tenzans, 44 Suiseis and 48 Zeros; this total included a number despatched to continue searching. However, as this wave departed it passed over the leading vessels of the Japanese surface force which mistook them for hostile and opened fire, two aircraft being shot down. At 1040 one of the American carrier groups was found and an attack was commenced at 1053. Defending fighters had been well-positioned to deal with this by the GCI operators aboard the American ships, and the pilots of these tore the attacking formations to shreds; 24 Tenzans, all 24 Suiseis and 31 of the Zeros were lost. The few survivors returned to report one hit on a carrier and claims by the defending escort fighters for 12 US aircraft shot down and six probables.

At 0900 the 2nd Air Flotilla's first wave went off, comprising seven Tenzans, 25 Zero fighter-bombers and 17 more Zeros as escorts. They sought initially to find the group of carriers already attacked by the first two waves, but they were then directed instead to the two other groups which had been spotted. In the event they failed to find these, but were themselves found by US fighters. Against these the escorting pilots were able to claim four victories, but at a cost of four fighter-bombers, one escorting fighter and one Tenzan.

While these attacks had been going on, the brand new large carrier *Taiho* was hit by a torpedo at 0810, while at 1120 the veteran *Shokaku* was hit by two such missiles; all three had been fired by US submarines. Meanwhile at 1015 the second wave from the 2nd Air Flotilla carriers (*Junyo*, *Hiyo* and *Ryujo*) had set course for the American task force. This strike contained three Tenzans, 27 Suiseis and 20 Zeros. Having got into the air bad weather caused the formation to split up into two groups. One of these failed to find the carriers, and due to the pilots' known inability to land back on their own vessels, headed instead for Guam. As they prepared to land here they were bounced by about 30 F6Fs which shot down 14 Zeros, nine dive-bombers and three Tenzans. The other group, comprising nine Suiseis and six Zeros, found a group of carriers but lost five dive-bombers and four Zeros when they attempted to attack these. Five minutes after this force had taken off, 1st Air Flotilla's second wave had also taken off from *Zuikaku*, this formation also splitting into two forces. Neither was able to find suitable targets, while in various engagements eight fighter-bombers and a Tenzan were lost.

The critically damaged *Shokaku* sank at 1410, followed only ten minutes later by *Taiho*. Nonetheless, throughout the day there had been no attacks on any of the Japanese carriers by US aircraft. Consequently those that had survived still had available between them 44 Zero fighters, 17 fighter-bombers, 11 dive-bombers and 30 torpedo aircraft. On 20 June supply vessels arrived to replenish the ships, allowing them to remain in the battle area. At 1615 the US carriers were again found by a search aircraft, but the only strike force to be launched was one of seven Tenzans. These failed to find any target, but all failed to return. Now it was that the US carrier air groups made their first attacks on their opposite numbers, their raid lasting an hour. 19 Zero fighters and seven fighter-bombers were launched from the surviving carriers of the 1st and 2nd Air Flotillas, the pilots of these claiming one F6F and seven TBFs shot down, plus one F6F and two TBFs probably so. In doing so, the defenders lost 14 of their number, including three which force-landed in the sea. 3rd Air Flotilla aircraft claimed two more shot down but eight Zeros were lost including two which had been launched only to get them off the decks and to allow them to take refuge in the air.

During this attack *Hiyo* was hit by one air-launched torpedo, but was later finished off by submarines. Three of the other carriers were also damaged during the air strike as were one battleship and one cruiser. All were still able to operate under full steam, and were thus able to withdraw northwards, making for Nakagusuku Bay, Okinawa.

iv) Aftermath

While the fighting between the two nations' carrier aircraft had been underway during 19 June, 34 land-based aircraft, including three fighter-bombers, had taken off from Guam to try and add their weight to the assault on the US vessels. A further eight Zeros took off to attack transport ships off Saipan, but all the latter were lost.

In an effort to reinforce the area, more aircraft – including nine Zeros – flew down to Guam from Yap, while 11 more Zeros from 201 Ku flew to Yap from Peleliu and 26 other aircraft, including ten Zeros, moved to Peleliu. By now the defeat of the Mobile Fleet was apparent, and from Tinian US fighters could be seen landing on Aslito airfield on Saipan. Fighting there became desperate for the Japanese and early in July all organised resistance there came to an end. US forces landed on Tinian on 24 July where savage ground fighting continued until early August, while on 21 July they had landed on Guam, where resistance continued until mid August. This meant that the whole of the Marianas group of islands was in American hands, allowing them to start creating bases from which B-29 Superfortresses could commence raids on the Japanese home islands more effectively than they had so far been able to do from China.

Despite the defeat of the fleet, an order was given for a unit known as the Hachiman Force to move to Iwo Jima. This force comprised the major part of the Yokosuka Kokutai, which despatched 27 aircraft to that island on 22 June, followed on 25th by 24 Zeros and five Gekkos. 23 Zeros of S302, 252 Ku, which had been ordered to amalgamate with Hachiman Force had also arrived on 21 June, 16 more following them in on 25th. S601 of 301 Ku was a J2M Raiden-equipped unit, but it now re-equipped with Zeros, nine of which also moved to Iwo Jima on 21 June, followed by another 31 on 25th.

While this was occurring, on 24 June US carrier aircraft raided Iwo Jima. All 59 Zeros of S302, S601 and Yokosuka Ku which had so far reached the island, were sent off to intercept and very heavy fighting ensued. S302 pilots claimed 19 victories, three each being credited to Wt Off Inui Ishida and PO1C Nasao Sugawara, but ten of the unit's Zeros were lost including that flown by Lt Cdr Nobuo Awa, the hikotai leader. The nine S601 pilots claimed five victories for the loss of four Zeros, but also lost their leader, Lt Katsumi Koda. Amongst the Yokosuka Ku pilots, claims were made for 11 shot down and six probables but the cost was a further nine Zeros including that of veteran leader Lt Sadao Yamaguchi. Wt Off Saburo Sakai, recently recovered from severe eye injuries suffered over Guadalcanal, claimed two while Wt Off Kaneyoshi Muto is reputed to have claimed four. That afternoon Hachiman Force attempted to attack the US carriers, 23 Zeros escorting torpedo aircraft and dive-bombers to undertake this task. Near Urakas they were intercepted by the ever-present Hellcats, losing ten Zeros and seven torpedo-bombers. The carrier air groups then raided Iwo Jima on 3rd and 4 July, during which period 252 Ku claimed 13 shot down but lost 14 of its own aircraft; three of the claims were again made by Sugawara. During the afternoon of 3rd 31 Zeros from S601 led by Lt Iyozo Fujita engaged 100 F6Fs, claiming seven and three probables, but for the loss of 17 Zeros. Next day S601 intercepted with 14 Zeros, claiming six for three losses. During these three days of attacks Ens Akio Matsuba had claimed five and one probable, while Wt Off Iwao Mita and PO2C Takayoshi Morita had both been credited with four victories; all three of these pilots were lost during the fighting on 4th. On this same date US aircraft also raided Chichijima where they were intercepted by nine A6M2s of Sasebo Ku. Only CPO Teruyuki Naoi was able to claim one shot down and two probables, but four of the float fighters failed to return. Meanwhile, US surface vessels bombarded Iwo Jima, destroying all the rest of the aircraft there on the ground. The surviving aircrews were subsequently returned to the homeland by transports.

Returning again to 19 June, from Truk on this date Lt Cdr Harutoshi Okamoto led 13 Zeros from 253 Ku towards Guam. Before they could land, these aircraft were attacked by a strong force of F6Fs, and while pilots were able to claim two shot down and one probable, four Zeros were lost including one flown by Lt Tatsuo Hirano and another by a veteran, Wt Off Yoshinao Tokuchi. 253 Ku retained a few aircraft on Truk, some of which intercepted B-24s over the atoll between 21 June-9 July. The only claim was for a probable on 7 July, but two Japanese pilots were lost, one of them Lt Masamichi Minokata. 253 Ku was then disbanded and the remaining pilots transferred to 201 Ku. On 17 August one of them, Lt Mitsuho Tanaka, intercepted 24 B-24s over Truk, but in doing so was shot down and killed, his aircraft falling outside the atoll.

Having lost 21 aircraft of S302 and S603, 202 Ku pilots also intercepted B-24s during the period 23 June-9 July, claiming five destroyed, four probables and 29 damaged, losing only one pilot killed in doing so.

THE PHILIPPINES CAMPAIGN

Reorganisation after the Marianas Campaign

When the fighting around the Marianas Islands and Guam came to a close the IJNAF was obliged to reorganise since it was now in a chaotic situation. Amongst the carrier flying units, 601 Ku became of hikotai composition, comprising S161 (48 Zeros led by Lt Hohei Kobayashi), S163 (48 Zero fighter-bombers led by Lt Masayuki Hida), plus K161, K262 and T61 equipped with dive-bombers and torpedo aircraft. However, it would take until the end of 1944 for these elements to recuperate and reform fully. 652 Ku was disbanded on 10 July, the majority of the flying personnel being transferred to 653 Ku. This latter unit now also became of hikotai composition, but apart from one dive-bomber element (K263), all other elements were to be equipped with 48 Zeros apiece. Lt Kenji Nakagawa became leader of both S164 and S165, while Lt Tetsuo Endo would command S166 with fighter-bombers. 634 Ku had equipped mainly with floatplanes and dive-bombers, but added S163 and S167, each with 48 Zeros; both these units were commanded by Lt Sumio Fukuda.

The land-based IJN units which had taken part in the Marianas Campaign had come close to total destruction. Consequently a number of veteran kokutais including 202, 252 and 253 Ku were disbanded on 10 July, while S901 was transferred to 153 Ku. The remaining crews of 253 Ku at Truk were posted to Higashi Caroline Kokutai. From the 1st

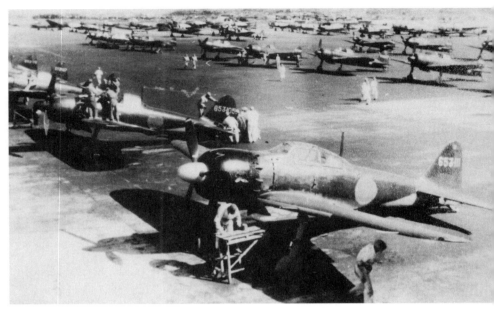

653 Ku Zeros at Oita in September 1944.

Air Fleet which had been expected to act as the mainstay of the defence of the Marianas, 261, 263 and 265 Ku were also disbanded on 10 July, such pilots as remained going now to 201 Ku. 501 Ku and S351 were also disbanded and their personnel absorbed by 201 Ku. Additionally, 10 July saw the disbandment of 301, 321, 343, 345 and 361 Ku.

1st Air Fleet
After this swingeing cull, 1st Air Fleet fighter units comprised just S901 (night fighters) of 153 Ku, S301, S305, S306 and S311 of 201 Ku. The greater part of this air fleet now moved to Davao on Mindanao Island in the Philippines. However, the headquarters had simply dissolved on Tinian by the end of July 1944. To take its place a new headquarters under the command of Admiral Kinpei Teraoka arrived at Davao on 12 August, moving to Manila a month later. As at 21 July the air fleet had 134 aircraft of which 96 were fighters. When the remaining aircraft at Yap and Truk were added, this total rose to 120 fighters and 56 other aircraft. At this stage the quality of the fighter pilots available was assessed, and all were divided amongst four categories. These were from A – excellent, to D – barely able to operate. Out of 54 pilots with S305, there were but 12 with A or B assessments; of the 57 in S311 12 were classed as B quality, and with S306's 46 pilots, only two reached B grading. S301 had only 22 pilots of whom just four were classed as A grade. However, by 1 September the 1st Air Fleet was brought up to a strength of 408 aircraft (249 serviceable) which total included 210 Zeros (130 serviceable) and 22 Gekkos (15 serviceable).

A buntai from 201 Ku led by Lt Naoshi Kanno was detached to Yap where between 16th-23 July formations of B-24s were intercepted each day. During this period eight of these bombers were claimed destroyed, plus nine probables and 46 damaged, However, five pilots were lost, including two who rammed their opponents, and after eight days of operations the detachment returned to Davao.

2nd Air Fleet
The forerunner of the 2nd Air Fleet, the 62nd Air Flotilla, had formed on 1 February 1944 with 11 kokutais including five with fighters. As the new Shiden fighter was too late

in arriving to become operational at this time, 265 Ku with Zeros was exchanged with the 61st Air Flotilla's Shiden-equipped 341 Ku, accompanying the 61st to the Marianas. On 15 June the 62nd Air Flotilla became the 2nd Air Fleet and following the reorganisation of fighter units on 10 July had under command the following units:-
    S804 (night fighters) of 141 Ku (ex 322 Ku);
    S308, S312, S313 and S407 (the latter ex 361 Ku) of 221 Ku;
    S401 (ex 341 Ku) and S402 (ex 345 Ku) of 341 Ku (these two hikotais equipped with Shidens).

Of the new units, S804 was based at Katori, S308, S312 and S313 were at Kasanbara, and S407 at Kagoshima.

On 1 October the 2nd Air Fleet fighter force had the following strength:-
    221 Ku – S308 and S313 with 35 Zero type 52s (23 serviceable), and six Zero type 21s (none serviceable); S312 at Hsinchu with 48 Zero type 52s (32 serviceable); S407 with 48 Zero type 52s (20 serviceable), and 13 Zero type 21s (eight serviceable).
    341 Ku – S401 with 32 Shidens (20 serviceable) at Takao and 11 Shidens (six serviceable) at Miyazaki; S402 with 30 Shidens (25 serviceable) and 22 Zeros (18 serviceable) also at Miyazaki.
    141 Ku – seven Gekkos (three serviceable) at Hsinchu and 26 Gekkos (18 serviceable) at Kanoya and Kushira.

Altogether, 2nd Air Fleet had 620 aircraft (361 serviceable), so the serviceability rate was 58%. Amongst the fighter units which accounted for 367 aircraft from this total, 198 were serviceable so that the rate amongst the fighter units was only 54%.

A Zero 52 (A6M5) of Sento 407, 221 Ku, about to take off. 221 Ku had four hikotais, and operated with 252 Ku as the core units of the IJN during the Philippines campaign.

3rd Air Fleet
The 3rd Air Fleet, formed on 10 July from the Hachiman Force which had operated over Iwo Jima, had one flyingboat, one reconnaissance, one attack and one fighter kokutais. The latter, 252 Ku, comprised S302 (Zeros, ex 252 Ku), S316 (Zeros, ex 301 Ku) and newly-formed S315 and S317 (both with Zeros). The newly-formed reconnaissance

kokutai had a new night fighter unit – S851. The elements of 252 Ku were stationed at Tateyama and Yokosuka, while S851 was at Katori. On 11 July 18 Zeros, three Gekkos and the air fleet's Rikkos were ordered to move to Iwo Jima. As of 1 October, therefore, S851 had 14 Gekkos (eight serviceable) at Katori and three (two serviceable) on Iwo Jima. The combined S315 and S317 had 25 Zero type 52s (20 serviceable) and 23 Zero type 21s (19 serviceable) at Tateyama, while the combined S316 and S302 had 101 Zeros (49 serviceable) at Mobara and 14 Zeros (eight serviceable) on Iwo Jima.

Other Units
Apart from the three air fleets detailed above, there was also the 12th Air Fleet in Chishima and Hokkaido. On 1 October this unit had:-
203 Ku – S303 with 42 Zero type 52s (40 serviceable) at Kagoshima; S304 with 52 Zero type 52s at the 1st Bihoro, nine Zeros at the 1st Shumushu, Chishima, and the newly-formed S812 with nine Gekkos at 1st Shumushu and the 1st Bihoro.
In the south was the 28th Air Flotilla, which had:-
331 Ku – S309 with 30 Zeros at Balikpapan. Five Zeros at Sandakan and other airfields.
381 Ku – S602 and S902 with 32 Zeros and eight Gekkos at Balikpapan, 32 Zeros, nine Raidens and two Gekkos at Kendari plus two Gekkos at Surabaya.
In China were 254 Ku (37 Zeros) and 256 Ku.

During September 1944, out of a total of 21 fighter kokutais (28 hikotais), five hikotais from 1st Air Fleet, seven of 2nd Air Fleet, five of 3rd Air Fleet and five hikotais of carrier fighter units were committed to the Philippines Campaign, together with detachments from two kokutais in South-East Asia and the two in China. Left behind for home defence were only five kokutais (Yokosuka, 210th, 302nd , 332nd and 352nd), plus two hikotais of 601 Ku.

US Carrier Aircraft Attack Southern and Central Philippines during September 1944
During September raids by B-24s by day on Davao, Mindanao, commenced. Interceptions by Zeros based on Cebu were not undertaken in order to conserve the aircraft for what was anticipated to come. However, at the suggestion of Lt Tadashi Minobe, the new hikotai leader of S901, on 2nd four Gekkos and two night Zeros of his unit took off at 1000, attacking a B-24 formation with 3-go phosphorus bombs. One Gekko and one Zero were shot down by escorting P-38s, but PO1C Yoshimasa Nakagawa flying a Type 2 land-based reconnaissance aircraft (forerunner of the Gekko night fighter) claimed to have shot down one of the P-38s with his front guns while following it in and out of clouds. During the night of 4 September Nakagawa, now flying a production Gekko, came upon a B-24 at very close range. Without an opportunity to get into a proper firing position, he allowed his aircraft's right-hand propeller to chew into the bomber, thereafter barely managing to land his badly damaged aircraft. Headquarters subsequently judged his attack to have destroyed his opponent.

On 9 September US carrier aircraft attacked Davao, achieving complete surprise due to a breakdown of the Japanese radar. Few aircraft were present at the time, however, and little damage was suffered. Next morning three Gekkos had just taken off from Tacloban, Leyte, when they were attacked by F6Fs and two were at once shot down. Those on the ground watched as the third aircraft, flown by CPO Takumi Tanaka, was engaged by four of the Hellcats, reporting that he had shot down two of these before falling himself.

Next day a rumour that US forces had landed on Mindanao caused 89 Zeros of 201

Ku to be moved from Manila to Cebu to be in a position to resist. However, the rumour proved to be a false one and about half the aircraft were flown back to the north. On 12th US aircraft swept through the central Philippines, including Cebu. Once more the radar had malfunctioned and a visual report from Suluan Island had not been received in time. As a result the 41 Zeros which took off found themselves in an unfavourable position, and whilst able to claim 20 F6Fs shot down and three more probably, 25 Japanese pilots were lost including Lt Hiroshi Morii, hikotai leader of S306, and CPO Shinsaku Tanaka, who had achieved earlier success over Palau; two more pilots did manage to bale out and survived. Moreover, a further 25 aircraft were either destroyed or damaged on the ground, leaving 1st Air Fleet with just 85 aircraft, 34 of which were Zeros.

Cebu, Tacloban and Legaspi were raided on 13th, six Zeros of 201 Ku led by Lt Kakichi Hirata representing the entire interception force that could be put up. Two F6Fs and two TBFs were claimed but three of the pilots were killed and a fourth wounded. Raids continued next day and on 15 September US forces landed on Morotai Island, just north of Halmahera, and Peleliu Island in the Palau group. Despite a valiant fight, the Japanese garrisons on both islands were soon captured and in early October USAAF aircraft moved to airfields there.

Despite radar warnings on this occasion, USN aircraft again achieved surprise when they attacked Manila on 21 September. 42 Zeros intercepted, their pilots claiming 27 of the raiders shot down, but again they lost 20 of their own number. Next day six fighter-bombers and nine escorting Zeros led by Lt Usaburo Suzuki, the S301 hikotai leader, attacked a US carrier group, making three bombing and strafing passes. Intercepted by F6Fs, they suffered the loss of six aircraft, surviving pilots claiming three of the US fighters shot down. Thus by 23 September the 1st Air Fleet in the Philippines was down to just 25 Zeros and 38 other types. Strenuous efforts to reinforce followed, and by 1 October numbers had increased to 230, of which 149 were serviceable; amongst these, 201 Ku had recovered to 132 Zeros (81 serviceable).

Lt Usaburo Suzuki, hikotai leader of Sento 301, who failed to return from a mission on 13 October 1944.

Attacks on Formosa

The air defence of Formosa at this time was in the hands only of the instructors at the training units on the island. In mid August 1944, therefore, seven Gekkos of S804 and 50 Zeros of S312 were sent to Hsinchu to take over this role. They were followed by seven Shidens of S401 which arrived at Takao at the end of the month, followed in mid September by 25 more of these new fighters.

US carrier aircraft attacked Okinawa on 10 October, but few interceptions were made by the defending forces. Indeed, the only claims were submitted by Wt Off Tomokazu

Kawanishi J1N1-J Shiden 11 ko with 20mm gun pack under the wing.

Ema of 254 Ku, who was credited with one shot down and one probable. However, two days later on 12th, USN aircraft appeared over Formosa in strength. Four Zeros of S312, 221 Ku, were on patrol at 0720 when some 20 F6Fs were encountered. One of these was claimed shot down but two of the Japanese fighters were lost. 30 more Zeros from the same unit were scrambled meanwhile, taking on an estimated 50 US aircraft over Touen. On this occasion the defenders believed that they had come off best, claiming 23 victories at a cost of 14 Zeros and 12 pilots. Five more Zeros intercepted F6Fs over Hsinchu, claiming two for one loss. Finally, three Zero pilots on their third sortie of the day and two on their fourth, intercepted ten F6Fs over Taoyuan during the afternoon, claiming three shot down for the loss of two Zeros and one pilot.

On this same date 37 Zeros were in combat with 30-40 F6Fs over Tainan, where the Japanese pilots claimed seven shot down and three probables – but at a loss of 17 Zeros. Seven Shidens of S401 were on patrol and were joined by 24 more which hurriedly took off. This force engaged about 60 US aircraft and during the Shiden's first combat of the war, their pilots claimed ten of the attackers shot down. PO1c Takeo Yamada and PO1c Hideo Hirakawa each claimed four F6Fs during this engagement, Hirakawa achieving the last of his four by ramming. However, 14 Shidens were lost.

By evening strength of the island's defences in terms of serviceable aircraft had fallen to six Zeros at Hsinchu, eight Shidens at Takao and 12 Zeros at Tainan. Raids on Formosa continued on 14th when 25 Zeros from the training units intercepted 20-40 F6Fs over the southern part of the island. Four victories were claimed but 13 Zeros were lost.

Meanwhile a special unit known as T Force had been formed to attack the US Task Force, striking by night on three consecutive dates beginning 12 October. The attackers suffered severe losses but the Japanese command announced that the major part of the US carrier force had been sunk. USN sources failed to confirm this, indicating that no carriers had been sunk and that only a few ships had suffered some damage. During the afternoon of 13 October 30 IJN attack aircraft escorted by 70 fighters took off from the Philippines to attack what were believed to be only the remaining few American carriers. However, nothing could be found and Lt Usaburo Suzuki, who was leading the fighter escort had to land in the sea off Taito with engine trouble and was lost. On this date the 254 Ku at Haikou despatched a detachment of 16 Zeros to Tainan, while two more of these fighters from 256 Ku at Shanghai were sent to the same base, followed two days later by another nine. This latter date also saw the arrival of the Shidens of S701 and S402 on Formosa.

During 11th and 12 October 634 and 653 Ku joined the 2nd Air Fleet, moving with 252 Ku and other units to Kyushu, and from there to Okinawa. This particular move was considerably interrupted by bad weather. During 13-14 October 634 Ku moved to Taichung. Following participation in the attacks on the US fleet by T Force, S303 moved to Luzon.

On 14 October 254 and 256 Ku were engaged over Changhua with about 40 hostile aircraft, the units claiming two shot down and one probable without loss. 50 Zeros from Ie Shima provided escort to torpedo-bombers, dive-bombers and fighter-bombers seeking to attack the US Task Force, but found nothing. During the return flight while making for Formosa, this formation was attacked by F6Fs which shot down three of the Zeros. By the night of 14th on Okinawa and Formosa were 221 Ku with 29 aircraft, 252 Ku with 50, S304 with 23, 624 Ku with 23 and 653 Ku with 20. At this time, however, one unit after another was sent from Kyushu, Okinawa and Formosa to the Philippines.

US carrier aircraft raided the Philippine capital, Manila, on 15 October, where intercepting Japanese fighter pilots claimed 15 shot down and ten probables, while major efforts were made to organise an attack from Formosa and the Philippines on the carriers. Ten Rikkos and nine Zeros from Tainan and Takao on such a raid were intercepted by eight F6Fs. Seven of the fighters, including Lt Masao Iizuka, commanding officer of S302, and nine of the Rikkos failed to return. From Hsinchu 16 Zeros, one Gekko and five Rikkos also took off, but 13 of the fighters had to turn back due to engine problems. The remainder of this small force was intercepted by 15 F6Fs which shot down two Zeros, the lone Gekko and two of the Rikkos. From Miyako Jima, meanwhile, 15 Zeros and three Tenzans also took off, but this formation too was intercepted, 30 F6Fs attacking at 0830. CPO Yoshinobu Nakamura claimed two victories against these, while Lt Eiji Muraoka who was leading in a fighter-bomber Zero, was believed to have dived his aircraft onto an American carrier. Five more of the Zeros and a Tenzan were lost.

From the Philippines seven fighter-bomber Zeros escorted by 19 fighter versions, attacked the same target, claiming a near miss on one carrier and a hit on a cruiser for the loss of six of the fighter-bombers, while the escorting pilots claimed seven F6Fs shot down. From Clark airfield nine Zeros, three Rikkos and 63 IJAAF fighters, accompanied by four Zeros and 12 Tenzans from Tuguegarao then sought to make a second attack on the shipping. Returning IJN fighter pilots claimed seven defending US aircraft shot down and three probables, but on this occasion the cost was three Zeros, three Rikkos, eight Tenzans and nine of the Army fighters.

Two formations from Formosa sought to attack the carriers again on 16 October. The first of these, launched from Tainan, included 40 Zeros which encountered at least 30 F6Fs at 1345. Pilots of 203 Ku claimed four destroyed and two probables without loss, while others from 254 and 256 Ku claimed four more but lost a veteran, Wt Off Satoshi Kano. 252 Ku claimed two victories but lost two pilots, but 634 Ku and 221 Ku lost six more Zeros, while seven Rikkos and 14 Tenzans were also missing. The second formation from Formosa departed Hsinchu, but failed to find the US fleet and returned to base. These were to be the last major attacks launched against the carriers for some time.

B-29s bombed the Formosan airfields on 17 October, but nine 252 Ku aircraft which attempted to intercept achieved no successes and lost two aircraft. 653 Ku which had lost the greater part of its strength during the fighting around Formosa now moved to Bamban on Luzon. Two days later eight 331 Ku Zeros led by Lt Kisuke Hasegawa and 15 from 381 Ku, led by Lt Keijiro Hayashi, flew to Mabalcat from Balikpapan. Pilots of 331 Ku's detachment returned to the homeland to collect new Zeros from the Mitsubishi factory.

The 'Sho-1-Go' Operation is Launched

During the morning of 17 October from a watchtower on Suluan, a small island east of Leyte in the Philippines, a telegram arrived at headquarters, announcing "Enemy is approaching". 90 minutes later the last message was received from this source – "Enemy landed"; nothing more was heard thereafter. Next evening HQ, Combined Fleet, issued the order "Sho-1-go Operation put into operation". Three Zeros at once took off from Cebu and flew to Leyte to attack shipping, claiming a near miss.

Before noon on 20 October US forces landed at Tacloban and Dulag on Leyte Island. At the time as this occurred the 1st Air Fleet had just 40 serviceable aircraft to hand, including 15 Zeros. The 2nd Air Fleet, which had been ordered to Formosa on the previous day, had, however, 395 aircraft of which 223 were currently serviceable. By 23rd 196 of this command's aircraft had reached the Philippines, of which 126 were fighters.

At this juncture 201 Ku underwent a change in its *modus operandi* which would become a pivotal point in the history of Japanese forces. Since the previous month, the kokutai's pilots had been undertaking training in the technique of skip-bombing which Allied attacks in the New Guinea area had shown to be extremely effective. This had been confirmed by the Suzuki unit of the kokutai on 22 September and by the Ibusuki unit on 15 October. However, Admiral Takijiro Ohnishi, who was about to take over as C-in-C of 1st Air Fleet from Admiral Teraoka, had become convinced that the most effective form of attack was the "organizational ram attack method" (i.e. suicide attack). The day before he arrived at the Air Fleet Headquarters, he summoned the senior personnel of 201 Ku and persuaded (or ordered) them to undertake such attacks. On 20 October as C-in-C, he ordered the formation of a ram attack unit with 24 pilots of 201 Ku, which he named *Jinpu Tokubetu Kogekitai* (Jinpu Special Attack Unit). This new unit he divided into four flights, each named after a verse written by a national patriot of the past. On the very next day the new unit took off, but in the event found nothing to attack, although two Zeros failed to return.

Kamikaze A6Ms of the tokko-tai (special attack units) 'Ouka Tai' and 'Seku Tai', about to take off from a road near Manila on 11 November 1944.

On the evening of 22 October at the Clark Field group of airfields on Luzon Island were the following units:

221 and 252 Ku with 62 Zeros;

341 Ku and S701 with 40 Shidens;

S303 with 20 Zeros;

1st Attack Unit with eight fighter-bombers and 32 Zeros of 221 Ku;

3rd Attack Unit with 50-60 Zeros of S304, 634 Ku and 653 Ku.

Eleven Gekkos of S804 flew in to Nichols Field, but when four or five of these aircraft were despatched on a search mission, only one returned. During the previous evening seven Zeros of 221 Ku had taken part in a night attack on Morotai Island, ten more repeating this raid on 31st. Thereafter, attacks on this target were left to JAAF heavy bombers and Ki 45kai twin-engined fighters.

The Americans' general offensive commenced on 24 October and at once 26 Zeros led by Lt Cdr Minoru Kobayashi, took off to attack carriers located at 090 degrees, 150 miles from Manila. 50 F6Fs were encountered and claims were made for six destroyed and one probable, Saburo Saito and Jiro Matsuda each accounting for two, and Yoshio Oh-ishi one. However, 11 Zeros failed to return including Kobayashi's. A further attack became split up in cloud, 51 Zeros led by Lt Oshibuchi engaging another 50 F6Fs at 075 degrees, 140 miles from Manila. The pilots of these claimed 11 shot down, three of them by CPO Kazuo Sugino, but amongst the units taking part in this mission, 634 Ku lost seven including their leader, Lt Sumio Fukuda, 203 Ku lost two, one of them flown by Wt Off Yukiharu Ozeki, while 341 Ku lost 11 of the 21 Shidens it had despatched.

The Attack by the Japanese Surface Fleet

The surface fleet sent out to challenge the US force included battleships, cruisers and destroyers, ordered to destroy the American forces on Leyte and the amphibious force off the coast supporting them. Just prior to the arrival in the area of this body of ships, a small formation of aircraft had been ordered to take part in the operation to act as a decoy to draw off the US carriers. During the early hours of 25th a small fleet including two battleships approaching from the south was annihilated by US battleships, while the main force, including the huge battleships *Yamato* and *Musashi*, which came under attack from 24th, was running behind time to rendezvous with the Nishimura Fleet in Leyte Bay. Nevertheless, the success of the decoy operation in drawing off the main US carrier force allowed the Japanese vessels to reach Leyte almost unopposed. However, here they were attacked by the air groups of a force of light carriers which had been supporting the landings. For no reason that has been discovered, this caused them to turn away and withdraw from the battle.

Meanwhile during the early hours of 24 October, the crew of a land-based reconnaissance aircraft reported sighting the main US carriers. On receipt of this information, several search aircraft took off from the Japanese carriers which had also been sent to the area, the crew of one of these finding these vessels at 1145. An attack force was then launched, divided into two parts. While one of these comprised only aircraft from *Zuikaku*, the other was composed of elements from *Zuiho*, *Chitose* and *Chiyoda*. The latter force, including 20 Zeros, nine fighter-bombers and four Tenzans, were intercepted by about 20 F6Fs at 1305. In the fight which followed Lt Mitsuo Ohfuji claimed two and one probable, while Lt Kenji Nakagawa, Wt Off Yoshinao Kodaira and three other pilots each claimed one. No US ships were seen, and only three aircraft returned to the carriers, two more landing on airfields in Luzon; six Zeros, one fighter-bomber and two Tenzans were lost.

The *Zuikaku* formation (ten Zeros, 11 fighter-bombers, one Tenzan and two Suiseis) had found the US Task Force at 1300. While the Zero escorts led by Lt Tai Nakajima provided cover, the fighter-bombers attacked, claiming two bomb hits on one carrier and one on another. Returning crew reported that they had seen the first of these sink rapidly, followed by the second. None of the attacking aircraft returned to the carrier; two Zeros, five fighter-bombers and the Tenzan were lost, the remainder landing at bases in northern Luzon. However, during the day no raids by US carrier aircraft were experienced.

However, at 0748 next morning US aircraft appeared in strength over the Japanese carriers. At 0815 the Combat Air Patrol dived on an incoming gaggle of aircraft. The pilots of five Zeros led by Lt Hohei Kobayashi claimed six fighters and four dive-bombers destroyed, three more pilots led by Ens Yoshimi Minami claimed three torpedo aircraft, while a further four Zero pilots led by Wt Off Harukichi Kubota claimed four more of the latter. All 12 Japanese pilots then had to force-land in the sea as their carriers were under attack. US aircraft kept on coming in waves and at 0947 *Chitose* was sunk. *Zuikaku*

Carrier *Zuiho* under attack off the Philippines on 25 October 1944.

followed her beneath the waves at 1414, *Zuiho* went down at 1526 and *Chiyoda* at 1647. Most of the pilots from these vessels were rescued by the destroyer *Hatsuzuki*, but this vessel was sunk by gunfire during the night and all the aircrews went down with her.

During the morning of 15th two waves of attackers were sent off from land bases to join the attack on the American carriers, but the position data supplied had been incorrect, and nothing was found. Low on fuel, both forces were obliged to land at Legaspi. At 1500 35 Zeros and 23 dive-bombers took off again, but became engaged by 35 F6Fs. Although claims were made for five of these shot down and two probables, six Zeros were lost including four from S304, as were two of the dive-bombers. Still failing to find the American warships, part of this force attacked Admiral Kurita's surface fleet. Kurita repeatedly reported this attack and called for it to desist. This caused a further raiding force which had taken off at 1730 to turn back without finding any US ships.

All the new ram attack units of 1st Air Fleet operated during the day, but the fate of seven Zeros of the Asahi and Yamazakura Units which set off and were not seen again, remained unknown. However, two Zeros of the Kikusui Unit gained one hit on a carrier at 0800 for the first success for what became known as the kamikazes. Five more Zeros of the Shikishima Unit found a group of carriers at 1040, two of them hitting the same carrier which was considered to have been sunk, one more also hit a carrier and the fourth hit a cruiser which was believed to have sunk rapidly. Four escorting Zero pilot led by the famous Wt Off Hiroyoshi Nishizawa, claimed two F6Fs shot down but lost one of their own to AA.

That evening, since standard attacks had achieved nothing but the suicide attacks appeared to have gained considerable success, Admiral Ohnishi pleaded with Admiral Fukutome, commander of 2nd Air Fleet, to form similar units. This Fukutome did, selecting dive-bombers for this task as the 2nd Jinpu Attack Unit. In consequence ram attacks (Tokko) took the place of standard operations from then until the end of the war. That evening the 1st and 2nd Air Fleets were combined to form the No 1 Combined Land-Based Flying Fleet; as the more senior officer, Fukutome was appointed commander.

Bomb-laden A6Ms of the Shikishima Unit, Jinpu Special Attack Unit (a suicide unit) prepare to take off from Mabalacat on 25 October 1944.

On 26 October two Zeros of the Yamato Unit were seen to ram the same carrier, while a third Zero rammed another. During the morning 14 Zero pilots claimed six F6Fs shot down for the loss of six of their own. 47 aircraft, including 18 Zeros, attacked Leyte, but nine were lost. Meanwhile, Wt Off Hiroyoshi Nishizawa, the IJN's top-scoring pilot of the war, had landed on Cebu Island following his sortie on the previous day. On seeking to return to his unit, he was ordered to leave his aircraft where it was and to fly out as a passenger in a Rikko. This was shot down by an F6F and he was killed.

The Leyte Battle and Afterwards
In an effort to try and force a decisive battle on Leyte, a convoy of Japanese Army troops and equipment sailed for Ormoc Bay where it was due to arrive on 1 November. In support the IJN undertook a maximum effort series of attacks on Leyte between 27th-31st October. During 25th-26th a number of fighters which had been sent to Formosa for service aboard the carriers, were despatched instead to Manila and Bambam; these included nine aircraft of 254 Ku and six of 256 Ku. Next day they moved to Nichols Field, and on 28th to Marcott.

On 27 October 17 Zeros being led by Lt Naoshi Kanno of 201 Ku, encountered 16 F6Fs over Marinduque Island whilst on the way to Cebu from Clark. The pilots of these Zeros claimed 12 victories over the American fighters for the loss of only one. During the evening of 28th 23 Zeros including three loaded with bombs, were led by Lt Kenji Nakagawa to attack Tacloban. Little AA fire was met and the jubilant pilots returned to claim 40 F6Fs and a single large aircraft burnt on the ground while one F6F and a lone P-38 were claimed shot down in the air; no losses were suffered. 13 of 221 Ku's aircraft were providing escort for this attack, and the pilots of these fought 30 defending fighters, claiming four destroyed and two probables for only one loss, while six accompanying Shidens of 341 Ku claimed two more without loss to themselves.

During an interception over Manila on 29 October the pilots of 133 fighters (including 31 Shidens), 14 pilots of 254 Ku and 32 of S304 claimed 33 victories for the loss of 22; six of the latter were from 254 Ku, including Wt Off Ema, and two from S304. On this same date ram attacks on shipping saw a dive-bomber hit a large carrier and another aircraft crash into a 'middle-sized' carrier.

On 30th carriers were again targetted by six Tokko fighter-bombers, three of which were reported to have hit a large carrier and one another of 'medium-size'. On the same day two Gekkos and three Zero night-fighters of S901 were led by Lt Minobe to Cebu to engage in night sorties against some troublesome torpedo-boats. Six of these would be claimed sunk during a period of four weeks. Cebu also saw the arrival of 252 Ku Zeros after they had undertaken a strafing attack on Tacloban on 30th. Next day they took off again to strafe Dulag airfield before returning to their base. In this same period Zeros of 653 and 634 Ku also landed at Cebu after strafing attacks on targets on Leyte. However from both these units and from 203 Ku, pilots were selected for service with the ram units without their having volunteered. Veterans such as Lt Haruta and Ens Tsunoda were retained to fly escorts to the Tokko units. Since Cebu and Mabalacat were the bases of 201 Ku, they became taboo for landing by the pilots of other fighter units because once they landed there, they immediately became candidates for Tokko.

On general offensive operations on 1 November, 91 aircraft including 37 Zeros, 11 Shidens and ten suicide aircraft took off from Clark and Nichols for Leyte. Having undertaken a strafing attack here, 22 of the Zeros led by Lt Nakagawa encountered B-24s, one of which was claimed shot down over Cebu. During the day over a reinforcement convoy four Zeros (including one flown by CPO Yoshio Oh-ishi) intercepted more B-24s, claiming damage to three of these. Next day eight 634 Ku Zeros led by CPO Kazuo Sugino strafed Tacloban airfield but found the AA defences much stronger than they had been a few days earlier, four of the Japanese fighters being lost. On 3rd Lt Nakagawa led 12 Zeros to this same target with the dire result that all were shot down by the ground defences.

Despite these crippling losses, a further attack on Tacloban was mounted on 4 November, one aircraft from 254 Ku and four from 252 Ku taking part. This time the attackers got away without loss and as they made for home, B-24s were spotted over Mactan Island, Ens Tsunoda and three of the other pilots claiming two destroyed and one probable.

Manila was raided by US aircraft on 5 November, 68 Japanese fighters being scrambled to intercept. 14 raiders were claimed shot down, but of the ten defending fighters lost in return, one was flown by Lt Torajiro Haruta, who had previously survived a Tokko mission from Cebu, but who was now killed. Next morning CPO Ki-ichi Nagano led seven other pilots on an interception mission over Luzon; they ran into about 60 F6Fs, Nagano and two others failing to return from this encounter.

6 November saw the arrival at Clark of 20 Zeros of 332 Ku, led by Lt(jg) Susumu Takeda. The unit would be annihilated in 20 days of fighting, Takeda himself falling on 27th. Meanwhile, on 11 November four Zeros from 254 and 256 Ku engaged nine P-38s over Ormoc, claiming three but losing two. This unit's remaining pilots then withdrew to China. Between them these two units had claimed 14 destroyed and four probables during a period of one month over Formosa and the Philippines, but had lost 24 pilots. On this same date seven Zeros of S304 led by Lt Mutsuo Urushiyama moved in to bases around Davao. From there they operated until 29 November, claiming two P-38s shot down and three B-24s damaged during three engagements, for the loss of three aircraft. On 13 November the pilots of nine Shidens operating over Manila, claimed three victories but suffered the loss of six of their own aircraft.

Attrition continued; on 15 Novembert 653 Ku was disbanded and 203 Ku left the Philippines for Japan. S901 also departed before the end of the month, but was replaced by S812, which arrived at Nichols. The first unit to take the Shiden into action, 341 Ku, continued to undertake strafing attacks on Leyte, interceptions and convoy escorts, but the unit's sortie rate was low due to the teething troubles still being experienced with the engines of these aircraft. Pilots also reported that the aircraft was outclassed by the USAAF's P-38s and P-47s. On 24 November the unit lost the leader of S701, the ace Lt Cdr Ayao Shirane, over Ormoc Bay. A number of the unit's pilots were then transferred to Tokko units and were all killed in action.

On 4 December a large US convoy was spotted in Surigao Strait and on 7th American forces landed on the west coast of Leyte from where they were able to bring the Japanese defenders under attack from both sides. The IJN at once attacked these landings with all available units. 12 Shidens challenged ten plus P-47s over the northern tip of Cebu, but lost five while their own pilots were able only to claim two of the US aircraft shot down. Eight Zero pilots claimed three of 15 P-38s shot down, but only two survived to return to base. Two more Zeros were also lost here, while an attack force from Cebu lost three Shidens and a Zero.

Another convoy was found on 13 December and during the early hours of 15th troops from this landed at San Jose on the southern tip of Mindoro Island, just south of Luzon. During 14th and 15th these ships and the landings underway were attacked by Tokko units, so that by 17 December the IJN had available only 28 operational aircraft; this total included four Zeros, four fighter-bomber Zeros, one night fighter Zero, four Shidens and seven Gekkos. Coincident with these latest landings, US carrier aircraft commenced raids on Luzon on 14 December, while during the latter half of the month Japanese floatplane bombers and Rikkos started making night attacks on San Jose.

The last IJN reinforcments for the Philippines arrived on 20 December when Lt Cdr Kawai arrived at Angeles with 20 S308 Zeros. Four days later 25 IJN fighters and 20 more from surviving IJAAF units, undertook an interception over Clark where eight victories were claimed. However, Lt Cdr Shiro Kawai of Tainan Ku fame was shot down, and while he was observed to bale out successfully near the base, he was not seen again. During two further interceptions next day two veterans, Wt Offs Tadashi Yoneda and Shoichi Kuwabara of S308, 221 Ku were amongst those lost.

A very large convoy entered Lingayen Gulf, western Luzon, in January 1945. On 4th 13 Shidens were strafed on the ground by two P-47s and eight were destroyed. Such aircraft as remained were used up in suicide attacks on 3rd and 9 January, bringing to an end IJN air operations over the Philippines. On 9 January US forces landed on Lingayen. Four remaining Shidens made attacks here, but all were lost during the next few days. Attempts were then made to evacuate the remaining pilots and aircrews to Formosa, but in doing so 13

Lt Cdr Shiro Kawai, hikotai leader of Sento 308, lost on 24 December 1944.

transports were lost by 10 February. The remaining men of the fighter force then remained on Luzon with insufficent food or arms for the rest of the war, most of them being killed during the land battle there.

Between 25 October 1944 and 5 January 1945 1,049 sorties were flown by Zeros, 303 by Shidens and 274 by Gekkos and Zero night fighters. By day the Zero and Shiden pilots claimed 120 US aircraft shot down and 60 destroyed on the ground.

AIR BATTLES IN DEFENCE OF THE HOMELAND

Fighting over the North-Eastern Frontier

The shortest route to Japan from US soil was via the northern route. Consequently, following the loss of Attu Island in the Aleutians, the Chishima Islands became the front line. 281 Ku (formed on 1 March 1943 at Tateyama as a Zero Fighter Kokutai) was despatched to Paramushir Island, the most northerly of the Chishima chain, during late May 1943. On 3 June 12 Zeros of 201 Ku which had been withdrawn from the central Pacific front were also sent to this base, but did not stay long before being ordered to the Solomons. Float fighters of 452 Ku which had departed Kiska Island after some severe fighting, had by now recuperated, and under the leadership of Lt(jg) Shunshi Araki moved from Yokosuka to a lake on Shumushu Island late in July.

The IJAAF's 54th Sentai was also stationed in northern Chishima, equipped with Nakajima Ki 43 fighters, and aircraft from this unit intercepted US bombers on 12 August and 12 September. Although 281 Ku fighters were not able to take part in either of these engagements, ten A6M2Ns of 452 Ku did join the Army fighters, their pilots

A6M2Ns and pilots of Sasebo Ku's float fighter unit on 22 June 1944, just prior to departure for Chichi Jima. Left, seated, WO San-ichi Hirano, leader of the detachment; 6th from left, standing, CPO Teruyuki Naoi, who claimed one destroyed and two probables over Chichi Jima on 4 July 1944; the other four pilots with him all failed to return.

claiming two B-24s shot down and a third probable on the second of these dates. No more action was seen, the float fighters returning to Yokosuka where the unit was disbanded on 1 October 1943, while 281 Ku was posted to the central Pacific at the end of November.

203 Ku had formed from the Atsugi Ku, a fighter operational training unit, on 20 February 1944, then moving to Chitose in Hokkaido late in March to await the melting of the snow on the northern islands. Late in April one buntai of A6Ms and three Gekkos flew up to Kataoka airfield, moving on in turn to Shumushu and Musashi, and to Paramushir, while the major part of the kokutai followed them to Kataoka in May.

During the evening of 13 May two Gekkos took off from 1st Shumushu airfield, CPO Yasuro Baba shooting down a PV-1 Ventura patrol bomber; another Gekko failed to return on this date, however. On the morning of 15 June seven Zeros from Paramushir intercepted three B-24s, one of which was claimed destroyed. On the same day Lt Oshibuchi led off 31 Zeros from Shumushu, these intercepting four PV-1s, claiming damage to two of these. It was later learned that two of these aircraft had force-landed on the Soviet Union's Kamchatka Archipelago.

24 Zeros were again scrambled on 24 July, 12 of these sharing in the destruction of another PV-1. The main part of 203 Ku then withdrew to Hokkaido, but 19 A6Ms of S304 which still remained in the area were led by Lt Mutuo Urushiyama to intercept US bombers over Chishima on several occasions during August. One was claimed shot down on 20th, but two Zeros were lost. Damage was claimed to other bombers on 28th, and on 18th and 25 September. Towards the end of September, however, the Urushiyama Unit withdrew to Bihoro and from there was sent to the Philippines. Gekkos of 203 Ku were also flown to Bihoro where they formed S812 and departed from Hokkaido. Consequently, no further IJN fighters were stationed in Chishima for the rest of the war.

Top: 2 Ku float fighter unit at Bettobinuma, Shushumu, on 2 September 1943. On the extreme left of the front row is Kiyomi Katsuki; 2nd row, 2nd from right, Shunshi Araki (leader) and 3rd row, 2nd from right is Teruyuki Naoi. The majority of these pilots later converted to land-based fighters.

Bottom: Pilots of Sento 303, 203 Ku, with Lt Cdr Okajima (2nd row, 3rd from left). '3-64' on the tail of the Zero behind them is the fundamental marking of S303.

The Costly Fighting over Iwo Jima

Iwo Jima was used as the staging base for attacks on Saipan during and after the Marianas Campaign. On 13 July 13 Zeros of S302, 252 Ku, arrived there, led by Lt Kunio Kimura. On 11 August 11 of these fighters intercepted 18 B-24s, claiming two damaged. Three days later 11 pilots again intercepted B-24s – 22 this time – claiming one shot down and one probable. The unit then moved to the Philippines to take part in the fighting there.

Genzan airfield, Iwo Jima, with a 252 Ku Zero present.

14 A6Ms of S317, 252 Ku, then moved to the island on 10 October to defend it from attacks by B-24s. Led by Lt Shigehiro Nakama, this unit mainly used the 3-go phosphorus bombs. Nakama personally damaged one bomber on 11th and shared in damaging another on 15th. On 21st ten Zeros intercepted 28 B-24s, one of which Nakama rammed, both aircraft falling into the sea. Lt Nakama was posthumously promoted two ranks and notice of this was made to the whole IJN. Following the death of their leader, the remaining pilots continued to defend the island, which on 5 November suffered its first raid by B-29 Superfortresses. On 18 November, however, P-38s appeared in the area for the first time and three of the unit's pilots were lost to these. Attempts to intercept B-29s were made from 27 November onwards, but without great success. Two had been claimed to have suffered some damage, but on 19 December the unit returned to Japan.

Admiral Teraoka, C-in-C 3rd Air Fleet since 17 November, had noted how ineffective attempts to intercept B-29s returning to Saipan after bombing targets in Japan had been. On 24 November, therefore, he ordered a strafing mission by Zeros on their base. 12 pilots of S317 were selected to undertake such an attack, to be led by Lt(jg) Kenji Ohmura.

These flew back to Iwo Jima on 26th, and next morning took off for Saipan, flying at very low level. Nothing further was heard, except that two of the fighters despatched landed on Pagan Island. One had flown so low that the propeller had struck the surface of the sea and become bent, requiring an emergency landing. The other had reached Pagan after taking part in the strafing attack, but had crashed on landing and the pilot had been killed. All those participating were posthumously promoted two ranks and the unit was renamed 1st Mitate Tokubetsu Kogekitai (First Shield Special Attack Unit). US records show that the Zeros of this unit arrived over Saipan just after the majority of the B-29s based there had taken off for an attack on Tokyo. Of those still present, three were destroyed and two others badly damaged by the strafing, but all the Zeros were then shot down, either by AA or fighters. One Zero force-landed, the pilot leaping from his cockpit and drawing his pistol at which point he was shot dead. This was Lt Ohmura, who was buried by the Americans with military honours.

US Marines landed on Iwo Jima on 19 February 1945. Admiral Teraoka immediately ordered 601 Ku to form another special attack unit to be called 2nd Mitate Tokubetsu Kogekitai, which comprised 12 Zeros, 12 dive-bombers and eight torpedo-bombers. The whole force took off from Katori on 21 February and after refuelling at Hachijo Jima, attacked shipping around Iwo Jima. None survived, but US sources admitted the loss to their attack of one escort carrier sunk, the veteran Fleet carrier *Saratoga* seriously damaged, one further escort carrier, one transport ship and two LSTs (Landing Ship Tank) damaged. Fighting to the last, the defenders held out for a month, following which Iwo Jima became a base for P-51 Mustang escort fighters and an emergency landing area for damaged B-29s returning from raids on the home islands.

B-29 Raids from Mainland China

As the Japanese Army began to prepare defences for the home islands, the IJN also undertook similar preparations in respect of the three major military harbours. In March 1944 302 Ku with 48 interceptors and 24 night fighters was formed at Atsugi for the defence of Yokosuka harbour. It was agreed that from 21 July this unit, together with the Kure Ku at Kure and the fighter unit of the Sasebo Ku at Sasebo would come under the operational control of the Army's defence system when the need arose. On 1 August 332 Ku was formed from the Kure Ku fighter unit while 352 Ku was similarly formed from the Sasebo unit. The establishment of each of these kokutais was 48 interceptors and 12 night fighters, requiring more J2M Raidens and Gekkos to be obtained.

During this period the USAAF sent B-29s to inland China across the Himalayan 'Hump' from bases in India. The first raid by these new very heavy bombers was made during the night of 15/16 June, the target being the Yahata Ironworks in northern Kyushu. It was during the night of 10/11 August that IJN fighters were first employed against these raiders, however, two 352 Ku Gekkos taking off in an effort to intercept the B-29s over Nagasaki and Yahata, although without success. On 20 August the bombers came by day, Yahata again being the target. This time eight A6Ms from the same unit achieved a successful interception, claiming one B-29 shot down and one probable. During the same raid Lt(jg) Sachio Endo in a 302 Ku Gekko took up the chase towards Jeju Do, being credited with shooting down two of the bombers, with a third as a probable and two more damaged.

A major attack occurred on 25 October when B-29s from Chengtu attacked the Ohmura Navy Plant. This time 71 fighters were sent up by 352 Ku and the Ohmura Ku, but against the bombers flying above 8,000 metres the interceptor pilots were plagued by cannon failures and other malfunctions so that successes were few. Lt(jg) Koichi Sawada, flying an A6M, claimed one shot down, Lt(jg) Koretake Ide in a Gekko claimed one probable, whilst other pilots could only achieve damage to the raiders, 18 of which were claimed to have been hit to some extent.

By day on 21 November the alarm was raised early by Japanese Army units in China, allowing 352 Ku to scramble six Gekkos, 16 Raidens and 43 Zeros which were waiting over Ohmura when the bombers approached. The Gekko pilots attacked first, claiming a single victory before continuing their attacks for an hour. On this occasion the B-29s were flying at a lower altitude than had been the case previously, allowing 352 Ku to submit claims for nine destroyed while the Ohmura Ku added three more. Amongst the claims made by 352 Ku was one for a ramming by Lt(jg) Mikihiko Sakamoto, who had already claimed three damaged; he did not survive this particular victory, however. One victory each were claimed by Lt(jg) Susumu Kawasaki (an ex-934 Ku float fighter pilot), and Ens Yoshimichi Saeki (a pilot with Tainan Ku during 1941-42). Successful Raiden pilots included another ex-Tainan Ku veteran, Wt Off Toshiyuki Ichiki, and another ex-Solomons veteran, Wt Off Yasunobu Nabara. The Gekko crews also claimed two victories.

On 19 December 47 fighters from 352 Ku took off, but were only able to claim three damaged, while on 6 January 43 pilots from this unit made two claims for the loss of Lt(jg) Sawada and a second Zero, the pilot of which survived. Thereafter raids by B-29s from Chinese airfields ceased.

The Start of B-29 Raids from Saipan

302 Ku provided the main air defence of the Kanto area for which purpose as at 1 November 1944 it had on hand 40 Raidens, 38 Zeros, 24 Gekkos, 20 Suisei dive-bombers modified for bomber interception duties and four Gingas – or more correctly Kyokkos, which were the fighter version of the new Yokosuka P1Y1 bomber; 60 of these various aircraft were serviceable. On that day a reconnaissance B-29 (known in the USAAF as an F-13) appeared over the Kanto area at very high altitude, but could not be caught by either IJN or IJAAF fighters. The same thing occurred on 7th when a further F-13 undertook a similar sortie. 302 Ku scrambled 38 fighters, but none could get within firing range.

Tokyo was attacked by B-29s for the first time on 24 November, the Nakajima Engine Plant at Mitaka being the main target. 109 fighters were sent up by 302 Ku, but the area in which they were held proved to be too far from the bombers' line of approach and the only claim that could be made was for one damaged, although in achieving this, two Japanese fighters were lost. The bombers returned to the same target on 3 December, but this time more was achieved, 302 Ku pilots claiming six shot down and three probables for the loss of two fighters and a third heavily damaged.

Raids followed on the Mitsubishi Engine Plant at Nagoya on 13th and 18 December, 210 Ku claiming two destroyed and one probable on the latter date, while Lt Sachio Endo of 302 Ku claimed one and two damaged during the same interception. As reconnaissance aircraft then appeared over the Hanshin area where Osaka and Kobe are located, 332 Ku at Iwakuni swiftly moved Zeros and Raidens to Naruo – a landing strip converted from an old horse-racing course, while Gekkos moved to Itami. On 22 December 332 Ku Raidens caught B-29s over Nagoya where CPO Akeshi Ochi claimed the unit's first victory, thereby winning for himself a bottle of whiskey; a second victory was claimed by 210 Ku.

The Nakajima Musashi Engine Plant was the target on 27 December, two Zero night fighter pilots of 302 Ku (Lt Shunshi Araki and CPO Yuji Komatsu) each being credited

Top: Zeros and Gekkos of 302 Ku line up at Atsugi. Suisei night fighters can be seen on the right.
Bottom: Battle-hardened NCO pilots of 332 Ku at Naruo. In the centre row, left, is Teigo Ishida, while in the centre is Susumu Ishihara.

with a victory, while Gekko pilot Lt Endo also claimed one. However, Komatsu was hit by the bombers' defensive fire and suffered severe wounds. The raiding force split on 3 January, one formation of B-29s bombing Osaka while the main force struck Nagoya. 27 fighters from 210 Ku took off from Meiji, claiming one and one shared. Meanwhile, Lt Jusaburo Mukai, a Zero night fighter pilot from 332 Ku, closed on one B-29 which he shot down with oblique cannon armament from very short range. During this period 302

Ku had despatched a detachment of four Gekkos to Hachijo Jima on 14 November, hoping that they would be able to intercept B-29s on their way to and from Saipan. However, the course they took proved to be too far from this location and on 10 January 1945 the detachment was withdrawn, having failed to make any interceptions.

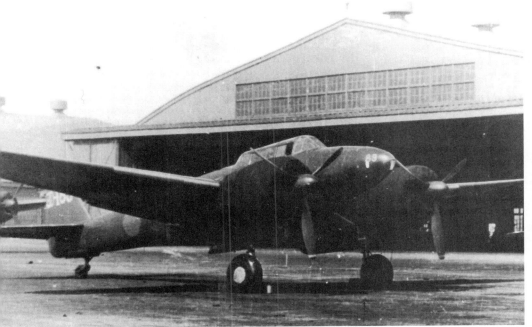

Top: A Mitsubishi J2M Raiden 21 of 302 Ku. Note one single cherry blossom and one double blossom on the tail, indicating credits for one destroyed and one probable.

Bottom: Although performance of the Gekko was below that of the B-29 Superfortress, pilots nonetheless attempted to intercept the bombers in these aircraft even by day.

The Nakajima Engine Plant suffered its fifth raid on 9 January, 302 Ku pilots claiming one bomber shot down, three probables and ten damaged – one of the latter by Lt Endo. On 14th the target was the Mitsubishi Nagoya Plant, and again claims were distinctly 'thin'. 210 Ku pilots shared a victory with Army fighters, adding claims for two damaged. Raiden pilots of 332 Ku claimed damage to five, while Lt Kisaku Hamano of 302 Ku commanding a Ginga, dropped 3-go bombs on the raiders, claiming one probably destroyed – the first success to be achieved by the crew of one of these aircraft. However, having reported shooting down one B-29 over the radio, Lt Endo failed to return. The bodies of him and his radio operator were subsequently found and it was assumed that they had baled out too late after their aircraft had been hit. Endo had claimed six destroyed and ten damaged at the time of his death.

Snow falls at Matsuyama in January 1945 as groundcrews prepare a Sento 301 Shiden for operations. The 'A' at the top of the rudder identifies the unit.

Over Nagoya on 23 January 210 Ku pilots claimed one shared destroyed and three damaged while the Gekko unit of 302 Ku added claims for two more shot down. The crew of a Ginga from this unit also claimed one victory, but from the Zero night fighter unit both Lt Hiroshi Morioka and his wingman were wounded. The night of 27th proved more profitable for 302 Ku when the Nakajima Plant in Tokyo was bombed, six destroyed and 15 damaged being claimed for the loss of two Gekkos and one Zero night fighter. On 10 February B-29s raided the Nakajima Ohta Plant in Gumma Prefecture in the northern Kanto area. Here 302 Ku claimed three destroyed and five damaged.CPO Yoshimitsu Naka, flying a Suisei which was commanded by Ens Hisao Kanazawa in the rear cockpit, dived behind and below one bomber, then attacking it with obliquely-mounted cannon. They then zoomed up and dived again, gaining hits during each of three such attacks and bringing the B-29 down for the first victory claimed by the crew of this night fighter adaptation of the dive-bomber.

US Carrier Aircraft hit the Kanto Area, while B-29s turn to Night Bombing

Prior to the Iwo Jima landings the US Navy had made plans to attack the Kanto area, located directly behind the target island. The first such raid was undertaken on 16 February, American fighters making a series of strikes throughout the day. All the IJN fighter units located in the Kanto area were engaged throughout the morning and afternoon. 302 Ku launched 18 Raidens and 30 Zeros, the pilots of these claiming nine victories for the loss of five aircraft and three pilots, including Lt Shunshi Araki. S310 of 601 Ku had been preparing to move to the Kanto area from Iwakuni, 16 Zeros led by Lt Katori flying in as planned. They at once became involved in the fighting, losing four aircraft and three pilots. 601 Ku's S308 put up eight Zeros, four being lost in achieving claims for one destroyed and one damaged. From the Yoko Ku Ens Matsuo Hagiri claimed one F6F shot down while flying one of the latest Shiden-kai fighters, while Wt Off Kaneyoshi Muto of this unit was reported to have shot down four single-handed.

Meanwhile, from 252 Ku 44 fighters took off, claiming 14 destroyed and nine probables for the loss of 11 aircraft and nine pilots. Ens Tetsuzo Iwamoto of S311 claimed one of each category. 48 Zeros and 11 Shidens from the Tsukuba Ku were also active, the pilots of these claiming six destroyed, three of them by Wt Off Takanobu Obata, and three damaged, but paid heavily, losing 12 of their number. A further 14 Zeros of 201 Ku flew over Hamamatsu where they engaged about 50 fighters. This unit was more successful, claiming 16 victories including three by PO1c Kunimichi Kato while suffering only a single loss.

Next day the attacks were resumed, this time the Tsukuba Ku launched 40 fighters to intercept, two victories being claimed for a single loss, although two Zeros which had been converted to become trainers, were also shot down. 601 Ku undertook three missions totalling 18 sorties during which six raiders were claimed, four by Lt Katori, while like the Tsukuba aircraft, only a single loss was suffered. 252 Ku also lost only one pilot from 13 fighters which took off, while 302 Ku reported ten victories for two losses; Four were claimed by Ens Sadaaki Akamatsu. Other 3rd Air Fleet units were also involved, claiming between them a further 12 shot down.

US carrier aircraft again raided the Kanto area on 25 February, but as it was snowing few interceptors got off the ground. 252 Ku managed to get off 20 of its fighters, three F6Fs being claimed but at a loss of five aircraft and three pilots. That afternoon IJN aircraft were also unable to intercept raiding B-29s.

Although Lt Hiroshi Morioka, leader of the A6M-equipped 302 Ku, lost his left hand in combat, he recovered to continue flying until the end of the war.

General Curtis LeMay who had made his name undertaking daylight precision bombing over Europe and who was now commanding the B-29 force, had decided following a period of trial and error, that night bombing with incendiaries was potentially the best way to destroy Japanese industry. Consequently, during the night of 9th/10th March 1945 more than 300 bombers unloaded incendiaries on Tokyo from low level and with considerable success. Nagoya was attacked in similar manner on 11th/12th, followed by Osaka on 13th/14th. LeMay considered all these raids to have been outstandingly successful, and this method was adopted for the majority of attacks on Japanese cities thereafter.

During these raids 302 Ku had managed to launch only four Gekkos on 9th /10th, just one damaged being claimed. Five of these aircraft from 332 Ku took off during the night of 13th, but two failed to return and a third was hit by mistake by AA fire, the crew having to undertake a force-landing. Kobe was raided during the night of 16th/17th, and Nagoya on 18th/19th to round off the first series of fire-bomb attacks.

The Okinawa Campaign Begins

US carrier air groups raided Kyushu on 18 March 1945, interceptors from 203 and 721 Ku rising to meet them. 22 Zeros from the former unit went off in small groups as a result of which they gained no successes at all, but lost seven aircraft and three pilots. The 721 Ku pilots were much more successful, claiming 24 destroyed, seven probables and four damaged; Lt Kanzaki alone claimed two shot down and three damaged, but again the cost was high. As the result of four separate engagements, 19 of the unit's Zeros were lost while eight more were destroyed on the ground.

Next day the Americans raided further to the east, targets including Kure harbour in Chugoku. From Matsuyama in the Shikoku area 54 brand new Shiden-kai fighters of 343 Ku were scrambled operationally for the first time and during this, their combat debut, their pilots claimed an incredible 52 victories for the loss of 16 of their fighters. Amongst the most successful of the unit's pilots were CPOs Katsue Kato, who claimed five, and Shoichi Sugita, who claimed three. 203 Ku also took part in the interceptions again, but this time without suffering any losses. The unit claimed six raiders shot down and three probables, CPO Takeo Tanimizu claiming one and one damaged.

The 5th Air Fleet, responsible for the defence of Kyushu, also attacked the US carriers on 18th and 19 March, believing that considerable damage had been inflicted. Their attacks were repeated on 20th and 21st in an effort to finish off the enemy carriers. Indeed, on the latter date the last and most deadly weapon was sought to be used. 18 Rikkos of 721 Ku, all carrying Ohka piloted rocket bombs (named Baka – 'Fool' – by the Allies) set off with a close escort of 19 Zeros from the kokutai's fighter unit, and an indirect escort of 11 more Zeros from 203 Ku, led by Lt Cdr Okajima. Before they could reach their target area, some 50 F6Fs pounced on them, shooting down all 18 Rikkos. The escorts tried to come to their aid, the 721 Ku Zero pilots claiming nine shot down and three probables, while those of 203 Ku claimed a further four and three probables. Despite these claims, 721 Ku lost ten Zeros including that flown by Lt Urushiyama, while two 203 Ku Zeros force-landed after being seriously damaged. At the end of the day's fighting, the 5th Air Fleet had lost almost all its attack aircraft and was down to just 50 fighters. So weakened, the defenders had to await the arrival of elements of 3rd Air Fleet, urgently moved to the area. Meanwhile, on 26 March US forces landed on Kerama Island, just to the west of Okinawa. To this the Japanese responded by launching their 'Ten-1-go' Operation, and the Okinawa Campaign had begun.

Lt Sadao Yamaguchi of Yokohama Ku, who was killed in action on 4 July 1944.

Veteran NCOs of Yokosuka Ku in summer 1944. Extreme left, kneeling, is Ryoji Oh-hara; standing, 2nd from left, is Masami Shiga; 3rd, Tomita Atake; 4th, Kiyoshi Sekiya.

Units of the 3rd Air Fleet began transferring to Kyushu; amongst the fighter units 252 Ku (S304, S313 and S316) moved to Tomitaka, 601 Ku (S308, S310) went to the 2nd Kokubu, and 210 Ku to Izumi. 343 Ku also moved up to Kanoya by section from 10 April onwards.

Amongst the defence units some 20 A6Ms from 352 Ku shifted from Ohmura to Kasanbara on 3 April where they were joined four days later by seven more Zeros from 302 Ku. The operational training units, Tsukuba Ku and Yatabe Ku, sent more Zeros to Tomitaka, 16 of these aircraft then going on to Izumi. From the Ohmura Ku and Genzan Ku further fighters were sent to Kasanbara, as were a buntai of A6Ms from Yoko Ku.

Sento 304, just before departure from Mobara, is addressed by Lt Cdr Hachiro Yanagisawa.

During the period 1st-5 April IJN units counter-attacked the landings on Okinawa, A6M fighter-bombers of 721 Ku attacking the supporting US carrier force on 2nd and 3rd, employing 30 aircraft on the first occasion and 22 on the second; respectively losses were four and six Zeros. On 3rd 36 Zeros and eight Shidens of 601 Ku set off as the escort for an attack by the remaining Japanese carrier forces. While so engaged, 25 Zero pilots engaged at least 30 hostile aircraft, claiming ten shot down and six or seven damaged;

CPO Yoshijiro Shirama claimed three of these victories and Lt Katori one, but the cost was high, eight Zeros and two Shidens failing to return.

On Formosa at this time were the remnants of units which had withdrawn from the Philippines, which included 201, 221, 341 and 153 Kokutais, although the actual strength of this group of units was precisely nil. On 5 February 205 Ku was formed at Taichung to include S302, S315 and S317 (with an establishment of 48 Zeros). In practice the core of 205 Ku comprised the Philippine survivors from the 1st Air Fleet amounting to 112 pilots, though with only about 20 fighters available.

The fundamental method of operation by the 1st Air Fleet by now was all-out Tokko attacks, so the normal basis of operations for 205 Ku became the despatch of a few bomb-laden aircraft on one-way missions, accompanied by one or two fighters, the pilots of which were to observe and report back on the results achieved. On 1 April the kokutai despatched three attack forces against the US carriers. Thereafter, from two to five bomb-laden Zeros took off every day from Hsinchu and Tainan, losing some of their number on each occasion but without achieving any results. With few reinforcements available from the homeland, unsurprisingly the number of aircraft remaining available shrunk rapidly during May.

The Kikusui Operation Commences

This new series of operations comprised a number of all-out attacks by the air forces of both the Army and the Navy, the first such being launched on 6 April. The job of the fighters was to obtain aerial superiority over the area between Kyushu and Okinawa to enable Tokko aircraft to undertake their missions. At 0900 on this first day 13 A6Ms from the Tsukuba and Yatabe Kokutais flew south to the area of Toku-no Shima but failed to meet any opposition; one aircraft was lost during the return flight. 20 more Zeros and ten Shidens took off at 1030, but five of each type of fighter returned due to mechanical problems, in each case including the formation leader. This time opposition was encountered, the Zero pilots becoming involved with 40 F6Fs of which they claimed three shot down for the loss of seven of their own aircraft. The Shidens did not take part in this engagement, although one aircraft was lost. The reason for this apparent lack of aggression by the Shiden pilots was actually due to the distance between Kanoya and Okinawa which was 650 km. While Zeros could fly such a distance, remain in the combat zone for up to 20 minutes, and then fly home, the Shidens (and Shiden-kais) did not enjoy such a range capability and were able only to fly as far south as Amami Oshima.

Seven 252 Ku Zeros followed at 1310, but five failed to return, the remaining two force-landing at Kikai Shima. They were followed by 23 more A6Ms of S312 at 1430, these fighting F6Fs over Toku-no Shima and Amami Oshima, where their pilots claimed three destroyed and two probables, but lost seven including the leader. 29 more Zeros which were despatched returned with claims of six US aircraft shot down without loss. From Formosa during the day four Zeros escorted Gingas to attack the carriers, while in a night interception the crew of a Gekko claimed one B-24 damaged.

Next day patrols of A6Ms were maintained over the battleship *Yamato* and other vessels, but after the fourth such patrol US carrier aircraft attacked this fleet and amongst the ships sunk was the mighty *Yamato* with its 18.1 inch guns.

More heavy fighting occurred on 11 April, commencing when 28 Zeros and six Shidens from 601 and 210 Ku took on F6Fs south-east of Kikai, suffering one loss. 22 more Zeros from 352 Ku and Ohmura Ku surprised four F6Fs and claimed two shot down; one A6M force-landed during the return journey. Two more F6Fs were claimed by six 252 Ku pilots over Kikai.

On 12th – the first day of Kikusui No 2 – 34 new Shiden-kais of 343 Ku led by Lt Kanno operated over Amami, claiming a massive 22 destroyed and three probables; these claims included three destroyed and one probable by Shoichi Sugita and one shot down by Kanno himself. However, 12 of the unit's fighters and 11 of its pilots were lost. 33 Zeros were then despatched but lost seven of their number while only able to submit one claim. Pickings were scarce for other units too. 16 Zeros from S312 and 352 Ku claimed three but lost three, while nine more Zeros from Tsukuba and Yatabe Ku, flying with 15 S311 aircraft and nine from Genzan Ku, shared between them just one and one probable for the loss of four.

US carrier aircraft attacked bases on Kyushu on 15 April, 36 A6Ms of 203 Ku rising to oppose the raiders, claiming four destroyed and two probables in three engagements. The cost was eight Zeros shot down and three more severely damaged. Kanoya was strafed during this attack, two Shiden-kais of 343 Ku being shot down as they tried to take off, the pilots being killed – one of whom was CPO Shoichi Sugita. 343 Ku moved to the 1st Kokubu on 17 April and a week later withdrew to Ohmura.

Next day (the first day of Kikusui No 3) brought further action for the fighters. At 0620 five Tsukuba Ku Zeros and six from S311 undertook an interception to claim three destroyed and one probable in return for two losses. 20 more Zeros followed at 0700, these 203 Ku aircraft suffering five losses for no results. Yet 12 more A6Ms, this time from 252 Ku, reached Amami where they encountered F6Fs, Lt Cdr Hachiro Yanagizawa, the formation leader, claiming two of these shot down and a third damaged. As the unit undertook the return flight, it was bounced over Kagoshima Bay and whilst the pilots were able to claim two of the attackers shot down, six Zeros and five of their pilots failed to return.

At 0730 30 Zeros and 11 Shidens from 601 and 210 Ku set off, becoming engaged between Amami and Kikai. Pilots of S308 claimed four shot down and four damaged for three losses, while four of the 210 Ku pilots claimed three shot down and one damaged (one of the former by PO1c Kunimichi Kato) for a single loss.

Finally, on 17 April, 16 601 Ku Zeros and ten from 252 Ku took off to take the long flight south again. Six of the latter turned back, but the remaining four became engaged with more than ten US fighters over Amami and three of them, including Lt Cdr Yanagizawa, failed to return.

From Kikusui No 4 to the end of the Okinawa Campaign

At this stage the Americans were able to repair the airfields on Okinawa, bomber and fighter units then moving up to this new base. Consequently, it was no longer necessary for the USN carriers to remain permanently on station. Until this point the main targets for the IJN's Tokko aircraft had been these carriers. With their departure, the IJN was obliged to change the targets for the self-immolators to other shipping around the island and the airfields thereon.

By now 601 Ku had lost 25 Zeros and one Shiden, and now moved back to the Kanto area. 210 Ku, having gained six victories but lost nine Zeros and a Shiden returned to Meiji airfield, while 252 Ku, after suffering the loss of 15 Zeros, and S304 also returned to the Kanto area. During late April Tsukuba Ku also moved to this area, followed in mid May by Yatabe Ku.

As a result of these departures, missions attempting to gain air superiority over the Okinawa area became restricted. Nonetheless, on 22 April near Okinawa 28 A6Ms from 352 Ku and other units fought F6Fs, pilots claiming four destroyed and eight probables for two losses. Six days later, on the day of Kikusui No 4, 29 S312 Zeros and aircraft from other units, claimed three F4Us over Ie Shima.

However, the main focus of attention now turned onto the interception of B-29s which were appearing in considerably increased numbers. On 21 April 35 A6Ms of 302 Ku joined with others in the interception of some of these high-flying bombers, claiming two destroyed and seven damaged for one loss. On the same day 32 of 343 Ku's Shiden-kais claimed two destroyed and two damaged, but of two lost in return, one was flown by Lt Yoshishige Hayashi, S407 hikotai leader, who was killed. On 28th 24 Zeros, drawn mainly from S312, made another interception of the bombers, PO1c Hirotomo Takakuwa ramming one, which was claimed as a probable. Next day 32 more Zeros, 16 of them aircraft of S312, claimed one B-29 shot down and 14 damaged for the loss of a single pilot.

The B-29s now commenced attacks on airfields in southern Kyushu, in response to which three defence units (302, 332 and 352) with a total of 46 Raiden interceptors moved to Kanoya between 22nd-30 April. The Raiden unit (the pilots called themselves the 'Tornado' unit) first engaged the bombers over this area on 27th, Lt Nakajima of 332 Ku claiming one success, Wt Off Takehiko Baba one probable, and other pilots 14 damaged; one Raiden was lost. Two days later Ldg Sea Shigeo Sakata of 302 Ku claimed two destroyed but was then shot down and killed himself. Ldg Sea Shoji Kuroda claimed one more destroyed while other pilots added claims for four damaged, two of the Raidens being seriously damaged in return. Six more damaged were claimed on 30th, while on 3 May claims for eight more damaged were made. Pilots continued to find it difficult to achieve more than some damage on their fast, high-flying and heavily-armed opponents; five damaged claims were made on each of 7th and 10th, during which period some of the interceptors were destroyed on the ground by bombing. Soon only eight serviceable aircraft remained available and on 16 May the pilots of 302 and 332 Kokutais returned to their bases, leaving the remaining Raidens behind. On 3 June the Raiden pilots of 352 Ku also departed, returning to their base at Omura.

Meanwhile, until the end of Kikusui No 4, 136 bomb-carrying Zeros, 116 dive-bombers and 77 torpedo-bombers were consumed on Tokko operations. Even floatplanes and Shiragiku crew trainers were now being used for these missions. 4 May saw the start of Kikusui No 5 when 45 A6Ms of 203 Ku were joined by other units in an air superiority sweep. Eight opposing aircraft were claimed shot down plus two probables, but at least three Japanese pilots failed to return, including a veteran, Wt Off Katsujiro Nakano. The 'crack' 343 Ku also sortied into the area with 36 Shiden-kais, engaging a substantial formation of F6Fs over Kikai. 13 of these were claimed shot down for the loss of six Shiden-kais. From Formosa 18 205 Ku Zeros escorted 21 fighter-bombers to attack shipping. Lt(jg) Tsunoda watched four of the latter undertake their suicide dives including one flown by Lt(jg) Tanimoto, which Tsunoda reported had impacted on a large aircraft carrier. In the evening Wt Off Yoshio Oh-ishi set off for an affirmation of the results achieved, but he never returned. Despite the efforts of both the Navy and Army air forces during these attacks, the Army's counter-offensive on Okinawa was halted, and from then onwards its defensive efforts declined steadily.

On 11 May (the day of Kikusui No.6) 65 Zeros from 203 Ku, 252 Ku and other units again sought to achieve air superiority over the route to Okinawa. On this occasion, however, 203 Ku lost six pilots and 252 Ku lost seven, while two more from this latter unit lost their lives in accidents. Two days later three waves of US carrier aircraft swept over the Japanese bases. 203 Ku launched 12 Zeros in an effort to intercept these raiders, but while six claims were made, four of the defenders were lost. On 14 May airfields on southern Kyushu were attacked, and the only two A6M pilots to try and get into the air were shot down as they were taking off.

During the closing days of May several days of bad weather coupled with a reduction in daylight Tokko missions, allowed a total of 216 fighters to be launched during two days. On 28th 343 Ku undertook two sweeps involving 28 individual sorties, but P-47s from Okinawa were encountered, five Shiden-kais and three pilots being lost to these. Zeros, meanwhile, were also in the area, their pilots claiming four victories but losing seven of their own.

343 Ku Shiden-kais led by Lt Keijiro Hayashi, the new hikotai leader of S407, fought 31 F4Us on 2 June, claiming 18 victories for just two losses. Next day saw the start of Kikusui No 9, but on this date 343 Ku was unable to claim a single success and lost one of its own aircraft. 28 A6Ms of S313 and S316 took off, but five returned early while the rest suffered eight losses, also without being able to claim anything in return. During the final stage of the fighting over and around Okinawa on 22 June 66, Zeros provided escort to Ohka-carrying Rikkos. 25 of the fighters returned early, while the rest claimed seven aircraft shot down but lost nine. 31 343 Ku Shiden-kais engaged Okinawa-based F4U Corsairs, claiming seven of these for the loss of four pilots, including Lt Keijiro Hayashi. This was the last aerial combat to take place during the fighting for Okinawa.

Fuyo Unit (131 Kokutai) at Kanoya in early April 1945. Second row, 3rd from left, Lt Cdr Tadashi Monobe; 3rd from right, Lt Eiichi Kawabati (leader of S804); third row, 5th from left, Wt Off Saburo Sue.

The activities of IJN fighters over Okinawa were fairly low after the end of April, but one unit which was heavily involved from start to finish was the Fuyo Unit. Sento 901, S812 and S804 had gathered at Fujieda, near Mount Fuji on the instigation of Lt Cdr Tadashi Minobe, and here they trained for night attacks – not, it should be emphasised, of a Tokko nature. Their task would be to attack both shipping and ground targets with single-engined aircraft. Becoming 131 Ku, the pilots named themselves the Fuyo Unit, Fuyo being the alternative name for Mount Fuji. By late in March the unit had gathered some 50 Suisei dive-bombers and 20 A6M Zeros. On 30 March a move was made to Kanoya with 25 of the dive-bombers and 16 fighters, search sorties commencing on 1 April.

During April the unit operated on 15 nights, undertaking a total of 115 Suisei sorties and 58 by Zeros involving searches, attacks on Okinawan airfields, and strikes against the US carrier force. During these operations ten Suiseis and six Zeros were lost. In mid May, following several raids by B-29s and carrier aircraft, Kanoya was abandoned by the unit, which moved back to Iwakawa, a secret base 30 km north-east of Kanoya, from where operations continued. Between 3 May and 21 June 158 Suisei sorties and 76 by Zeros were flown during 20 nights. These raids were concentrated on the Okinawan airfields, costing a further 11 Suiseis. Following the termination of the Okinawa Campaign a further 22 nights saw the unit operating, 148 Suisei sorties and 73 by Zeros being made until the end of the war. During this phase only five Suiseis and one Zero were added to the unit's losses.

Raids against airfields and shipping in and around Okinawa faced anti-aircraft fire so fierce that a second pass meant certain death. Nonetheless, the unit's crews flew on, led by veteran leaders like Wt Off Mi-ichi Hirtano (and ex-float fighter pilot who then flew 12 missions on the Zero), Lt(jg) Akira Saito and Ens Masanori Kawamura (each of whom undertook more than 11 sorties in Suiseis). Although these raids were always pursued by US night fighters, CPO Yoshimasa Nakagawa, who had claimed night victories over the south-west Pacific, claimed an F6F-5(N) shot down during the early hours of 10 June, whilst in August he claimed to have brought down three B-24s with a single 3-go phosphorus bomb.

The War Ends

During the morning of 7 April 1945 fighters from 302 Ku, 601 Ku and 252 Ku rose to intercept B-29s approaching the Kanto area. This time there was a difference, however, for unknown to the Japanese pilots for the first time the bombers were escorted by P-51 Mustang fighters from new bases on Iwo Jima. In fact more decisive successes were achieved on this occasion by the use of desperate methods. A dozen 601 Ku pilots engaged the bombers over Choshi, one victory being credited to Ldg Sea Tetsuya Ishida who died while making a ramming attack, a second victory being claimed by more normal methods by Ldg Sea Yoshiharu Nishioka. Another ramming victory was claimed by 252 Ku for PO1c Yohei Watanabe, who also failed to survive. Lt Kyoji Takahashi of this unit was shot down by fire from the B-29s and baled out, but no escort fighters were encountered by either of these units. The 302 Ku pilots were less fortunate, a Suisei, three Gekkos and a Raiden being shot down by P-51s.

That same afternoon P-51s appeared over Nagoya before the B-29s reached the target area and as a result IJN units were not even able to take off. Worse was to come, for on 12 April 17 IJN and IJAAF aircraft were lost to P-51s, and night fighter units were prevented from making any interceptions by day. The Mustangs undertook a pure fighter sweep over the area on 19th, encountering 302 Ku aircraft; three Japanese fighters were shot down, two were badly damaged and two more were strafed and destroyed on the ground. A month later on 29 May, just before noon 60 Zeros challenged P-51s over Kisarazu, but lost three aircraft and one pilot without success. This same kokutai was engaged with P-51s twice in June, losing two of its fighters on 10th and four on 23rd.

Concluding their offensive against Kyushu airfields, the B-29s then reverted to night incendiary raids against large cities. The night fighter unit of 210 Ku had been disbanded on 5 May, but during the early hours of 14th 332 Ku and a detachment aircraft from 302 Ku intercepted B-29s over Nara, and while each unit suffered the loss of one aircraft, CPO Yoshimitsu Naka flying a Suisei, claimed one bomber shot down and a second damaged. A week later the 302 Ku detachment withdrew to Atsugi. From here before dawn on 24 May the unit intercepted B-29s over Tokyo with considerable success, the crews of six out of eight Gekkos each claiming one victory, while two more were claimed by the pilots of other types for a total unit claim of eight destroyed and six damaged – all achieved for the loss of but one Suisei.

During the night of 25/26 May 302 Ku claimed 16 destroyed and eight damaged for the loss of two aircraft. Of these, the Gekko unit claimed five of the destroyed and two damaged, two shot down and one damaged being credited to Ldg Sea Hajime Yamashita. Ginga crews claimed two and one damaged, Suisei crews claimed two and one damaged, one of the former by CPO Yoshimitsu Naka, while the pilots of Zero night fighters claimed three and two damaged. More Gekkos from Yoko Ku also took part in the night's interceptions. Lt Cdr Masaji Yamada, this unit's leader, was recorded as shooting down one and damaging a second bomber, but crashed while force-landing his damaged aircraft, he and his crew being killed. Most notably, CPO Juzo Kuramoto attacked bomber after bomber, claiming five shot down and a sixth damaged; for this achievement he was commended and promoted at once to Wt Off. This raid did indeed see the heaviest casualties of the war suffered by the B-29s, which lost 26 aircraft to all causes, with 100 more suffering damage.

By day on 29 May Yokohama was attacked by escorted B-29s. Three Raidens and eight Zeros were sent up by 302 Ku, but three of the former and two of the latter were lost to P-51s. At last the unit was able to claim a reverse against the Mustangs on 23 June when Lt(jg) Sadaaki Akamatsu and another pilot, both flying Raidens, managed to force two P-51s into a turning combat, Akamatsu claiming both shot down. Before that, on 1 June, 332 Ku had also managed to intercept B-29s over Osaka, where Lt Tota Hayashi claimed one destroyed and two damaged before having to bale out; his and three other Zeros falling to return fire from the bombers.

Following the end of operations over Okinawa, the Navy ordered its units to restrict flying in preparation for the anticipated landings on the home islands. Despite this order, 343 Ku continued to engage quite frequently in combats during July. On 2 July 24 Shiden-kai pilots engaged F4Us over Mount Kaimon, but results were unfavourable. While Lt Kanno claimed a single victory, five Japanese aircraft and four pilots were lost. Three days later eight of the unit's fighters fought P-51s but suffered four more losses. One more pilot was lost on 9 July while intercepting B-24s.

More success was achieved on 24 July when 21 Shiden-kais led by Lt Oshibuchi attacked carrier aircraft on the way back to their ships following an attack on Kure. On this occasion 16 victories were claimed, including one each by Ens Kazuo Muranaka and Lt Oshibuchi, but six 343 Ku pilots were lost. These included Lt Takashi Oshibuchi himself, and Lt Kaneyoshi Muto who had just been transferred to the unit from Yoko Ku. Pilots from 343 Ku attacked B-24s over Yaku Shima on 1 August, but as they did so, they were bounced by P-51s. As a result claims were limitated to one for a B-24 probably shot down and one for a fighter (misidentified as a P-47). In return two pilots were lost; Lt Naoshi Kanno's Shiden-kai was seen to have been damaged apparently by a 20mm cannon shell. After escorting him for a while, his wingman had to leave him, and he was not seen again. B-29s, this time escorted by P-47s, were intercepted by 24 Shiden-kais on

8 August, CPO Shigeo Suzaki losing his life when he rammed one of the bombers. The engagement was again ill-fated, for apart from Suzaki's aircraft, eight more failed to return. On 12 August three 343 Ku aircraft were lost to friendly fire while one was shot down during what was the unit's final engagement of the war.

Other units were finally permitted to undertake some sorties in the days immediately prior to the surrender, but the atomic bombs had been dropped without any resistance from the fighters of either the Army or the Navy, and it was already too late. 15 August saw the final sorties of the war on the day on which the fighting finally ceased.

SECTION TWO

Unit Histories

Explanation of JNFU Plane Markings

1 銀 is Silver

2 黒 is Black

3 ライト グレー is Light Grey

4 黄 is Yellow

5 赤 is Red

6 白 is White

7 ダークグリーン is Dark Green

8 青 is Blue

9 On Hashiguchi's tail (X-183 3rd Ku) there are 11 PINK victory marks

10 茶 is Brown

11 The upper surface of 251 Ku's Zero was painted mottled green

12 332 Ku Zeros had a yellow circle and white numbers

NB: In this section the pilots' names are given family first.

AIRCRAFT CARRIERS AND SEAPLANE TENDERS

Akagi

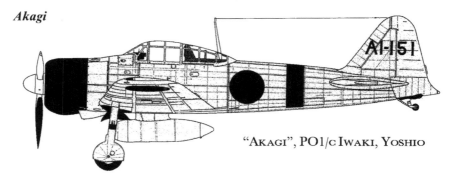

"AKAGI", PO1/c IWAKI, YOSHIO

Akagi was laid down in 1920 as a battlecruiser, but because of the Washington Treaty it was reconstructed as an aircraft carrier. Completed in March 1927, it was admitted to the Combined Fleet. From November 1931 it received a major reconstruction which took 18 months to complete. This was followed by further substantial works which took a further three years. With these works it was altered from having three flight decks to a single bow to stern flight deck to allow more practical operation of aircraft on an all-through basis. Following this second reconstruction, the vessel became of 41,000 tons displacement, and was possessed of a speed of 32.5 knots. It now carried 12 A5M fighters (plus four spares), 19 D1A dive-bombers (plus five spares) and 35 B4Y torpedo aircraft (plus 16 spares).

From January to mid-February 1939 she cruised around south China and Hainan Tao, then undertaking training in the waters around Japan. From October 1941, maintaining hull and arms, the vessel joined the 1st Air Flotilla of the 1st Air Fleet. The numbers of aircraft aboard now incorporated 12 A6Ms (plus four), 18 D3As (plus six) and 48 B5Ns (plus 16).

Akagi in its final form with one all-through flight deck and a small island structure on the port side.

In the first attack wave of the Hawaiian operation (Pearl Harbor), Lt Cdr Itaya, the hikotai leader of *Akagi*, led 43 A6Ms including nine from this carrier. *Akagi's* fighters shot down three aircraft and then strafed Hickam and Ewa airfields, claiming 25 destroyed on the ground; one A6M was lost to AA. In the second wave Lt Saburo Shindo led nine Zeros to escort 18 dive-bombers, strafed Hickam Field, and all nine returned safely.

After returning to Japan on 23 December, the ship departed on 8 January 1942 to participate in the occupation of the Bismarck Islands. Stopping at Truk, it then attacked Port Darwin with nine Zeros, sharing four victories with other fighters and claiming eight aircraft destroyed on the ground. After attacking Tjilatjap in Java, *Akagi*'s air group attacked Ceylon in early April, nine A6Ms fighting intercepting fighters over Colombo on 5th, claiming nine destroyed and seven probables without any losses. Six further victories were then claimed over Trincomalee on 9th.

At the end of April the vessel returned to Japan again, but left the Inland Sea on 27 May for Midway. During the morning of 5 June (Tokyo Time) Lt Shirane led nine Zeros to Midway, claiming 11 enemy aircraft shot down, then strafing and destroying a B-17; one A6M was lost. Over *Akagi* during 12 interceptions, the carrier's fighter pilots claimed 21 shot down and shared in the destruction of 30 more with other fighter units. However, *Akagi* was hit at 1030 by US dive-bombers and set on fire, sinking during the night. Defending fighters (CAP) landed on *Hiryu* instead, and continued to operate in defence of the fleet. When *Hiryu* too was hit, they landed in the sea near destroyers, and all the pilots were safely picked up.

Lt(jg) Masao Sato (right) and Lt(jg) Masao Asai on *Akagi*.

Hiroshi Ohara (left) and Masao Taniguchi in front of one of *Akagi*'s A5Ms. Identity code 'AI' is visible on the tail of the aircraft.

Akagi's fighter unit in June 1941. Second row, 2nd from left, Masanobu Ibusuki; 3rd, Lt Cdr Shigeru Itaya; 4th, Saburo Shindo; 3rd row, extreme left, Sadamu Komachi; 3rd, Masao Taniguchi; 4th, Yoshio Iwaki.

Hikotai leaders

Dec 1938 – Nov 1939	Lt Cdr Mohachiro Tokoro
Apr 1941 – June 1942	Lt Cdr Shigeru Itaya

Buntai leaders

Dec 1938 – Oct 1940	Lt Takahide Aioi
Dec 1938 – Nov 1939	Lt Masao Yamashita
Apr 1940 – Oct 1940	Lt Kenjiro Notomi
Apr 1941 – Dec 1941	Lt Saburo Shindo
Dec 1941 – Jun 1942	Lt Ayao Shirane
Dec 1941 – Jun 1942	Lt Masanobu Ibusuki

A6Ms on *Akagi* about to take off for the attack on Pearl Harbor on 7 December 1941.

Hiryu

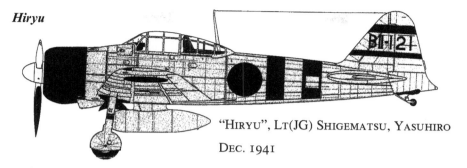

"Hiryu", Lt(JG) Shigematsu, Yasuhiro
Dec. 1941

Completed on 5 July 1939 at Yokosuka Navy Yard, *Hiyo* joined the 2nd Air Flotilla on 15 November. In September 1940 the flying unit from the carrier moved to Hainan Tao in preparation for the occupation of northern Indochina. In February 1941 the vessel supported the blockade of southern China. It then took part in the Hawaiian Operation, six of its Zeros being included in the first wave led by Lt Okajima, these aircraft strafing Barber's Point where 22 aircraft were claimed to have been set afire. In the second wave nine Zeros were led by Lt Nono to Kaneohe and Bellows airfields, and after strafing here the unit claimed two shot down, although one A6M failed to return. During the return journey to Japan, *Hiryu* attacked Wake Island. Here PO3c Isao Tawara shot down the defenders' two remaining F4Fs, making it easy for marines to land.

Hiryu.

On 5 April 1942 nine of the vessel's fighters, again led by Lt Nono, took part in the attack on Colombo, Ceylon, where the pilots claimed 16 Hurricanes and eight Swordfish shot down.

During the Battle of Midway nine of the ship's Zeros led by Lt Shigematsu, formed part of the first wave to attack the island, claiming 18 interceptors shot down. After *Akagi*, *Kaga* and *Soryu* had been hit by US dive-bombers, *Hiryu* despatched two waves of attackers against the American task force, badly damaging USS *Yorktown*. Four Zeros led by Lt Shigematsu covered the Kobayashi dive-bomber unit, claiming seven aircraft shot down but losing three of their own. Four more Zeros were led by Lt Mori to escort the Tomonaga torpedo-bomber unit, joined by two A6Ms from *Kaga*. The fighter pilots claimed 11 defending aircraft shot down but lost two of their own aircraft, including that flown by Lt Mori. Over the Japanese fleet 33 defending A6Ms (including aircraft from other carriers) claimed 43 victories, but five of *Hiryu*'s aircraft were lost while so engaged. During an afternoon attack *Hiryu* was hit and set on fire; she sank that night.

Hiryu pilots keep warm by the fire, including Lt(jg) Moriyasu Hidaka (extreme right).

Hikotai leaders

Nov 1940 – Apr 1941   Lt Cdr Shigeru Itaya

Buntai leaders

Jan 1940 – Nov 1940   Lt Shigeru Itaya
Nov 1940 – Apr 1942   Lt Sumio Nono+
Nov 1940 – Apr 1941   Lt Takumi Hoashi
Jan 1942 – Jun 1942   Lt Shigeru Mori+

n.b. Pilots marked + were killed whilst serving with the ship or unit.

Hiyo

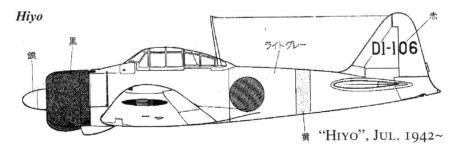

黒   ライトグレー   赤

銀   黄   "HIYO", JUL. 1942~

Hiyo fighter unit in April 1943. Front row, 3rd from left, Lt Kiyokuma Okajima; 2nd row, 3rd from left, Hideo Morinio.

Nihon Yusen's passenger liner *Izumo Maru* when under construction, was purchased by the IJN and reconstructed as an aircraft carrier. These works were completed at the Kawasaki Shipyard in Kobe on 13 July 1942, and the vessel was named *Hiyo*. The established number of aircraft to be carried by her was set at 21 fighters, 19 dive-bombers and nine torpedo-bombers. On completion, she joined the 2nd Air Flotilla while her air group trained at Iwakuni, Kagoshima and Kasanbara, completing training at Saeki on 14 September 1942. The ship left Saeki Bay on 4 October, reaching Truk five days later. By 17 October she was approaching northern Guadalcanal, launching an attack on the island on that date. On 20th the carrier again sought to head south but suffered an engine failure. In consequence, part of the air group was transferred to *Junyo* on 22nd, and *Hiyo* returned to Truk for repairs. The remaining unit aboard (16 A6Ms and 17 D3As) flew down to Rabaul on 23rd and from here next day the Zeros escorted Rikkos to Guadalcanal. Part of the unit then moved to Buin on 1 November, taking part in the Third Battle of the Solomons. On 13th and 14th, patrolling over a convoy to Guadalcanal, Lt Cdr Kaneko was killed and other pilots force-landed. The Rabaul detachment returned to Truk on 11 November while the other detachment at Buin remained until 14 December, then returning to Japan. On 22 March 1943 *Hiyo* left Saeki for Truk from where the flying unit (27 A5Ms and 12 D3As) were again detached to Rabaul. From here during 2nd-17 April it took part in the 'I-go' operation during which *Hiyo* fighters claimed 15 destroyed and 11 probables, although the 2nd Air Flotilla lost seven fighters and six dive-bombers.

After returning to Tokyo Bay on 22 May, *Hiyo* left Yokosuka on 10 June, but was torpedoed east of Miyake Jima and was forced to turn back. The flying unit then flew to Buin via Iwo Jima, Tinian and Truk during early July, but all were transferred to *Ryuho* on 15th. The *Hiyo* flying unit was reconstructed with an establishment of 24 Zeros, 18 D3As and nine B5Ns and was trained at Singapore before moving to Truk once more on 21 December. On 25 January 1944 the fighter unit flew to Rabaul where a period of extremely violent action followed. Led by Lt Hohei Kobayashi, the hikotai leader, the unit's pilots claimed some 80 victories by 20 February, losing 12 aircraft in return. A return to Truk then followed. Leaving the flying unit at Truk, *Hiyo* then sailed home to Japan. In March 1944 the carrier was posted to the 1st Mobile Fleet and took aboard the aircraft of 652 Ku for participation in the Battle of the Philippine Sea. Here, however, the vessel was sunk by US carrier aircraft on 20 June 1944.

Hikotai leaders

Jun 42-Dec 42	Lt Tadashi Kaneko+
Dec 42-Jan 43	Lt Yoshio Shiga
Jan 43-Jul 43	Lt Kiyokuma Okajima
Nov 43-Mar 44	Lt Hohei Kobayashi

Buntai leaders

Jun 42-Jun 43	Lt Iyozo Fujita
Jun 43-Jul 43	Lt(jg) Keigo Fujiwara
Nov 43-Jan 44	Lt(jg) Fujikazu Koizumi+

Hosho

Hosho was the first Japanese aircraft carrier, completed on 20 September 1922 at Yokosuka Navy Yard. In February 1923 an experimental landing of the Type-10 carrier fighter was carried out as the first carrier landing in Japan. Early in the Showa era the vessel carried nine fighters, three reconnaissance aircraft and three torpedo carriers. As a result of the Shanghai Incident which began on 29 January 1932, *Hosho* and *Kaga* were stationed on the coast off the city to support the fighting ashore. Three Type 3 fighters led by Lt Mohachiro Tokoro escorted two of *Hosho*'s torpedo aircraft to the area, engaging several times with a total of nine Chinese fighters – the first aerial combats involving Japanese aviation.

Part of the unit moved forward to Kunda airfield, Shanghai, on 7 February. From here on 26th six fighters escorted an attack on Kangzhou airfield, fighting five defending fighters and claiming three of them shot down. In July 1937 the China Incident broke out and *Hosho*, together with *Ryujo*, left Sasebo on 12 August for the Maan Islands off Shanghai. From 16th support was provided to actions ashore, but only on 22 August were the carrier's fighters involved. On that date three Type 90 fighters led by Lt(jg) Harutoshi

A badly damaged Type 3 carrier fighter is lifted aboard *Hosho*. Note the inscription 'Ro-221' on the tail; Katakama Ro was the identification of *Hosho* at this time.

Okamoto intercepted two Martin 139 bombers, claiming one shot down. *Hosho* returned to Sasebo, then taking part in a series of attacks on Chinese air bases around Canton from 21 September onwards. On the morning of the opening day six *Hosho* fighters took part in attacks on Tianhe and Baiyun airfields, engaging with ten Curtiss Hawk biplanes, six of which were claimed shot down. However, five of the Japanese fighters then had to force-land in the sea due to fuel shortage, although all the pilots were rescued by IJN destroyers. That same afternoon a strike force drawn purely from *Hosho* repeated the attacks, this time ten Hawks being encountered by the nine escorting Type 90s, the pilots of which claimed five Chinese fighters destroyed. No further combats were to take place before the ship returned to the homeland on 17 October. The vessel went onto the reserve on 1 December 1937 and spent the next three years undergoing repairs and reconstruction. In November 1940 it joined the 3rd Air Flotilla, but as new carriers were completed it became a second line vessel, providing aircraft for defensive patrols over major warships, and anti-submarine sorties. When the 3rd Air Flotilla was disbanded in April 1942, the ship was attached to the 1st Fleet, undertaking a similar range of duties. However, from June of that year *Hosho* became the training ship for carrier deck take-off and landings, but without her own air group. During 1945 the ship was again placed on the reserve, in which category she survived the war.

Buntai leaders

Jan 37 – Oct 37	Lt Kyoto Hanamoto
Nov 40 – Sep 41	Lt Harutoshi Okamoto
Sep 41 – Dec 41	Lt Ayao Shirane
Dec 41 – Apr 42	Lt(jg) Moriyasu Hidaka

Kaga

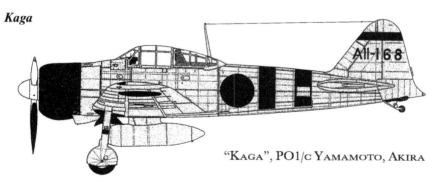

"KAGA", PO1/c YAMAMOTO, AKIRA

The keel of this ship was laid down as a battleship in 1920 at Kawasaki Shipyard (Kobe), but following the Washington Treaty which limitated the number of capital ships for the IJN, she was reconstructed as an aircraft carrier and was completed in March 1928. She entered service in November 1929. At this time the air group comprised 12 fighters (plus three spares), six reconnaissance aircraft (plus two spares) and 18 torpedo-bombers (plus six spares); the fighters in use at the time were the Type 3 carrier fighter. Following one year of further works, she became the flagship of the 1st Air Flotilla in December 1931, but at the outbreak of the Shanghai Incident the following month the flotilla was amalgamated into the 3rd Fleet and sailed for Shanghai. Here the air group went ashore at Kunda airfield, Shanghai, to support the ground fighting. On 22 February three of *Kaga*'s fighters encountered and shot down a Boeing P-12 for the first aerial victory ever to be gained by the Japanese armed forces.

Top: Kaga, unlike her sister ship *Akagi*, was operated from 1932 throughout the China Incident of 1937-1941.
Middle: Type 3 carrier fighter from *Kaga*, with Katakana 'Ni' and '203' on the tail and fuselage. Note that the torpedo-bombers in the background carry similar identity.
Bottom: Kaga fighter unit in 1936. Front row, 5th from left, Lt(jg) Kofukuda; 6th, Lt Cdr Koike; 7th, Lt Minowa.

Fighter unit on *Kaga* in October 1938. 2nd row, extreme left Jiro Chono; 2nd, Chujiro Nakano (hikotai leader); extreme right Osamu Kudo; 3rd row, extreme right, Yoshio Fukui; 2nd from right Chitoshi Isozaki.

The carrier then underwent two years of major reconstruction from October 1933, the three small flying decks being converted into a single clear-through deck. When posted to join the 2nd Air Flotilla in November 1935, the ship had aboard 16 Type 90 fighters, 16 Type 94 dive-bombers and 28 Type 89 torpedo-bombers. On 7 July 1937 the China Incident commenced, but the ship's air units then undertook training operations from 23 July-8 August. They then carried out patrols over Japanese convoys to north China, 8-12 August, while on 15 August operations commenced over central China, although for this phase the number of torpedo-bombers aboard was reduced by about half. *Kaga*'s first aerial combat saw the pilots of six Type 90s engage four Chinese aircraft and claim three shot down. Shortly thereafter the first Type 96 (A5M) fighters arrived, the first engagement involving these new aircraft occurring on 4 September. From 15 September a detachment to Kunda involving six Type 90 and six Type 96 fighters, 18 dive-bombers and 18 torpedo aircraft provided air support for ground forces for about ten days. Then from 19th the detachment took part in attacks on Nanking. A return to Sasebo on 26 September saw the ship's air complement changed to 16 Type 96 fighters, 16 Type 96 dive-bombers and 32 Type 96 torpedo-bombers. Operations were resumed over southern China during early October, requiring the carrier to move to the South China Sea where it operated until the end of 1938 when Canton was captured. During the period of one year from December 1937 Kaga sailed 29,048 miles, her aircraft repeatedly attacking important targets in southern China and becoming involved in aerial fighting over Nangxiong. During this period six fighters were detached to Shanghai and Nanking from December 1937-mid January 1938, while nine were sent to Nanking March-April 1938.

During the two years beginning in December 1938 the vessel underwent major reconstruction which altered its subsequent establishment of aircraft to 12 fighters (plus

A5M1 of *Kaga*, flown by Lt Hideki Shingo during the early period of the China Incident. Tail marking is 'R-125'.

four), 18 dive-bombers (plus six) and 48 torpedo-bombers (plus 16), the ship's total displacement rising to over 40,000 tons. During October 1940 she joined the 1st Air Flotilla, which in April 1941 became a part of the 1st Air Fleet. During October 1941 her complement of fighters was raised to 18 (plus six spares), while the total number of aircraft she was able to carry rose to 63 – plus 21 spares.

PO2c Nakajima with an A5M on *Kaga*. Note 'K' on the tail which was employed after December 1937.

A5M in the process of taking off from *Kaga*. Katakana 'Ni' on the tail replaced 'K' from July 1938.

During the attack on Hawaii on 8 December (Japanese Time) 1941, nine A6Ms led by Lt Shiga took part in the first attack, nine more led by Lt Nikaido following in the second wave. Pilots claimed one aerial victory and 20 or more destroyed on the ground, but lost four Zeros. In January 1942 *Kaga* took part in operations in the South-Eastern area, including the attack on Port Darwin when damage to the ship's hull required a return to Japan for repairs.

Following the completion of these works, *Kaga* departed the Inland Sea to take part in the Battle of Midway. On the morning of 5 June 1942 nine A6Ms were involved in the first attack on Midway Island, 12 victories being claimed. Subsequently during the fighting above the ships, 32 more victories were claimed, but the carrier was hit by USN dive-bombers and sunk; six of the vessel's fighter pilots were killed during the action.

Hikotai leaders

Oct 35 – June 36	Lt Cdr Motoharu Okamura
Jun 36 – Nov 36	Lt Cdr Sadao Koike
Nov 36 – Dec 37	Lt Cdr Takeo Shibata
Dec 37 – Dec 38	Lt Cdr Chujiro Nakano
Nov 40 – Sep 41	Lt Cdr Tadao Funaki

Buntai leaders

Nov 36 – Mar 38	Lt Chikamasa Igarashi
Nov 36 – Mar 38	Lt Tadashi Nakajima
Dec 37 – Sep 38	Lt Hideo Tashima
Sep 38 – Dec 38	Lt Masao Yamashita
Sep 38 – Dec 38	Lt(jg) Kashira Ikeda
Apr 41 – Apr 42	Lt Yoshio Shiga
Apr 41 – Sep 41	Lt Kiyokuma Okajima
Sep 41 – May 42	Lt Yasushi Nikkaido
Apr 42 – Jun 42	Lt Masao Iizuka
May 42 – June 42	Lt Masao Sato

Zero of *Kaga*.

Kamikawa Maru

A6M2Ns on the floatplane tender *Kamikawa Maru*. Code on the tail 'YII' was used from August-December 1942.

On 18 September 1937 this ship became a floatplane tender, the air element from which was engaged in the China Incident. From December 1941 it became involved in operations over South-East Asia, in June and July 1942 taking part in the Aleutians operations. During August 1942 a float fighter unit was attached to *Kamikawa Maru* with 11 A6M2Ns and two F1Ms with which the ship arrived at Shortland Island, south of Bougainville, on 4 September. Float fighters were at once engaged in operations, while a base for their use was rapidly set up at Rekata Bay on the east coast of Santa Isabel Island. From here reconnaissance sorties and patrols were at once commenced. On 13 September a Grumman F4F Wildcat was claimed shot down for the unit's first victory, but generally the F4F proved to be a difficult opponent. Consequently, during the early morning hours of 14 September three A6M2Ns were lost, while that afternoon when two of the unit's aircraft encountered two F4Fs, although Lt Jiro Ono, the buntai leader, was able to claim one shot down, his wingman was killed for the fourth loss of the day. On 12 October the float fighter unit of the 14th Kokutai arrived at Rekata Bay, joining with the remnants of *Kamikawa Maru*'s to provide cover for convoys during those hours when land-based fighter protection could not be given. However, on 7 November the only remaining pilot was killed and on 7 December the float fighter unit was disbanded. While operating in the Solomons area the pilots of the two units had claimed 14 aircraft shot down plus one more probably so, but had lost nine pilots. On 25 May 1943 *Kamikawa Maru* hit a mine and sank.

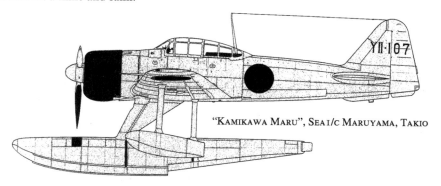

"KAMIKAWA MARU", SEA I/C MARUYAMA, TAKIO

Buntai leader

Aug 42 – Dec 42   Lt Jiro Ono

Junyo

Junyo was launched at Mitsubishi Nagasaki Shipyard on 26 June 1941 and completed fitting out on 3 May 1942. She then joined the 4th Air Flotilla of the 1st Air Fleet and commenced training of her air group at Saeki. Her establishment was set at 16 A6M fighters and 24 Type 99 (D3A) dive-bombers. The 4th Air Flotilla was to attack the Aleutian Islands at the same time as the Midway operation was being undertaken. On 18 May *Junyo* departed Saeki, also carrying 12 Zeros and pilots of 6th Ku which were to be stationed on Midway following the Japanese invasion of that island. Leaving the Inland Sea on 22 May, the air group launched 13 fighters and 12 dive-bombers to attack Dutch Harbour in the Aleutians during the morning of 4 June (Japanese Time). During the day's second strike, US aircraft were met for the first time and three were claimed shot down. Next day, despite very adverse weather conditions, a further attack was made during which six P-40s were claimed shot down.

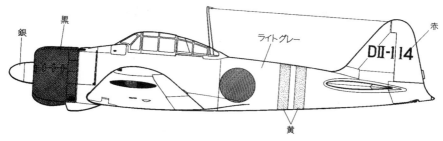

銀　黒
ライトグレー
赤
D Ⅱ-1 14
黄

"Junyo", May-Sep., 1942

Following the return to the homeland which followed, the 4th Air Flotilla was renamed 2nd Air Flotilla with which the carrier would now carry 21 fighters, 18 dive-bombers and nine torpedo aircraft. In early October the carrier group to which *Junyo* now belonged, sailed for the seas around Guadalcanal. Here, initial duty involved patrolling over battleships until the Battle of Santa Cruz on 26 October. During the latter date the carrier despatched 12 fighters to escort dive-bombers attacking the US carriers, the pilots of these claiming nine American aircraft shot down and five probables. Six of the unit's A6Ms then escorted the third attack to be launched during the afternoon.

Junyo fighter unit, June 1942. Front row, 3rd from left, Masao Sasakibara; 2nd row, extreme left, Ichiro Yamamoto; 2nd from left, Saburo Kitahata; 4th, Lt Yoshio Shiga; 5th, Kiichi Oda; 3rd row, 2nd from left, Masashi Taniguchi; 4th, Takashi Okamoto.

In the period 18-20 December 1942 *Junyo*'s fighters patrolled above an Army convoy making for Wewak, New Guinea, while from 17-25 January 1943 the air group was detached to Wewak for further such duties, and while here interceptions of B-24 heavy bombers were also made. In late January and early February the vessel also took part in the evacuation of Guadalcanal, then sailing home to Saeki. A new sortie from this base commenced during March, and on 2 April the air group again left the carrier, this time to fly down to Rabaul for further operations over the Solomons. Here the aircrews took part in the 'I-go' operations before withdrawing to Truk on 17 April. Following another brief visit home, the carrier sailed to Roa in the Marshall Islands on 15 June. Immediately

Top: Junyo fighter unit, January 1943. Front row, extreme right, Masaichi Kondo; 2nd from right, Sumio Fukuda; 2nd row, extreme left, Shizuo Ishii; 2nd from left, Tomita Atake; 4th row, extreme right, Wataru Nakamichi.

Bottom: Junyo fighter unit at Roi, Kwajalein Atoll in summer 1943. 2nd row, extreme left, Masaichi Kondo; 3rd from left, Lt Naoyoshi Miyajima; extreme right, Sumio Fukuda; 3rd row, 2nd from left, Shizuo Ishii; 6th, Takeo Banno; 4th row, 2nd from left, Tomita Atake; 6th, Wataru Nakamichi.

after the US landings on Rendova Island, all aircraft from the 2nd Air Flotilla carriers were detached to Buin and from 2 July were engaged in daily air battles until the end of August. During this period the *Junyo* fighter pilots claimed 37 victories plus 13 probables, but lost nine aircraft; disbandment followed on 1 September.

The air group was reformed on 1 November 1943, training in the Singapore area. With an establishment of 24 fighters, 18 dive-bombers and nine torpedo-bombers, the carrier next sailed to Truk again on 1 December. Following a brief detachment to Kavieng, the air group moved to Rabaul on 25 January 1944. Until 20 February fighting was constant, claims being made for 40 destroyed and 30 probables, but the flying units of the 2nd Air Flotilla were virtually annihilated. By this time the *Junyo* air group was without its carrier, for the vessel had returned home for urgent repairs. Following participation in the Battle of the Philippine Sea, she was to spend the rest of the war in the Inland Sea, suffering damage whilst at Saeki just before hostilities ceased at which time she was virtually the sole remaining Fleet carrier remaining to the IJN.

Hikotai leaders

May 42 – Dec 42	Lt Yoshio Shiga
Nov 43 – Mar 44	Lt Moriyasu Hidaka

Buntai leaders

Jun 42 – May 43	Lt Yasuhiro Shigematsu
May 43 – Aug 43	Lt Naoyoshi Miyajima
Jul 43 – Aug 43	Lt Keigo Fujiwara
Nov 43 – Jan 44	Lt Todoroki Mae

Ryuho

Substantial reconstruction of the submarine tender *Yaigei* to convert it into an aircraft carrier commenced on 20 December 1941 and was completed on 30 November 1942. Now named *Ryuho*, on 12 December while on her way to Truk carrying replacement aircraft for the carriers then based there, she was torpedoed by a submarine off Hachijo Jima and had to return to base. Here she was under repair until February 1943. From the end of March the vessel was then involved in providing deck landing and torpedo dropping training.

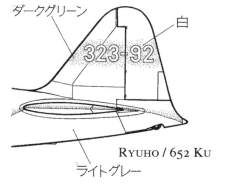

ダークグリーン

白

RYUHO / 652 KU

ライトグレー

On 12 June the ship, now able to carry 21 fighters and nine torpedo-bombers, joined the 2nd Air Flotilla, taking aboard the air group from *Hiyo* which had returned to Japan after being hit by a torpedo. *Ryuho* arrived at Truk on 21 June, but on 30th Allied forces landed on Rendova Island as a result of which the air groups from the 2nd Air Flotillas were despatched to Buin on 2 July, taking part in attacks on Rendova two days later. From then on these units were involved in daily battles over the central Solomons which proved very costly, the remaining aircrews being transferred to the 26th Air Flotilla on 1 September. By this time *Ryuho*'s fighter pilots had claimed some 30 victories.

During September 1943 the 2nd Air Flotilla began rebuilding the various air groups, moving with these to Singapore to train where there was abundant fuel available. On 10 December the flotilla moved to Truk, and from here between 27 December - 9 January air cover was provided to a convoy carrying army units from Kavieng. US carrier air attacks were driven off successfully on 1st and 4th January. On 25 January the air groups were again detached to Rabaul where they were involved in heavy fighting once more. *Ryuho* fighter pilots claimed 40 victories during this period, but suffered such heavy losses that by mid February only four or five aircraft could be put up for the continuing defence of Rabaul. Finally, on 4 March the 2nd Air Flotilla units were disbanded, although 652 Ku was then attached to the Flotilla to take their place. *Ryuho* was involved in carrying aircraft to Saipan for ten days from the end of March, but then departed for the Inland Sea. From here it moved to Tawi Tawi, taking part in the Battle of the Philippine Sea. During this action on 20 May the carrier was damaged by air attack, but was still able to navigate and fight. Thereafter it was employed only for transporting aircraft and in April 1945 departed from the combined fleet. In this guise it survived the end of the war.

Hikotai leader
Jul 43 – Sep 43   Lt Kiyokuma Okajima

Buntai leaders
Jul 43 – Sep 43   Lt(jg) Keigo Fujiwara
Sep 43 – Mar 44   Lt(jg) Hiroshi Yoshimura

Ryujo

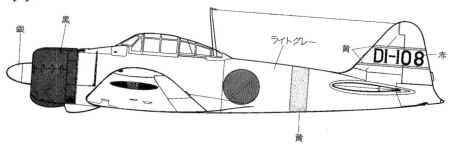

銀   黒   ライトグレー   黄 DI-108 赤

黄

"RYUJO", JUNE 1942. PO1/C KOGA, TADAYOSHI

The keel was laid at Yokohama Dock on 9 May 1933 for a small aircraft carrier of 8,000 tons displacement. On completion the vessel, which achieved a top speed of 29 knots, joined the 2nd Air Flotilla. Initially, the air component comprised six Type 3 carrier fighters and nine Type 13 torpedo-bombers. However, by the time the China Incident broke out there were 12 Type 95 fighters present. *Hosho* and *Ryujo* now formed the 1st Air Flotilla and on 6 August 1937 joined the 3rd Fleet, arriving off the coast of China near Shanghai on 14th of that month. *Ryujo's* air group fought its first air combat over Paoshan during the afternoon of 22 August when four fighters led by Lt Tadashi Kaneko spotted 18 Curtiss Hawks at an altitude of 3,000 m. Taking these by surprise, the Japanese pilots claimed six shot down without any of their aircraft receiving so much as a bullet hole. Next day another four, led this time by Lt(jg) Minoru Suzuki, engaged 27 fighters in the same area, including this time some Boeing 281 (P-26) aircraft. Nine were claimed on this occasion, including three by Suzuki himself.

The carrier then returned to Sasebo for resupply before sailing for southern China where operations were undertaken for ten days from 20 September. On 21st Canton was raided, nine *Ryujo* fighters led by Lt Cdr Kozono providing the escort; six more claims were submitted on conclusion of this mission. A second attack at once followed, and again nine of the ship's fighters were the escort, their pilots on this occasion claiming five shot down and one probable. A return to the Shanghai area followed on 3 October, and on 5th the air group went ashore to Kunda to provide support for the ground forces.

During 1938 the vessel joined *Soryu* as the 2nd Air Flotilla, returning to south China both in March and October. A year undergoing overhaul followed from November 1939 to November 1940, on conclusion of which the ship joined *Hosho* to form the 3rd Air Flotilla, attached to the 1st Fleet. By the time of the outbreak of the Pacific War *Ryujo* had joined *Kasuga Maru* to form the 4th Air Flotilla and had aboard 16 A5M fighters and 18 B5N torpedo-bombers. On 27 November 1941 the ship departed Saeki Bay and during the early morning of 8 December nine A5Ms led by Lt Takahide Aioi escorted B5Ns to attack Davao, Mindanao in the Philippines. During a second attack three fighters provided cover, but one was shot down by AA fire. The vessel and its air group then took part in operations over the southern Philippines before being diverted to Malaya. There targets around Singapore and in the Dutch East Indies were attacked, while during April a sortie into the Indian Ocean followed during which many attacks were made on shipping in the Bay of Bengal. On 23 April the ship returned to Japan where the composition of the air group was changed to include 16 A6Ms and 20 B5Ns.

On 26 May *Ryujo* left Ominato in company with *Junyo* to strike Dutch Harbour in the Aleutians during 4-5 June. On the latter date a PBY flyingboat was shot down, but one A6M was hit by AA and force-landed; this aircraft was later salvaged by the Americans, providing them with their first almost intact example of the Zero. In July 1942 the carrier joined *Junyo* and *Hiyo* to form the 2nd Air Flotilla which sailed for the Solomons area on 16 August. Here *Ryujo* separated from the main group and on 24th launched an attack on Guadalcanal, 15 Zeros led by Lt Notomi escorting the bombers to the target where 15 intercepting aircraft were claimed shot down. However, the carrier then came under a concentrated counter-strike during which ten defending A6M pilots claimed 11 of the attackers. Despite this defence, *Ryujo* sank that evening and all her fighters had to be landed in the sea.

Lt(jg) Minoru Suzuki's Type 95 fighter on *Ryujo*; in this aircraft he claimed three Chinese aircraft shot down on 23 August 1937. Katakana 'Ho' was used as the recognition identification on this carrier.

Ryujo fighter unit in 1941. Front row, centre, Lt Takahide Aioi; back row, 4th from left, Teruo Sugiyama.

Hikotai leaders

Dec 33 – Oct 35	Lt Cdr Ryutaro Yamanaka
Oct 35 – Dec 37	Lt Yasuna Kozono
Nov 41 – Feb 42	Lt Takahide Aioi
Jun 42 – Aug 42	Lt Kenjiro Notomi

Buntai leaders

Nov 36 – Oct 37	Lt Shigeru Itaya
Oct 37	Lt Manbei Shimokawa
Oct 37 – Dec 37	Lt Kiyoto Hanamoto
Dec 37 – Jun 38	Lt Mitsugu Kofukuda
Nov 40 – Nov 41	Lt Masaji Suganami
Mar 42 – Jul 42	Lt Minoru Kobayashi
Jul 42 – Aug 42	Lt Masao Iizuka

Shoho

The fast tanker *Kenzaki* was launched at the end of December 1935, but was subsequently rebuilt as a submarine tender. On completion of these works on 15 January 1939 the vessel was attached to the combined fleet. In November 1940 a further rebuild commenced to create an aircraft carrier, and in this guise she was completed as *Shoho* on 22 December 1941. With a displacement of 11,200 tons and a length of 180 m, she possessed a maximum speed of 28 knots and was fitted to operate 12 fighters and 12 torpedo aircraft. On 3 February 1942 she sailed from Yokosuka to transport A6M fighters of the 4th Ku from Truk to Rabaul. Further Zeros were carried to this base during March, but on 8 April *Shoho* returned to Yokosuka.

During May as part of the 6th Flotilla with the 4th Fleet, the carrier accompanied the force aiming to occupy Port Moresby, New Guinea. At this time the ship was carrying six Zeros, some A5Ms and 12 B5Ns. During the morning of 3 May her air group covered landings on Tulagi in the southern Solomons, while two days later escort was provided

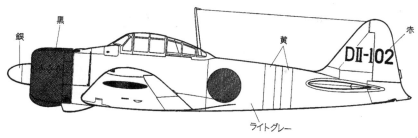

"SHOHO", DEC.1941-MAY 1942

over the convoy heading for New Guinea. Whilst so engaged during the early morning of
7th, at 0900 four A6Ms and two A5Ms were confronted with an attacking force estimated
to be 93 strong. The defending pilots claimed four shot down and one probable, but the
carrier was struck by seven torpedos and 13 bombs, sinking in 20 minutes. Three of the
defending fighters were lost, while the other three force-landed on Devonsea Island.

Buntai leader
Nov 40 – May 42   Lt Kenjiro Notomi

Shokaku

The carrier was completed at Yokosuka Navy
Yard on 8 August 1940, becoming part of the
5th Air Flotilla with *Kasuga Maru* on 1 May
1941; the latter vessel was transferred to the
4th Air Flotilla, *Shokaku* being joined instead
by *Zuikaku* in the 5th. At this time its
establishment included 12 A6Ms, with six
spares, 18 D3As with six spares and 18 B5Ns,
also with six spares. The fighter unit

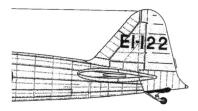

"SHOKAKU", PO1/c MATSUDA, JIRO

commenced training at Ohmura on 7 October, and the carrier sailed from the Inland Sea
on 18 November with an enhanced air group comprising 18 fighters, 27 dive-bombers
and 27 torpedo-bombers for the strike on Pearl Harbor. Six of the A6Ms took part in the
first wave, led by Lt Tadashi Laneko, strafing Kaneohe and Bellows Fields where 33
aircraft were claimed destroyed on the ground. No participation in the second wave was
involved, the ship's fighters flying CAPs. On return to base the strength of the air group
was again altered, this time to 21 fighters, 21 dive-bombers and 21 torpedo-bombers.
With this complement, she sailed again on 8 January 1942 to take part in attacks on
Rabaul and Lae. On 2 February she returend to Yokosuka, leaving once more on 17
March, this time for the Indian Ocean, where the attacks on Ceylon took place early in
the following month. On 9 April Lt Kaneko led ten fighters to this target, the pilots
claiming 23 British aircraft shot down for the loss of a single Zero. During 7-8 May 1942
the 5th Air Flotilla was involved in the first carrier-versus-carrier battle in history in the
Coral Sea. During 8th Lt Hoashi led nine *Shokaku* fighters over the US Fleet, claims for
21 victories being made. However, the carrier was hit and set ablaze, although emergency
fire-fighting teams were able to douse the flames.

Shokaku then entered Kure harbour for repairs, her flying unit flying to Kanoya and
Saeki for rest and replenishment. The air group composition was then changed yet again,
this time to 27 fighters, 27 dive-bombers and 18 torpedo-bombers – the importance of

Shokaku fighter pilots in November 1941; front row, extreme left, Ichiro Yamamoto; 2nd from left, Masao Sasakibara; 2nd row, 2nd from left, Masao Iizuka; 3rd, Lt Takumi Hoashi; 3rd row, 3rd from left, Jiro Matsuda; 4th, Kenji Okabe; 5th, Sadamu Komachi; 4th row, 2nd from left, Yoshimi Minami.

the fighters yet again being emphasised. At the same time the 3rd Fleet was also re-organised, now comprising *Shokaku*, *Zuikaku* and *Zuiho* as the 1st Air Flotilla. This fleet sailed on 24 August, participating in the Battle of the Eastern Solomons. Here four of *Shokaku*'s fighters took part in providing escort for the first wave, led by Lt Shigematsu, to attack USS *Enterprise*. Over the US vessels four victories were claimed for one loss. Following this engagement, 15 Zeros were led to Buka Island by Lt Shingo, and from here, despite vile weather, took part in attacks on Guadalcanal on 29th and 30 August, and 2 September. During these raids 15 F4Fs were claimed shot down for the loss of six pilots; Lt Shingo force-landed on Guadalcanal itself, and Lt Ibusuki also force-landed, but both pilots were able to return safely later.

During the Battle of Santa Cruz on 26 October 1942, Lt Miyajima led four fighters in the first wave attack on the US carriers, while Lt Shingo led five in the second wave, five victories being claimed by these pilots. Then, during the American counter-attack 24 defending fighter pilots intercepted the incoming raiders, claiming six and three shared with the fighters of other carriers. One Dauntless dive-bomber was rammed by PO1c Ohmori just before it released its bombs, but despite this sacrifice, *Shokaku* was hit and badly damaged. While only three fighters were lost, the losses of dive-bombers and torpedo aircraft were great and replenishment was again essential.

By the end of February 1943 repairs were completed and on 15 July the ship sailed to Truk. The air group, comprising 27 fighters, 18 dive-bombers and 27 torpedo-bombers continued training at which time the fighter unit changed its fundamental three aircraft formation for that involving four aircraft, as used by that time by American, British and German fighter units. In order to take part in the 'Ro-go' operation, the air group flew south to Rabaul on 1 November, where 25 *Shokaku* fighters intercepted an incoming raid

Top: NCO pilots of *Shokaku* in October 1942. 2nd row, extreme right, Ichiro Yamamoto; 2nd from left, Tetsuzo Kikuchi; 3rd, Jiro Matsuda; 5th, Yoshinao Kodaira; 3rd row, 3rd from left, Masao Sasakibara; 4th, Sadamu Komachi.

Bottom: Lt Hideki Shingo's Zero 'EI-111' about to take off in October 1942.

next day, assessed as being comprised of 200 fighters and bombers. 40 of these raiders were claimed shot down plus seven more as probables, and amongst the pilots involved, Wt Off Hitoshi Sato claimed eight, Wt Off Kazuki Miyabe six and Lt Hohei Kobayashi four. Thereafter, the carrier's air group, together with aircraft from other carriers and land-based units took part in attacks on Bougainville and interceptions of raids on their own base, until 13 November when the unit moved back to Truk. By this time *Shokaku*'s fighters had claimed 84 destroyed and 23 probables during the month for the loss of eight. The 1st Air Fleet carriers then moved to Roi to asssist in operations over the

Shokaku fighter unit at the end of October, 1943. 2nd row, 3rd from left, Hohei Kobayashi; 3rd row, 5th from left, Hitoshi Sato; 3rd from right, Kenji Okabe; 2nd, Takao Tanimizu; 4th row, 5th from left, Kazuo Sugino.

Marshall Islands on 26 November. Following heavy losses on 5 December, the remaning fighters withdrew to Truk, and from there to the homeland. Here the air group was disbanded on 15 February 1944 and formed into the new 601 Kokutai. On 19 June 1944 *Shokaku* was sunk during the Battle of the Philippine Sea.

Hikotai leaders

Jul 42 – Oct 42	Lt Hideki Shingo
Mar 43 – Nov 43	Lt Masuzo Seto

Buntai leaders

Sep 41 – Apr 42	Lt Tadashi Kaneko
Nov 41 – June 42	Lt Takumi Hoashi
Apr 42 – Jul 42	Lt Shigehisa Yamamoto
Jul 42 – Nov 42	Lt Naoyoshi Miyajima
Jul 42 – Nov 42	Lt Masanobu Ibusuki
Feb 43 – Oct 43	Lt Hohei Kobayashi
Oct 43 – Dec 43	Lt(jg) Yasuho Masuyama
Oct 43 – Dec 43	Lt(jg) Ikuro Sakami

Soryu

In August 1937 *Soryu* was completed at Kure Navy Yard, joining the 2nd Air Flotilla. The initial air group was to include 18 Type 96 fighters, 27 Type 96 dive-bombers and 12 Type 96 torpedo-bombers. Initially, however, Type 96 fighters (A5Ms) were in short supply and the fighter unit had to use the older Type 95. On 25 April 1938 nine fighters, 18 dive-bombers and nine torpedo aircraft from the carrier flew to Nanking to co-operate with forces advancing up the Yangzhe river. After undertaking a number of defensive patrols and sorties in support of the ground forces, the group moved to Wuhu in early June, and then to Anking, the main duty continuing to be air defence. Polluted drinking water caused numerous of the unit's personnel to fall ill, but on 25 June, having shot down one Chinese aircraft, Wt Off Sakae Kato crashed and was killed. It was assumed that he had fainted due to the illness being suffered by so many members of the group.

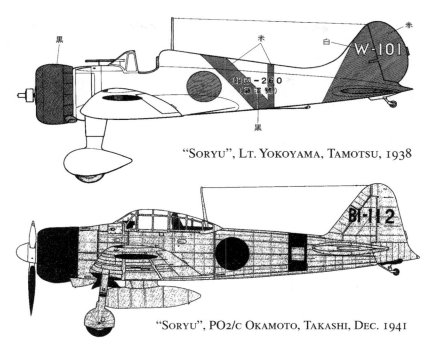

黒 赤 白 赤

W-101

靖国-260
(藤澤號)

黒

"Soryu", Lt. Yokoyama, Tamotsu, 1938

BI-112

"Soryu", PO2/c Okamoto, Takashi, Dec. 1941

On 10 July the air group returned to the carrier apart from one fighter buntai which remained behind as the nucleus of a new 15th Kokutai.

From late September 1938 *Soryu* took part in operations over Canton, but without any air combat occurring, and in December returned to the homeland. During 1939 and 1940 the carrier was engaged solely on fleet exercises, but as so many fighters had been sent to China, only nine were maintained aboard during this period. From mid September and through October 1940 the carrier moved to Hainan Tao to co-operate with the occupation of northern Indochina. From February 1941 came a move to the waters around Formosa to blockade southern China. In April 1941 the 1st Air Fleet was formed, 2nd Air Flotilla becoming a part of this, and in mid July *Soryu*'s air group moved to Sanya, Hainan Tao, to co-operate this time with the occupation of southern Indochina.

In December 1941 eight of the carrier's A6Ms led by Lt Suganami, took part in the first wave attack on Hawaii, strafing Boiler and Barber's Point where 27 aircraft were claimed to have been destroyed on the ground. Some of the fighters attacked aircraft which were believed to be carrier-borne, and five of these were claimed shot down. Nine more Zeros took part in the second wave, led by Lt Iida, and these strafed Kaneohe airfield, but here Iida's aircraft was hit by AA fire and dived into the ground. During the return flight American aircraft were encountered, two victories being claimed for the loss of two A6Ms. On 12 January 1942 *Soryu* sailed from the Inland Sea, attacking Port Darwin on 19 February. The fleet then sailed into the Indian Ocean, where on 5 April Colombo naval base in Ceylon was attacked, nine of the group's Zeros taking part led by Lt Fujita. Here 11 aircraft were claimed shot down plus three probables, for the loss of a single Zero. *Soryu* next participated in the Midway battle in June. During the initial wave of attack on the island nine Zeros took part, led by Lt Suganami, the pilots of these claiming six intercepting fighters shot down. Then when US carrier aircraft attacked the

Top: Lt Tamotsu Yokoyama taking off from *Soryu* in his A5M.
Bottom: Soryu's A5Ms during a manoeuvre from a base airfield. Note B5N torpedo-bomber on left.

Japanese carriers, *Soryu* put up 18 fighter sorties during six missions, claims against the raiders totalling 32, but this could not prevent the carrier being hit and sunk.

Hikotai leaders
Dec 37 – Mar 38	Lt Cdr Mohachiro Tokoro
Mar 38 – Jul 38	Lt Mochifumi Nango
Nov 40 – Aug 41	Lt Cdr Ryosuke Nomura

Buntai leaders
Dec 37 – Jul 38	Lt Mochifumi Nango
Dec 37 – Oct 40	Lt Tamotsu Yokoyama
Dec 39 – Oct 40	Lt Kiyoto Hanamoto
Nov 41 – Jun 42	Lt Masaji Suganami
Nov 41 – Dec 41	Lt Fusata Iida +
Dec 41 – Jun 42	Lt(jg) Iyozo Fujita

Soryu's fighter NCOs in 1941. Front row, extreme right, Yoshio Iwaki; back row, 2nd from right, Jiro Tanaka.

Zuiho

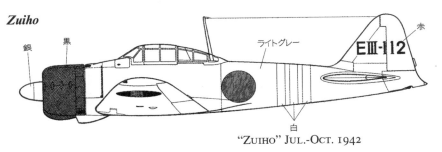

"ZUIHO" JUL.-OCT. 1942

A high-speed tanker, named *Kenzaki*, had further construction halted following its launch, and was then reconstructed as an aircraft carrier during September 1939, being completed in December 1940 as *Zuiho*. Initially the vessel was employed for deck-landing training, but on 10 April 1941 joined *Shoho* to form the 3rd Air Flotilla. By December 1941 the ship's establishment of aircraft was 12 fighters and 12 torpedo aircraft, and during the period 17 February-2 March 1942 the vessel was employed to transport aircraft from Yokosuka to Davao. During the Midway battle, the carrier accompanied the fleet's battleships and saw no action. *Zuiho* became a part of the 1st Air Flotilla on 14 July 1942, her air group now operating 21 fighters and six torpedo-bombers. During early September the vessel cruised around the Solomon Sea but again saw nothing of the enemy, returning to Yokosuka on 23rd. From here personnel and equipment for 6th Ku were transported to Rabaul on 8 October. The ship now formed a part of 1st Air Flotilla again, engaging in the Battle of Santa Cruz where it at last saw action. Nine *Zuiho* Zeros took part in escorting the first wave on its way to attack the US carriers, but on the way

Top: Zuiho, although only a small reconstructed carrier, became one of the mainstays of the carrier force from the middle of the Pacific War.
Bottom: Zuiho fighter unit in October 1942. Front row, 2nd from left, Moriyasu Hidaka; 3rd, Lt Cdr Sakuma Minowa; 4th, Masao Sato.

a US force was seen heading in the opposite direction on a similar mission. The Zero pilots at once attacked, claiming 14 shot down for the loss of four of their own fighters. With the second wave attack 14 Zeros led by Lt Sato claimed four shot down over the US carriers in return for two losses. During the US attacks *Zuiho* was hit by one small bomb and returned to Truk on 28th for repairs, after which it sailed for the homeland.

Reverting to her more usual employment, the vessel then transported aircraft from Kure to Truk on 17 January 1943. Then forming part of the 2nd Air Flotilla, she departed Japan on 29th to cover the withdrawal from Guadalcanal, returning to Truk on 9 February. The air group flew to Wewak for ten days, 18-28 February, then moving to

Top: Zuiho fighter unit, February 1943. Front row, 2nd from left, Moriyasu Hidaka; 5th, Masao Sato; 6th, Akira Yamamoto; extreme right, Tsutomu Iwai.

Kavieng 2-13 March, undertaking patrols over convoys carrying reinforcements for the army. On 2 April the air group moved to Rabaul, taking part in the 'I-go' operation until 18th, during which time 18 victories were claimed. Another return to Japan followed in early May from where between July and 5 November aircraft were transported to Truk three times. After being detached briefly to Kavieng from late August-early September, the air group moved to Rabaul on 1 November 1943 with 18 fighters and eight torpedo aircraft, to take part in the 'Ro-go' operation. Here in about ten days of interceptions and attacks on Bougainville, the unit claimed 25 shot down and ten probables for the loss of eight pilots including Lt Sato. On 1 December 1943 the *Zuiho* air group was disbanded, the carrier being once again engaged as a transport, making four trips between then and mid February, carrying aircraft and other equipment. On 1 February she was transferred to the 3rd Air Flotilla with *Chitose* and *Chiyoda*. On 15 February 653 Ku was formed, coming under the control of the 3rd Air Flotilla. Meanwhile, *Zuiho* continued to operate as a transport to Guam until the end of March. Following renewed operational training, she left the Inland Sea to take part in the Battle of the Philippine Sea, again escaping any damage. However, during the Battle of Engano, *Zuiho* was hit by US aircraft and sank on 25 October 1944.

Hikotai leaders

Jun 42 – Nov 43	Lt Masao Sato+

Buntai leaders

Apr 41 – Nov 41	Lt Takumi Hoashi
Feb 42 – Apr 42	Lt Shigeru Mori
Apr 42 – Jun 43	Lt Moriyasu Hidaka
Jun 43 – Nov 43	Lt Kanji Nakagawa

Zuikaku

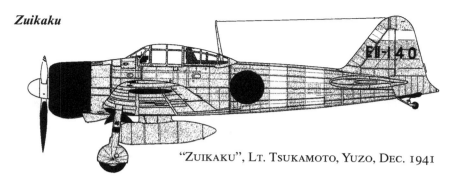

"Zuikaku", Lt. Tsukamoto, Yuzo, Dec. 1941

Zuikaku was completed at Kawasaki Shipyard (Kobe) on 25 September 1941, and with her sister ship *Shokaku* formed the 5th Air Flotilla, the air group commencing training at bases on Kyushu. During the attack on the Hawaiian Islands on 7 December 1941 six of the ship's Zeros took part in the first wave attack to strafe Kaneohe, but took no part in the second attack, being retained as CAP. From January 1942 the vessel co-operated with operations in the southern Pacific area, taking part in April in the attacks on Ceylon where during the attack on Trincomalee on 9th ten pilots led by Lt Makino to escort dive-bombers, claimed 20 victories for the loss of two pilots and their aircraft.

The next role for the 5th Air Flotilla was to support the planned occupation of Port Moresby, resulting in it taking part in the Coral Sea battle during 7-8 May. During the second day Lt Tsukamoto led nine fighters to escort torpedo-bombers attacking the US carriers. Over the enemy ships in a fierce battle the Japanese pilots claimed 26 destroyed and three probables, while over their own carriers ten more Zeros led by Lt Okajima resulted in claims for a further 22 shot down plus two probables; only one Zero was lost from each formation.

Following the Battle of Midway, in which the carrier did not take part, a new mobile fleet (the 3rd Fleet) was formed on 14 July, *Zuikaku* becoming part of the 1st Air Flotilla, an air group of 27 fighters, 27 dive-bombers and 18 torpedo-bombers training at Saeki and Kanoya. During August the carrier left the Inland Sea, then taking part in the Battle of the Eastern Solomons during which her fighters in the first wave of attackers claimed six victories for the loss of three aircraft. Between 28 August-4 September 30 Zeros from the 1st Air Division flew to Buka Island to take part in attacks on Guadalcanal, but details of what the *Zuikaku* air group claimed to have achieved have not survived. On 11 October the carrier departed from Truk, fighting the US carriers again at the Battle of Santa Cruz. Here eight *Zuikaku* fighters led by

Lt(jg) Yuzo Tsukamoto on *Zuikaku* in the Indian Ocean.

Top: Zuikaku fighter unit at December 1941. 2nd row, 2nd from left, Yuzo Tsukamoto; 4th, Masao Sato; extreme right Tatsuzo Iwamoto.

Bottom: Zuikaku, which after Midway became the core of the IJN carrier force until she was sunk on 25 October 1944.

Zuikaku fighter unit in October 1943; 2nd row, 4th from left, Lt Kenjiro Notomi; 3rd row, extreme right, Saburo Saito; 4th row, 6th from left, Yoshio Oh-ishi.

Lt Shirane took part in the first wave attack against the American carriers, claiming 14 US aircraft shot down, while four more Zero pilots in the second wave claimed another nine, but losses amongst the bomber aircraft were considerable. During 27 sorties over the Japanese fleet, meanwhile, six dive-bombers were claimed shot down. Total losses of fighters for *Zuikaku* during the battle amounted to five.

Returned to Kure, the ship was repaired and replenished, sailing again on 17 January 1943 for Truk. From here 36 fighters led by Lt Notomi moved to Rabaul on 29 January, and from there to Buin where patrols were flown over the convoy withdrawing troops from Guadalcanal. Three such missions were undertaken during which some 40 victories were claimed against US aircraft attacking the vessels. The air group then returned to Truk on 17 February. The 3rd Fleet's air groups returned to Rabaul for the period 2-18 April to take part in the 'I-go' operation with attacks on Guadalcanal, Oro Bay, Port Moresby and Rahi. During this operation *Zuikaku* fighter pilots claimed eight destroyed and five probables for the loss of three. The carrier then returned to Japan on 8 May where training took place before she sortied again on 15 July, returning again to Truk. With the announcement of the 'Ro-go' operation of 28 October, the air groups of the 1st Air Flotilla, including 24 *Zuikaku* fighters, moved to Rabaul on 1 November. Next day ships off Cape Torokina were attacked, after which attacks and interceptions were made until 13 November. By that time the *Zuikaku* fighter pilots had claimed 28 destroyed and 19 probables, but had lost eight of their own pilots, including their leader, Lt Notomi. At that stage, the unit returned to Truk.

Part of the unit continued to operate over Rabaul under the command of Lt Nakagawa until the end of January 1944, while 26 other fighters of 1st Air Flotilla moved to Roi in the Marshall Islands on 26 November. Here a raid by US carrier aircraft on 5 December destroyed ten of these aircraft, the rest returning to Truk on 7th. On 12

December *Zuikaku* and *Shokaku* returned to the homeland where on 15 February the *Zuikaku* air group was disbanded, being amalgamated into the new 601 Ku. *Zuikaku* took part in the Battle of the Philippine Sea, and then on 25 October 1944 in the Battle off Engano, she received a concentrated attack by many US carrier aircraft and was sunk in the sea north-east of Luzon.

Hikotai leaders
Jun 42 – Jul 42	Lt Kiyokuma Okajima
Nov 42 – Nov 43	Lt Kenjiro Notomi +
Nov 43	Lt Kenji Nakagawa

Buntai leaders
Sep 41 – Jan 42	Lt Masao Sato
Nov 41 – Apr 42	Lt Masatoshi Makino +
Jan 42 – Jun 42	Lt Kiyokuma Okajima
Apr 42 – Jun 42	Lt Yuzo Tsukamoto
Jul 42 – Nov 42	Lt Ayao Shirane
Jul 42 – Nov 42	Lt Shigeru Araki +
Nov 42 – May 43	Lt Naoyoshi Miyajima
Oct 43 – Nov 43	Lt(jg) Takeo Sekiya

NAMED KOKUTAIS (LAND-BASED FIGHTER UNITS)

Chitose Kokutai

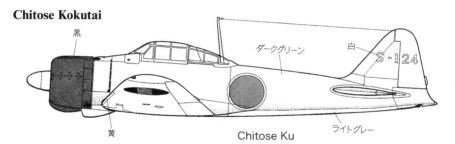

黒　ダークグリーン　白　S-124
黄　ライトグレー
Chitose Ku

On 1 October 1939 the kokutai was formed as a Rikko unit at Chitose, Hokkaido, and a fighter unit led by Lt Cdr Kiyoji Sakakibara, the hikotai leader, was attached. The unit took no part in operations over China, but on 25 January 1941 became a part of the 24th Air Flotilla with the 4th Fleet. Training took place at Saipan during June-August, while in October 36 Type 96 fighters (A5Ms) flew to Roi in the Marshall Islands. 12 of these fighters were then detached to Taroa at the eastern end of the Marshalls. As the Pacific War started, Rikkos attacked Wake Island, but due to their short radius of action, the A5Ms did not escort them, remaining to patrol over the bases. The senior buntai leader, Lt Harutoshi Okamoto, led a buntai of fighters to Truk, then undertaking patrols over this atoll. When Rabaul was occupied by Japanese forces on 23 January 1942, Okamoto's unit moved to Lakunai airfield here on 31 January, flying in via Kavieng, joining the Kawai unit there. On 10 February these detachments formed the fighter unit of the newly formed 4th Kokutai.

Meanwhile, the main part of the unit in the Marshalls (18 A5Ms at Roi and 15 at Taroa) met raids by US carrier aircraft. On 1 February 12 A5Ms intercepted over Roi, their pilots claiming five shot down, while at Taroa a total of 22 A5M sorties brought claims for 12. In March 1942 most of the remaining fighters in the Marshalls moved to Wake where they gradually converted to A6Ms. At the end of May part of the unit, together with 1st Ku sent 15 fighters on attachment to Tainan Ku. Veteran pilots were

Top: Chitose Ku fighter unit detachment on Wake Island in April 1942. Front row, 2nd from left, Lt Yoshitami Komatsu; 3rd, Lt Cdr Ryutaro Yamanaka; back row, extreme left, Masami Shiga.
Bottom: 'S171' of Chitose Ku, which was flown by Hideo Watanabe.

then selected for the newly-formed 1st and 2nd Ku. The Rikko unit subsequently operated over New Guinea and the Solomons until the end of October, but the fighter unit remained at the Marshall bases until 1 December 1942 when they were renumbered as the 201st Kokutai.

Commanding Officers

Nov 39 – Jul 40	Capt Kozo Matsuo
Jul 40 – Sep 41	Capt Kaoru Umetani
Sep 41 – Dec 42	Capt Fujiro Ohashi

Hikotai leaders

Nov 39 – Sep 41	Lt Cdr Kiyoji Sakakibara
Sep 41 – Apr 42	Lt Cdr Chikamasa Igarashi
Apr 42 – Nov 42	Lt Yoshitami Komatsu

Genzan Kokutai

Formed as a mixed unit of Rikkos and fighters at Wonsan, on the east coast of Korea on 15 November 1940, the unit was to have a fixed establishment of 27 Rikkos (plus nine spares) and 18 Type 96 fighters (plus six spares). It was attached initially to the 2nd Combined Kokutai, but in January 1941 it joined the 22nd Air Flotilla of the 11th Air Fleet, moving to Hankow at the end of April. From here the Rikkos took part in bombing raids on targets in Sichuan Sheng while the fighters patrolled over the bases, meeting no opposition. The unit then returned to Wonsan where the fighter unit was omitted from September.

黒

ライトグレー

GENZAN KU

On the outbreak of the Pacific War Genzan Ku was engaged in operations over South-East Asia and around the Indian Ocean, before moving to Rabaul in May. During April 1942 a fighter unit with 36 aircraft was again attached to the kokutai, training commencing at Kisarazu in May with pilots transferred from Tainan and 3rd Kokutais. On 20 September 1942 the fighter unit of Genzan Ku was renamed 252 Kokutai.

Kanoya Kokutai

The Kanoya Kokutai was formed at Kanoya on 1 April 1936 as a Rikko unit to which 12 Type 95 fighters were attached. The Rikkos moved to Taipei, Formosa, in August 1937

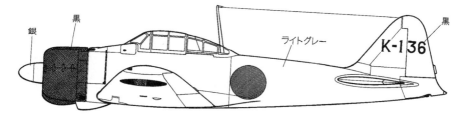

KANOYA KU. APR.-SEP. 1942

as the China Incident broke out, attacking targets on the seas around Shanghai. The unit then moved to Shanghai, being engaged over central China until the end of March 1938. The fighter unit meanwhile, provided air defence at Taipei until disbanded at the end of March.

Rikkos of Kanoya Ku operated over South-East Asia from the beginning of the Pacific War, and on 1 April 1942 a fighter unit with 36 aircraft was attached. Most of the fighters which had been attached to the 22nd Air Flotilla Headquarters, were transferred to this new unit. Around 16 September nine Zeros and 23 Rikkos (G4Ms) moved to Kavieng, New Britain, the fighters then taking part in operations over Guadalcanal from 21st of the month. Here their pilots claimed the unit's first four victories in the new area on 29 September. A few days later on 1 October 1942 Kanoya Ku was renamed 751st Kokutai

Zero 21 of Kanoya Ku in summer 1942. 'K-112' indicates the parent unit of this aircraft.

Tainan Kokutai

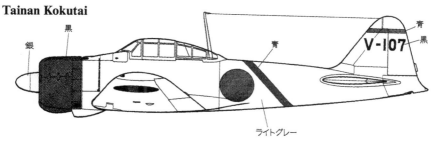

TAINAN KU, POI/C SAKAI, SABURO

The kokutai was formed on 1 October 1941 as a fighter unit at Tainan, Formosa, with an establishment of 54 fighters (plus 18 spares), and six reconnaissance aircraft. At once high pressure training commenced prior to operations over the Philippines as soon as the Pacific War started. Just prior to the outbreak, because there were no escort fighters available to cover Rikkos which would be operating over the Malayan peninsula, Tainan Ku and 3rd Ku each sent 14 fighters to the 22nd Air Flotilla. This left Tainan Ku at the outbreak of war with 45 A6M Zero 21s, 16 Type 96s (A5Ms) and six Type 98 (C5M) reconnaissance aircraft. On 8 December Lt Hideki Shingo led 44 A6Ms to attack Iba and Clark Fields in Luzon. Here nine US aircraft were claimed shot down and 60 destroyed on the ground by strafing; five Zeros failed to return. After that the unit attacked airfields around Manila every day, but resistance was minor. One buntai led by Lt Masuzo Seto

Tainan Ku officers on Jolo Island, January 1942. Front row, 2nd from left, Gitaro Miyazaki; 2nd row, 2nd from left, Yasuna Kozono; 3rd, Capt Masahisa Saito (CO); extreme right, Hideki Shingo; 3rd row, 5th from left, Junichi Sasai.

PO2c Gisuke Arita with 'V-141' on Bali Island. The legend on the fuselage reads 'Hokoku Dai 439 Go' (upper) and 'Dai 6 Ohbayashi Go'.

moved to Legaspi on Luzon on 14th, to cover the advance of the Japanese Army on the island. At the end of December with the occupation of the Philippines almost complete, 41 Zeros moved to Jolo Island, south-west of Mindanao. The unit then took part in operations over the Dutch East Indies. Tarakan was attacked on 30 December, the unit then moving there on 16 January 1942 immediately following its occupation. Attacks on Balikpapan and Bandjarmasin followed, and then a move to the former of these locations.

From 3 February the unit commenced operations over eastern Java, gaining air superiority here by defeating Dutch and US fighters over Surabaya and Malang. Part of the unit moved to Makassar on 18 February to cover landings on Bali Island, then moving to Wyndham airfield on that island on 26th. The rest of the unit gradually followed, operating from there until the close of fighting on Java on 9 March. Nine fighters were then detached back to the Philippines to escort bombers raiding Corregidor Island where US forces continued to resist for some weeks.

Top: 'V-117', an A6M of Tainan Ku, demonstrates by the two diagonal and two horizontal bands that it is the aircraft of the buntai leader (Lt Akira Wakao) or hikotai leader (Lt Hideki Shingo).
Bottom: Tainan Ku/251 Ku pilots just prior to leaving Rabaul.

On 1 April establishment was reduced to 45 fighters and six reconnaissance aircraft, and the unit was transferred to the 25th Air Flotilla in the south-east area (which was mainly to be New Guinea). At this stage about half the pilots were sent home, the remainder boarding a transport vessel which arrived at Rabaul two weeks later. Here ex-members of 4th Ku were transferred to Tainan Ku, having been engaged in operations over New Guinea while awaiting the arrival of the main unit. Thus on 28 April the kokutai had on hand 24 Zeros at Lae, but only eight plus six A5Ms at Rabaul. The supply of new aircraft had now slowed, and until August only about 20 aircraft could be put up for each mission. During this period only Tainan Ku operated over eastern New Guinea; during the period April to July 1942 it undertook 51 offensive missions totalling 602 sorties. Claims during these latter operations amounted to 201 destroyed and 45 probables, while victories claimed during interceptions raised the total to about 300. The cost had been just 20 Zeros.

Following the landing of US Marines on Guadalcanal the situation changed dramatically. On 7 August with the news of the landings, Tainan Ku suspended a planned attack on Rabi, Lt Cdr Nakajima, the hikotai leader, heading a force of 18 Zeros escorting Rikkos the 560 miles (1,000 km plus) to reach Guadalcanal. Here they engaged US carrier aircraft, claiming 36 shot down and seven probables for the loss of two (actual USN losses were 11 F4Fs and one damaged). Following this day's fighting the kokutai was engaged over both the southern Solomons and New Guinea, claiming many victories but with accumulating fatigue, even the veteran Zero pilots were killed one after another.

Lt Cdr Tadashi Nakajima, hikotai leader of Tainan Ku.

During the three months from August to the end of October, a further 164 destroyed and 37 probables were claimed, but for the loss of 32 pilots killed. On 1 November 1942 the unit was renamed 251st Kokutai.

Commanding Officers

Oct 41 – Dec 42	Capt Masahisa Saito

Hikotai leaders

Oct 41 – Apr 42	Lt Hideki Shingo
Apr 42 – Oct 42	Lt Tadashi Nakajima

Toko Kokutai

Formed as a flyingboat unit on 15 November 1940, the kokutai was initially engaged over the Philippines and Dutch East Indies following the start of the Pacific War. It then took part in the occupation of Atsu and Kiska in the Aleutian Islands on 8 June 1942, six flyingboats moving to Kiska. On 5 July the float fighter unit was formed at Yokosuka and carried aboard the floatplane tender *Chiyoda* to Kiska where it was incorporated into Toko Ku. As there were no airfields on Kiska, the float fighters began patrolling instead of reconnaissance floatplanes which had previously been undertaking this role. Enemy aircraft were encountered for the first time on 8 July when B-24s raided the island. On 18 July one B-24 was shot down for the float fighter unit's sole victory. This was because on 5 August a new kokutai, the 5th, was formed and all Toko Ku float fighters were transferred to it. On 1 November 1942 Toko Ku was renamed 851st Kokutai.

TOKO KU

Buntai leader of the float fighter unit

Jul 42 – Aug 42	Lt Kushichiro Yamada

A6M2N float fighter of Toko Ku; note 'D-107' on the tail.

Yokohama Kokutai

The kokutai was formed on 1 October 1936 as a flyingboat unit, becoming a part of 24th Air Flotilla for operations mainly in the Marshall Islands. When Japanese forces occupied Rabaul in February 1942, the unit moved there. On 1 April 1942 a float fighter unit was attached to Hama Ku (as Yokohama Kokutai was abbreviated to), but initially on paper only. It was to be the middle of May before the unit was actually formed at Yokosuka with 12 A6N2N aircraft, which arrived at Rabaul on 3 June, at once beginning patrols. Initially, hostile aircraft were only met on one occasion before the unit moved to Tulagi where the flyingboat unit had already moved in. Patrolling here, the float fighter pilots claimed their first victory on 10 July when a B-24 was claimed shot down. Thereafter heavy bombers were intercepted every day until early August, a few claims being made and losses suffered. On 7 August, however, as US forces landed on Guadalcanal, strafing US carrier fighters destroyed all the remaining float fighters at their moorings. All the personnel of the unit

Lt Ri-ichiro Sato, leader of the float fighter unit of Yokohama Ku.

were recorded as being killed in action
when the island was invaded. One record
indicates that there were a few float
fighters further north which patrolled from
Shortland from the end of August, but
without any contact with US aircraft; the
float fighter unit was omitted from the
kokutai immediately after this, and on 1
November 1942 was renamed 801st
Kokutai.

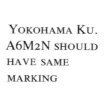

YOKOHAMA KU.
A6M2N SHOULD
HAVE SAME
MARKING

Yokosuka Kokutai

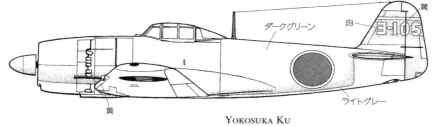

YOKOSUKA KU

The oldest land-based unit of the Imperial Japanese Navy, this kokutai was formed on 1
April 1916 for the purposes of educating and training aircrews. The unit then undertook
research aerial warfare and tested new aircraft, engaging also in advanced aviation
education. For the 'A-go' operation in June 1944 the major part of the Yoko Ku, as it was
known, and the 27th Air Flotilla formed the 'Hachiman Force' which was sent to Iwo
Jima. At this time the establishment of Yoko Ku for its fighter unit was 48 carrier fighters,
48 interceptors and 12 night fighters, the hikotai leader being Lt Cdr Tadashi Nakajima.
On 22 June Nakajima led 27 Zeros to Iwo Jima to join a further 24 Zeros and five Gekkos
which had already arrived there. On 24 June Zeros of Yoko Ku fighter unit intercepted
US carrier aircraft in company with 252 Ku and 301 Ku, claiming ten F6Fs shot down
with six more as probables, for the loss of nine aircraft. That afternoon nine Yoko Ku
Zeros took part in escorting an attack on the US carrier force but were intercepted over
Urracas Island. Here 17 Japanese fighters were lost, including four Yoko Ku aircraft, and
the planned attack had to be aborted. During 3 and 4 July American carrier aircraft
raided Iwo Jima while USN ships bombarded the island, but there were no Japanese
aircraft left, and on 6th the remaining pilots were transported back to the mainland.
From November 1944 B-29s began raids on targets in the Kanto area, including on
Tokyo, and in February 1945 US carrier aircraft also raided the area. On 16 February at
least ten fighters (a mixed force of Zeros, Shiden-kais and Raidens from Yoko Ku and
Investigation Section) led by Lt Yuzo Tsukamoto, sought to intercept. Next day Yoko Ku
fighters were able to surprise a mixed formation of F6Fs and F4Us, identified as 19-
strong, 13 of these being claimed shot down with the remaining six as probables. During
the Okinawa fighting a Zero buntai led by Lt Kunio Iwashita took part until the middle
of April when they were withdrawn to Yokosuka. During day interceptions after P-51s
started accompanying the bombers, Yoko Ku pilots were seldom able to make claims, but
late in May a few Gekko crews were able to register a number of claims against B-29s by
night. After that date the unit was seldom engaged except during the last few days before
the war actually ended.

WO Shigetoshi Kudo taxying the J5N1 Tenrai twin-engined fighter prototype for a test flight at Yokosuka Ku, where all such tests of new aircraft were undertaken.

Officers of Yokosuka Ku, from left to right: Lt Kunio Iwashita, Lt Cdr Masanobu Ibusuki and Lt Yuzo Tsukamoto. All were veterans who then took command of hikotais.

Shiden 11 of Yokosuka Ku in flight in the hands of CPO Ohara.

NUMBERED KOKUTAIS (FIRST NUMBERING)

1st Kokutai

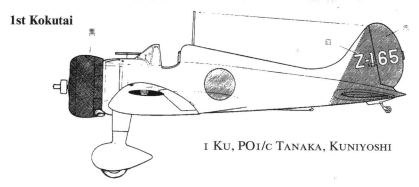

I KU, PO1/C TANAKA, KUNIYOSHI

The unit was formed as a mixed kokutai of fighters and Rikkos at Kanoya on 10 April 1941. Initial establishment was set at 36 Rikkos and 24 fighters. From July 1941 the Rikko unit repeatedly attacked targets in Sichuan Sheng, operating from Hankow. During this period the fighter unit's Type 96 fighters (A5Ms) flew patrols over Hankow without meeting anything. In September the fighter unit was omitted, following which the Rikkos took part in operations over the Philippines and Dutch East Indies during the opening phase of the Pacific War. Raids over New Guinea between late February and late March followed, but on 10 April the kokutai moved to Taroa in the Marshall Islands and was transferred to the 24th Air Flotilla.. On 1 April 1942 a fighter unit was again added, which was intended to have 27 fighters plus nine spares. In practice, all that was available at this time was 13 A5Ms, the pilots for which were drawn mainly from Chitose Ku. At the end of May a 15 aircraft mixed formation from 1st and Chitose Ku were led by Lt Joji Yamashita to Rabaul where the pilots and their aircraft were all transferred to Tainan Ku, taking part in fighting over New Guinea and seeking to complete the training of new pilots as these gradually arrived at Rabaul. In the middle of August a US force landed on Makin Island in the Gilbert Islands. Lt(jg) Heitaro Morita led the remaining 1st Ku fighters with those still with Chitose Ku, to the area but found nothing. One buntai then moved to Mille Island to co-operate in the occupation of Nauru and Ocean Islands. On 1 November 1942 1st Ku was renamed 752nd Kokutai, the fighter unit then being amalgamated into Chitose Ku which became 201 Ku a month later.

2nd Kokutai

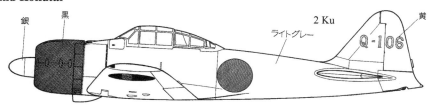

The unit was formed on 31 May 1942 at Yokosuka as a mixed force of fighters and dive-bombers with an establishment of 16 of each type of aircraft. On 29 July the unit boarded the specially reconstructed carrier *Hachiman Maru* for transport to New Britain. On 6 August 15 A6Ms and 16 D3As took off to land at Lakunai airfield. The unit's A6Ms were of the A6M3 variant (Zero 32), which had an increased maximum speed, but a resulting reduced range. As a consequence they could not fly from Rabaul to Guadalcanal and back, so instead the unit was employed for operations over eastern New Guinea, for which they moved to Buna. From here they attacked Rabi on 24 August, encountering P-39s, claiming seven shot down and two probables without loss. Offensive and defensive missions were flown until 8 September when the unit was thrown into the attack on Guadalcanal, the availability of a landing ground at Buka Island mitigating the range problem.On 28 September Cdr Yamamoto, the unit commander, led 21 fighters to Buka from where several attacks were launched. On 1 November 2nd Ku became 582nd Kokutai.

3rd Kokutai

Formed as a Rikko unit on 10 April 1941, the bombers moved to Hanoi during July, and in September operated against the interior of China. In mid September the kokutai was re-formed as a fighter unit with an establishment of 54 aircraft (plus 18 spares) and nine reconnaissance aircraft (C5Ms). The kokutai despatched a detachment of 13 A6Ms and three C5Ms to operate over Malaya just before the outbreak of the Pacific War, leaving available 45 Zero 21s, 12 A5Ms and six C5Ms. The distance between Takao, where the unit was based, and Manila, Luzon, was 500 miles so that it would be necessary for the pilots to be able to fly a return distance of 1,000 sea miles and to allow for 20 minutes combat over the target. For this requirement, tests of long flights were undertaken with satisfactory results, obviating the need for aircraft carriers to have to operate in the area.

Capt Gaifu Kamei and Lt Cdr Takeo Shibata, respectively the commanding officer and executive officer of 3 Ku, watch as the first elements of their unit take off for the attack on Luzon on 8 December 1941.

On 8 December 53 Zeros led by Lt Tamotsu Yokoyama, the hikotai leader, took off from Takao, escorting Rikkos to attack Iba and Clark airfields in the Manila area. Here the Japanese pilots claimed more than ten intercepting fighters shot down for the loss of two Zeros, then going on to claim a further 20 plus aircraft destroyed on the ground by strafing. Two days later 24 pilots returned to the area, claiming 44 victories in the air and 42 more on the ground. (USAAF losses on this date to both the 3rd and

3 Ku, PO2/c Hashiguchi, Yoshiro

Tainan Kokutais included 11 P-40s shot down, 12 P-35s destroyed on the ground with six more damaged, three B-18s and a considerable number of miscellaneous aircraft; two US Navy PBY flyingboats were also shot down and one more destroyed at anchor.)

Top: 3 Ku detachment at Rabaul. 2nd row, 8th left, Lt Cdr Sasakibara; 9th, Lt Aioi; 6th row, 5th left, Kiyoshi Ito.
Bottom: Zero 21 'X-182' of 3 Ku at Lakunai, Rabaul.

A move was made to Davao on 23 December, and from here Dutch Brewster Buffalos defending Tarakan were overwhelmed on 28th. A move to newly-occupied Menado followed on 12 January, from where Ambon was attacked. From 25th the kokutai began moving to Kendari, launching an attack on Koepang on Timor Island next day. On 2 February the greater part of the unit moved to Balikpapan, and next day Lt Yokoyama led 27 Zeros to Surabaya, Java, joined there by 27 more Zeros from Tainan Ku. Here 3rd Ku claimed 39 aircraft shot down plus 21 on the ground by strafing (Dutch and US losses). Co-operating with forces landing in eastern Java followed, and stock was taken of the kokutai's achievements. Between 12 January-3 March 1942, 86 aerial victories had been claimed together with 90 more on the ground. From the outbreak of the war on 8 December 1941 to the completion of the occupation of South-East Asia, 150 aircraft had been claimed shot down and about 170 destroyed on the ground; the cost had been 11 pilots killed.

On 3 March 17 Zeros led by Lt Miyano attacked Broome and Wyndham airfields in north-western Australia, claiming to have destroyed more than 20 flyingboats there. While part of the unit then returned to Japan on 10 March, the main element stayed at Koepang, escorting Rikkos to attack Port Darwin in the Australian Northern Territories on several occasions. Here the unit became regularly involved with the P-40s of the 49th

Zero 'X-128' of 3 Ku.

Kunimori Nakakariya of 3 Ku with Nakakariya's Zero, 'X-138'.

Fighter Group, claiming some victories against these but losing a number of pilots, Lt Tokaji included, when the American pilots included 'dive-and-climb' attacks which were difficult for the Zeros to combat. During September 1942 Lt Cdr Sakakibara, the Hikocho, led 21 Zeros and four C5Ms to Rabaul, taking part from here in attacks on Guadalcanal, and on convoy patrols and interceptions until early November. Claims during this period amounted to 48 destroyed and 20 probables for the loss of eight pilots. On 1 November 1942 the unit was renamed 202nd Kokutai.

4th Kokutai

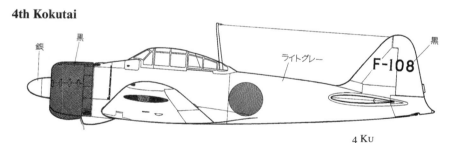

4 KU

The unit was formed on 10 February 1942 with both Rikkos and fighters, establishment being set at 27 of each type of aircraft. The fighter unit was composed of one chutai of the Chitose Ku led by Lt Harutoshi Okamoto, which moved from Truk to Rabaul, and one led by Lt Shiro Kawai which arrived there from Palau. Initial equipment was the A5M, but these were gradually replaced by A6Ms. Surumi and Gasmata in the western part of New Britain were occupied and part of the new unit moved there at once, claiming three RAAF Lockheed Hudsons shot down on 11 February. From the latter part of the month newly-acquired Zeros escorted Rikkos to Port Moresby, New Guinea. In early March Japanese forces occupied Lae and Salamaua in eastern New Guinea, seven A6Ms being moved to Lae on 11th. On 14 March eight Rikkos and 12 Zeros attacked Horn Island at the north-eastern tip of Australia, encountering 49th Fighter Group P-40s and claiming six shot down and two probables for the loss of two pilots. On 1 April 1942 the 4th Ku became an all-Rikko unit, the fighters being transferred to Tainan Ku.

5th Kokutai

Top: Pilots of 5 Ku engaged in helping create facilities at the exposed Kiska base in the Aleutians.

Bottom: 5 Ku float fighter pilots at Kiska in August 1942. Back row, extreme left, Sea2c Hachiro Narita; 2nd from left, PO2c Giichi Sasaki. Note 'R-106' on the tail of the A6M2N. Several of the unit's pilots, including the leader, are not present in this photograph.

On 5 August 1942 the 5th Ku was formed as a
float fighter unit with 12 aircraft. Initially the
float fighter unit of Toko Ku which had six
A6M2Ns and which had been operating over
Kiska in the Aleutians, led by Lt Kushichiro
Yamada, became the 5th Ku, being removed
administratively from Toko Ku. Interceptions
over Kiska continued, but half a unit of three-
seat reconnaissance floatplanes were then
attached, and during August five E13As were sent

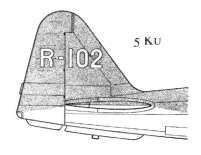

5 KU

to Kiska to undertake search and patrol sorties. The new kokutai's first success was
recorded on 8 August when the float fighters shot down two USN floatplanes which had
been launched from a cruiser. Subsequently opponents came to include heavy bombers,
flyingboats and fighters, engagements usually taking place at low altitude and in dreadful
weather conditions. A particularly heavy combat occurred on 15 September when four
A6M2Ns challenged a formation of 12 heavy bombers and 28 escorting fighters. Five of
the US aircraft were claimed shot down, four of them by PO2c Giichi Sasaki, plus one
probable, but two float fighters failed to return and Sasaki's aircraft overturned as he
landed, reducing available A6M2Ns to just one. The floatplane tender *Kimikawa Maru*
arrived on 25 September to deliver six replacement A6M2Ns and two E13As. These were
immediately in action from the next day, but by 4 October no float fighters remained
serviceable. On 1 November 1942 the unit became 452nd Kokutai.

6th Kokutai

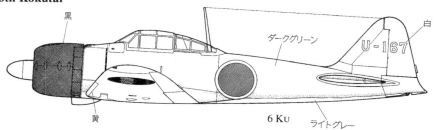

黒   ダークグリーン   白

U-167

黄   6 KU   ライトグレー

Formed at Kisarazu on 1 April 1942 as a fighter unit with an establishment of 60 fighters
and eight reconnaissance aircraft, the 6th Ku at the time of its formation had in fact only
six fighters available to it, although it did have a core of veteran pilots who had just
arrived from the Tainan and 3rd Kokutais. On 18 April an attempt was made to raid the
B-25s launched from USS *Hornet* on the famous 'Doolittle Raid', but no interceptions
were achieved. By 30 May the unit had 33 Zeros, 24 A-class pilots and 31 of C-class. For
the Midway operation in June the unit's aircraft and personnel were divided between five
carriers, including *Junyo*, the intention being that they should be stationed ashore
following the occupation of the island. On 4 June seven 6th Ku Zeros led by Lt Miyano
took part in the attack on Dutch Harbour, while next day over the carriers some of the
pilots were able to claim the kokutai's first victories while engaging the USN raids on the
Japanese carriers, three pilots led by Lt Kaneko from *Akagi* claiming five of the attackers.

    At the start of the operations over the Solomon Islands, Lt Mitsugu Kofukuda, the
hikotai leader, led 18 of the veteran pilots from the homeland to Rabaul on 21 August.
However, 6th Ku had what were known as 2-go Zeros (Type 22 and 32) which had a
shorter range capability. These could only be used to take part in raids on Rabi

Top: Mainstay of 6 Ku was the Zero 32 like this one, photographed here.

Bottom: 6 Ku detachment at Kisarazu, just before leaving for Rabaul. Back row, extreme left, Lt Mitsugu Kofukuda; 3rd from left, Tomoichi Ema; 4th, Momoto Matsumura; 8th, Saji Kanda.

and Moresby. Following the completion of Buin airfield on Bougainville Island, the unit moved there on 8 October, allowing them to operate further south. However, on 11 October on the final patrol mission of the day in very bad weather, five Zeros were lost when their pilots had to force-land in the sea. Meanwhile, on 7 October the carrier *Zuiho* brought down 27 more Zeros to Rabaul, these then moving to Buin to reunite the whole unit. On 1 November 6th Ku was renamed 204th Kokutai.

12th Kokutai

At the outbreak of the China Incident the 12th Kokutai was formed at Saeki Kokutai on 11 July 1937. With an establishment of 12 Type 95 fighters, 12 Type 94 dive-bombers and 12 Type 92 torpedo aircraft, it became a part of the 2nd Combined Kokutai. On 7 August it moved to Zhoushuizi airfield at Darian, engaging in convoy patrols before returning to the homeland at the end of the month. With the situation at Shanghai critical, on 5 September 12th Ku was amalgamated into the 3rd Fleet, moving to Kunda airfield, close to the city. The Type 95 fighters possessed too short a range to allow them to take part in attacks on Nanking, and consequently were employed for ground support and air defence of the base area. Between October and November conversion took place to the Type 96 (A5M), and following the occupation of Nanking the unit moved to

Dajiaochang airfield at Nanking. From here the unit joined aircraft of 13th Ku in attacks on Nanchang and Hankow. On 22 March 1938 the 12th Ku was designated a fighter unit only, and took over the fighters of the 13th Ku, which became a Rikko unit. The 12th now comprised 2.5 tai (five buntais) with an establishment of 30 A5Ms, and it had on strength more than 50 fighter pilots; it also still retained a torpedo aircraft unit.

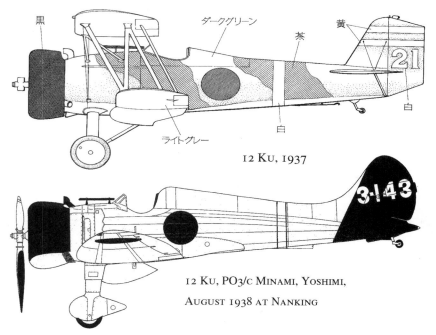

黒   ダークグリーン   茶   黄

ライトグレー   白   白

12 Ku, 1937

12 Ku, PO3/c Minami, Yoshimi,
August 1938 at Nanking

12 Ku at Hankow in summer 1938. Front row, 3rd from left, Shigema Yoshitomi; 4th, Yasuna Kozono; 5th, Tadashi Nakajima; 6th, Takahide Aioi; 2nd row, 2nd from left, Toshio Kuroiwa; 7th from left, Kiyonobu Suzuki; 8th, Shigetaka Ohmori; 3rd row, extreme left, Tetsuzo Iwamoto.

Top: A5M1 flown by Lt Shigema Yoshitomi at Nanking. Note two fuselage bands indicating the leader's aircraft. (A white band showed that it was an operational aircraft.)

Middle: An A5M2 which was fitted with an enclosed cockpit, was also supplied to 12 Ku for trials.

Bottom: 12 Ku pilots being debriefed after a mission to Chengtu on 4 October 1940. Vice-Admiral Shigetaro Shimada, C-in-C China Area, inspects his pilots who are, front row, left to right, Lt Tamotsu Yokoyama, Matsuo Hagiri, Ichiro Higashiyama, Saburo Shindo, unknown and Ayao Shirane.

Zero 11 ('3-112') of 12 Ku in August 1941 with officers of the unit. Twenty-eight victory marks (kite in a circle) indicate victories claimed by the unit during a certain period.

Based at Anking, on 29 April 1938 27 A5Ms escorted Rikkos to Hankow where the pilots claimed 28 Chinese aircraft shot down and seven more as probables. Repeated attacks on the two cities continued and in the period from the end of April-19 July the unit claimed 100 destroyed and 12 probables for the loss of five aircraft.

After the fall of Hankow the 12th Ku moved there, but the majority of the pilots returned to Japan, the unit being reduced to a much smaller size. Now the Rikkos were raiding targets in inland China, mainly at Chengtu and Chungking in Sichuan Sheng. The A5Ms lacked the range to be able to accompany the bombers all the way to these targets, but in summer 1940 the A6M Zero appeared on the scene, and this did have the range capability. On 21 July 1940 six A6Ms were led by Lt Yokoyamato Hankow, nine more soon following. On their fourth escort mission on 13 September, 13 Zeros led by Lt Saburo Shindo caught 30 Chinese fighters in the air and claimed 27 of them shot down. From then until summer 1941 the 12th and 14th Kokutais claimed 103 shot down and 163 destroyed on the ground for the loss of just three aircraft. The 12th Kokutai was disbanded on 15 September 1941.

Commanding Officers

Jul 37 – Nov 37	Capt Osamu Imamura
Nov 37 – Dec 38	Capt Morihiko Miki
Dec 38 – Oct 39	Capt Shun-ichi Kira
Oct 39 – Jun 40	Capt Takasue Furuse
Jun 40 – Mar 41	Capt Kiichi Hasegawa
Mar 41 – Sep 41	Capt Ichitaro Uchida

Hikotai leaders

Jul 37 – Mar 38	Lt Cdr Motoharu Okamura
Mar 38 – Dec 38	Lt Cdr Yasuna Kozono
Mar 38 – Aug 38	Lt Cdr Mohachiro Tokoro
Dec 38 – Oct 39	Lt Cdr Takeo Shibata
Oct 39 – Nov 40	Lt Cdr Sakuma Minowa
Oct 39 – Jan 40	Lt Cdr Shigema Yoshitomi
Nov 40 – Sep 41	Lt Cdr Sei-ichi Maki

13th Kokutai

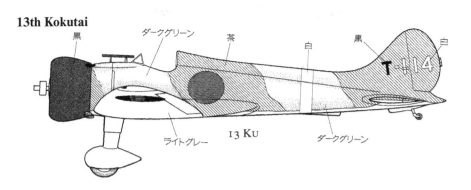

黒　ダークグリーン　茶　白　黒　白

T-114

ライトグレー　I3 Ku　ダークグリーン

Formed at Ohmura on 11 July 1937, the unit was amalgamated with the 2nd Combined Kokutai and moved to Zhoushuizi airfield in Darian, China, where it undertook convoy patrols until the end of August when it returned to Japan. It was ordered to Shanghai on 5 September, moving on 9th to the nearby Kunda airfield. Strength at this time was 12 A5Ms (two buntais), six Type 96 dive-bombers and 12 Type 96 torpedo-bombers. Taking part in the first air raid on Nanking on 19 September, Lt Shichiro Yamashita led 12 A5Ms to escort dive-bombers, engaging at least 20 fighters, 12 of which were claimed shot down, plus a further three probably so, all without loss. Further raids were led by Lt Nango, and on 2 December six A5M pilots became engaged with 30 Chinese aircraft, claiming 13 shot down on this occasion. Following the fall of Nanking the unit moved forward to Dajiaochang airfield from where Nanchang and Hankow were attacked. Although many claims were made, two buntai leaders were shot down during this period. During a reorganisation of units on 15 November 1938 the fighter unit was removed from 13th Ku which was subsequently disbanded two years later on 13 November 1940.

A5Ms of 13 Ku at Kunda, Shanghai, in September 1937. 'T' was the unit identifying code until that month.

Top: 13 Ku A5Ms at Kunda. An aircraft on the left carries '4-134' on the tail, while one on the right is '4-127'. From October 1937 '4-' became the unit's code.

Bottom: 13 Ku at Oita just after its formation. Front row, 6th left, Kuniyoshi Tanaka; 8th, Momoto Matsumura; 3rd from right, Kan-ichi Kashimura; extreme right, Shigetaka Ohmori; 2nd row, 2nd from left, Sadaaki Akamatsu; 4th, Toshio Kuroiwa; 7th, Lt Shigema Yoshitomi; 5th from right, Mitsugu Mori.

Commanding Officers

Jul 37 – Mar 38	Capt Sadao Senda
Mar 38 – Dec 38	Capt Kanae Kosaka
Dec 38 – Nov 39	Capt Kikuji Okuda +
Nov 39 – Nov 40	Capt Rinosuke Ichimaru

Hikotai leaders

Jul 37 – Sep 37	Lt Cdr Tsuguo Ikegami
Sep 37 – Dec 37	Lt Cdr Chujiro Nakano
Aug 38 – Nov 38	Lt Cdr Mohachiro Tokoro

14th Kokutai

The unit was formed on 6 April 1938 with an establishment of 12 fighters (1 tai), six dive-bombers (0.5 tai) and 18 torpedo aircraft (1.5 tais). It became a part of the 5th Fleet, moving to Sanzaodao in southern China to support operations in Canton during the autumn of 1938. It took part in attacks on Kweilin (Guilin) and Liuchow, but without seeing any aerial combat. On 15 December 1938 the kokutai was re-organised to comprise only Rikkos and torpedo aircraft, but a further re-organization in November 1939 altered it to incorporate 18 fighters and nine dive-bombers. During the closing days of the year the unit joined the 12th Ku and *Akagi*'s air group to undertake attacks on Kweilin and Liuchow. On 20 December 13 of the unit's fighters together with others from 12th Ku were led by Lt Aioi to Kweilin where 14 Chinese aircraft were claimed shot down. On 10 January 1940, 14 A5Ms of 14th Ku and 12 from 12th Ku claimed another 16 shot down plus nine destroyed on the ground.

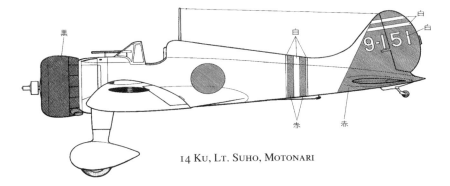

14 KU, LT. SUHO, MOTONARI

14 Ku in January 1939; seated in the centre are Lt Hideki Shingo (left) and Lt Takumi Hoashi.

Type 95 carrier fighter (A5N) of 14 Ku at Weizhoudao.

During the period May-September 1940 the unit was engaged in air defence sorties over Hankow, receiving nine A6Ms on its return to southern China. On 7 October seven of these fighters led by Lt Kofukuda reached Kunming from Hanoi, claiming 14 aircraft shot down here, while four were claimed strafed on the ground, all the Japanese aircraft returning safely. Subsequently the unit attacked targets in Yunnan Sheng, while in late July 1941 the greater part of the unit moved to Hanoi, detaching 12 fighters to Saigon. However, on 15 September 1941 the whole unit was disbanded where it stood.

14 Ku A5Ms being refuelled at Weizhoudao, Tonkin Bay, during November-December 1939.

Commanding Officers

Apr 38 – Dec 38	Capt Hiroki Abe
Dec 38 – Nov 39	Capt Hiroshi Higuchi
Nov 39 – May 40	Capt Tameki Nomoto
May 40 – Oct 40	Capt Shigematsu Ichii
Oct 40 – Apr 41	Capt Toshiyuki Yokoi
Apr 41 – Sep 41	Capt Gin Nakase

Hikotai leaders

Dec 38 – Nov 39	Lt Cdr Tadao Funaki
Nov 39 – May 40	Lt Chikamasa Igarashi
May 40 – Nov 40	Lt Cdr Asaichi Tamai

A5M4 of 14 Ku seen from behind.

Early model A6Ms of 14 Ku undertaking a long-range sortie deep into south-west China.

A5Ms of 14 Ku at Nanning following a combat on 27 December 1939.

14th Kokutai (second)

The second 14th Ku began life on 1 April 1942 as a flyingboat unit which became a part of the 24th Air Flotilla, operating initially from Jaluit. In August a detachment was sent to the Solomon Islands. In early September a float fighter unit with nine aircraft plus three spares was formed since the situation in the Solomons was becoming critical, and this was sent to the south-east area, becoming a part of 14th Ku on 20 September 1942. On 2 October nine A6M2Ns and 97 men went aboard the floatplane tender *Kiyokawa Maru* at Yokosuka, arriving at Shortland on 12th where they came under the command of

14 Ku (2ND),

PO3/c Jitoh, Hisao

Kamikawa Maru of R-area flying unit. The first aerial encounter occurred on the morning of 13 October against B-17s, the float fighters using Rekata Bay as an advanced base. From here convoy patrols and standing patrols over the base were flown, during which Wt Off Gi-ichi Minami claimed one B-17 shot down and a second probably so, while PO3c Hisao Jito claimed three Douglas SBD Dauntless dive-bombers. On 1 November 1942 the unit was renamed 802nd Kokutai.

15th Kokutai

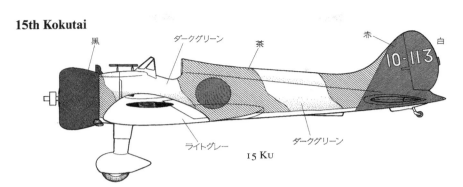

黒　　　ダークグリーン　　　茶　　　赤　　　白

10-113

ライトグレー　　　ダークグリーン

15 Ku

Formed at Ohmura on 25 January 1938, the unit moved to Anking on 10 July. Establishment was 12 fighters, 12 dive-bombers and six torpedo-bombers, but in fact actual strength comprised nine Type 95 fighters, nine Type 96 fighters (A5Ms), 18 dive-bombers and nine torpedo aircraft and with these it provided support for operations around Hankow. At this base from May was the air group from *Soryu*, the major part of its fighter unit including Lt Nango being transferred to 15th Ku of which Nango then became hikotai leader. On 18 July Nango led six A5Ms to escort dive-bombers and torpedo-bombers to Nanchang. Enemy aircraft were encountered over Poyanghu, one of which Nango shot down, but he then collided with another and was killed. During September the unit's strength was altered to include 18 fighters, 18 dive-bombers and nine torpedo-bombers with which it moved to Poyanghu (Poyang Lake) to take part in the Hankow offensive, but on 1 December 1938 the unit was disbanded.

Commanding Officer
Jun 38 – Nov 38   Capt Kazutari Kabase

Hikotai leader
Jun 38 – Jul 38   Lt Mochifumi Nango +

Type 96 (A5M) and Type 95 (A5N) carrier fighters of 15 Ku at Anking in summer 1938. '10-' on the tail identifies this unit.

Fighter Unit attached to HQ, 22nd Air Flotilla

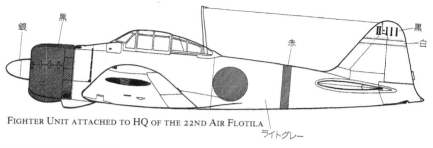

FIGHTER UNIT ATTACHED TO HQ OF THE 22ND AIR FLOTILA

Top: Zero of the Fighter Unit attached to the 22nd Air Flotilla at Kota Baru, Malaya, in December 1941.
Bottom: Takaji Buntai from 3 Kokutai formed part of the 22nd Air Flotilla attached Fighter Unit. Front row, 3rd from left, Tadatsune Tokaji; 4th, Morio Miyaguchi; 3rd row, 2nd from right, Ken-ichi Abe.

Just before the outbreak of the Pacific War two British battleships (HMS *Prince of Wales* and *Repulse*) arrived at Singapore, and to counter these three Rikko kokutais were attached to the 22nd Air Flotilla in the area. There were, however, no fighters and on 22 November it was decided these were necessary, so detachments were sent to join the Flotilla. These comprised 14 Zeros and three C5Ms from Tainan Ku, led by Lt Kikuichi Inano, and 13 Zeros with a further three C5Ms led by Lt Tadatsune Tokaji from 3rd Ku, plus nine A5Ms. The commanding officer was Cdr Yutaka Yamada, the executive officer of Takao Ku, while the hikotai leader was Lt Cdr Mohachiro Tokoro. These elements all reached Saigon by 1 December, although two Zeros were lost along the way. The greater part of the unit then moved to Soc Trang.

Following the outbreak of war, these fighters flew both offensive and defensive missions. Starting on 12 January 1942 escorts were provided to Rikkos raiding Singapore, and by 29th of that month 150 sorties had been flown on these duties, 40 aircraft being claimed shot down by the fighters and bombers jointly, with 30 more claimed destroyed on the ground. Two Zeros were lost together with one C5M, but whilst under escort by the Zeros, only two Rikkos were shot down. On 5 February the unit moved to Kuching from where it was involved in fighting over western Java, following which it moved to Bangkok. On 1 April 1942 it was absorbed into the Kanoya Kokutai.

Inano Buntai from Tainan Ku also formed part of the 22nd Air Flotilla attached Fighter Unit. Front row, centre, Lt Kikuichi Inano; right, Kozaburo Yasui; back row, extreme left, Minoru Honda; extreme right, Sadao Yamashita.

NUMBERED KOKUTAIS (SECOND NUMBERING)

131st Kokutai (131 Ku)

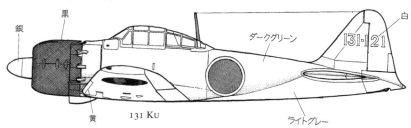

The kokutai was formed at Yokosuka on 10 July 1944 as a reconnaissance and night fighter unit. It comprised Teisatsu 11 Hikotai (abbreviated to T11) with 24 reconnaissance aircraft and Sento 851 Hikotai (S851) with 24 night fighters. Training began at Katori, T11 with C6N Saiun aircraft and S851 with Gekkos. By 1 August eight of the latter were on hand and already from the end of July two or three Gekkos had been flying sorties over Iwo Jima, intercepting B-24s. On 15 November 131 Ku became a dive-bomber unit instead, S851 being transferred to Hokuto Ku. In March 1945, however, it was again reformed as a night fighter unit, this time with S804, S812 and S901, each with 24 aircraft. Moving to Kanoya, the unit then came under the command of Kanto Kokutai where it effectively became a night attack unit with single-engined aircraft, using D4Y Suisei dive-bombers and Zeros. It was classed as an ordinal unit, meaning that it was not destined to undertake Tokko (suicide attacks), at the direction of Lt Cdr Tadashi Minobe, the commander at Kanoya, and Lt Masashi Tokura, the hikotai leader of S812. Training took place at Fujieda, Shizuoka, following which pilots and aircraft were sent to Kanoya. Beginning with a dawn search mission on 1 April, the unit took part in the Okinawa fighting. The first attack on US carriers was undertaken during the night of 4 April, eight Suiseis and four Zero night fighters being involved. Thereafter attacks were made on shipping off Okinawa and airfields on the island practically every night. During these attacks 93 sorties were made by Suiseis and 60 by Zeros, more than ten of the unit's aircraft being lost. By this time Kanoya itself was coming under almost daily attack. Consequently, a move was made to a secret airfield in the mountains to the north-east of Kanoya from which the unit operated until the end of the war. By that time it had made some 600 sorties during which more than 60 aircraft were lost.

153rd Kokutai (153 Ku)

The unit was formed on 1 January 1944 with 12 reconnaissance aircraft, becoming a part of the 23rd Air Flotilla, based at Kendari, Celebes. When US forces invaded western New Guinea in April, it moved to Wakde Island, undertaking patrols from there for about a month. On 5 May S311 was transferred from 381 Ku to 153 Ku and the 23rd Air Flotilla was amalgamated into the 1st Air Fleet. Although S311 had an establishment for 48 carrier fighters, at this stage it was only at about 50% strength. During May Allied bombers began attacking Sorong and Manokwari, S311 becoming engaged in intercepting these attacks from Sorong, employing about ten Zeros at this base. 21 victories were claimed, including several B-24s, but by 26th of the month six of its own aircraft had been lost. On that date S311 had five Zeros at Sorong and 15 more at Kendari, but next day Allied troops landed on Biak Island. S311, joined by 202 Ku, made attacks on these forces as well as maintaining the defence of Sorong. On 29 May, however, Wt Off Shinji Ishida led four Zeros to attack the airfield on Biak, but all failed to return. Three from a formation of six were lost on 2 June, and three days later, down to only one aircraft, the unit withdrew to Kendari to recuperate.

US forces then landed on Saipan on 15 June, the unit at once sending ten Zeros to Peleliu, Yap and Truk, but these were not to be involved in any aerial fighting whilst there. On 10 July S311 was transferred to 201 Ku, and 153 Ku received instead S901 with 24 night fighters from the disbanded 251 Ku. Thus the kokutai now had two hikotais, S901 and T103, with which it moved to Davao on Mindanao in early August. Coincidentally, night bombing of Davao commenced at this time. However the lack of ground radar coupled with the low maximum speed of the Gekko meant that the only successful interception occurred on 2 September when CPO Yoshimasa Nakagawa rammed a B-24 and survived to land back at base. Attacks by US carrier aircraft were anticipated at this time, and S311 moved its nine Gekkos and two Zero night fighters to Tacloban, Leyte. From here on 9 September three Gekkos were despatched on a search and strike mission, but these were intercepted by F6Fs, two being shot down and the third damaged. In mid September the greater part of S901 moved to Nichols Field, Luzon. During the first raids on Manila on 21 and 22 September two Gekko crews spotted the US carrier task force, CPO Saburo Sue claiming to have obtained a hit on one carrier. In mid October S901 was ordered to hinder the activities of torpedo boats operating in Ormoc Bay, Leyte, and to achieve this Lt Minobe, the hikotai leader, led four Gekkos and four Zeros to Cebu from where the unit was able to claim six such vessels sunk in four weeks. However, during a period of two months in the Philippines, the unit lost 14 aircraft including three buntai leaders. Following this, S804, S812 and S851 were all transferred to 153 Ku for brief periods before departing for Formosa or Japan. Ultimately the headquaters staff and ground crew ended up fighting in the mountains west of Clark Field until the end of the war, by which time the majority had been killed, including the commanding officer, Capt Wada.

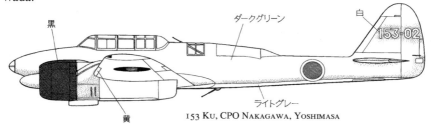

黒　ダークグリーン　白

153-02

黄　ライトグレー

153 Ku, CPO Nakagawa, Yoshimasa

Commanding Officers

Jan 44 – Feb 44	Cdr Moriyoshi Yamaguchi
Feb 44 – Jul 44	Cdr Rikihei Inoguchi
Jul 44 – Nov 44	Capt Nobukichi Takahashi
Nov 44 – ?	Capt Tetsujiro Wada+

201st Kokutai (201 Ku)

On 1 December 1942 201 Ku was formed from Chitose Ku and 752 Ku, becoming a part of the 24th Air Flotilla stationed in the Marshall Islands and at Wake Island. In February 1943 the unit was ordered to return to Japan, arriving at Kisarazu during March. It then moved to Matsushima for recuperation and training, and to defend the northern area. For the latter 12 Zeros were moved to Paramushir in the Kuriles (Chishima) on 3 June. This detachment ended after a short spell, following which 45

Zeros were loaded aboard the carrier *Junyo*, which reached Rabaul on 15 July. Another 18 Zeros led by Cdr Nakano flew the whole 5,000 km to Rabaul at the same time, arriving on 12th. Here the kokutai was assigned to the 21st Air Flotilla, although at this stage it had only eight A-class pilots, the rest of the unit's strength being made up of 20 B-class and 24 C-class pilots.

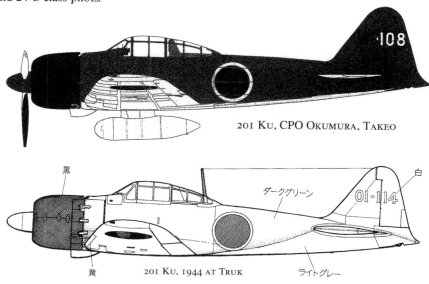

201 Ku, CPO Okumura, Takeo

黒   ダークグリーン   白

201 Ku, 1944 at Truk

黄   ライトグレー

A detachment of 25 aircraft was sent to Buin, from where on 21 July Lt(jg) Tomokichi Arai led a raid on Rendova where eight victories were claimed. Next day over the Special Transport *Nisshin*, US dive-bombers and escorting fighters were engaged. The unit was then heavily involved in severe fighting for the next two months. On 14 September a total force of 117 Zeros intercepted a raid estimated to be over 200 aircraft strong against which claims were made for 29 fighters and four bombers shot down. CPO Okumura made three sorties, claiming nine 'small aircraft' shot down plus one 'large aircraft' shared.On 22 September US forces landed at Finschhafen in eastern New Guinea, Lt Shiro Kawai leading 12 Zeros in an attack on the shipping off the landing beaches. Here they were bounced by a large force of American fighters, losing four veterans including CPO Okumura and Wt Off Hongo. Following the withdrawal of 204 Ku from Buin on 8 October, 201 Ku continued to operate from there alone. On 22nd many of the unit's aircraft were damaged on the ground during raids, leaving only nine Zeros to retreat to Buka Island. Soon both Buka and Ballale became unusable due to constant raids, and a further withdrawal was made to Rabaul. From late October until early January 1944 the kokutai was engaged in attacks on Torokina, on Cape Marcus, and in defence of Rabaul itself, finally withdrawing to Saipan to rest and re-equip. Since July 1943 the unit's pilots had claimed some 450 victories.

On 11 February the unit received 23 Zeros at Saipan, eight of which were then sent to Truk, led by Wt Off Masao Taniguchi. However on 17th all these were lost, shot down by US carrier aircraft. At once 20 more Zeros were led to Truk by Lt Kawai, but on 23rd the remaining four aircraft at Saipan were wiped out by a raid there, and the aircraft which had been sent to Truk, had to return. As at 4 March 1944 201 Ku was reorganised to comprise two hikotais, S305 (with an establishment of 48 fighters, led by Lt Shiro

Kawai) and S306 (also with an establishment of 48 fighters, led by Lt Torajiro Haruta). S305, which had been the flying unit of 201 Ku before the reorganisation, moved to Peleliu in mid March. From here on 30th 20 Zeros intercepted US carrier aircraft, the pilots claiming 17 shot down, but nine failed to return, nine more were badly damaged and the last two both force-landed. 201 Ku then withdrew to Davao to recuperate during April and May.

A6M 'WI-111' of 201 Ku in flight in August 1943.

Meanwhile the formation of S306 had been continuing at Kisarazu and during mid April the unit began moving to Cebu and Davao where 32 aircraft concentrated during the middle of May, continuing training there. When orders for the 'A-go' operation were issued, 11 Zeros were flown to Yap while a detachment also moved to Guam. The latter, after undertaking attacks on US shipping off Saipan, returned to Cebu in July. One buntai formed from the remaining pilots of 263 and 343 Kus, together with other units that had been disbanded, moved to Yap in mid July. Here on 16th and 23rd Lt Kanno led this unit to undertake interceptions of B-24s, eight of which were claimed shot down, seven probably so, and 46 damaged for the loss of five pilots including two who rammed.

After the battle for the Mariana Islands, remaining fighters were concentrated in 201 Ku, which effectively became a new unit. It was now composed of S305 (Lt Masanobu Ibusuki), S306 (Lt Hiroshi Morii, followed by Lt Naoshi Kanno in September), S301 (Lt Usaburo Suzuki) and S311 (Lt Takeo Yokoyama). The new commander was Capt Sakae Yamamoto, executive officer was Cdr Asaichi Yamai, and hikocho was Lt Cdr Tadashi Nakajima. In early August B-24s started raids on Davao, but when these began to be escorted by P-38s during September, the Japanese fighters were held back and conserved for the defence of the Philippines, being sent to Cebu, Mactan Island, Legaspi and Nichols Field. By 1 September 201 Ku had no less than 210 Zeros to hand of which 130 were immediately serviceable. To enhance the strike power of the defending units, from late August 201 Ku pilots were trained in the art of skip-bombing, using bomb-carrying Zeros by a dive-bomber instructor from Yokosuka Ku. Just as the basic training was completed US carrier aircraft raided the Philippines and it was not this skill that came to the fore, but the success of Tokko – the suicide attacks. On 9 September the carrier air groups attacked Davao, 201 Ku fighters being moved to Clark Field instead. On 10th came a false report that landings were taking place at Davao, causing the fighters to be moved to Cebu in the afternoon. Next day 60 Zeros moved to Mactan and Nichols, while on 12th there were about 100 Zeros in the area when US aircraft raided Cebu. 41 took off to intercept from an unfavourable position, their pilots claiming 23 victories. However, 25 Zeros were shot down, including that flown by Lt Hiroshi Morii, 14 more force-landed, 25 more were destroyed or badly damaged on the ground, 30 more suffering minor damage. Then on 21 and 22nd September the carrier aircraft raided Manila. 42 Zero sorties were launched in defence, and while 27 of the attackers were claimed shot down, 20 Japanese aircraft were lost in the air and ten more destroyed on the ground. When account is taken of the true US losses as opposed to the claims made by the Japanese pilots, the full extent of the disaster may be appreciated.

Nonetheless, the morale of the pilots remained high, and on 22nd 15 bomb-carrying Zeros led by Lt Suzuki in company with ten Suisei dive-bombers attacked the US carrier force east of Lamon Bay, hits on five ships being claimed, as well as three strafing attacks on carrier decks. On 15 October Lt Ibusuki led 25 Zeros in another strafing attack on the carriers. Then on orders from Admiral Takijiro Ohnishi, the Jinpu Special Attack Unit was formed within 201 Ku, commencing 'special' missions on 20th. The first success was achieved on 25th when the Shikishima Unit (five aircraft including Lt Seki) reportedly sank one carrier, set a second on fire and also sank a cruiser, these results being reported by Wt Off Nishizawa who was leading the direct escort. As a result of this proven effectiveness, the special attack, ramming attack, or Tokko as was at once named, became the basic method of attack, employed on a daily basis, and with a considerable wastage of Zeros. In January 1945 the remaining pilots were evacuated to Formosa, while the ground crew stayed on in the Philippines, fighting in the mountains west of Clark until the end of the war, by which time over 200 men had been killed.

Commanding Officers

Dec 42 – Mar 43	Cdr Ryutaro Yamanaka
Mar 43 – Jul 44	Cdr Chujiro Nakano
Jul 44 – Nov 44	Capt Sakae Yamamoto
Nov 44 – Jan 45	Cdr Asaichi Tamai
Jan 45 – Mar 45	Cdr Zen-ichi Nakamura

Hikotai leaders

Apr 42 – May 43	Lt Yoshitami Komatsu
May 43 – Mar 44	Lt Shiro Kawai

202nd Kokutai (202 Ku)

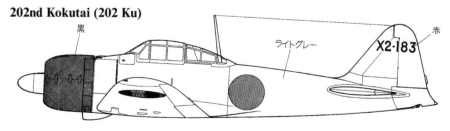

黒   ライトグレー   X2-183   赤

202 Ku

On 1 November 1942 the 3rd Ku was renamed as 202nd. At this time the greater part of the kokutai was operating over the Solomons and New Guinea, but almost immediately after being renamed, it moved back to the south-west area. Most of the unit then took up station at Kendari, Celebes, but detachments were maintained for air defence at Ambon, Tual, Koepang, Makassar and Surabaya. On both 3rd and 15 March 1943

Elements of 202 Ku visiting Kendari for 'rest and recreation'. Lt Cdr Minoru Suzuki, hikotai leader, is third from left, wearing a tropical helmet.

Rikkos were escorted to Darwin, meeting Spitfires and veteran Australian fighter pilots for the first time on the latter date when four Spitfires were claimed shot down.

Following the arrival of a new hikotai leader, Lt Cdr Minoru Suzuki, the unit trained in long distance, over-water navigation, and in formation fighting. On 2 May the kokutai took off from Koepang to fly to Darwin where an estimated 33 Spitfires were encountered, 21 of these being claimed shot down (four of them probably so) without loss. (Actual RAAF losses amounted to five Spitfires.) From then until early September the unit was engaged on six occasions over Darwin and Brooks Creek, raising its claims in the six months since March, to a total of 79 destroyed and 22

202 Ku NCO pilots including Kiyoshi Ito (2nd row, 2nd from right, in flying overalls).

probables for the loss of three Zeros and two Rikkos. On 9 September 27 Zeros from Tual escorted Rikkos to Merauke. Here seven Zeros became separated from the main formation and were pursued by P-40s, losing three pilots including Ens Miyaguchi. During December 20 pilots led by Wt Off Tadao Yamanaka were detached to 204 Ku to aid in the fighting over Rabaul.

As a result of the reorganisation of units on 4 March 1944, 202 Ku became composed of S301 (48 fighters led by Lt Cdr Minoru Suzuki) and S603 (48 fighters led by Lt Cdr Hideki Shingo). 43 S301 A6Ms were then moved to Moen, Truk, via Tinian, where they were engaged in intercepting B-24s until late May. During this period one buntai was detached to Ponape, which on 30 April was subjected to an attack by US carrier aircraft. Seven Zeros led by Lt(jg) Kawakubo escorted dive-bombers to attack the ships, claiming 16 interceptors shot down for the loss of four aircraft. On 23 May 28 Zeros led by Lt Cdr

Lt(jg) Toshio Shiozuru and his 202 Ku Zero 'X2-172'.

Suzuki left Truk for Sorong on 1 June where rendezvous was made with S603 to take part in the Biak operation until 12th. Next day a move was made to Peleliu, while on 19th attacks were made on shipping off Saipan, which was reached via Yap.

S603, which had been formed from the flying unit of 331 Ku, on being transferred to 202 Ku moved 33 aircraft from Surabaya to Meleyon Island during early March to provide air defence. At the end of the month the major part of the hikotai moved to Truk, joining the main body of the kokutai. On 30 April US carrier aircraft attacked Truk again, 15 S603 Zeros being led by Lt(jg) Hiroshi Suzuki to escort dive- and torpedo-bombers against the task force. 11 hostile aircraft were claimed shot down for four losses. In late April 18 Zeros accompanied dive-bombers to Halmahera, from here taking part in convoy escorts and interception of B-24s. 14 more Zeros left Truk on 23 May, arriving at Sorong on 1 June under the leadership of Lt Cdr Shingo, and here they joined forces with S301. With a combined total of about 50 A6Ms the kokutai took part in the Biak operation until 12 June, losing 21 aircraft. The remaining 16 serviceable Zeros moved to Peleliu on 17th, taking part in a few attacks on Saipan. At the end of June interceptions of B-24s over Yap and Peleliu were made, but the unit then withdrew to Davao where it was disbanded on 10 July 1944.

Commanding Officers

Apr 41 – Aug 42	Capt Tokio Kamei
Aug 42 – Oct 42	Capt Kaoru Umetani
Oct 42 – Aug 43	Cdr Motoharu Okamura
Aug 43 – Mar 44	Cdr Sadagoro Uchida
Mar 44 – Jul 44	Cdr Shigeki Negoro

Hikotai leaders

Sep 41 – Mar 42	Lt Tamotsu Yokoyama
Mar 42 – Mar 43	Lt Takahide Aioi
Mar 43 – Apr 43	Lt Minoru Kobayashi
Apr 43 – Mar 44	Lt Cdr Minoru Suzuki

203rd Kokutai (203 Ku)

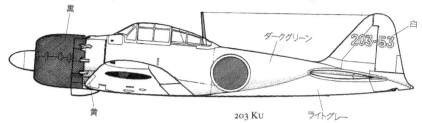

203 Ku

Atsugi Kokutai had originally been formed in April 1943 as an operational training unit for carrier fighter pilots, but on 20 February 1944 was renamed 203rd Kokutai. Intended to have an establishment of 96 fighter aircraft, at the beginning it had on hand 59 Zeros and 12 Gekkos. The latter were based at Kisarazu, and on 4 March these were transferred to 302 Ku.

203 Ku joined the 51st Air Flotilla in the 12th Air Fleet, the main body of the unit (46 aircraft) departing Atsugi for Chitose. With the overall reorganisation which occurred on 15 April, 203 Ku became comprised of S303 (led by Lt Cdr Kiyokuma Okajima) and S304 (Lt Takashi Oshibuchi), each unit to comprise 48 aircraft. While waiting for the winter snows to

Inset: Lt Cdr Kiokuma Okajima (leader, Sento 303), with 203 Ku in northern Japan.

Bottom: Zero '3-71' of Sento 303, 203 Ku, at Kagoshima in early 1945 with the Sakurajima volcano in the background.

finish melting, one buntai of Zeros and three Gekkos were sent to Kataoka airfield on Shumushu Island and Musashi airfield on Paramushir Island at the end of April; the main body followed to Kataoka during the latter half of May. On 13 May the first victory was claimed against US heavy bombers, which continued to be the main target to interception thereafter. During the night of 26 June USN cruisers bombarded Musashi, destroying six Zeros and damaging seven more.

On 11 August the main body withdrew to Bihoro, but a detachment of 19 Zeros remained at Shumushu, defending the northern Kuriles. On 18 September S303 with 27 Zeros was sent to join T Force, moving to Mobara, and then to Kagoshima, where it was ordered to wait. Orders came for a move to the Philippines on 12 October, the flight being made by way of Okinawa and Formosa. S304 with 29 Zeros had moved to Katori on the previous day, and on 14th flew on to Yontan airfield. That same afternoon the unit escorted an attack force bound for the US task force. Next day fighters were despatched from Formosa, while on 16th 14 Zeros led by Lt Oshibuchi joined other units in attacking the American fleet, sharing in the claimed destruction of six aircraft. All units rendezvoused over Taichung on 22nd, 29 Zeros then moving to Bamban, Luzon.

From here 16 Zeros of S304 took part in the general all-out attack launched against the US carriers on that date, following which the two hikotais took part in raids almost every day at the time when the main method of anti-shipping attack was being shifted to Tokko (suicide). 203 Ku was one of the units selected for such methods, 16 of the unit's pilots, including Lt(jg) Ichiro Watanabe, were to be killed undertaking Tokko attacks. On 15 November 203 Ku withdrew to Kasanbara where S303 was transferred to 201 Ku and S304 to 221 Ku. At the end of October S812 with 24 Gekkos had flown to the Philippines and been absorbed into 203 Ku, undertaking night bombing raids on targets in Luzon and night interceptions until the end of December by which time the hikotai had no aircraft left.

203 Ku then became composed of S303 and S312 which had also returned from the Philippines. As at 18 March 1945 S303 had been rebuilt to 32 Zeros and 57 pilots at Kagoshima and Izumi, while S312 had 31 Zeros and 51 pilots at Kasanbara. During combats over Kyushu on 18-19 March, 12 victories were claimed at the cost of 11 Zeros. On 21 March 11 Zeros led by Lt Cdr Okajima escorted the first raid attempted by Rikkos carrying Ohka piloted glider bombs. However, all the Rikkos were shot down despite the escort, before they could reach their targets. During April 1945 203 Ku joined the 5th Air Fleet to take part in the fighting over Okinawa, but in May was transferred to the 72nd Air Flotilla. By late July it had S303, S309, S311, S312 and S313 under command, but was ordered to preserve aircraft and pilots for the anticipated invasion of the homeland, and so it ended the war.

Commanding Officer
Feb 44 – May 45   Capt Ryutaro Yamanaka

204th Kokutai (204 Ku)

On 1 November 1942 6th Ku became 204th Kokutai with an intended operating strength of 60 fighters and eight reconnaissance aircraft. At the time the unit was based at Buin on Bougainville Island, taking part in attacks on Guadalcanal, escorts to convoys heading for that island, and interceptions of US raids on Buin. Until the end of the year nine pilots were lost in combat during convoy patrols, but 11 more went missing in bad weather, or in other types of accidents. During the 'I-go' operation of April 1943, attacks were made on Guadalcanal, Port Moresby and Milne Bay, 20 victories being claimed.

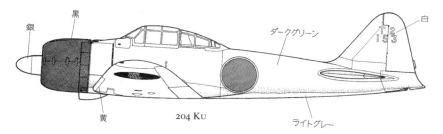

銀　黒　　　　　　　　　　　　　ダークグリーン　　　　T2 白
153

黄　　　204 Ku　　　　　　　ライトグレー

However, on 18 April six Zeros led by Lt(jg) Takeshi Morisaki escorted Rikkos carrying Admiral Isoroku Yamamoto and his staff to Buin. Following a long and carefully planned interception flight, USAAF P-38s attacked, and while the escorting fighter pilots claimed four of the attackers shot down and two more as probables, both the Rikkos were shot down and the Admiral was killed. Capt Sugimoto, the kokutai commanding officer, then planned to use fighters as fighter-

204 Ku Zero 32, 'T2-190', at Buin.

bombers, employing this method in an attack on Russell Island on 7 June. Losses were heavy, however, and the method was not used again. At this time the losses being suffered by dive-bomber units had become so heavy that a new method of protecting these was devised, whereby the fighters joined them at the point where they came out of their dives.

NCO pilots of 204 Ku at Buin. Kneeling on the left is Hideo Watanabe; centre, Yoshio Nakamura; standing, 3rd from left, Saji Kanda.

Zero 'T2-188' of 204 Ku at Buin.

This was also tried on 16 June, but Lt Miyano, the hikotai leader, was lost, and at one point the unit had no officer pilots left at all. Allied forces landed on Rendova Island in the central Solomons on 30 June, several days of intense aerial activity following as a result of which the kokutai amalgamated with the remains of 582 Ku and 2nd Air Flotilla. Even after this reinforcement, strength was exhausted, and early in October the unit withdrew to Rabaul. The unit then fought over Rabaul and undertook an attack on Torokina, Cape Marcus. On 26 January 1944 the remaining 12 Zeros were evacuated to Truk. Here the unit began training new personnel, but on 17 February US carrier aircraft attacked. 31 Zeros were scrambled to intercept, the pilots claiming about 30 victories, but 18 pilots were killed. This was the end, and the survivors were withdrawn home to Japan where on 4 March 204 Ku was disbanded. Including its time under the title 6 Ku, the unit had claimed in total about 1,000 victories.

Commanding Officers

Apr 42 – Mar 43	Capt Chisato Morita
Mar 43 – Sep 43	Capt Ushie Sugimoto
Sep 43 – Mar 44	Cdr Takeo Shibata

Hikotai leaders

Apr 42 – May 42	Lt Hideki Shingo
May 42 – Jul 42	Lt Tadashi Kaneko
Jul 42 – Mar 43	Lt Mitsugu Kofukuda
Mar 43 – Jun 43	Lt Zenjiro Miyano +
Jul 43 – Sep 43	Lt Cdr Saburo Shindo
Sep 43 – Dec 43	Lt Kiyokuma Okajima
Dec 43 – Jan 44	Lt Yoshio Kuragane
Jan 44 – Mar 44	Lt(jg) Torajiro Haruta

205th Kokutai (205 Ku)

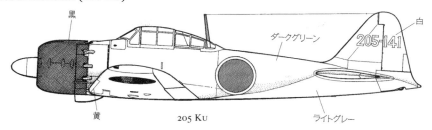

205 Ku was formed at Taichung, Formosa, on 5 February 1945, joining the 1st Air Fleet. Allocated to the kokutai were S302, S316 and S317, each with 48 fighters. The core of the pilot strength was provided by 1st Air Fleet survivors from the Philippines, and initially the unit had just 23 Zeros. By 10 March 112 pilots were to hand, located with S302 at Taichung, S315 at Tainan and S317 at Hsinchu, but available aircraft strength had fallen to only 20. As the fundamental tactic of 205 Ku was Tokko to which all its flying personnel were nominated, when the battle for Okinawa commenced the kokutai moved to Ishigaki Shima and Miyako Shima. From these bases one or two Zeros flew with each formation of bomb-carrying Tokko Zeros to report back on the results achieved. Few new aircraft were received, and by the end of May regular attacks could no longer be sustained. During June one or two attacks a month was all that was possible, and this continued until the end of the war.

Commanding Officer
Feb 45 – end   Cdr Asaichi Tamai

Pilots of 205 Ku; 2nd row, 7th from left is Ens Kazuo Tsunoda.

210th Kokutai (210 Ku)

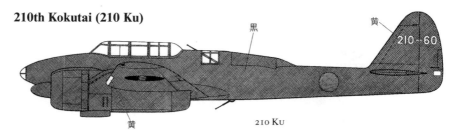

黒

黄

210--60

黄

210 KU

Formed at Meiji on 15 September 1944, the kokutai was a composite training unit with many different types of aircraft. Establishment was 48 fighters, 48 interceptors, 12 night fighters, 24 reconnaissance aircraft, 24 dive-bombers and 24 torpedo-bombers. In fact by 31 December the unit had 29 Zeros, 31 Shidens, 12 Gekkos, eight fighter trainers, 47 Suiseis, 11 D3As, 18 Tenzans and three B5Ns. Administrative composition of the unit was two tais of fighters, two tais of interceptors, half a tai of night fighters, one tai each of the other main types. Buntai leader of the Shidens was Lt Kazumasa Mitsumori and that of the Zeros was Lt Toshio Shiozuru.

B-29s started raiding the homeland islands from the Marianas late in 1944 and instructors of 210 Ku began undertaking interceptions of these bombers over the Nagoya area. The first combat was by day on 13 December when 14 Zeros, four Shidens and three Gekkos took off, but these were only able to claim two bombers damaged between them. During January 1945 a detachment from Tokushima equipped mainly with Shidens, took part in intercepting raids on the Hanshin area, while the main part of the unit operated over the Nagoya area. When US carrier aircraft attacked Kanto and Tokai for the first time on 16 February, 14 210 Ku Zeros engaged about 50 raiders, claiming 12 shot down and four probables, while next day two more were claimed for the loss of only one Zero. Late in March 32 Zeros and 14 Shidens moved to Kokubu No.1 airfield where they came under the command of the commanding officer of 601 Ku. From 6 April 210 Ku took part in the fighting over Okinawa as a support unit, but in mid month returned to Meiji, having claimed six victories but lost ten aircraft. On 5 May 210 Ku became an all-fighter unit with 95 aircraft, held ready for the invasion of the homeland, as a result of which it saw no major operations prior to the end of the war.

Gekko of 210 Ku at Meiji.

A Shiden 11a of 210 Ku.

221st Kokutai (221 Ku)

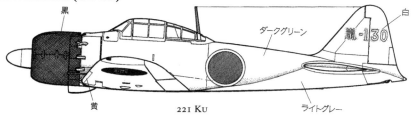

黒 ダークグリーン 白

嵐-130

黄 221 KU ライトグレー

The kokutai was formed on 15 January 1944 at Kasanbara with a planned establishment of 54 fighters plus 18 spares. By the end of February it had just ten Zeros on strength, but more aircraft arrived in gradual stages until by the end of June it had 14 Zero 21s and 24 Model 52s. With the reorganistation of 10 July, the unit took under its wing S308, S312, S313 and S407, each due to have a strength of 48 aircraft, and the kokutai became part of 2nd Air Fleet. In mid August 51 Zeros of S312 moved to Hsinchu in three waves, there taking part in interceptions and convoy patrols. On 12 October US carrier aircraft raided Formosa, Lt Keiji Kataki leading 43 S312 aircraft to intercept. 16 American aircraft were claimed shot down plus seven probables, but 15 Zeros were lost. 11 out of 15 available Zeros escorted Rikkos on 15 October, but the major part of the formation turned back when no target was found. Some Zeros continued, but these were attacked by F6Fs and lost two of their number.

With the start of the 'Sho-go' operation, 221 Ku departed Kyushu on 14 October for Taichung and Hsinchu, while on 23rd 60 aircraft from all four hikotais moved to the Philippines, flying in to Angeles, Luzon. Next day these aircraft took part in the attacks on the US carrier forces east of Luzon, then undertaking attacks on targets on Leyte, and interceptions there. Some pilots were then selected for Tokko units, and gradually the unit was worn down. In December 221 Ku was selected to remain as the main fighter unit in Luzon, having added to it S303, S304, S315 and S317, but all these units were exhausted, and on 20 December 20 Zeros of S308 flew in as the final reinforcements. Many of these were lost during an interception on 24th, including Lt Cdr Kawai, the remnants then being lost over the next two days. On 8 January the remaining pilots were ordered to return to Formosa, the ground crews remaining to take part in the land fighting where many were to be killed before the war ended.

Commanding Officers

Jan 44 – Jul 44	Cdr Ichiro Himeno
Jul 44 – Nov 44	Capt Masahisa Saito
Nov 44 – Jan 45	Cdr Katsutoshi Yagi

251st Kokutai (251 Ku)

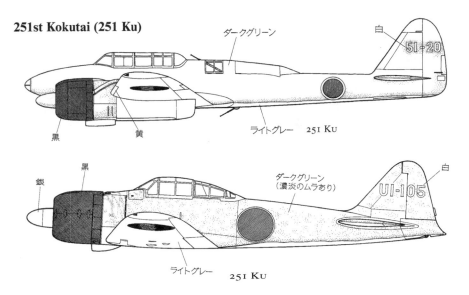

ダークグリーン

白

51-20

黒　　　黄

ライトグレー　251 Ku

銀　　黒

ダークグリーン
(濃淡のムラあり)

白

UI-105

ライトグレー　251 Ku

The Tainan Ku was renamed 251 Ku on 1 November 1942 and in the middle of that month the unit returned to Japan to recuperate. Established strength was to be 60 fighters, eight reconnaissance aircraft and four transports. Initially the core of pilots consisted of a little over ten of those who had returned from Rabaul, training at Toyohashi then emphasising formation fighting. On 1 May 1943 the unit was ordered to return to Rabaul, 58 Zeros and seven reconnaissance aircraft led by hikotai leader Lt Ichiro Mukai, arriving at Lakunai airfield on 10th. On 14th 32 Zeros escorted Rikkos to Oro Bay, their pilots claiming eight shot down and five probables without loss. On 7 June 35 of the unit's fighters and on 12th 32 of them, jointly with 204 Ku and 582 Ku, claimed 18 victories and five probables over Russell Island on the first occasion and 11 shot down on the second of these dates, for the loss of eight Zeros. During an attack on ships off Lunga Point on 16 June 30 Zero pilots claimed ten but lost seven.

Allied forces landed on Rendova Island on 30 June, 24 Zeros from 251 Ku attacking the invading force, but on this occasion lost eight pilots including Lt Mukai and Lt(jg) Takeyoshi Ohno. The unit then moved to Rabaul airfield and led by Lt(jg) Oshibuchi, engaged in attacks on Rendova and interceptions over Buin. At the instigation of Cdr Kozono, some of the unit's reconnaissance aircraft had been fitted with obliquely-mounted cannons both above and below the fuselage, and two of these were sent down to Rabaul. Using one, CPO Kudo claimed two B-17s shot down over Rabaul on 21 May. This modified aircraft which became known as the Gekko, was employed at night, and on 10 June Wt Off Ono also claimed two victories flying one, while during the nights of 11th and 13th Kudo was able to make one claim each night; on 15th he added two more, and then another during the night of 30 June. After these successes, raids by night on Rabaul ceased. However, the night attacks were then turned against Buin and consequently two of the night fighters were moved to Ballale in early July, six more nocturnal victories being claimed here, while the aircraft were also used for attacking torpedo-boats around Rendova.

On 1 September 1943 251 Ku became entirely a night fighter unit, the unit's Zero pilots being transferred to 201 Ku and 253 Ku. Prior to this, during the period from May to the end of August 1943, these latter pilots had claimed about 230 victories for the loss of 34.

From September the Gekko unit engaged in interceptions by night, reconnaissance sorties by day, and night attacks on ground targets. Gradually daylight raids around Buin became heavier, the unit operating from Rabaul via Ballale, but in October a full return to Rabaul took place, but with a detachment being sent to operate from Kavieng. As the main activities of the US heavy bombers were now reverting to daylight, and as counter measures were being employed at night, the Gekko pilots were unable to gain many successes at this time. During February 1944 the main part of the unit withdrew to Truk, but during raids here on 17th and 18th nine Gekkos were destroyed on the ground. With reorganisation on 1 April 1944 the flying unit of 251 Ku became S901 with an establishment of 24 night fighters. Two or three Gekkos remained at Rabaul to undertake night interceptions until May, when this detachment too moved to Truk, joining the rest of the unit there. Until the end of June the Gekkos intercepted night raids by B-24s, but with little success. The unit was disbanded on 10 July, S901 being transferred to 153 Ku.

Commanding Officers

Nov 42 – Sep 43	Cdr Yasuna Kozono
Sep 43 – Mar 44	Cdr Ikuto Kusumoto
Mar 44 – Jul 44	Cdr Takeo Shibata

Hikotai leaders

Nov 42 – Feb 43	Lt Cdr Tadashi Nakajima
Feb 43 – Jun 43	Lt Ichiro Mukai +
Jun 43 – Jul 44	Lt Hiroshi Sonokawa

252nd Kokutai (252 Ku)

On 20 September 1942 the fighter unit of Genzan Ku became independent and was named 252 Ku. Equipped with 60 fighters, the greater part of the unit was transported to the Solomons on board *Taiyo*, arriving at Rabaul on 9 November. Two days later the first operation was flown when Lt Shigehisa Yamamoto led 11 Zeros with others from 263 and 582 Ku to Guadalcanal. The unit claimed one victory and returned safely. On 12th Lt Suganami led 12 A6Ms to escort Rikkos carrying torpedos, and on this occasion eight victories were claimed. Next day six of the unit's pilots patrolled over the damaged battleship *Hiei*, but in appalling weather conditions all had to force-land at Rekata Bay, Lt Yamamoto being badly injured. During escort to a convoy to Guadalcanal on 14th, 14 US aircraft were claimed shot down, but Lt Suganami failed to return. Actions like these continued, and by February 1943 when the kokutai was ordered to move to the Marshall Islands, the unit had claimed 109 destroyed and 36 probables for the loss of 16 pilots.

ダークグリーン 白

252-108

ライトグレー
252 Ku

Moving to the Marshalls during February and March, the headquarters became based at Roi, two buntais were detached to Wake, and one buntai each were sent to Taroa (Maloelap) and Nauri. Patrols were then flown over the islands, but until September all that was encountered was sporadic raids by B-24s, no action resulting. On 19 September, however, US carrier aircraft raided the Gilbert Islands, resulting in 12 Zeros led by Lt Tsukamoto being despatched to Tarawa. Here B-24s were intercepted until 14th, when with the carriers gone, the Zeros returned to Taroa. On 6 October about 100 aircraft

252 Ku detachment on Maloelop Atoll, Marshall Islands. Front row centre is Lt(jg) Sumio Fukuda.

raided Wake, 26 Zeros being put up by 252 Ku's detachment. The pilots of these claimed 14 of the raiders shot down, but lost 16 of their own number, including Wt Off Sueda. Six A6Ms were then sent off to escort seven Rikkos to Wake to reinforce the garrison, but en route the formation was bounced by F6Fs and only three Zeros survived to reach the island. US Marines then landed on Makin and Tarawa Islands on 21 November, opening up a new major action. On 24th 19 Zeros were led by Lt Suho from Taroa to attack the American troops at Makin, each fighter carrying a pair of 60kg bombs. Again, this force was intercepted by F6Fs and six were shot down; returning Japanese pilots claimed that they had managed to destroy five of the Hellcats with five more as probables. Next day 30 more Zeros set off for Makin, but on this occasion six failed to return.

From late December until 30 January 1944 252 Ku sought to intercept B-24s over Taroa, about 50 of these bombers being claimed shot down, although suffering some considerable losses both in the air and on the ground in doing so. On 1 February a US force landed on Kwajalein Atoll where 120 pilots of 252 Ku were rescued from this and other islands by flyingboats and returned to Truk. From here they were sent home to Tateyama to recuperate.

In late March 1944 the kokutai moved to Misawa, and on 1 April became of hikotai composition. Thus the flying unit of 252 Ku became S302, which by early May had 55 Zeros available. On 15 June 252 Ku was incorporated into the Hachiman Force, 23 A6Ms moving to Iwo Jima on 21st, followed by 16 more four days later. Meanwhile, on 24th US carrier aircraft attacked Iwo Jima, the fighters which had reached the island intercepting in strength. In the hard fighting which followed, 252 Ku claimed 19 victories but lost ten aircraft and their pilots, including Lt Awa, the hikotai leader. During further engagements on 3rd and 4 July, the kokutai claimed 13 destroyed but lost 14 Zeros. With no aircraft left, the pilots were evacuated to Tateyama.

Reorganised again on 10 July, 252 Ku was now to control S302, S315, S316 and S317, each of which were intended to have 48 fighters. At least ten S317 Zeros were led to Iwo Jima in late September by Lt Nakama where they were to undertake 19 interceptions which brought claims for just two shot down and three damaged. Lt Nakama personally rammed a B-24, but was killed in doing so. He was included in an All Forces Bulletin and received posthumous promotion of two ranks. On 27 November 11 bomb-laden Zeros

Top: Sento 302, 252 Ku, at Misawa in May 1944. Note the Zeros sub-type 52 in the left row and sub-type 21 in the row to the right.
Bottom: Zero 52c of 252 Ku, mainstay of the unit in spring 1945.

were led by Lt(jg) Kenji Ohmura to Saipan to bomb, and then to land there to destroy parked B-29s.

Meantime, during October the greater part of 252 Ku had been despatched to Clark Field, Luzon, on 22nd as the US invasion threatened. On 24th 26 Zeros led by Lt Cdr Minoru Kobayashi, leader of S317, took part in the general all-out attacks on the approaching carrier forces. Intercepted before reaching their target, the kokutai's formation lost 11 aircraft including that flown by Kobayashi. Following that date, the kokutai was involved in launching attacks on targets in Leyte, search and attack missions against the American carriers, and interceptions. By mid November the unit had lost three hikotai leaders. At that stage some pilots were transferred to Tokko units (suicide ram units), the rest being sent back to Japan.

By February 1945 the kokutai headquarters were at Mobara, while S304, S308, S311, S313 and S316 were training at either Mobara or at Tateyama; S308 would leave the unit during March on transfer to 601 Ku. During 16 February 45 Zeros from S308 and S311 intercepted 30 F6Fs, claiming 14 shot down and ten probables for the loss of nine A6Ms. Further interceptions on 17th and 25th brought more losses, reducing the number of aircraft available to the kokutai to 39. At the end of March 144 Zeros of S304, S313 and S316, together with Suisei dive-bombers of K3, moved to Kokubu to take part in the battle for Okinawa, where 39 pilots were killed making Tokko attacks. On 16 April 12 Zeros led by Lt Cdr Yanagizawa were engaged over Amami Oshima with F6Fs. Here the pilots claimed 13 victories but lost five of their own. Next day Lt Cdr Yanagizawa failed to return. During the period 1-17 April 252 Ku lost 15 Zeros and five Suiseis, then returning to the Kanto area. From there S304 was sent to Koriyama and S316 to Mobara where only part of these hikotais were involved in interceptions, the majority seeing no further action until the end of the war.

Commanding Officers
Sep 42 – Feb 44	Capt Yoshitane Yanagimura
Feb 44 – Jul 44	Cdr Tadao Funaki
Jul 44 – Nov 44	Capt Masahisa Fujimatsu
Jun 45 – end	Cdr Kiyoji Sakakibara

Hikotai leaders
Sep 42 – Nov 42	Lt Masaji Suganami +
Dec 42 – Mar 44	Lt Motonari Suho

253rd Kokutai (253 Ku)

A Rikko and fighter unit, Kanoya Ku became 751 Ku, but on 1 November 1942 the fighter unit was separated to form 253 Ku. With an establishment of 48 fighters and four reconnaissance aircraft, the kokutai joined the 21st Air Flotilla with the main part of the unit based at Kavieng, but with detachments at Rabaul, Buka

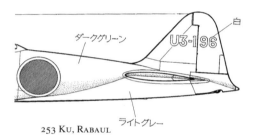

253 KU, RABAUL

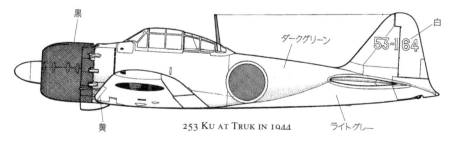

253 KU AT TRUK IN 1944

Island and Surumi. Here the unit was engaged over both New Guinea and the Solomon Islands. On 1 April 1943, 26 A6Ms led by Lt Masao Iizuka were engaged in a fight over Russell Island during which 28 opposing aircraft were claimed shot down in 30 minutes. The unit continued to operate here until mid May when it moved to Saipan for recuperation. By this time about 110 victories had been claimed for the loss of about 20 pilots during missions to Guadalcanal and Oro Bay, and during convoy patrols and interceptions.

With recuperation, the unit was increased in strength to 96 fighters and moved to Tobera airfield during September where it became the nucleus of the Rabaul fighter force in place of 201 and 204 Ku. These units had been worn out by daily operations over eastern New Guinea and the northern Solomons. From mid-October daylight raids by US aircraft became very heavy, 253 Ku intercepting on every occasion with 30-40 Zeros.

253 Ku at Kavieng in January 1943. 2nd row, 4th from left, Masao Iizuka; 5th, Lt Toshitaka Ito; 6th, Lt(jg) Saburo Saito; 7th, Tadashi Torakuma; 3rd row, 6th from left, Shigeru Shibukawa; 4th row, 5th from left, Minoru Honda.

By late January 1944 there remained only 253 Ku and the fighter units of the 2nd Air Flotilla, and losses were mounting. By mid-February the number of victories claimed had exceeded 500, but the numbers available to the defence had fallen to no more than 20 aircraft.

On 17th and 18 February US aircraft had attacked Truk and consequently the main body of 253 Ku was moved here with 23 Zeros. They rested and consolidated on Moen, then from mid March became engaged in intercepting B-24s, attacking these with 3-go phosphorus bombs. Meanwhile, nine A6Ms remained at Rabaul, undertaking interceptions, bombing and strafing attacks on targets on Green Island, and reconnaissances of the Admiralty Islands. These duties continued with small numbers of repaired Zeros until the end of the war.

In the Marshalls, on 1 April 1944 the reorganised 253 Ku now comprised S309 and S310, each with an intended establishment of 48 fighters, continued to intercept attacks by the US carriers when these occurred. On 30 April the kokutai lost 20 aircraft during one such raid, the number of available aircraft shrinking to five. By 16 June the numbers available had grown to 13 when the 'A-go' operation was ordered. Consequently, on 19th Lt Cdr Okamoto led all 13 from Truk to Guam, but as they were landing at their destination they were surprised by F6Fs which shot down five which were still in the air, then going on to destroy the rest which by then were on the ground. With no aircraft left, the survivors were withdrawn by transports and a submarine, returning to the homeland. There on 10 July 1944 253 Ku was disbanded, seven Zeros which had survived being transferred to Higashi Caroline Ku, where they continued to be used for interceptions until October.

Commanding Officers

Nov 42 – Jul 43	Cdr Yoshito Kobayashi
Jul 43 – Jun 44	Cdr Taro Fukuda
Jun 44 – Jul 44	Cdr Sho-ichi Ogasawara

Hikotai leaders

Nov 42 – Aug 43	Lt Toshitaka Ito
Aug 43 – Jan 44	Lt Cdr Harutoshi Okamoto
Jan 44 – Mar 44	Lt Tatsuo Hirano

254th Kokutai (254 Ku)

Formed on 1 October 1943 with a planned strength of 24 fighters, four torpedo aircraft and one transport, the unit moved to Sanya, Hainan Tao where it became engaged with China-based B-24s and B-25s over here and Hong Kong. On 5 April 1944 fighters from the unit escorted bomb-carrying Zeros of Sanya Ku and Haiko Ku to Nanning. The escorts became separated from their charges by cloud and the fighter-bombers were intercepted, many being shot down including that flown by Lt(jg) Nakahara. During October 1944 US carrier aircraft raided Formosa and 16 of 254 Ku's Zeros moved forward to Tainan on 13th. Next day the pilots fought 40 raiders, claiming three F6Fs shot down, while on 16th six Zeros took part in an attempted attack on the carriers, acting as escort and claiming two more F6Fs. On 27 October the kokutai moved to Nichols Field, Luzon, to take part in attacks on Leyte, also undertaking interceptions and convoy patrols until mid November. It then returned to Hainan Tao to recuperate, but on 1 January 1945 the unit was disbanded, the fighters being amalgamated into 901 Ku.

254 Ku pilots; in the front row, seated in chairs, are Lt Akira Sugiura and Lt(jg) Takeshi Inoue.

Commanding Officer

Oct 43 – Jan 45   Capt Kuro Hori

Hikotai leader

Oct 43 – Jan 45   Lt Isamu Matsubara

256th Kokutai (256 Ku)

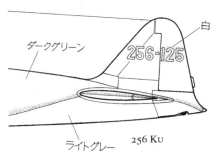

256 Ku

Formed at Longhua airfield, Shanghai, on 1 February 1944, the unit was established for 24 fighters and eight torpedo-bombers, becoming engaged in the air defence of Shanghai, and on anti-submarine and convoy patrols. Young pilots were also trained by the unit which had an actual initial strength of 15 Zeros and five B5Ns. When US carrier aircraft raided Formosa, Lt Keizo Yamazaki led ten Zeros from Longhua on 13 October to rendezvous with 254 Ku, taking part in interceptions on 14th and 15th. On 16th five out of nine available Zeros provided part of the escort for an attack force seeking the carriers, the pilots claiming two F6Fs shot down. Six Zeros were incorporated into the 254 Ku detachment, moving to Nichols Field, Luzon, on 27th. From there they moved to Marcott, engaging in attacks on Leyte and in air defence. By early November the unit had no aircraft left. On 7th Lt Yamazaki arrived with some new aircraft and operations continued until 15th when there were only three survivors remaining, these withdrawing back to Tainan from where the detachment was ordered to return to Shanghai on 17th. At Longhua meanwhile, Lt Chiyoyuki Shibata had led the remainder of the unit which had not gone forward to Formosa. Here on 1 November there were still to hand 11 Zero 21s, 13 Zero 52s and three Raidens, these aircraft attempting without success to intercept B-29s from bases in China raiding western Japan. On 15 December 256 Ku was disbanded, the remaining aircraft being amalgamated into 951 Ku.

Commanding Officer
Feb 44 – Dec 44   Capt Ichitaro Uchida

Hikotai leaders of the fighter unit
Feb 44 – Mar 44   Lt Hiroshi Mori
Apr 44 – Dec 44   Lt Keizo Yamazaki

261st Kokutai (261 Ku)

261 Ku

The kokutai was formed at Kagoshima on 1 June 1943 with an establishment of 36 fighters; this was rapidly increased to 72. On 1 July it joined the newly-formed 1st Air Fleet, on 19 February 1944 moving to Katori from where on 22nd 68 Zeros flew to Iwo Jima. When US carrier aircraft raided the Mariana Islands, 261 Ku moved up to Saipan during the next few days. On 30 March the Americans attacked Palau and at once 32 aircraft were led by Lt Masanobu Ibusuki to escort dive-bombers of 523 Ku sent to attack the US carrier force in this area. This could not be found and the formation landed on Peleliu. Next morning early, a large force of carrier aircraft raided this island, being intercepted by fighters from 201, 261 and 263 Ku, which had still not been able to climb very high when the enemy aircraft were met. 15 of the raiders were claimed shot down and three more probably, but 20 Zeros from a force of 28 failed to return. The other eight were all badly damaged, or were destroyed on the ground. Survivors returned to Saipan.

While 261 Ku began to recuperate, PBYs and B-24s raided Saipan and Meleyon(Woleai) frequently during April, the kokutai despatching detachments in turn to the latter base, especially on 23 April when 27 Zero pilots intercepted B-24s, claiming four destroyed, two of them by ramming, for the loss of four A6Ms. By mid May the unit had 42 Zeros on strength, 34 of which were immediately serviceable. On 27th of that month Amercian forces landed on Biak Island, following which Lt Ibusuki led 27 Zeros

Zero 'Tora-110' of 261 Ku, badly damaged in a training accident.

Cdr Taketora Ueda, commanding officer of 261 Ku.

to Wasile, Halmahera, on 2 June. From here on 12 June 12 Zeros departed to attack Biak, but all failed to return. Landings on Saipan followed and the remaining aircraft withdrew to Guam and Yap. One buntai remained here and on Saipan, but there were few pilots available due to sickness, while most aircraft were now worn out by the long flights they had undertaken, only a few remaining in a fit state to undertake interceptions over Guam, or to make attacks on the shipping around Saipan. By 21 June only two aircraft were left available on Guam and the main body of the unit attempted to leave either as passengers in Rikkos, or in a submarine. However, the latter was sunk and all the pilots on board were lost.

A small number of pilots, including CPO Minpo Tanaka, were left on Guam, undertaking attacks on targets on Saipan until 15 July. During the period 27 May–15 July 261 Ku pilots had claimed 76 aircraft shot down, but had lost 28 pilots. On 10 July 1944 the kokutai was disbanded, the remaining pilots being absorbed into 201 Ku.

Commanding Officer
Jun 43 – Jul 44   Cdr Taketora Ueda

Hikotai leader
Jun 44 – Jul 44   Lt Masanobu Ibusuki

263rd Kokutai (263 Ku)

Formed at Wonsan, Korea, on 1 October 1943, the unit was initially to have a strength of 36 fighters, but this was increased to 72, following which it joined the 1st Air Fleet and commenced training at Matsuyama. Ordered to the Marianas, 18 of the unit's Zeros left Katori on 20 February 1944 to fly to Tinian, arriving next day. On 23rd US carrier aircraft raided the island, and although 263 Ku took off to intercept, 11 aircraft were lost, including that flown by Lt(jg) Wajima, while six more were destroyed on the ground. Surviving pilots, including Lt Yasuhiro Shigematsu, the hikotai leader, returned to Matsuyama in transports. In mid March the unit

TAIL MARK OF 263 KU

despatched 49 Zeros to Guam No.1 airfield, and on 30th of that month 25 of the kokutai's Zeros together with others from 261 Ku moved to Peleliu Island. Next day US carrier aircraft made an anticipated raid on Palau, 18 263 Ku Zeros intercepting, their pilots claiming five shot down. However, 15 of these failed to return and three more

263 Ku pilots at Matsuyama in March 1944. 2nd row, 6th from left, Yasuhiro Shigematsu.

aircraft were destroyed on the ground. Again, the survivors returned to Japan, travelling via Saipan. On arrival they were supplied with new aircraft and ordered back to Guam. At the end of May 28 of the Zeros moved to Wasile as American forces were landing on Biak Island. The eight which remained at Guam sought to intercept carrier aircraft over Guam on 11 June, but four of these were lost. On 15 June US landings on Saipan followed, and the greater part of the unit returned to Guam. Here during the period 15-18 June these were engaged in air defence and in attacks on the shipping around Saipan, but these operations cost at least 20 Zeros shot down. Yet more replacements arrived from the homeland, so that at least ten aircraft could be available, but on 8 July F6Fs attacked six of these, only one of which survived to reach Peleliu. On 10 July 1944 263 Ku was disbanded, but eight A6Ms remaining on Peleliu moved to Davao, Philippines, where they were absorbed by 201 Ku. A handful of aircraft remained on Guam, undertaking sorties until none remained flyable by 18 July. Remaining pilots were then flown out in Rikkos.

Commanding Officer
Oct 43 – Jul 44   Cdr Asaichi Tamai

Hikotai leader
Oct 43 – Jul 44   Lt Yasuhiro Shigematsu

265th Kokutai (265 Ku)

265 Ku was formed at Kagoshima on 15 November 1943; its initial establishment of 36 fighters was quickly increased to 72, and it trained at Kasanbara. During January 1944 the unit moved to Hsinchu, Formosa, and from here on 1st and 2 May 32 of the unit's Zeros were sent via Kasanbara, Katori, to Saipan No.1 airfield to relieve 341 Ku. 16 more A6Ms followed just before the US attack on the Marianas, but the kokutai's remaining aircraft moved to Iwo Jima, but then no further. Interceptions of heavy aircraft were made on 28 May, but then on 5 June 32 Zeros were moved to Kau, Halmahera, to take part in operations over Biak where US landings had been made. After taking part in the fighting in the Marianas, the unit returned to Peleliu during 14-15 June. Six aircraft led by Lt Yasuhiko Ukimura then moved to Yap, taking part in attacks on shipping off Saipan on 15th. Here a single claim for a victory was made, but two aircraft were lost in return, the remaining fighters landing on Guam. During the Marianas operations Lt Suzuki led the main part of the unit to intercept B-24s, operating from Yap and Peleliu No.2 on several occasions, one further claim resulting. Part of the unit also operated from Guam, but details of this are not available. It seems that the remaining elements of 265 Ku were engaged over Iwo Jima on 15 June, but apparently suffered heavy losses.

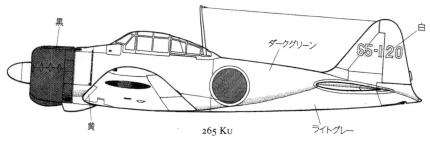

黒

ダークグリーン

白

65-120

黄   265 Ku   ライトグレー

During early July the unit's remaining aircraft retreated to Davao in the Philippines where the kokutai was disbanded; many of the surviving pilots, including Lt Suzuki, were then amalgamated into 201 Ku. The ground crew, meanwhile, reached Rota Island about 11 June, where they remained.

Commanding Officer
Nov 43 – Jul 44   Cdr Terujiro Urata

Hikotai leader
Nov 43 – Jul 44   Lt Usaburo Suzuki

281st Kokutai (281 Ku)

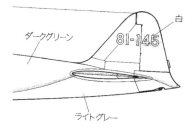

Formed on 20 February 1943 at Tateyama, 281 Ku had an establishment of 48 fighters; it commenced training during May, joining the 12th Air Fleet. A move was made to Paramushir in northern Chishima on 23 May, via Misawa and Bihoro. B-24s raided the area sporadically from August 1943, but at this stage the unit took no part in any interceptions. On 1 October 42 Zeros were to hand, and in November 16 of these were detached to Rabaul. Here, on arrival, they were transferred to 204 Ku. On 24 November the kokutai transferred to the 24th Air Flotilla, moving to the Marshall Islands to reinforce this area. 21 Zeros moved to Roi on 3 December, followed by 18 more on 6th, the remainder of the unit following on 8th. Roi was raided by US carrier aircraft on 5 December, 27 of the kokutai's fighters being delayed in take-off as they tried to scramble; four were lost. On 14 December Lt Takaichi Hasuo, the hikotai leader, led 13 Zeros on detachment to Mille, from here several interceptions being made of heavy bombers. Following attacks on Makin Island in the Gilbert Islands, they returned to Roi. Roi and Kwajalein were raided by heavy bombers late in December, but on 30th and 31 January 1944 US carrier aircraft attacked, destroying 25 Zeros on the ground. American landings here followed on 1 February; all the pilots and ground crews of 281 Ku were then killed in the ground fighting by 6th, at which time the islands were completely occupied, there being no Japanese survivors.

Commanding Officer
Feb 43 – Feb 44   Cdr Mohachiro
                         Tokoro +

Hikotai leader
Feb 43 – Feb 44   Lt Takaichi Hasuo +

Cdr Mohachiro Tokoro, commander of 281 Ku, seen here earlier while serving with the 22nd Air Flotilla attached Fighter Unit.

301st Kokutai (301 Ku)

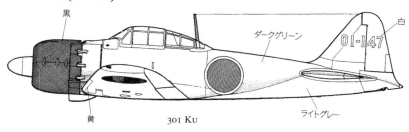

黒

ダークグリーン

白

01-147

黄   301 Ku   ライトグレー

301 Ku was formed on 5 November 1943 at Yokosuka to operate the new Mitsubishi J2M Raiden interceptor. It joined the 26th Air Flotilla on 1 February 1944, but transferred to the 22nd Air Flotilla on 20th of that month. On 4 March, following the reorganisation of units, the kokutai became comprised of S316 with 48 A6M fighters and S601 with 48 J2M interceptors. On 5 May the 22nd Air Flotilla became a part of the 1st Air Fleet, 301 Ku then moving to Truk. From there, in order to take part in operations over Biak, where a landing by US forces had occurred, 20 Zeros and 49 Raidens moved to Tateyama on 29 May. However, S316 had not been set up as an ordinary fighter unit, but was to fly Zero night fighters for which pilots of two-seat floatplanes had been gathered and trained only for night flying; they had received scarcely any air combat training before leaving the homeland. Despite this handicap, 18 of the hikotai's Zeros were led to Tinian on 2 June by Lt Shigeji Torimoto, and from here on 11th ten rose to intercept US carrier aircraft, suffering heavy casualties in so doing. On the same day 19 more Zeros moved to Iwo Jima, and from here during the afternoon of 15th 18 were sent up to intercept a formation of about 60 aircraft from the US carriers. The predictable result was that 16 failed to return, one survived to force-land, and only one landed back at base.

Meanwhile, S501 had been awaiting an improvement in weather conditions, which had been very bad, before risking a long-range flight with the Raidens. In desperation the unit was switched to Zeros, nine of which were sent to Iwo Jima on 21 June, followed by 31 more on 25th. The initial batch intercepted carrier aircraft on 24th, claiming five shot down but losing four A6Ms, including that flown by Lt Katsumi Koda. Following the

Lt Ryoichi Koga, maintenance officer of 301 Ku, in the cockpit of a J2M Raiden interceptor.

arrival of the second batch of 31, all were led by Lt Iyozo Fujita to intercept on 3 July, these pilots claiming seven shot down and three probables – but for the loss of 17 Zeros. Next day 14 pilots were off again, this time claiming six victories for three losses. All remaining aircraft were then destroyed on the ground by naval bombardment, surviving pilots being evacuated to Japan in transports. On 10 July 1944 301 Ku was disbanded.

Commanding Officer

Nov 43 – Jul 44   Cdr Katsutoshi Yagi

Hikotai leader

Nov 43 – Mar 44   Lt Iyozo Fujita

302nd Kokutai (302 Ku)

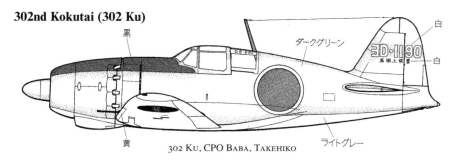

302 KU, CPO BABA, TAKEHIKO

Formed as an air defence unit on 1 March 1944 at Kisarazu, 302 Ku was initially intended to have 48 interceptors and 24 night fighters, one tai with J2M Raidens commencing training at Yokosuka and one tai with Gekkos at Kisarazu; late in March these elements moved to Atsugi. By November 1944 302 Ku was composed of the 1st hikotai (two buntais of Raidens and one of Zeros), the 2nd hikotai (two buntais of Gekkos and one of Gingas) and the 3rd hikotai (two buntais of Suisei night fighters); between them, these units had a strength of ten Raidens, 27 Zeros, 15 Gekkos, two Gingas and six Suiseis. However, during 1945 the main strength of the kokutai would be represented by about 30 Raidens. In the meantime one buntai of Gekkos led by Lt(jg) Endo was detached to northern Kyushu, where several victories were claimed against China-based B-29s. During November Lt Endo led three of the Gekkos on detachment to Hachijo Jima, but no interceptions were to result from this move, and the aircraft returned to the main islands.

Trained for high altitude combat against B-29s, on 3 December 74 aircraft, including 24 Raidens, intercepted incoming bombers to claim six destroyed, three probables and eight damaged. During February 1945, however, 302 Ku suffered losses to US carrier aircraft operating over the home islands, while from April P-51s from Iwo Jima began escorting the B-29s, and it became impossible to employ the night fighter types by day any more. At this time the losses of Zeros and Raidens also increased, although veterans such as Lt(jg) Akamatsu continued to achieve some successes. When the battle for Okinawa began in early April, one buntai of Zeros was sent to southern Kyushu to operate over the air corridor to the island. As B-29s were raiding the main airfields in Kyushu at this time, 302 Ku's Raidens, together with those of 332 and 352 Ku, were moved to Kanoya from where they continued to undertake interceptions. During some ten days from 12 May, 12 Gekkos and eight Suiseis were despatched to Itami to assist in the air defence of the Hanshin area. During May the kokutai achieved a few successes by

Top: Gekko unit of 302 Ku on 1 January 1945 at Atsugi. 2nd row, 7th from left, Lt Sachio Endo.
Bottom: Zero night fighters of 302 Ku at Atsugi on 1 January 1945. Front row, 3rd from left, Lt Shunshi Araki; 4th, Lt Hiroshi Morioka (leader).

night against low-flying B-29s carrying incendiary bombs, although the number of night fighters available soon decreased. Detachments were sent to Komatsu and Maebashi to protect numbers for the final battle, and on the last day of the war Lt Morioka led eight Zeros into a fight with six F6Fs, one of these being claimed shot down, but four 302 Ku fighters were lost.

Commanding Officer
Mar 44 – end   Capt Yasuna Kozono

The Ginga night fighter version of the Yokosuka P1Y1 bomber. This type of aircraft served with 302 Ku, the only unit to employ it in intercepting the B-29s.

312th Kokutai (312 Ku)

Top: The forerunner of 312 Ku was the Hyakurigahara Detachment of Yokosuka Ku. 312 Ku then flew Zero 21s of the latter kokutai. Note the 'Yo-192' marking on the tail of this aircraft, which was soon changed to '312-'.
Bottom: Zero 21 'Yo-191' in the foreground and Zero Trainer 'Yo-190' behind – both aircraft of 312 Ku.

Formed at Yoko Ku on 5 February 1945, the unit was to use the 'Shusui' fighter; this was a rocket-propelled Japanese version of the German Messerschmitt Me 163. Training began at Kasumigaura with two buntais of Zeros (24 aircraft) and engine-less glider versions of the rocket fighter. On 7 July the first powered prototype with Lt Toyohiko Inuzuka, a buntai leader of 312 Ku, took off, zoomed upwards and stalled into the ground. The war ended before the second prototype could make its maiden flight.

Commanding Officer
Feb 45 – end   Capt Takeo Shibata

Hikotai leader
Feb 45 – end   Lt Cdr Yorio Yamagata

321st Kokutai (321 Ku)

The kokutai was formed at Mobara on 1 October 1943 as the first night fighter unit for the IJN, with 18 Gekkos plus six spares. In February 1944 this establishment was increased to 54 Gekkos plus 18 spares and training began at Matsuyama. The unit joined the 61st Air Flotilla and on 21st 12 aircraft led by Lt Ichiro Shimoda, the hikotai leader and an ex-dive-bomber pilot, moved to Tinian where they undertook anti-submarine patrols. On 23 February US carrier aircraft raided the Marianas for the first time and five Gekkos took off on a reconnaissance mission, but two were shot down and Lt Shimoda's aircraft force-landed after being heavily damaged. On the ground six Gekkos were destroyed and two were badly damaged by the attack, leaving

Lt Ichiro Shimoda, hikotai leader of 321 Ku. Earlier he had taken part in the attack on Pearl Harbor as a dive-bomber pilot.

the unit with no serviceable aircraft. Replacements gradually arrived, and by 1 March 11 Gekkos were based on Tinian No.3 airfield from where they were quickly dispersed to a number of other bases. At this time they were rarely used for night interceptions, continuing anti-submarine and convoy patrols, and dawn patrols. On 15 May there were six at Guam, ten at Tinian and 15 at Katori, while early in June four were despatched to Yap; of six sent to Peleliu, four were destroyed on the ground by a B-24 raid. The two detachments then joined up to intercept night raiders and late in June moved to Davao. On 10 July 1944 321 Ku was disbanded, but details of its activities up to this date are unknown, as the headquarters at Tinian was annihilated on 2 August.

Commanding Officer
Oct 43 – Jul 44   Cdr Tokutaro Kubo

Hikotai leader
Oct 43 – Jul 44   Lt Ichiro Shimoda

331st Kokutai (331 Ku)

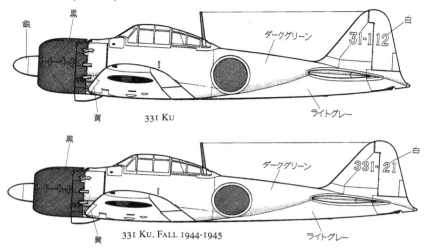

331 Ku was formed at Saeki on 1 July 1943 as a composite unit with 24 fighters and 24 torpedo-bombers. 36 fighters and 18 torpedo aircraft went aboard *Junyo*, but left the ship on 27 August, flying to Sabang on the northern tip of Sumatra. On 1 September the torpedo-bomber unit was separated from the kokutai, forming 551 Ku, following which the fighters undertook patrols and air defence sorties. A detachment was sent to Car Nicobar, Port Blair, Mergui, to intercept B-24s and here some victories were achieved. On 5 December 1943 27 Zeros escorted Rikkos to Calcutta via Burma, claiming four RAF Hurricanes shot down and two probables. To strengthen the defences of the inner Pacific area, 331 Ku moved there in February 1944, but on 4 March the unit was reformed as S603 and joined 202 Ku, 33 Zeros moving to Meleyon on 24 March. At the same time a new 331 Ku was formed at Iwakuni with 24 Zeros, one buntai led by Lt Akira Tanaka then moving to Penang, Malaya, via Davao. In October this 331 Ku was composed of S309 (48 fighters) and X253 (48 torpedo-bombers), at the start of the month S309 having 30 Zeros (20 serviceable) at Balikpapan, Borneo, five (four serviceable) at Penang, Cuching and Sandakan. The main body of S309 was engaged in the air defence of Balikpapan together with 381 Ku, but in February 1945 it moved to Singapore. In May 13 Zeros returned to Omura and on 15th the unit was disbanded.

Commanding Officer
July 43 – May 45   Cdr Hisao Shimoda

Hikotai leader
Jul 43 – Mar 44   Lt Cdr Hideki Shingo

332nd Kokutai (332 Ku)
Formed on 1 August 1944 from the fighter unit of Kure Ku at Iwakuni, the unit was created for the air defence of the Kure area. Initially, it was to have 48 interceptors and 16 two-seat floatplanes, but on formation it actually had 45 Zeros (28 serviceable), 12 Gekkos (two serviceable) and two unserviceable Raidens. However, by 1 November it had improved upon this with 27 Zeros, 15 Raidens and six Gekkos available. On 6 November

黒　　　　　ダークグリーン　　　白

茶

332 KU. CPO OCHI, AKESHI

黄　　　ライトグレー

the main body of the Zero unit was ordered to the Philippines where next day 20 fighters led by Lt(jg) Susumu Takeda arrived at Clark Field where they were transferred to 201 Ku, and before long they had been annihilated. Also on 6th eight Zeros and six Gekkos had been detached to Atsugi to defend the Kanto area, these returning to Iwakuni on 15 December. As 332 Ku was also to defend the Hanshin area, nine Zeros and 11 Raidens moved to Naruo, and the Gekko unit to Itami. The kokutai's first operations had been flown on 25 October without any results, but on 22 December B-29s were intercepted over the Hanshin area, CPO Ochi claiming the unit's first victory over one of these. One more was claimed on 3 January 1945 and a probable on 14th, but results were not to get much better than this. During the battle for Okinawa, B-29s began bombing bases in Kyushu, and to oppose these attacks 17 Raidens moved to Kanoya during 23-25 April, joining other Raidens of 302 Ku and 352 Ku in interceptions until 12 May. 332 Ku joined the 72nd Air Flotilla on 25 May with a new establishment of 48 interceptors and 24 night fighters, but it was then transferred to the 53rd Air Flotilla on 3 August, days before the war ended.

Commanding Officers

Aug 44 – Feb 45　　　　　Cdr Takeo Shibata
Feb 45 – end　　　　　　 Cdr Katsutoshi Yagi

A force-landed Raiden, '32-101' of 332 Ku. In this aircraft CPO Ochi had shot down a B-29 for the unit's first victory on 22 December 1944, for which his CO presented him with a bottle of whisky.

Top: A Raiden in front of a backdrop of smoke and dust from another bombing raid.

Bottom: 332 Ku pilots around a stove awaiting a mission .

341st Kokutai (341 Ku)

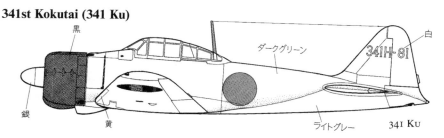

黒　ダークグリーン　白
341H-81
銀　黄　ライトグレー　341 KU

341 Ku was formed at Matsuyama on 15 November 1943 to fly the Kawanishi N1K1 Shiden fighter which had been developed from the Kyofu float fighter. Initially intended to operate 36 interceptors, this figure was increased to 72 in February 1944. A move was made to Kasanbara on 5 December, while on 14 January 1944 the unit moved again, this time to Tateyama. Due to delays in production, it was mid February before the first Shidens finally arrived, and initially Zeros were employed for training. The Shiden suffered considerable initial teething troubles; due to the wings being mounted at the mid point on the fuselage, the undercarriage legs were long, and frequently collapsed, while the Homare engines were also subject to numerous breakdowns. As a result training with these aircraft for formation fighting only commenced in June. In the interim one buntai of the unit's pilots, led by Lt Motoi Kaneko, were ordered to deliver Zeros to the Marianas, arriving initially at Iwo Jima. Here, however, they were intercepted by US carrier aircraft on 15 June, ten being shot down; the rest then returned to Japan. At last, by 10 July 341 Ku had under command S401 and S402, each with 48 interceptors, the unit then moving from Tateyama to Meiji. A third hikotai, S701, had meanwhile been formed at Yokosuka with more Shidens; intended to act as an escort unit for T Force, it continued its training here. On 31 August 17 Shidens of S401 moved to Takao, followed in mid September by 25 more. Here they undertook interceptions of USAAF aircraft from China, but the serviceability of the aircraft remained poor during this period, and strength was maintained only by continuing to employ Zeros as well. On 1 October S401 had 32 Shidens at Takao of which 20 were serviceable, augmented by 11 Zeros at Miyazaki, of which six were serviceable. At these same airfields, S402 had 25 of 30 Shidens serviceable, and 18 of 22 Zeros.

On 12 October 1944 US carrier aircraft attacked Formosa and Lt Masaaki Asakawa led the Shidens from Takao to intercept, ten victories being claimed but for the loss of 14 Shidens. The hikotais here were now joined by S701 which arrived in Formosa to undertake its duties in connection with T Force. The centre of attention then shifted to the Philippines, the greater part of 341 Ku moving there by 23 October, arriving at Marcott, Luzon, with 36 Shidens, 21 of which were immediately available. Next day these all took off to provide escort for the general attack launched against the US carrier and invasion force approaching; this operation cost the kokutai 11 aircraft. During the attack on Tacloban four days later six Shiden pilots engaged F6Fs, claiming two victories, while over Manila next day (29th) 11 Shiden pilots accompanied by five in Zeros claimed eight victories for the loss of six Shidens. During November S701 was removed from T Force and also joined 341 Ku, following which S701's hikotai leader, Lt Cdr Ayao Shirane, led all three hikotais. The kokutai was then engaged in attacks on targets in Leyte, air defence, convoy patrols, and attacks on torpedo-boats. Due to its high speed, the Shidens were also employed dropping urgent supplies to Japanese forces on Leyte. In combat over Clark Field on 13 November nine Shiden pilots claimed three victories, but lost six of their number. They were finding that despite having been equipped with a new type, their

341 Ku at Tateyama in March 1944. 2nd row, 3rd from left, Takashi Okamoto; 5th, Lt Ayao Shirane (hikotai leader).

Shidens enjoyed few advantages over the American fighters they were encountering, whilst their servicability remained very poor.

On 24 November Lt Cdr Shirane was killed over Leyte, following which Lt Iyozo Fujita, leader of S402, took over the leadership of all three hikotais. During December at least seven pilots were killed during interceptions over their own base, and from mid month the Shidens became mainly used for high speed reconnaissance. During these flights Lt Takuo Mitsumoto frequently spotted the US carriers. Even in 341 Ku, some pilots were selected for Tokko missions, but during the morning of 2 January 1945 as 12 Shidens were lined up preparing to take off on a mission, they were suddenly strafed by P-47s and most were destroyed. By 9 January only four Shidens were left, and these were led by Lt Iwashita to Tuguegarao. From here they made attacks on shipping in Lingayen Gulf, but after a few days all the aircraft had been lost. Lt Fujita then led the remaining pilots on foot to Tuguegarao where they were picked up by transports and taken to Formosa. The groundcrews were left in Luzon, subsequently fighting in the mountains around Clark Field, the majority being killed by the time the war ended.

Commanding Officers
Nov 43 – May 44	Cdr Shoichi Ogasawara
May 44 – Oct 44	Capt Motoharu Okamura
Oct 44 – end	Cdr Tadao Funaki

343rd Kokutai (first), (343 Ku)

Formed at Kagoshima on 1 January 1944, the kokutai was intended to be equipped with Shiden interceptors, but by the end of February only a single example had reached it, so it was decided that A6Ms should be employed instead. On 27 March 12 Zeros led by Lt Shinya Ozaki left Katori for Tinian No.1 airfield. They were to be followed by 3 May by the kokutai headquarters and further elements of the unit. By 15 May 53 Zeros were to hand, of which 42 were immediately available, and at the end of the month Ozaki

ダークグリーン　白
ライトグレー

343 Ku (THE FIRST)

led 37 of these to Palau, to the newly-completed Airai airfield there. Here the unit became involved in the air defence of Palau. On 11 June eight Zero pilots intercepted US carrier aircraft over Tinian, then moving to Guam; details of the fate of this detachment are not available. On 17 June Ozaki led 12 Zeros from Palau to Yap and that evening these

attacked ships off Tinian, losing two aircraft,
Two days later Ozaki claimed one victory over
Guam, but then had to force-land and was
killed. Thereafter 343 Ku aircraft were engaged
in interceptions and in attacks on shipping,
flying from Yap and Guam, but lost Lt
Kawamura and other HQ staff. They moved
back to Airai on 28 June, but on 10 July the
kokutai was disbanded.

Commanding Officer
Jan 44 – Jul 44   Cdr Masao Takenaka

Hikotai leader
Jan 44 – Jul 44   Lt Shinya Ozaki +

Inset: Lt Shinya Ozaki, hikotai leader of the first 343 Ku.
Bottom: 343 Ku at Kagoshima in early 1944. 2nd row, 8th from left, Cdr Masao Takenaka (CO); 9th, Lt Shinya Ozaki (hikotai leader).

343rd Kokutai (second), (343 Ku)

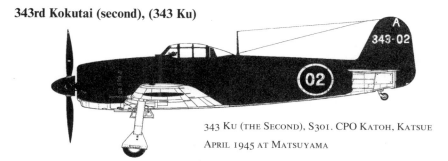

343 KU (THE SECOND), S301. CPO KATOH, KATSUE
APRIL 1945 AT MATSUYAMA

A new 343 Ku was formed on 25 December 1944 at Matsuyama, composed of S301, S701 and S407. The unit's equipment comprised 48 new N1K2J Shiden-kais – a low-wing development of the N1K1 Shiden, which was not to be identified by US aircrews or their intelligence service prior to the end of the war. Training began at Matsuyama, Oh-ita and

Izumi, but all elements gathered at Matsuyama at the end of January. At this point S401 and S402 which had returned from the Philippines, were transferred to 343 Ku and began recuperation. S401 was then employed for the training of newly-qualified pilots at Tokushima, while S402 was transferred to 601 Ku during March.

Pilots of Sento 407, 343 Ku, at Matsuyama in February 1945. 2nd row, 3rd from right, Capt Minoru Genda (CO); extreme right, Yoshishige Hayashi (hikotai leader); 3rd row, 3rd from right, Minoru Honda.

For information gathering, T4 hikotai (24 Saiun reconnaissance aircraft) was also attached to the kokutai. Initially, 343 Ku trained with existing N1K1 Shidens, then re-equipping with Shiden-kais during March, by the end of which month this process was mainly completed. On 19 March 1945 56 Shiden-kais and seven Shidens intercepted US carrier aircraft over the Inland Sea, the Japanese pilots claiming a massive 52 victories for the loss of 15 Shiden-kais and a single Shiden. In April the battle for Okinawa commenced, and on 8th 343 Ku moved to Kanoya to take part in the 'Kikusui-go' operation from 12 April. From then until 22 June the kokutai flew six escort missions to Tokko aircraft flying to Okinawa. However, since the range of the Shiden-kai was less than that of the Zero, the unit's pilots flew only as far as Kikai Shima. Nonetheless, during this period they were able to claim 106 victories in 165 sorties, losing 29 Shiden-kais. During the same period, but particularly between 18 April and 11 May, 343 Ku pilots also intercepted B-29s over Kyushu, but even the Shiden-kai, armed with four 20mm cannons, found the Superfortresses difficult to bring down, In 120 sorties against these raiders only 12 were claimed shot down, while three Shiden-kais were lost and seven had to force-land. In one of the former, Lt Yoshishige Hayashi was one of those killed.

343 Ku flew back to Omura on 3 May from where four flyingboats were shot down. Despite the policy of seeking to preserve aircraft for the anticipated invasions of the home islands, the kokutai was to engage in aerial combats on several occasions. On 2 June 18 F4U Corsairs were claimed shot down over Kagoshima Bay, although Lt Keijiro Hayashi, the new leader of S407 was lost. (The US units involved were VF-85 and VBF-85 – relatively little-known F4U squadrons flying from USS *Shangri La*. Nine of the unit's Corsairs were lost as a result of this day's operation; two were definitely shot down by the Japanese fighters, two ditched during their return flight, possibly due to damage; one ran out of fuel; two were hit by anti-aircraft fire, one had crashed on take-off, and one was written off due to damage on its return.)

16 more victories were claimed over Bungo Strait on 24 July, this time for the loss of Lt Takashi Oshibuchi, leader of S701, and of Ens Kaneyoshi Muto, one of the IJN's top-scoring pilots of the war. On 1 August Lt Naoshi Kanno, leader of S301, was killed, but by the end of the war the unit had claimed about 170 victories – by far the highest total achieved by any unit during the later stages of the conflict. The cost had been high, however, 85 pilots being lost including top aces like Muto and CPO Shoichi Sugita.

Commanding Officer
Dec 44 – end   Capt Minoru Genda

n.b. For a detailed account of this unit, see *Genda's Blade: Japan's Squadron of Aces; 343 Kokutai* by Henry Sakaida & Koji Takaki; Air War Classics 2003.

Shiden-kai of 343 Ku (the 2nd) with rails for rocket projectiles. In the cockpit is CPO Tokokazu Kasai.

345th Kokutai (345 Ku)

345 Ku was formed on 15 January 1944 at Naruo with an establishment of 72 interceptors, plus 18 spares. Joining the 62nd Air Flotilla, the intention had been to equip the unit with Shidens, but due to delays in manufacture, A6Ms were supplied instead. These arrived too late to take part in the Marianas battles, so the 62nd Air Flotilla was disbanded. It was initially intended to transfer 345 Ku to the 2nd Air Fleet, but in the event it too was disbanded.

Commanding Officer
Jan 44 – Jul 44   Cdr Korokuro Tatemi

Hikotai leader
Jan 44 – Jul 44   Lt Cdr Miyoshi Sonoda

352nd Kokutai (352 Ku)

The unit was formed on 10 August 1944 for the air defence of Sasebo, Nagasaki and Omura. Based at Omura, it was to have 48 interceptors and 12 night fighters, but subsequently was divided into three units – one of Zeros, one of Raidens and one of Gekkos. By the end of September the kokutai had 33 Zeros, eight Raidens and five Gekkos, Already by this time, raids by B-29s from China had commenced in June on targets in Kyushu, and eight Gekko crews led by Lt(jg) Sachio

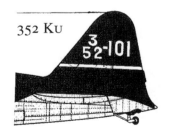

352 Ku

Endo had been detached to the kokutai by 302 Ku. The first engagement occurred during the afternoon of 20 August when 33 Zeros and four Gekkos took off, 352 Ku pilots claiming one destroyed and one damaged. Lt(jg) Endo, meanwhile, pursued the bombers as far as Jeju Do, claiming two shot down, one probable and two damaged. On 25

October 50 Zeros, 13 Raidens and six Gekkos took part in interceptions, claiming nine destroyed, ten damaged and burning, seven damaged, trailing black smoke, and 14 more hit, but results not observed. Lt Mikihiko Sakamoto had rammed a B-29 and been killed. By this time, when raids on the Kanto area by bombers from the new airfields in the Marianas began, Endo had returned to his own unit.

Raids by China-based B-29s then ceased, and on 1 March 1945 352 Ku had 39 Zeros (21 serviceable), 39 Raidens (18 serviceable), eight Gekkos (four serviceable) and six Suisei night fighters (three serviceable). As the fighting for Okinawa began, one buntai of Zeros led by Lt Manae Uematsu moved to Kokubu, taking part in several escort missions and combats before returning to Ohmura. The Raiden unit, led by Lt(jg) (reserve) Yoshihiro Aoki, then moved to Kanoya with other Raiden-equipped units, 302 Ku and 332 Ku, spending the next three weeks intercepting B-29s here. Reorganisation occurred on 25 May, S902 transferring into 352 Ku which then had a strength of 28 Zeros, 25 Raidens, 12 Gekkos and seven Suiseis, plus a detachment of nine Raidens at Naruo, and thus the unit remained until the end of the war.

Commanding Officers
Aug 44 – Dec 44	Capt Ryuji Terasaki
Dec 44 – Jul 45	Capt Bunzo Shibata
Jul 45 – end	Capt Tatsuto Yamada

361st Kokutai (361 Ku)

CPO Minoru Honda, leader of the NCO pilots of Sento 407, addresses them. Meanwhile, Lt Yoshishige Hayashi, the hikotai leader, stands by.

The kokutai was formed at Kagoshima on 15 March 1944, becoming a part of the 62nd Air Flotilla. It was joined initially by S407 which was to have 48 interceptors, but as no Shidens were available, the unit commenced training with A6Ms. It had 11 of these available and ten in maintenance on 1 May, but on 10 July 1944 the kokutai was disbanded and S407 was transferred to 221 Ku.

Commanding Officer
Mar 44 – Jul 44	Lt Cdr Kiyoji Sakakibara

Force-landed Zero of Sento 407. Note Japanese hanji 'Akiru' on the tail.

381st Kokutai (381 Ku)

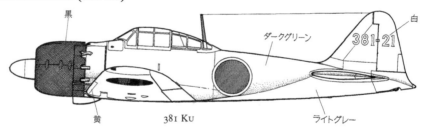

黒　　　ダークグリーン　　　白
381-21
黄　　　381 Ku　　　ライトグレー

381 Ku was formed with 36 A6N fighter-bombers at Kendari, Celebes, on 1 October 1943. On 1 April 1944 it comprised S602 led by Lt Takao Kurosawa, now with 48 fighter-bombers, and S902 with 24 night fighters, led by Lt Hideo Matsumara. Also now included was S311, formed to include pilots from 202 Ku with 48 fighters led by Lt Kunio Kanzaki, which was also formed at Kendari, and was based there. Meanwhile, S602 trained at Toyohashi on Zeros, but during February it was to incorporate ten Raidens as well as 35 Zeros. Due to the small numbers of Raidens available, only the advanced party of the unit departed in February for Celebes. The main body of the hikotai left in March for Balikpapan, but this still had Zeros, although four Raidens followed soon afterwards. The headquarters was by then based at Laikan airfield, Makassar, with a detachment at Balikpapan for air defence. S311 was spread around the Indies at Surabaya, Tarakan, Ambon, Waingapu, Sorong and Babo, but on 5 May this hikotai was transferred to 153 Ku. Meanwhile, S902 had been joined by a number of ex-float fighter pilots, but even by

mid April it still had only two Gekkos, although numbers did slowly increase.

On 1 September 1944 S602 had 40 Zeros (31 serviceable) at Balikpapan, 32 Zeros (15 serviceable) and nine Raidens (seven serviceable) at Kendari, while S902 had eight Gekkos (five serviceable) at Balikpapan, two at Kendari and two at Surabaya. During September USAAF B-24 bombers drawn from both the 5th and 13th Air Forces began raids on Menado where the detachment led by Wt Off Kamihura undertook some effective interceptions using 3-go air-burst bombs. From the end of the month B-24s undertook four heavy raids on Balikpapan which were intercepted by 381 and 331 Ku which between them claimed more than 80 victories. (During these attacks US B-24 losses to defending fighters amounted to 17, while during the two

Lt Cdr Takeo Kurosawa, hikotai leader of 381 Ku, seen here during his earlier time with 12 Ku in China.

later attacks six escorting fighters were also lost.) During the two raids mounted on 10th and 14 October, American fighters had accompanied the bombers all the way and losses of the defending units increased. It became impossible to employ the Gekkos by day, and thereafter they were used for night bombing attacks on Morotai Island. S602 moved to Singapore in mid March 1945, engaging in convoy patrols from there, while S902 moved to Bali Island, and then to Koepang for similar work during the withdrawal from Timor Island. In early April the fighters were ordered home to Japan. With the disbandment of 11th Ku (a fighter training unit), 12th Ku (dive-bombers) and 13th Ku (Rikkos), 381 Ku was reorganised to use the aircraft from these units which included 24 fighters, 12 Rikkos, 12 torpedo-bombers and 24 advanced trainers to take part in Tokko operations. 381 Ku was preparing to move to Sumatra when the war ended.

Commanding Officers

Oct 43 – Oct 44	Capt Katsuji Kondo
Oct 44 – end	Capt Daizo Nakajima

452nd Kokutai (452 Ku)

452 Ku was formed by the renumbering of 5th Ku which had been operating over Kiska in the Aleutians, on 1 November 1942. In the month prior to this renumbering there were no aircraft available, but in early November *Kimikawa Maru* brought float fighters and their pilots to Attu Island. However, these were all destroyed by strafing P-38s and in bad weather accidents in a very short time. Consequently in late December *Kimikawa Maru* returned to deliver seven more A6M2Ns and their pilots who flew them

452 Ku. PO2/c Naoi, Teruyuki

to Kiska on 26th. On the last day of the year these were engaged in the first aerial combat to occur over the area for some considerable time. On this date their pilots claimed one B-25, one PBY and one P-38 shot down, while next day when five float fighters engaged

Float fighter unit of 452 Ku at Bettobi Numa, Shumushu. Centre, seated, is Lt Shunshi Araki (leader); 2nd row, 2nd from right, is Teruyuki Naoi; 3rd row, 2nd from left, is Kiyomi Katsuki.

six P-38s, two of the latter were claimed destroyed. Towards the end of January 1943 transport vessels and escorting ships were spotted off nearby Amchitka Island, pairs of float fighters attacking these on three consecutive days in early February. Six float fighters and a two-seat floatplane were engaged with US aircraft on 14th, one P-39 being claimed shot down. On 19 February PO1c Gi-ichi Sasaki, the only ace in the area, failed to return from a strafing sortie. After this, supply to these islands became increasingly difficult and 452 Ku only operated when conditions were favourable to it. On 27 March two P-38s were claimed as the unit's final victories in the Aleutians, following which the surviving pilots were evacuated from Kiska in a submarine. Recuperation took place at Yokosuka, and in July one buntai of A6M2Ns were led to Bettobi Numa, Shumushu Island, by Lt(jg) Shunshi Araki. Here bombers were intercepted on 12 August and 12 September, on the latter date the unit claiming two B-24s shot down and a third probably. Autumn weather then froze the marshy waters from which the unit was operating, and on 1 October the float fighter unit was eliminated from the kokutai's establishment.

453rd Kokutai (453 Ku)
When Sukumo Ku was formed on 1 April 1943 a float fighter buntai was included. On 1 January 1944 Sukumo Ku became 453 Ku, based at Ibusuki. On 20 February 1944 the float fighter unit was removed from the unit.

501st Kokutai (501 Ku)
501 Ku was formed in July 1943 as a dive-bomber unit; in October it moved to Rabaul where it was engaged over New Guinea and the Solomon Islands, but here it was nearly wiped out, withdrawing to Truk at the end of January 1944. In February a fighter-bomber unit was added to the kokutai, using Zeros which had been held ready for use, and training began on Eten Island. During 17th and 18 February US carrier aircraft

raided Truk for the first time, 501 Ku undertaking a total of 25 sorties during three missions. During these operations PO1c Taizo Maki claimed three victories, but was killed during the third sortie. In total, the unit's pilots claimed ten F6Fs and three TBFs shot down, but lost 11 Zeros including that flown by Wt Off Sadao Matsumoto, the unit leader. Since the pilots had all been trained originally for dive-bombing, five of them attacked the American task force, LdgSea Shigeyoshi Takemura claiming a near miss on a battleship with a 60 kg bomb; one of the F6Fs was claimed by CPO Isaku Mochizuki.

The kokutai returned to Japan on 4 March and here it was brought up to a new established strength of 48 fighters and 48 dive-bombers, then joining the 26th Air Flotilla. The fighter unit then became comprised of S351 (led by Lt Takeo Yokoyama), which formed at Kisarazu, then despatching ten Zeros to Peleliu on 20 March, soon followed by others. Ten days later a large force of carrier aircraft raided this island, and although 12 of the unit's fighters got off to claim four victories, five of their own number were lost, including the leader, Lt Tomojiro Yamaguchi. The remaining seven aircraft were damaged on the ground during the attacks, but five of these were subsequently repaired and flown to Davao in early April, then returning to Japan. 17 more A6Ms moved to Davao where training began. Late in May the unit joined with fighter-bombers of 201 Ku to operate together, and on 10 July this arrangement was formalised when both 501 Ku and S351 were disbanded, the pilots all being transferred to 201 Ku.

582nd Kokutai (582 Ku)

582 Ku at Buin, June 1943. Front row, seated in chairs, extreme left, Usaburo Suzuki; 2nd from left, Cdr Sakae Yamamoto (CO); 3rd from left, Lt Cdr Saburo Shindo; 5th, Kazuo Tsunoda; 2nd row, 3rd from left, Ki-ichi Nagano; Tomezo Yamamoto (behind Shindo); 3rd row, 5th from right, Kiyoshi Sekiya.

On 1 November 1942 2nd Ku was renamed
582 Ku, the unit at this time being engaged in
convoy patrols over transports sailing to
Guadalcanal. Following these duties,
operations were flown over Buna until the
end of December. The unit was next involved
in escorting a convoy to Lae, after which it
helped to provide cover for the withdrawal
from Guadalcanal, from late January
operating from bases at Buin and Munda.

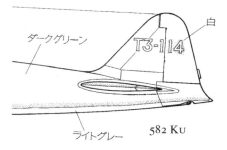

ダークグリーン
T3-114
白
ライトグレー   582 Ku

Staff of 582 Ku at lunch (CO on right).

Cdr Sakae Yamamoto (right), CO of 582 Ku, and
Lt Usaburo Suzuki.

During the 'I-go' operation in April, the kokutai took part in attacks on shipping off
Guadalcanal, and on targets at Moresby and Milne Bay. 28 victories were claimed on 13
May and 7 June, while in defence of Buin on 5 June 17 aircraft were claimed shot down.
On 16 June 16 of the unit's fighters and 24 dive-bombers, led by Lt Cdr Shindo,
rendezvoused with other units for an attack off Lunga, where four victories were claimed
but at a loss of four Zeros and eight dive-bombers.

Following US landings on Rendova Island on 30 June, very heavy fighting developed
over the central Solomons, 582 Ku being much involved until 12 July when the unit's final
mission was flown. On 1 August it was disbanded, its pilots either being sent home to
Japan or transferred to 201 or 204 Ku. 582 Ku's fighter unit had claimed about 220
victories by this time.

601st Kokutai (601 Ku)

Formed on 15 February 1944 from the ex 1st Air Flotilla flying unit, 601 Ku was intended
to serve aboard carriers, and on 10 March was attached to the new 1st Air Flotilla (*Taiho,
Shokaku* and *Zuikaku*) for that purpose. For such duties the establishment was large,
including 81 A6Ms, 81 D4Y Suisei dive-bombers, 54 B6N Tenzan torpedo-bombers and
nine reconnaissance versions of the D4Y. Completing land-based training at Iwakuni, the
unit moved down to Singapore to undertake carrier take-off and landing qualification
during April. The carriers then moved from their Lingga anchorage to that at Tawi Tawi,

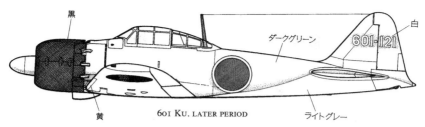

黒　ダークグリーン　白　601-121

黄　601 Ku. LATER PERIOD　ライトグレー

but there were so many US submarines around that no further training was possible here. On 15 June the 1st Air Flotilla set off to the west of the Marianas, becoming engaged in the Battle of the Philippine Sea on 19-20 June. The first wave sent off to attack the US carriers incorporated 48 fighters, 53 dive-bombers and 27 torpedo aircraft. Spotted by the USN's radars, the formations were hit by defending F6Fs so effectively that only one hit on a carrier could be claimed while 32 fighters, 41 dive-bombers and 23 torpedo-bombers were lost. Consequently, the second attack wave comprised just four A6M fighters, ten A6M fighter-bombers and four B6Ns. Worse, these aircraft failed to find the American carriers, but were found by the defending fighters, eight fighter-bombers and one B6N failing to return. Next day eight Zeros rose to intercept US aircraft, and jointly with fighters of the 2nd Air Flotilla, claimed 15 shot down. The majority of the Japanese fighters then had to force-land in the sea as darkness had fallen.

On 10 July 601 Ku joined the 1st Air Flotilla (*Zuikaku* and *Ryuho*), which on 10 August came under the direct command of the 3rd Fleet. At this time 601 Ku became of hikotai composition, comprising S161 (48 fighters led by Lt Hohei Kobayashi), S162 (48 fighter-bombers led by Lt Masayuki Hida), K162, K262 and T61 with dive-bombers and torpedo aircraft, and training began at Oh-ita. During the battle off Engano, an attack force including 12 fighters of 601 Ku took off from four carriers on 24 August, while 13 fighters, including eight of 601 Ku, intercepted US aircraft attacking the Japanese carriers, claiming 12 shot down. They were unable, however, to stop all four carriers being sunk. The remaining pilots were picked up from the sea by destroyers, but these too were sunk at night by American warships, the pilots going down with them.

The flying unit was again reformed with 24 fighters, 12 dive-bombers and 12 torpedo-bombers, the hikotai leader of the fighters being Lt Akio Katori. Training began at Matsuyama, but on 1 February 1945 the 1st Air Flotilla was disbanded, 601 Ku joining instead the 3rd Air Fleet. The unit then comprised S310 with 48 Zeros led by Lt Katori, K1 (dive-bombers) and K254 (torpedo aircraft). During March K254 was transferred elsewhere and 601 Ku was joined by S308 (48 A6Ms led by Lt Kakichi Hirata) and S402 (48 N1K1 Shidens led by Lt Iyozo Fujita). Meanwhile, during the morning of 16 February 20 Zeros of S310 had taken off from Iwakuni to Katori, then becoming engaged in a fight with F6Fs during which four of the Japanese fighters were lost. Next day seven S310 fighters led by Lt Katori claimed six US aircraft shot down – four of them by Katori personally. When US forces landed on Iwo Jima the 2nd Mitate Unit (a suicide unit with 20 Tokko aircraft and 12 direct escort fighters) was formed from within 601 Ku's aircraft and personnel. This unit attacked ships around Iwo Jima on 21 February, gaining a number of hits on vessels.

As the invasion of Okinawa became imminent, the 3rd Air Fleet in the Kanto area moved to Kyushu in late March, 601 Ku going to Kokubi No.1 airfield with 38 Zeros, eight Shidens and 18 Suiseis, where all had arrived by 1 April. On 3rd 32 Zeros and all eight Shidens escorted dive-bombers, becoming engaged in a combat over Kikai Shima. The unit's pilots claimed 11 destroyed and five probables, but lost eight. The unit then became involved in the fighting over Okinawa, on 16 April 26 Zeros and four Shidens again becoming engaged over Kikai where this time four victories were claimed but four losses were sustained. On this same date four bomb-carrying Zeros attempted to attack the US carriers, but none returned. By 17 April 601 Ku had lost 26 fighters and 23 dive-bombers, and at this stage returned to the Kanto area. Recuperation and training of new pilots at Hyakurigahara got underway, and by June the unit had about 70 D4Ys and 100 A6Ms available. During July S402 was transferred to Tsukuba Ku, while on 18th of that month about 50 aircraft of S308, led by Lt Takao Hirose, moved to Yamato, Nara Prefecture, while about 60 of S310 went to Suzuka, Mie Prefecture. These units remained at these bases until the end of the war, preserving their strength for the anticipated invasions of the home islands.

Commanding Officers

Feb 44 – Feb 45	Cdr Toshiie Irisa
Feb 45 – end	Capt Ri-ichi Sugiyama

634th Kokutai (634 Ku)

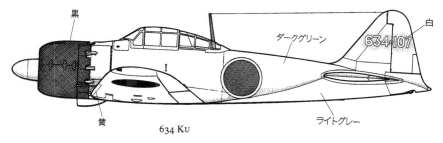

黒　　　　　　　　　　　　　　　　　　　　　　　　　　　　　　　　　　　白

ダークグリーン

黄　　　　　　　　　　　　　　　　　　　　　　ライトグレー

634 Ku

To provide an air unit for service on the carrier battleships *Ise* and *Hyuga*, which together formed the 4th Air Flotilla, 634 Ku was formed on 1 May 1944 with an initial establishment of 22 Suisei dive-bombers and 22 E16A Zuin floatplane bombers. During July and August *Junyo* and *Ryuho* were added to the flotilla and establishment was raised to 24 dive-bombers, 24 float bombers, plus S163 and S167, each with 48 fighters, the two hikotais to be led by Lt Sumio Fukuda. When US carriers launched attacks on Formosa on 12 October, 60 Zeros of S163 and S167 which had been training at Tokushima, together with the dive-bombers and float bombers, moved to Kanoya where they joined the 2nd Air Fleet. During 13th and 14th they moved to Taiching via Okinawa, most aircraft then gradually being fed into the Philippines, based at Clark Field, Luzon, by 23rd. The kokutai's fighters then took part in the general attacks on the US task force and invasion fleet, but lost seven pilots including Lt Fukuda. Here the unit took part in several interceptions, while a part moved to Sebu following the US attack on Tacloban on 1 November. Undertaking interceptions over Cebu and attacks on Leyte, half the unit's aircraft were lost in a couple of weeks. Then on 4 November Lt(jg) Korekiyo Otsusji was transferred to lead Baika Unit, a Tokko unit, leaving the 634 Ku fighters with no officer pilots. On 8 January 1945 the kokutai became a floatplane unit, the fighter element being eliminated.

Commanding Officer

May 44 – Jan 45	Capt Takahisa Amaya

652nd Kokutai (652 Ku)

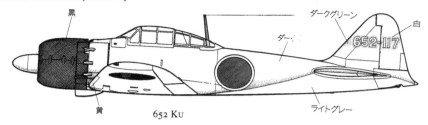

黒　　　　　　　　　　　　　　　ダークグリーン　　　　白

ダー

652-117

黄　　　652 KU　　　　　　　　　　ライトグレー

Fighter pilots of 652 Ku on *Ryuho*; 2nd row, 3rd from left, Moriyasu Hidaka; extreme right, Tetsuo Kikuchi.

652nd Kokutai was formed on 10 March 1944 from the ex-2nd Air Flotilla flying units, and was attached to the new 2nd Air Flotilla (*Junyo*, *Hiyo* and *Ryuho*). As the units incorporated in the new kokutai had been exhausted in the Rabaul fighting, the remaining aircraft were left at Truk and the pilots returned to the homeland on 2 March. Here as these men represented the last of the three carrier units, urgent training was undertaken during the period of two months following the end of March. The new unit was intended to have 81 fighters, 36 dive-bombers and 27 torpedo-bombers, but as at 1 April it had no more than 30 Zero 21s, 13 Zero 52s and four D3A dive-bombers. However, when it departed for Tawi Tawi in May it had been brought up to 135 aircraft, including 27 Zero 21s and 53 Zero 52s; Zero 21s were now to be used as fighter-bombers, fitted to carry 250 kg bombs. On 11 May the 2nd Air Flotilla departed Japan to reach Tawi Tawi on 16th, where rendezvous with other ships of the fleet took place. The area was alive with US submarines and there were no airfields available in the vicinity, so no further flying training could be undertaken, nor training of the fleet itself. On 15 June it moved to the west of the Marianas, on 19th and 20th taking part in the Battle of the Philippine Sea. During the first wave of aircraft despatched to attack the US carriers on 19th were 15 fighters, 25 fighter-bombers and seven torpedo-bombers of 652 Ku. These failed to find the carriers, but were discovered themselves by F6Fs, losing two fighters, four fighter-bombers and one torpedo aircraft. The second wave found the carriers near Rota Island, but lost one of six Zeros and five of nine D4Y dive-bombers without being able to witness the results of their attack. A second group of 20 Zeros, 27 D3As and two B6Ns also failed to find the carriers. However, just before landing on Guam, they were intercepted by about 30 F6Fs. Six of the latter were claimed shot down, but only one Zero, seven dive-bombers and a single B6N Tenzan managed to land safely. Next day the remaining dive-bombers and torpedo-bombers were ordered to fly to safety out of the battle zone, while all the remaining Zeros – 19 Type 52s and seven Type 21s – sought to intercept raids on their own carriers. The pilots claimed seven destroyed and four probables, but 11 Zeros were shot down and three more force-landed on the sea. On 10 July the 2nd Air Flotilla and 652 Ku were disbanded, many of the aircrew being absorbed into 653 Ku.

Commanding Officer

Mar 44- Jul 44   Cdr Sho-ichi Suzuki

653rd Kokutai (653 Ku)

653 Ku on "Zuiho"

Formed on 15 February 1944, 653 Ku took over the ex-3rd Air Flotilla flying units from *Chitose*, *Chiyoda* and *Zuiho* at Iwakuni. The proposed strength was supposed to be 63 fighters and 27 torpedo-bombers, the major part being Zero 21s which were to be used as fighter-bombers. On 11 May the three carriers sailed from the western sector of the Inland Sea, heading for the area west of the Marianas, where they arrived on 15 June. During 19-20 June the ships and their aircraft were involved in the Battle of the Philippine Sea, commencing during the morning of 19th when 14 fighters, 45 fighter-bombers and eight B6Ns set off to launch the initial attack on the American carriers. The kokutai's aircrews claimed one hit on a carrier and one on a cruiser, but lost eight fighters, 32 fighter-bombers and two B6Ns. A further attack was launched on 20th, comprising two B6Ns, ten fighter-bombers and four fighters, but soon after their departure came an attack on the Japanese carriers by US aircraft. Fighters and fighter-bombers were scrambled to defend the fleet, joining with aircraft of 652 Ku in claiming about 20 of the attackers shot down.

Following the Marianas fighting 653 Ku comprised S164 and S165, each unit with 48 fighters and both led by Lt Kenji Nakagawa, S166 with 48 fighter-bombers led by Lt Tetsuo Endo, and K263 (torpedo-bombers). The main body of 653 Ku joined the 2nd Air Fleet as US carrier aircraft raided Formosa, losing half its strength in the fighting here, while Lt(jg) Manabu Ishimori led 24 fighters to Bamban. The more experienced pilots were sent aboard four carriers, on 24th together with 601 Ku, 56 aircraft attacking the US carrier force. They were intercepted on their way by F6Fs, and only *Zuikaku*'s air group reached the opposing carriers. Here the six fighters, 11 fighter-bombers and single torpedo aircraft achieved little, the remaining aircraft, including that flown by Lt Nakagawa, reaching Manila where they rendezvoused with the part of the unit which had moved directly there. Under Nakagawa, at least 20 pilots assembled at Bamban on 26th, took part in two interceptions, and then moved to Cebu from where they undertook convoy patrols and attacks on Tacloban. During an early morning strafing attack on Tacloban on 3 November, Nakagawa and others were shot down and killed, leaving the unit with no officer pilots. Lt Yasuo Masuyama and other pilots who had been left behind at Oh-ita then moved to the Philippines to rejoin the unit, but on 15 November it was disbanded.

Commanding Officer

Feb 44- Nov 44   Cdr Gunji Kimura

721st Kokutai (721 Ku)

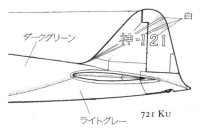

The kokutai was formed on 1 October 1944 at Konoike with Ohka-piloted rocket bombs and their parent Rikkos. Fighter escort was perceived to be essential, and the unit was therefore to include S306 (48 Zeros) for this purpose, led by Lt Cdr Hachiro Yanagizawa, and then Lt Kunio Kanzaki, followed by Lt Daihachi Nakajima. At one period during the battle for Okinawa S305 and S307 were also attached. The first drop test with an Ohka was successfully undertaken on 23 October 1944, and by the spring of 1945 750 Ohkas had been produced. 721 Ku became known as the Jinrai (Divine Thunder) unit, training beginning at Konoike in mid November. On completion of this, the kokutai became a part of the 5th Air Fleet and moved to Kanoya during February 1945. By 18 March S305, with 27 Zeros, was at Kanoya too, while S306 and S307 were at Tomitaka with 64 Zeros. On that date US carrier aircraft raided Kyushu, S306 and S307 being sent up to intercept. While their pilots claimed 22 victories, 23 Zeros were lost. So, on 21 March when 721 Ku made its first attack with Ohkas only 19 of the unit's fighters were available to accompany the Rikkos, joined by 11 of 203 Ku. The heavily-laden Rikkos were intercepted by about 50 US fighters, and although ten of these were claimed shot down, ten Zeros and all 18 Rikkos were destroyed before the Ohkas could be launched.

Fighter unit of 721 Ku; 2nd row, 2nd from left, Wataru Nakamichi.

In addition to the Ohkas, 721 Ku then also formed a fighter-bomber unit for Tokko raids, and during the Okinawa campaign this lost 48 fighter-bombers. Meanwhile, the first success for an Ohka was claimed on 12 April when one hit on a battleship was achieved. From then until 22 June when the Okinawa actions came to an end, 40 Ohkas were expended.

Commanding Officer
Oct 44 – end   Capt Motoharu Okamoto

751st Kokutai (751 Ku)

On 1 October 1942 Kanoya Ku became 751 Ku, and from 3rd of that month the fighter
unit flew nine attacks to Guadalcanal and three interceptions over Kavieng. On 1
November the fighter unit left 751 Ku to form 253 Ku.

802nd Kokutai (802 Ku)

On 1 November 1942 the 14th Ku was renamed 802nd Kokutai
at which time the unit was operating float fighters over
Guadalcanal, using Rekata Bay on the east coast of Santa
Isabel Island as its base. On 7 November five float fighters led
by Lt Hidero Goto, together with one more from *Kamikawa
Maru*, flew a convoy patrol mission. During this, they were
bounced by F4Fs led by Capt Joe Foss, USMC, and all were
shot down. From then on only CPO Eiji Matsuyama and

802 Ku, 1943 in Marshals

LdgSea Shinkichi Ohshima were left to undertake patrols over the base. On 10 December
Ohshima engaged P-38s, claiming one shot down and one probable, while on 19 December
Matsuyama claimed another of these fighters. Both pilots again fought P-38s on 5 January
1943, Matsuyama claiming one and one shared, while Ohshima added a probable.

802 Ku. Lt(JG) Yamazaki, Keizo

Float fighter unit of 802 Ku at Shortland in February 1943. 2nd row, extreme left, San-ichi Hirano; 3rd from left,
Lt Takeo Yokoyama (leader); 4th, Lt(jg) Keizo Yamazaki; 3rd row, extreme right, Shinkichi Ohshima.

Late in December, meanwhile, a new float fighter 802 Ku was formed at Yokosuka, beginning operations on 14 January 1943. Four days later nine A6M2Ns encountered six P-39s, claiming two shot down but losing two float fighters from which the pilots baled out, although one was killed. Until 14 February the unit continued to fly patrols and air defence sorties over Shortland, during which period no enemy aircraft were engaged. On 18 March the unit moved to Jaluit in the Marshall Islands, but here no opposition was encountered, and on 15 October 1943 the float fighter unit was

Lt(jg) Keizo Yamazaki's 802 Ku A6M2N in February 1943. Note the three hatchets, indicating victories.

removed from 802 Ku. During the period 13 October 1942-14 February 1943, the unit's pilots had claimed 13 aircraft shot down plus one shared with another unit, together with eight probables. 13 aircraft and seven pilots had been lost.

902nd Kokutai (902 Ku)

902 Ku was a floatplane unit located at Truk to which on 15 October 1943 the float fighter unit which had been serving with 802 Ku, was transferred. On 21st seven float fighters led by Lt(jg) Keizo Yamazaki arrived at Dublon Island, Truk Atoll from where on 3 November they moved to Greenwich (Kapingamarangi) Atoll, south of Truk. Here patrols

902 Ku

Float fighter pilots of 902 Ku on 3 November 1943 at Dublon Island, Truk Atoll. Seated in chairs, left to right, are Lt(jg) Juji Torimoto, Lt Keizo Yamazaki and Lt(jg) Michiji Kawano.

were commenced, but next day a low-flying B-24 made a strafing attack which damaged all the Japanese aircraft. Where possible, these were repaired, and together with reinforcement aircraft, efforts were made to intercept B-24s over the next few days, but without success, causing the unit to return to Truk. On 17 February 1944 eight float fighters managed to take off just before a raid by US carrier aircraft. Without any chance to form up, the pilots of these aircraft attacked the raiders individually, LdgSea Mitsuo Suzuki and PO1c Takeji Ito each claiming two shot down; both had their aircraft hit, Suzuki force-landing and Ito baling out. PO1c Sotojiro Demura also force-landed after claiming one victory, but four other pilots were shot down and killed. One pilot managed to take off a second time, but he too was hit and had to force-land, leaving the unit with no serviceable aircraft. On 4 March the float fighter element of 902 Ku was omitted.

934th Kokutai (934 Ku)

A floatplane unit was formed as the 36th Kokutai at Balikpapan on 30 June 1942, and on 1 November of that year was renamed 934th Kokutai. On 28 February a float fighter unit for this kokutai was formed at Yokosuka, arriving at Ambon by the special transport ship *Sagara* on 18 March, joining the parent unit in the air defence of the area. At Maikoor on Aru Island a floatplane base was formed in late April to which the floatplane unit of 934 Ku moved with a few of the float fighters. On the day of arrival, 24 April, the first aerial combat was experienced, while next day PO2c Hidenori Matsunaga and others claimed a Beaufighter shot down. Until the early days of May the unit suffered no losses but claimed four aircraft shot down and one probable. Following a E13A reconnaissance floatplane being shot down off northern Australia, however, float fighters were allocated to escort such aircraft. On one such sortie on 20 June Wt Off Takeru Kawaguchi claimed a twin-engined aircraft shot down, while on 10 August Lt Toshiharu Ikeda fought three Spitfires, claiming one of these shot down before the floatplane he was escorting was also brought down. The air defence of Maikoor continued until 21 November, on which date Kawaguchi himself was shot down and killed. Up to this point the unit had claimed 21 shot down and five probables for the loss of four pilots. However, on 10 December 934 Ku stopped using Maikoor.

A6M2N of 934 Ku marked with a lightning flash. It is believed to have been the aircraft flown by Hidenori Matsunaga.

Float fighter unit of 934 Ku at Ambon in December 1943. Front row, 2nd from left, Lt Toshiharu Ikeda (leader); 2nd row, 4th from left, Hidenori Matsunaga.

The float fighters of the kokutai were then employed in the defence of Manokwari and Ambon, the pilots going to Singapore to collect and fly down new N1K1 Kyofu float fighters which had been delivered to that base. The first victory for a Kyofu was claimed in January 1944 during an interception over Ambon when PO1c Kiyomi Katsuki shot down a B-24 on 16th. Earlier in the war this pilot had brought down a B-17 over the Solomons when he rammed it in an F1M biplane floatplane. During interceptions in January, 934 Ku pilots claimed in total three B-24s shot down and a fourth probable for the loss of only one pilot. However, following an engagement with PBY flyingboats on 23 February, the fighter unit of 934 Ku was removed on 1 March 1944.

Sento Hikotais

Sento 161st Hikotai (S161)

Formed on 10 August 1944 under the command of 601 Ku, the unit had an authorised strength of 48 carrier fighters. It was to serve on carriers, led by the highly-experienced Lt Hohei Kobayashi, and began training at Oh-ita. Elements went aboard four different carriers, taking part in the battle off Engano on 24 October, following which the aircraft all flew ashore to Luzon Island. On 25th eight S161 Zeros led by Lt Kobayashi intercepted US carrier aircraft, sharing in the claimed destruction of 12 aircraft. The aircraft and pilots on Luzon continued to operate until 15 November, when the hikotai was disbanded.

Sento 162nd Hikotai (S162)

Also formed on 10 August 1944 for service with 601 Ku, the unit comprised 48 fighter-bombers for carrier operations, led by Lt Masayuki Hida. The unit commenced training at Oh-ita, but as this was not completed, the unit remained in the homeland and was disbanded on 15 November 1944.

Sento 163rd Hikotai (S163)

On 10 August 1944 the unit was formed under the command of 634 Ku, intended to have 48 carrier fighters to be led by Lt Sumio Fukuda, a battle-hardened veteran. Training took place at Tokushima, and in October the unit joined the 2nd Air Fleet, moving to Kanoya. It then took part in the fighting over Formosa. From Taichung the unit then moved to Clark Field, Luzon, from where on 24 October it took part in the general initial attack launched against the US carriers approaching the Philippines. During this operation it lost seven aircraft, including that flown by Lt Fukuda. During the following days it steadily exhausted its supply of pilots and aircraft during attacks on Leyte, and was disbanded on 15 November.

Sento 164th Hikotai (S164)

The hikotai was formed on 1 September 1944 as a 48 carrier fighter unit under the command of 653 Ku. It was led by Lt Kenji Nakagawa, who also commanded S165. Trained at Oh-ita, it joined the 2nd Air Fleet for the fighting off Formosa, then serving on the various carriers involved in the battle off Engano. Subsequently elements operated from Bamban and Cebu during the battle for Leyte, but it was disbanded on 15 November 1944.

Sento 165th Hikotai (S165)

Formed on 1 August 1944 under the command of 653 Ku, the unit was to have 48 carrier fighters and to be led by Lt Kenji Nakagawa. It trained at Oh-ita, then providing air contingents for four carriers with which it took part in the battle off Engano.It then joined S164 on Leyte where it continued to operate until disbanded on 15 November.

Sento 166th Hikotai (S166)

Formed under the command of 653 Ku on 1 August 1944, the hikotai was to have 48 fighter-bombers under the command of Lt Tetsuo Endo. Part of the unit was serving on carriers at the time of the battle off Engano, while subsequently it joined three other hikotais in operations over Leyte. It was disbanded on 15 November 1944.

Sento 167th Hikotai (S167)

On 10 August 1944 the unit was formed under the command of 634 Ku with 48 carrier fighters led by Lt Sumio Fukuda, who also led S163. With that unit it moved from Tokushima to Kanoya, then becoming engaged in the fighting around Formosa. Then moving to Clark Field, Luzon, it took part in the initial attack on the US carrier task force on 24 October, then operating over Leyte until 15 November, when it was disbanded.

Sento 301st Hikotai (S301)

The hikotai was formed from the flying unit of 202 Ku on 4 March 1944. With an establishment of 48 fighters, the unit was led initially by Lt Cdr Minoru Suzuki, the hikotai leader of 202 Ku. 43 aircraft of S301 moved to Truk via Tinian, engaging in interceptions of B-24s while also maintaining detachments at Mortlock and Ponape. On 30 April US carrier aircraft raided Truk and seven Zeros led by Lt(jg) Kawakubo escorted dive-bombers to attack the US ships. Intercepted, the pilots claimed 16 victories for a loss of four aircraft. On 23 May 28 Zeros led by Lt Cdr Suzuki departed Truk, flying to Sorong via Peleliu on 1 June. With S603 they operated over Biak on 2nd and 12 June, while on 13th they moved back to Peleliu from where on 19th 20 Zeros attacked ships off Saipan, reaching their targets via Yap.

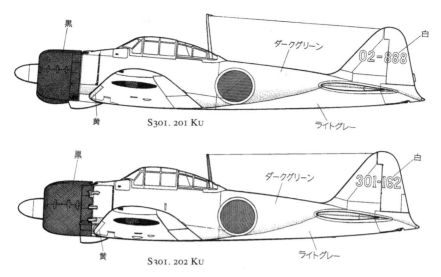

黒　ダークグリーン　白

02-888

S301. 201 Ku

黄　ライトグレー

黒　ダークグリーン　白

301-162

S301. 202 Ku

黄　ライトグレー

On 10 July 202 Ku was disbanded, S301 transferring to 201 Ku, Lt Usaburo Suzuki becoming the leader. Training followed at Davao, but here losses were suffered during mid and late September when US carrier aircraft raided the area. During the morning of 22 September Lt Suzuki led 15 bomb-carrying Zeros to attack carriers east of Lamon Bay, five bomb hits and three strafing attacks being claimed. On 20th a special ram attack, or Tokko unit was formed in 201 Ku and all pilots of every hikotai were designated as personnel for Tokko units. Lt(jg) Yoshitaka Kuno of S301 was the first pilot to be killed on such a sortie.

Sento 301 at Matsuyama in January 1945. Front row, 2nd from left, Lt Naoshi Kanno (hikotai leader); 4th, Minoru Genda (CO 343 Ku); 2nd row, 2nd from left, Mitsuo Hori; 4th, Isamu Miyazaki; 3rd from right, Shoichi Sugita; 5th row, 5th from left, Tomoichi Kasai.

S301 was transferred to 252 Ku in November, but on 25 December 1944, when S343(the second) was formed, S301 was incorporated into this new unit, which was led by Lt Naoshi Kanno. It equipped with the new N1K2J Shiden-kai fighter. From 19 March 1945 until Lt Kanno was killed on 1 August, the unit fought over Matsuyama, in the battle for Okinawa, and in the interception of B-29s over Kyushu until the end of the war.

Sento 302nd Hikotai (S302)

On 1 April 1944 the flying unit of 252 Ku was formed into S302 by the separation of the flying and ground personnel. It at once returned to 201 Ku command, and was led by Lt Nobuo Awa, its strength being established as 48 fighters. During training it moved from Tateyama to Misawa where by early May it actually had 55 Zeros on hand. On 15 June the Hachiman Force was formed and under the command of this force, it moved to Iwo Jima. Here on 24 June its pilots claimed 19 victories, but lost ten, Lt Awa included. Further engagements on 3rd and 4 July saw 13 more victories claimed but 14 aircraft lost.

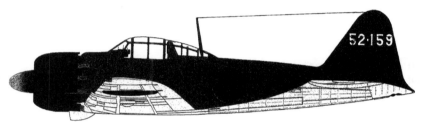

S302. 252 KU. CPO KUWABARA, SHOICHI

The remaining pilots were withdrawn back to Tateyama to rebuild under a new hikotai leader, Lt Masao Iizuka. On 10 July three more hikotais were placed under 252 Ku's command. During October, with the battle for the Philippines imminent, the unit participated in the aerial battles over Formosa, but here on 15th Lt Iizuka was amongst several lost. A move to Clark Field, Luzon, followed, the unit taking part in the general attack launched against the US carrier task force as it neared the area on 24 October. 11 of the 26 Zeros launched by 252 Ku were lost on this date, following which the remaining aircraft took part in operations over Leyte until mid November when it was transferred to 201 Ku and continued operating. At the end of the Philippines battles the remaining pilots were withdrawn to Formosa and transferred to the newly-formed 205 Ku on 5 February. During March this kokutai had only about 20 aircraft available for its three hikotais. The invasion of Okinawa then occurred and the aircraft available were divided into those to be used for Tokko operations, and those which were to escort them and observe the results. With few supplies from the homeland available, the units steadily wasted away, and following the fall of Okinawa during May, ceased to operate.

Sento 303rd Hikotai (S303)

The reorganisation of units on 15 April 1944 led to the formation of S304 from the flying unit of 203 Ku, which initially retained command of the new hikotai. Intended to have a strength of 48 fighters, the new unit was led by Lt Cdr Kiyokuma Okajima. On 30 March, just before the reorganisation, 46 Zeros moved from Atsugi to Chitose, and at the end of April one buntai was detached to Shumushu and Paramushir in northern Chishima. The main body then followed to Kataoka airfield on Shumushu Island, a Lockheed PV-1 Ventura being claimed shot down on 13 May, followed by a B-24 on 15 June. Several more

Sento 303 in July 1945. 2nd row, 5th from left is Lt Osamu Kurata; 3rd row, extreme left, Takeo Tanimizu.

interceptions followed, but without result. On 11 August the main part of the unit moved to Bihoro, and on 18 September S303 was attached to T Force, moving to Mobarta and then to Kagoshima. As the battle around Formosa commenced on 12 October, the unit moved to the Philippines via Okinawa and Formosa. Joined by S304, which had followed S303, the two units took part in attacks on US carriers and the battle for Leyte. On 15 November S303 was transferred to 201 Ku, continuing to operate until the end of December by which time the unit was exhausted and returned to Japan. Here it rejoined 203 Ku, recuperating at Kagoshima and Izumi; by 18 March it had been rebuilt to a strength of 32 Zeros. On this date it intercepted US carrier aircraft, undertaking 30 sorties during four scrambles, claiming three shot down and three damaged, but losing five of its own pilots killed, three missing and two baled out. Next day 12 Zeros with four more from S312, claimed four shot down and three probables. Lt Cdr Okajima led 11 Zeros (five of them from S303) to cover the first mission to be flown by Rikkos carrying Ohka-piloted rocket bombs. This was intercepted, and while the escorting pilots were able to claim four destroyed and three probables, all the Rikkos and their charges were lost. With the start of the US invasion of Okinawa, the unit was engaged in providing escort for Tokko aircraft, and following the loss of the island, S303 moved to Tsuiki in late July under the command of Lt Osamu Kurata, and here it ended the war.

Sento 304th Hikotai (S304)

The hikotai was formed on 15 April 1944 jointly with S303 from the air group of 203 Ku. With an establishment set at 48 fighters, the unit was led by Lt Takashi Oshibuchi, and had already moved from Atsugi to Chitose just prior to the reorganisation. From May-July the unit was involved in intercepting American bombers and on 11 August moved to Bihoro. However, 19 Zeros under Lt Mutsuo Urushiyama remained at Shumushu, and from here interceptions were made on eight occasions between 20 August and 25 September, but were only able to claim one probable and two damaged. On 11 October 23 Zeros led by Lt Oshibuchi moved to Katori, covering an attack force seeking US carriers as it moved to Izumi, Okinawa and Formosa. On 16th Lt Oshibuchi led 14 Zeros to cover this force, sharing in the claimed destruction of six raiding aircraft. 29 of the unit's Zeros moved to Bamban, Luzon, on 22 October and from here two days later 15 were led by Oshibuchi to take part in an attack on the US carrier force with other units. Subsequently they operated over Leyte, where by the end of the month 17 raiders had

Sento 304 in spring 1945 at Mobara. Front row, 5th from left, Lt Cdr Hachiro Yanagizawa (leader); 2nd row, 2nd from left, Masami Shiga; 8th, Katsuyoshi Yoshida.

been claimed shot down during interceptions, five more being claimed as probables, the unit losing eight of its own. Interceptions increased during November, and on 11th Lt Urushiyama led seven Zeros to Davao, Mindanao, to undertake convoy patrols and interceptions there.

On 15 November S304 was transferred to 221 Ku, continuing to operate until early in January 1945, 16 pilots, including Lt(jg) Ichiro Watanabe, being killed while undertaking Tokko suicide missions. During March the hikotai was attached to 252 Ku under the command of Lt Cdr Yanagizawa. It moved to Kokubu on 16 April to undertake escorts, but on 17th three aircraft were lost which included the new commander. Shortly thereafter the unit withdrew to the Kanto area, moving to Koriyama where the new leader, Lt Cdr Moriyasu Hidaka, preserved its strength until the end of the war.

Sento 305th Hikotai (S305)

Formed from the flying unit of 201 Ku on 4 March 1944 with a strength of 48 fighters under Lt Shiro Kawai. The unit took part in the interception of a raid by US carrier aircraft during which the 20 Zero pilots involved claimed 17 aircraft shot down but lost nine of their own, while 11 more were destroyed on the ground. A withdrawal to Davao followed in order to recuperate. On 10 July Lt Masanobu Ibusuki, the ex-hikotai leader of 261 Ku, became the leader of S305. During September 201 Ku was exhausted during three raids by US carrier aircraft, but on 15 October Lt Ibusuki led 25 Zeros to attack a carrier force east of Luzon, some of the carriers being strafed. S305 was then transferred to 252 Ku on 15 November, and then in March 1945 to 721 Ku. With this unit the hikotai took part in the first mission with Ohka-piloted rocket-propelled glider bombs. In April it transferred to 701 Ku, but on 10 May 1945 was disbanded.

Sento 306th Hikotai (S306)

Formed on 4 March 1944 from the flying unit of 201 Ku with 48 fighters, the unit was led by Lt Torajiro Haruta, ex-hikotai leader of 204 Ku. Immediately after formation it trained at Kisarazu, and during mid April-May it moved to Cebu and Davao. In July there was a change of leadership, Lt Ken-ichi Ban taking over until replaced in August by Lt Hiroshi Morii, and then by Lt Naoshi Kanno on 25 September. During this period Lt Kanno led one buntai to Yap to intercept B-24s. Between 16th-23 July eight interceptions were made, claims being submitted for eight destroyed, nine probables and 48 damaged. The unit was joined here by other hikotais, making further interceptions during September, but in October the detachment returned to Japan where the hikotai was transferred to 721 Ku. In February 1945 721 Ku became part of the 5th Air Fleet, the fighters moving to Moyazaki. On 19 March S306 and S307 were at Tomitaka with 64 Zeros between them and during the day the two hikotais jointly claimed 22 victories but lost 23 Zeros. On 21 March they escorted the first mission by Ohka-carrying Rikkos, but although some claims were made against interceptors, all the Rikkos were lost and the escorts also suffered some casualties. From then onwards escorts to Ohka-carriers and Tokko aircraft followed. During May S307 was disbanded, the aircraft and pilots being passed to S306 which rose to a strength of 96 fighters, and remained virtually intact it ended the war.

Sento 307th Hikotai (S307)

Formed in February 1945 with 48 fighters, the unit came under the command of 721 Ku. On 19 March it was at Tomitaka, joining S306 in intercepting raids by US carrier aircraft on three occasions from 0500 onwards. Two units jointly claimed 24 shot down and seven probables, but lost 19 pilots. 32 Zeros took off on 21 March to provide direct escort to the Jinrai Unit, being engaged by F4Us and F6Fs against which nine were claimed shot down plus three probables. However, several Zeros were lost including that of Lt Musuo Urushiyama, a buntai leader. 721 Ku was then engaged in the fighting over Okinawa, but on 5 May S307 was disbanded and its remaining pilots and aircraft transferred to S306.

Sento 308th Hikotai (S308)

Formed under the command of 221 Ku on 10 July 1944, the unit was established for 48 fighters and was commanded by Lt Hisaya Hirusawa. After training at Kasanbara, 221 Ku moved to the Philippines via Okinawa and Formosa during October, but S308 was left behind as last reserve. On 20 December, therefore, 20 Zeros from this hikotai were led by Lt Cdr Shiro Kawai to Angeles, on Luzon. On 24th Lt Cdr Kawai baled out over Clark Field but was not seen again. The fighting which followed over the next

ダークグリーン 白

白

S308. 221 KU

ライトグリー

two days almost annihilated the unit, which lost many of its pilots. In February 1945, following withdrawal from the Philippines, the hikotai was transferred to 252 Ku. With this unit 23 S308 Zeros intercepted US carrier aircraft in the Kanto area, pilots claiming eight destroyed and five probables for the loss of six pilots killed, one seriously wounded and one baled out. In March the unit was transferred again, this time to 601 Ku with which it took part in the Okinawan fighting. A move back to the Kanto area followed, and then to Yamato on 18 July, where with some 50 Zeros still on hand, the unit was still based when the war ended.

Sento 309th Hikotai (S309)

The hikotai was formed on 1 April 1944 from the flying unit of 253 Ku and with the standard 48 fighters. Led by Lt Tatsuo Hirano, it took part in the air defence of Truk. During the Battle of the Marianas the unit tried to move 13 Zeros to Guam on 19 June, but these were intercepted and lost five, Lt Hirano included. On 10 July with the disbandment of 263 Ku, S309 was also disbanded.

In October it was formed again, this time from the flying unit of 331 Ku, and with Lt Kisuke Hasegawa as leader. On 1 October it had 30 Zeros available, 20 of which were serviceable, at Balikpapan, plus five at Sandakan, Kuching, and Penang. The main part of the unit was engaged in the air defence of Balikpapan, but in February 1945 this was withdrawn to Seletar, Singapore. Ordered home to Japan during May, 13 Zeros reached Ohmura. Here on 15 June Lt Hideo Matsumura, ex-leader of S902, became leader until the end of the war.

Sento 310th Hikotai (S310)

Also formed from the flying unit of 253 Ku on 1 April 1944, S310 was led by Lt Cdr Harutoshi Okamoto with 48 fighters. It took part in the defence of Truk, but on 19 June all its 13 remaining Zeros sought to move from Truk to Guam, but just before landing they were attacked and five were shot down, all remaining aircraft then being destroyed on the ground. Survivors were returned to Japan, where the unit was disbanded on 10 July.

Sento 310, 601 Ku, in March 1945. Front row, 2nd from left, Tsutomu Iwai; 5th, Lt Hideo Katori (leader); 3rd row, 2nd from left, Kunimori Nakakariya; 4th row, 3rd from left, Yoshijiro Shirahama.

On 1 February 1945 the 1st Air Flotilla was disbanded, 601 Ku joining the 3rd Air Fleet instead, and from the flying unit of this kokutai, S310 was reformed, once more to have 48 fighters and to be led by Lt Akio Katori. On the morning of 16 February 20 Zeros took off to move from Iwakuni to Hyakurighara. Although flying with care, they were caught en route by F6Fs and suffered the loss of four. Next day, however, seven of the unit's Zeros intercepted dive-bombers, claiming six shot down, four of them claimed

Sento 310 in May 1945 at Matsushima. Front row, 7th from left, Lt Akio Katori; 2nd row, 6th from right, Yoshijiro Shirahama; 5th, Kunimori Nakakariya.

by Lt Katori personally. US forces landed on Iwo Jima on 19 February, and within 601 Ku the 2nd Mitate Unit (comprised of 20 Tokko aircraft and 12 escort fighters) was formed, on 21st attacking ships off Iwo. In late March the 3rd Air Fleet moved to Kyushu, 601 Ku going to Kokubu No.1 airfield until 1 April. From 3-17 April it took part in the fighting over Okinawa. It then returned to Hyakurigahara to preserve its strength for the anticipated invasions of the home islands. On 18 July S310 moved to Suzuka with 80 aircraft on hand, and here it ended the war.

Sento 311th Hikotai (S311)

The unit formed at Kendari, Celebes, on 1 April 1944 with pilots transferred from 202 Ku and 381 Ku. With the usual fixed establishment to be 48 fighters, the hikotai was led by Lt Kunio Kanzaki. On 5 May it was removed from the command of 381 Ku, moving to 153 Ku at which stage it received a new leader, Lt Shin Yamauchi. With about ten Zeros at Sorong, the unit engaged in interceptions of Allied bombers, by 26 May recording 21 victories over B-24s and other types. At this time S311 had five aircraft at Sorong and 15 at Kendari, but on 27 May American forces landed on Biak Island. Aerial fighting followed on every day until 5 June, the Sorong-based part of the hikotai losing nine aircraft and pilots, including Ens Kikunasa Fujita and Wt Off Shinji Ishida. On 7th the unit withdrew to Kendari; between 5 May-16 June, the unit had claimed 32 aircraft shot down for the loss of 15 Zeros.

As the fighting for the Marianas began, on 18 June ten Zeros and one D4Y moved to Truk and Meleyon, via Wasile, Peleliu and Yap, but there was no opportunity for aerial combat. On 10 July S311 was transferred to 201 Ku, moving to Davao to begin training under the leadership of Lt Takeo Yokoyama. Pilots of 201 Ku were trained in skip-bombing with bombs carried by their Zeros, but during September fighter units suffered considerable damage from raids on southern and central Philippines and on Luzon. At this point the new commander of the 1st Air Fleet, Admiral Ohnishi, ordered that 201 Ku become the parent unit for Tokko units. These were to achieve considerable results but at the cost of many pilots and aircraft. In December 1944 S311 was transferred to 252 Ku and returned to Japan under Lt Kosuke Tabuchi, to recuperate at Mobara and Tateyama. In February 1945 S311 intercepted the first raids by US carrier aircraft on the

Kanto area, following which in March it was transferred to 203 Ku to take part in the fighting over Okinawa. The hikotai was then ordered to conserve its strength under the command of Lt Kisuke Hasegawa, and thus it remained until the end of the war.

Sento 312th Hikotai (S312)

Formed on 10 July 1944 from the flying unit of 221 Ku, the hikotai had a fixed establishment of 48 fighters under the leadership of Lt Toshio Shiozuru, ex-hikotai leader of 221 Ku and a veteran fighter leader who had been on operations since the beginning of the Pacific War. After training at Kasanbara, in mid August the unit moved its 51 Zeros in three waves to Hsinchu,

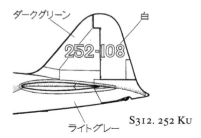

S312. 252 Ku

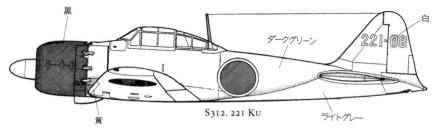

S312. 221 Ku

Formosa, to undertake air defence and convoy patrols. On 12 October US carrier aircraft raided Formosa, S312 putting up 43 aircraft to intercept. 16 aircraft were claimed shot down plus seven probables, but 15 Zeros were lost. On 15th 16 Zeros, including 11 from S312, escorted four Rikkos, with Gekkos pathfinding. The greater part of the formation turned back after finding nothing, but the remainder of the force was engaged by F6Fs and two Zeros were lost. With other 221 Ku hikotais, the unit then moved to Angeles, Luzon, by 23rd, next day taking part in the general attack on the approaching task force, during which some of the unit's aircraft were lost. Attacks on Leyte and air defence sorties followed until the unit withdrew from the Philippines. Led now by Lt Yoshihiro Hayashi, in March 1945 it joined 203 Ku, by mid month having 31 Zeros and 51 pilots at Kasanbara. On 18 March US carrier aircraft were intercepted over Kyushu, while part of the unit escorted Ohka-carrying Rikkos on their first mission. From 5 April the hikotai was engaged in the fighting over Okinawa, suffering some losses including Lt Kunio Kanzaki on that date. During July the unit withdrew to Tsuiki under Lt Kosuke Tabuchi to conserve its strength, and there it remained until the war ended.

Sento 313rd Hikotai (S313)

Formed on 10 July 1944 under the command of 221 Ku with a fixed complement of 48 Zeros, the hikotai was commanded by Lt Shiro Kawai who had been the leader of S305. Training took place at Kasanbara following which the unit took part in the general attack of 24 October on the US carrier task force, having moved to Angeles, Luzon. It then took part in attacks on Leyte and on air defence scrambles until the US landings on Luzon in January 1945, when it returned to Japan.

S313. 221 Ku

Sento 313 at Mobara in April 1945. 2nd row, 3rd from left, Lt Yasuo Masuyama (leader).

In February came a transfer to 252 Ku and the commander became Lt Yasuo Masuyama who had been operating from carriers. Following brief recuperation at Mobara and Tateyama, it moved to Kokubu at the end of March, taking part in the fighting over Okinawa. During these operations a part of the unit formed the 3rd Mitate Unit on 3 April to undertake Tokko missions. On 17 April, however, 252 Ku withdrew to the Kanto area where in May S313 was transferred to 752 Ku at Konoike. Here it undertook only occasional interceptions until transferred again just before the end of the war to 203 Ku with which it saw out the conclusion of hostilities.

Sento 315th Hikotai (S315)

Formed on 10 July 1944 as part of 252 Ku with 48 fighters, the unit was led by Lt Masuzo Seto who had been a carrier pilot. The unit began training at Tateyama, and as the 2nd Air Fleet moved to the Philippines, S315 flew to Clark Field, Luzon via Okinawa and Formosa. It engaged in attacks on the US carrier fleet on 14th and 15 October, losing two pilots. During general attack operations on 24th four more pilots were killed, while during attacks on Leyte three were lost on 29th over Tacloban, followed by six more during an interception on 5 November. On 4 December Lt Seto was killed while on his way to the Philippines. The unit was transferred to 201 Ku, while on 25th even veterans like CPO Tadashi Yoneda and CPO Sho-ichi Kuwabara were lost whilst involved in an interception. Transfer to 221 Ku followed and the unit moved back to Formosa where it was transferred to the newly-formed 205 Ku. Further losses of men and aircraft followed during the fighting over Okinawa, the unit ending the war without being reformed.

Sento 316th Hikotai (S316)

The unit was formed under the command of 252 Ku on 10 July 1944 with 48 fighters led by Lt Torajiro Haruta, commencing training at Tateyama. With other hikotais it transferred to the Philippines, on the way taking part in attacks on US carrier task forces on 15th, 16th and 17 October. It was then engaged in the interception of B-29s over Formosa, while 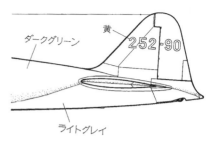 during the big operation on 24th three pilots were lost, including buntai leader Lt Iwao Akiyama. The unit was then involved in attacks on US forces on Leyte, while during an interception over Clark Field, Luzon. On 6 November four more pilots were lost including the hikotai leader. He was replaced on 15 November by Lt Tetsuo Endo who had previously led S166, the unit then returning to Japan to recuperate at Tateyama. Leadership passed to Lt Yasujiro Abe in February. The following month under the leadership of this veteran Yokaren course pilot, the unit moved to Kokubu, but at half strength. Here it took part in the fighting over Okinawa until late in April when a move was made to the Kanto area. Rebuilding of the unit began again at Mobara, while some pilots undertook interceptions, and this was how the hikotai remained until the war ended.

Sento 316 at Mobara, spring 1945. Front row, seated in chairs, from left to right: Lt(jg) Hachitara Hayashi; Lt Yasujiro Abe; Lt(jg) Susumu Kawasaki; 3rd row, 3rd from right, Shigeru Takahashi.

Sento 317th Hikotai (S317)

The sento was formed on 10 July 1944 from 252 Ku, equipped with 48 fighters and commanded by Lt Kazumasa Mitsunori. Training commenced at Tateyama, 14 Zeros led by Lt Hidehiro Nakama, a buntai leader, engaged in interceptions against B-24s on Iwo Jima during mid October-mid December. During this period Lt Nakama was killed and others were later shot down by escorting P-38s. The first Mitate Unit was formed in S317 under Lt(jg) Kenji Ohmura, but on 27 November the unit was strafed and all were killed on Saipan. The main body of the unit moved to Clark Field, Luzon, via Okinawa and

Formosa, on the way taking part in attacks on US carrier forces off Formosa. During the attack on 24 October, Lt Cdr Minoru Kobayashi, by then leader of S317, headed one of the attack forces with 26 Zeros, but he and one other pilot were killed. The unit then took part in attacks on Leyte and in the air defence of Luzon. On 15 November S317 was transferred to 201 Ku, Lt Minoru Kawazoe becoming the leader. A move back to Formosa followed, and in February the unit was again transferred, this time to the newly-formed 205 Ku. Following the Okinawan fighting the sento was exhausted and was sent to Hsinchu to recuperate and in this form it ended the war.

Sento 318th Hikotai (S318)
Formed on 5 April 1945 under 381 Ku, the unit was led by Lt Shoichi Kusakari, and received personnel from 11th Ku. However, on 25 May it was disbanded and both personnel and aircraft were attached directly to 381 Ku.

Sento 351st Hikotai (S351)
The fighter-bomber unit of 501 Ku had been badly damaged during the first major raid by US carrier forces and was reformed as S351 with 48 fighters on 4 March 1944; leader was Lt Cdr Bunto Inoue, ex-hikocho of 501 Ku. Only four days later he handed over to Lt Takeo Yokoyama, an ex-float fighter leader. The unit began training at Kisarazu, but ten A6Ms were sent to Peleliu on 20 March. These intercepted US carrier aircraft on 30th, sustaining heavy losses resulting in the return of the survivors to Kisarazu. A second detachment of 17 Zeros were sent to Davao on 10 July, but the sento was then disbanded and the pilots were absorbed into 201 Ku.

Sento 401st Hikotai (S401)
The unit was formed on 10 July 1944 as part of 341 Ku with an establishment of 48 Shidens led by Lt Ayao Shirane, the ex-hikotai leader of the kokutai. Training commenced at Tateyama, but this base proved to be too small and a move was made to Meiji. Production and deliveries of Shidens proved inadequate, only 20 having been received by August despite 341 Ku having been formed in November 1943. Nonetheless, 17 of the Shidens were sent forward to Takao at the end of August where they became engaged in air defence. On 11 September Lt Shirane was transferred to S701, Lt Masa-aki Asakawa, a bunti leader, taking over the hikotai. On 1 October S401 had 32 Shidens at Takao but only 20 were serviceable. However, when on 12th American carrier aircraft raided Formosa, 31 Shidens were launched by the unit, the pilots claiming ten of the raiders shot down at a cost of 14 of their own aircraft. The hikotai then moved to Marcott on Luzon, taking part in the general attack on 24th, undertaking interceptions and reconnaissance flights until early January. Having withdrawn to Formosa, the unit then returned to Japan where it was transferred to 343 Ku. It then moved to Tokushima to become a training unit for new pilots until the end of the war.

Sento 402nd Hikotai (S402)
Like S401, this unit was formed on 10 July 1944 as part of 341 Ku, to be equipped with 48 Shidens and led by Lt Iyozo Fujita of Midway fame. By 1 October 30 Shidens were to hand, of which 25 were serviceable, together with 22 A6Ms (18 serviceable), all based at Miyazaki. With other units of 2nd Air Fleet it moved to Marcott, Luzon, where it was mainly engaged in air defence and convoy patrol activities.

ダークグリーン 白

341S-52

S402. 341 Ku

ライトグレー

Together with S701, the three Shiden units suffered many mechanical breakdowns which gradually evaporated their strength. When US landings on Mindoro Island took place, S402 attempted to attack the landing force with a Tokko (suicide) attack, but as the aircraft took off they were bounced by American fighters; Lt Seiya Nakajima, Lt Sumio Akikawa and several others being killed. By early January the unit had no aircraft left and withdrew to Japan where it was transferred to 343 Ku. A further transfer to 601 Ku followed in March. With a number of Zeros, eight Shidens were flown to Kokubu No.1 airfield, and from here on 3 April seven Shidens took off to cover Tokko aircraft. However, the distance to be covered to Okinawa was 650 km and as a result only two were able to accompany their charges all the way, three turning back due to shortage of fuel, while the other two were not seen again. The hikotai continued to operate until mid April when 601 Ku withdrew to the Kanto area. Here the unit was transferred to Tsukuba Ku with which it ended the war.

Sento 403rd Hikotai (S403)
Formed on 5 May 1945 under Tsukuba Ku with 48 interceptors, the pilots came from the parent kokutai and from Yatabe Ku and were commanded by Lt Kazumasa Mitumori. However the war ended before any operations were undertaken.

Sento 407th Hikotai (S407)

Sento 407 of 361 Ku in May 1944. Front row, 2nd from left, Minoru Honda; 6th Lt Yoshishige Hayashi (leader).

Formed under 221 Ku on 10 July 1944, the unit was led by Lt Yoshishige Hayashi and established to have 48 fighters. Following training at Kagoshima, the unit moved to Angeles, Luzon, on 23 October, taking part in the initial attack on the US invasion fleet and losing one pilot. With the other hikotais of 221 Ku it then took part in attacks on Leyte, and on convoy patrols and interceptions. Four aircraft were lost on 11 November, and in December the unit returned to Japan. With the formation of 343 Ku, the hikotai was transferred to this unit where it converted to Shiden-kai aircraft. Training on the new fighter took place at Izumi and Matsuyama under the combined leadership of Lt Hayashi and the leader of the NCOs, Wt Off Minoru Honda. The unit took part in 343 Ku's first aerial combat on 19 March, moving then to Kanoya where it was involved in providing cover for Tokko aircraft flying to Okinawa. It then moved to Ohmura where Lt Hayashi was killed whilst attempting to intercept B-29s. On 2 June 1945 over Kagoshima Bay 21 Shiden-kais led by Lt Keijiro Hayashi, the unit's new leader,

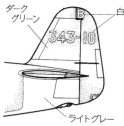

S407. 343 Ku (II)

bounced F4Us, claiming 18 shot down. However, on 22 June Lt Hayashi was also shot down and killed, the unit then seeing little more action before the war ended.

Sento 601st Hikotai (S601)

The unit was formed as part of 301 Ku with 48 interceptors on 4 March 1944; the leader was Lt Iyozo Fujita, ex-hikotai leader of 301 Ku. When the Marianas invasion commenced, it was planned that the unit should move to Iwo Jima, but there was concern regarding the over-water performance of the Raiden fighters with which the unit was equipped. As a result, a switch in equipment was made to the reliable Zero. The first nine of these fighters were sent to Iwo Jima on 21 June, where three days later their pilots claimed five victories but lost four aircraft. A further 31 fighters arrived on 25th, all taking off on the afternoon of 3 July. However, they were attacked from the side while still climbing for altitude and in the fight which followed 17 Zeros were lost, surviving pilots claiming ten victories in return. More aircraft and pilots were lost during a further combat next day and those remaining were destroyed by bombardment by US ships. The surviving personnel returned to Japan in transports and the unit was disbanded on 10 July.

Sento 602nd Hikotai (S602)

S602 was formed on 1 April 1944 within 381 Ku, and was initially equipped with 48 fighter-bombers under the leadership of Lt Takeo Kurosawa, the ex-hikotai leader of the kokutai; its base was Kendari. Main aircraft operated was the Zero, but the unit also had a few J1M Raidens. By 1 September it had 40 Zeros at Balikpapan, plus 32 Zeros and nine Raidens at Kendari. During the month a detachment led by Wt Off Keishu Kamihira claimed victories over heavy bombers over Menado. Between 30 September and 14 October four raids by B-24s were intercepted and together with 331 Ku more than 80 victories were claimed. However the bombers were accompanied by fighter escorts from 10 October onwards, the unit losing six pilots on this date and four on 14th. In March 1945 the main body of the unit withdrew to Singapore from where in early April it returned to Japan; here it was disbanded on 25 May.

Sento 603rd Hikotai (S603)

Formed on 4 March 1944 from the flying unit of 331 Ku under Lt Cdr Hideki Shingo, ex-hikotai leader of 331 Ku, the unit became part of 202 Ku. Initially 33 Zeros moved to Meleyon Island, flying patrols here until 1 April when the unit moved to Truk. Here it became engaged in the interception of heavy bombers raiding the island, eight claims being made including one and two probables on

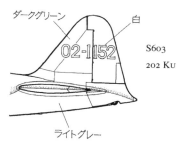

S603

202 Ku

Sento 701 at Omura in July 1945. Front row, extreme right, Kazuo Muranaka; 2nd row, 5th from left, Masao Sasakibara; 3rd from right, Hiroshi Okano.

3 April. When on 30 April US carrier aircraft attacked the atoll, 11 victories were claimed but four aircraft failed to return. In late May the unit moved to Sorong via Guam and Peleliu, led by Lt Toshiharu Ikeda, who became commander on 2 June for action during the Biak operation. When US forces landed on Saipan, the unit moved to Yap for the defence of that island against B-24s, but in July it moved again, this time to Davao. It was disbanded on 10 July 1944.

Sento 701st Hikotai (S701)

Formed at Yoko Ku on 10 July 1944 with Shidens, the unit was led for just a month by Lt Cdr Hideki Shingo, who then handed command over to Lt Kunio Iwashita. The unit had a fixed establishment of 48 Shidens, and soon became a part of T Force, being transferred on paper to 762 Ku. Lt Iwashita then fell sick, and taken over by Lt Shirane, the unit moved with other units of T Force to the south to take part in the battle off Formosa. It then moved to Marcott where the other two Shiden-equipped hikotais were based. Lt Shirane, the senior fighter leader at this base, then led the whole Shiden force to take part in the fighting around Leyte. Here the Shidens were mainly involved in air defence and convoy patrols. On 24 November Shirane, now promoted Lt Cdr, was killed, and on 5 December Lt Akira Takeda took over, leading the unit until it withdrew to the homeland.

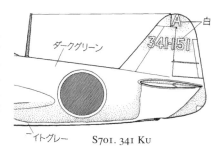

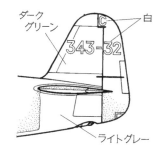

On 1 January 1945 Lt Takeshi Oshibuchi became leader and the unit was transferred to the new 343 Ku and converted to Shiden-kais, the pilots recuperating and becoming familiar with their new aircraft at Oh-ita, and then at Matsuyama. On 19 March Lt Oshibuchi led three hikotais to engage US carrier aircraft, a total of 50 victories being claimed including one by Oshibuchi himself, and two by Ens Akio Matsuba. As the battle for Okinawa began, 343 Ku moved to Kanoya to provide cover for Tokko aircraft, then moving to Ohmura to intercept B-29s. On 24 July Oshibuchi led 21 Shiden-kais against a sweep by carrier aircraft during which he was shot down and killed, although his pilots managed to claim 16 victories. Thereafter the unit saw no further action before the war ended.

Sento 804th Hikotai (S804)

The unit formed on 15 March 1944 as part of 322 Ku with 24 night fighters under the command of Lt Hideo Kodama. Training began at Katori, but the arrival of the unit's Gekkos was delayed and in the event was too late to allow any part to be played in the fighting for the Marianas. With the disbandment of 322 Ku, S804 was transferred instead to 141 Ku, moving in August to Hsinchu via Kushira. By 1 September it had six aircraft at Hsinchu and 16 at Kanoya, while within a month its strength had risen to 32 Gekkos. On 13 October command changed to Lt Fumio Shigeta, while on 13 October when the unit moved to Nichols Field on Luzon, it was led to this new base by Lt Eiichi Kawabata.Until the opening stages of the Philippines invasion the unit was involved mainly upon reconnaissance, also joining with S901 in night convoy patrols and attacks on Tacloban. When S901 returned to the homeland, it operated instead in co-operation with S812. During this period Lt Kawabata led four aircraft to Cebu, undertaking night attacks on targets on Leyte. Lt Kawabata became the unit leader on 15 November at the same time as the unit was transferred from 141 Ku to 153 Ku. In January the hikotai withdrew to Japan, commencing recuperation at Fujieda where it was to become a single-engined night attack unit under the command of 131 Ku. When the fighting for Okinawa began it moved to Kanoya to take part in its new role, attacking airfields and ships on and around Okinawa by night. These operations continued until the end of the war, although Lt Kawabata was killed on 12 April 1945, Lt Sadahiko Ishida then becoming leader.

Sento 812th Hikotai (S812)

S812 was formed on 1 November 1944 from the night fighter unit of the 51st Air Flotilla with an establishment of 24 aircraft, and led by Lt Masashi Tokukura, an ex-floatplane pilot. On 15 November it was transferred to 153 Ku and moved to Nichols Field, Luzon, in the Philippines. Here, operating with S804, the unit engaged in night attacks on various targets, including torpedo boats. By 5 December it had 22 aircraft on hand, of which 12 were serviceable, but was exhausted by attacks on Mindoro Island and withdrew to the homeland, becoming part of the Fuyo Unit. Here Lt Tokukura and other pilots trained new pilots at Fujieda, sending them subsequently to Kyushu to take part in attacks on Okinawa.

Sento 851st Hikotai (S851)

On 10 July 1944 the unit was formed within the Yoko Ku with 24 night fighters. The first commander was Lt Atsushi Sugawara, who had gained fame as the back seat man for CPO Kudo at Rabaul. Part of 131 Ku, the unit sent detachments of a few Gekkos to Iwo Jima from 28 July. Each of these in turn undertook patrols, but one Gekko was shot down by day, ending such activity. In August the unit moved to Katori, and in mid

November to Chitose. Four aircraft were then sent to Etorofu Island for anti-submarine patrols. In January it was transferred to 153 Ku for service in the Philippines, but in fact only reached Formosa where it was transferred to the newly-formed 133 Ku on 15 February. From Takao it then engaged in night defence, searches and anti-submarine patrols, also becoming involved in the fighting for Okinawa, losing ten aircraft. On 15 June 133 Ku was disbanded and the unit was transferred to 132 Ku, but on 15 July S851 was also disbanded.

Sento 901st Hikotai (S901)

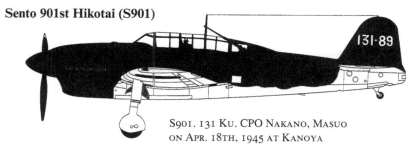

S901. 131 Ku. CPO Nakano, Masuo
on Apr. 18th, 1945 at Kanoya

The unit was formed at Truk on 1 April 1944 from the flying unit of 251 Ku, equipped with 24 Gekko night fighters led by Lt Atsushi Sugawara. Two or three Gekkos which had remained at Rabaul continued to operate until May when they returned to Truk, joining the main body of the unit. At Truk night operations against B-24s were undertaken. On 10 July 1944 251 Ku was disbanded and the unit was transferred to 153 Ku. Leadership also changed, Lt Tadashi Minobe who had been hikotai leader of 301 Ku

Officers of Sento 901, 153 Ku, at Nichols Field, Luzon, in late October 1944. Front row, 2nd from right, Lt Tadashi Minobe.

and who had been engaged in the night defence of Davao, taking over. In early September nine Gekkos and two Zero night-fighters were available. On 9 September when US carrier aircraft first attacked targets in southern Philippines, two out of three Gekkos involved in a search mission were lost. During mid September the unit moved to Nichols Field on Luzon, and from here on 21 September two Gekkos on a search and strike sortie attacked a carrier, CPO Saburo Sue (with Wt Off Takeaki Shimizu in the back seat) claimed to have achieved a hit on this vessel with one bomb. In mid October in support of Japanese landings at Leyte, the unit undertook attacks on torpedo boats (PT boats) claiming to have sunk six of these during a four-week period. The Gekko unit of S804 then arrived from Japan to take part in operations, but on 15 November S901 was transferred to 752 Ku in the Kanto area, returning to the homeland.

At Fujieda the unit received Suiseis and Zeros to train for night attacks by single-engined aircraft and became known as the Fuyo Unit. It now came under the control of 131 Ku with two other hikotais, continuing to undertake training. Lt Minobe who had proposed the methods of attack to be employed, became the hikocho of 131 Ku, and Lt Susumu Eguchi became leader of S901. In April 1945 it moved to Kanoya, engaging in night searches, attacks on airfields and on ships off Okinawa. In May, however, it moved back to Iwakawa in the mountain district of Miyazaki where it continued to operate until the end of the war with considerable losses.

Sento 902nd Hikotai (S902)

Formed on 1 April 1944 from the night fighter 381 Ku, the unit's initial establishment was 24 night fighters. The unit was led by Lt Hideo Matsumura, previously a float fighter pilot who had served with 934 Ku. As had been the case with 381 Ku, the unit's main base was Kendari, and detachments were sent to Surabaya, Makassar and Balikpapan. Fierce combats were fought over Balikpapan on four occasions against B-24s, the Gekkos also taking part in daytime interceptions. However, on 10 October escort fighters accompanied the bombers and S902 lost three aircraft and five members of the crews. Raids on Balikpapan then ceased and late in October the unit operated from Wasile, making night attacks on Morotai. From December until April 1945 anti-submarine patrols were undertaken, the unit claiming to have sunk four submarines. During March and April sorties were flown in support of the Army units withdrawing from Timor Island. It then gathered in all detachments to Surabaya from where the aircraft flew north to Ohmura in mid May. On 25 May it was transferred to 352 Ku and on 15 June a new leader, Lt Fujito Hoshiko, took over. Towards the end of June a detachment of six Gekkos led by Lt Koretake Ide moved to Tsuiki to intercept mine-laying B-29s. Five of these aircraft were claimed shot down, but in doing so all six Gekkos were lost. Meanwhile, the main body of the unit managed to claim a single B-29 before the war ended.

Aces

AN ALPHABETICAL LISTING OF BIOGRAPHICAL NOTES

CPO Ken-ichi Abe

Born in 1923 in a fishing village in Oita Prefecture, Abe graduated in October 1941, just before the outbreak of war from Otsu 9th Yokaren Class. Initially, he was trained as a dive-bomber pilot, but was then transferred to fighters. After completing advanced training in February 1942 at the Oita Kokutai, he was assigned to the Fighter Unit attached to the 22nd Air Flotilla Headquarters. With this unit he flew in South-East Asia, but had no opportunity to engage in aerial combat. In April the unit was transferred to the Kanoya Kokutai (later 253 Ku); in September he arrived with this unit at Rabaul. Abe's first combat experience was during an attack on Guadalcanal on 29 September. On that day PO3c Abe shared four victories (including a probable) with PO1c Tsumoru Ohkura. Abe continued to engage in combat from then on, in the Solomon Islands and over eastern New Guinea, adding further victories. On 1 April 1943 in a fight over the Russell Islands, he claimed two F4Fs shot down for his final victories. On 6 May he suffered injuries during a combat which included broken bones, as a result of which he was returned to Japan. Here the war ended before he had fully recovered. The total number of his claims were five confirmed shot down, two probables and five shared.

Cdr Takahide Aioi

Born in Hiroshima Prefecture in 1912, Aioi graduated from the Naval Academy as a member of the 59th class in 1931. He completed the 25th Aviation Cadet course in July 1934 and became a fighter pilot. After being posted to the Tateyama Ku, Ryujo, and then to Saeki Ku, he joined the 12th Ku in July 1938 and moved with it to Zhoushuizi. Shortly thereafter he returned to Ohmura. In August, as a result of the outbreak of the Shanghai Incident, he was sent to Kunda airfield. Flying a Type 95 fighter, he was engaged primarily in patrols and in supporting naval ground forces. In December he was promoted Lt and returned to the homeland as a buntai leader in the Kasumigaura Ku. However, during the following March he was transferred as a buntai leader to the 12th Ku again, and took part in combats over central China.

Aioi's first aerial combat occurred during an attack on Hankow in 29 April 1938. During the confused air fighting that ensued, he claimed to have shot down two I-15 fighters. Following this, during the attack on Nanchang on 26 June, he led three aircraft of his shotai away from the main force, but these pilots then found themselves surrounded by about 20 enemy fighters. After a hard struggle the three each shot down two of their attackers. Lt Aioi himself barely managed to escape, but was able to get back to Anking. After taking part in a number of air battles over Nanchang and Hankow, he was transferred to *Akagi* in December, again as a buntai leader. For a time in October

1939 he returned to the 12th Ku, serving with it in south China. In January 1940 he returned to *Akagi*. Postings followed to Oita and Yokosuka Kokutais, but the start of the Pacific War found him a hikotai leader on *Ryujo*. On 8 December he led a group of nine carrier fighters off Mindanao to escort torpedo aircraft attacking Davao, where no hostile opposition was encountered. In February 1942 he was transferred to the 3rd Ku as a hikotai leader and in November that year was promoted Lt Cdr. Until April 1943 he continued to serve in this role in what had become 202 Ku, and was greatly involved in the attacks on Darwin, and in the counter-offensive against Guadalcanal. In August 1944 he was posted to 601 Ku as hikotai leader and also commanded the unit on the ground. In October, during the Battle of Cape Engano, *Zuikaku* was sunk at a time when Aioi happened to be aboard, but he was rescued by a destroyer. Having been landed in the Philippines, he returned from there to Japan in the spring of 1945. When the war ended, he was executive officer of 343 Ku.

Renowned as a typical commanding officer of a naval fighter unit, subsequent to the war he joined the Maritime Self-Defence Force where he rose to the rank of admiral. After serving as OC, Self-Defence Fleet, he retired, dying on 6 February 1993. He has been credited with ten victories.

Lt(jg) Sadaaki Akamatsu

Akamatsu was born in 1910 in Kochi Prefecture, the son of the head of a weather observatory station. Following graduation from Kainan High School, he volunteered for the Navy and entered Sasebo Naval Barracks in 1928. He is among the veteran fighter pilots who graduated in March 1932 from the 17th Pilot Training Class.

After serving on *Akagi*, *Ryujo* and *Kaga*, as well as in the Yokosuka and Ohmura Kokutais, Akamatsu was assigned to the 13th Ku in December 1937 following the outbreak of the China Incident. At the time he engaged in air

battles over central China he was already 27 years of age and a senior PO1c aviator. He was known for his bravery in action, such as the occasion when he claimed four enemy aircraft shot down in one sortie during an attack on Nanchang on 25 February 1938. By the time of his transfer to serve on board *Soryu*, he had a record of 11 aircraft destroyed.

Promoted Wt Off in April 1941, Akamatsu was posted to the 3rd Ku during World War II and served in the Philippines and Dutch East Indies. In May 1942 he returned to the homeland where later, in July 1943, he was transferred to 331 Ku and took part in the December attack on Calcutta. Soon after this he again returned to Japan where until the end of the war he served with 302 Ku at Atsugi, piloting a J2M Raiden fighter with distinction, taking part both in the defence of the nation's capital and, for a brief period, in the Okinawa Campaign.

With a background of 14 years in fighter aviation and a flight time exceeding 6,000 hours, Akamatsu was well-known amongst his colleagues for both his heroic acts and for his eccentricities. Although he himself claimed a score of 350 enemy aircraft shot down, it is judged that the actual number achieved was around 27.

Wt Off Tomita Atake

Completing the 47th Pilot Training Class in October 1939, Atake was posted to Chitose Ku the following year. Immediately before the outbreak of war, he was assigned to air defence duties in the Marshall Islands. During the early dawn of 1 February 1942, PO3c Atake intercepted a raid by US carrier aircraft. Of two A5Ms which were the first to take off to intercept, Lt Kurakane's aircraft was hit and he had to ditch. Piloting the other, Atake flew at very low altitude to challenge the enemy aircraft to combat despite his own disadvantageous position. However, one of the ailerons on his aircraft was damaged when it struck the wing of an F4F, and he was forced to land at once.

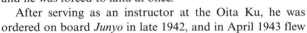

After serving as an instructor at the Oita Ku, he was ordered on board *Junyo* in late 1942, and in April 1943 flew down to Rabaul. During an attack on a large formation of enemy aircraft off Guadalcanal on 7th, he claimed two Grummans shot down (one as a probable). In mid-July his unit flew down to Buin, but just before landing he was bounced by enemy fighters and his A6M was shot up, obliging him to ditch; he was rescued by the crew of a destroyer. He continued to take part in operations over the Solomons until the end of August, while in September he was transferred to 204 Ku. Before returning to Japan at the end of the year, Atake claimed five aircraft shot down while attached to 204 Ku. In February 1944 he was posted to the Yokosuka Ku and in June moved to Iwo Jima. During a large-scale air battle here on 24 June his aircraft was again shot up and he made an emergency landing. Wounded, he was sent back to the homeland. On recovery, he was posted to 1001 Ku, a transport unit, but he ended the war with S313 at Konoike.

In total he claimed ten victories.

PO2c Takao Banno

Born in Aichi Prefecture in 1921, Banno entered the Navy in 1939; he became a pilot, graduating top of the 53rd Pilot Course in August 1940. He was assigned to *Junyo*, going with this vessel to the Solomons. Here on 17 October 1942 he shared two F4Fs destroyed with Lt Yoshio Shiga and a third pilot during his first air combat. He remained with the carrier during the Wewak operations of January 1943, and then took part in the 'I-go' operation of April, when his unit flew from land bases. *Junyo*'s fighter group continued to take part in the central Solomons fighting until the end of August when it was transferred to 204 Ku. Banno claimed one fighter and one dive-bomber shot down over Buin on

16 September, and as an element leader had claimed five destroyed by 3 October. During the morning of 7 October he took part in an attack on a torpedo boat base on Vella Lavella, where he claimed two P-38s shot down in quick succession. However, while chasing a third out to sea he was shot down in flames, having been the only member of his unit to achieve any success during the attack on this day. With 204 Ku he had claimed seven destroyed and one probable, and while with *Junyo* he had shared in 25 victories and nine probables with other members of the unit; more than ten of his victories were officially recognised.

Ens Jiro Chono

Born in Ehima Prefecture in 1907, Chono joined the Navy in 1927, graduating from the 15th Pilot Training Class in April 1930 as a fighter pilot. When the China Incident broke out he was an element leader with the *Kaga* fighter unit. On 11 November 1937 he and PO3c Hidaka pursued three Northrop bombers, claiming one shot down each. On 13 April 1938 during a raid on Canton the lead fighter element turned back, but PO1c Chono led two A5Ms and two Type 95 fighters to continue providing cover for 18 dive-bombers, becoming engaged against more than 20 Chinese fighters. In this fight Chono claimed two shot down, the pilots with him claiming a further four. As leader of the second element escorting six dive-bombers in an attack on Nanxiong on 30 August, the Japanese fighters again found themselves up against about 20 interceptors. This time Chono was able to claim three of these shot down and a fourth probable despite his own aircraft suffering 15 hits. Returning to the homeland later that year, he was promoted Wt Off, but returned to China during 1940 to fly with 14 Ku. On 21 February 1941 his aircraft was hit by AA over Kunming and he was killed. His official score at this time was seven.

Wt Off Tomokazu Ema

Born in Nara in 1912, Ema claimed two victories over China whilst serving with the 14th Ku. Following the outbreak of the Pacific War he was posted to the newly-formed 6th Ku, being sent to Rabaul. From here he claimed an F4F probably shot down over Guadalcanal, while on 2 October he claimed to have caused two of these fighters to collide with each other and crash, also claiming two shared victories. Subsequently he was transferred to 254 Ku in China where he claimed a probable victory over a P-38 over Sanya on 29 July 1944. As the war neared the Philippines, the kokutai sent a detachment to Formosa, and thence to Luzon. Ema collected a new aircraft from the Suzuka Plant in which he later intercepted US carrier aircraft over Oroku, Okinawa, on 10 October. He then joined the detachment at Nichols Field, Luzon, where he claimed one aircraft shot down and one probable to raise his total to eight victories. He was shot down and killed over Manila on 29 October 1944.

PO1c Masuaki Endo

Graduating with the Otsu 9th Yokaren Class in October 1941, just before the outbreak of hostilities, Endo was posted to the Tainan Ku in February of the following year. In April he served at Rabaul and Lae, where he took part in the air battles over eastern New Guinea and the Solomons. Surviving these, he returned to Japan with the unit in November 1942. In May 1943 he joined 251 Ku which was heading to the south-east area of operations. Based on Rabaul, he then took part in many missions in the area. On 7 June during an air battle in which 36 fighters under the command of Lt Ichiro Mukai undertook an attack on the Russell Islands, Endo was seen to shoot down a P-38, but his aircraft was then shot up by a P-39. Apparently realising the hopelessness of his position, he rammed the P-39 and fell to his death. He had claimed 14 victories.

Cdr Sachio Endo

Endo became notable as the "King of the B-29 Killers" while operating against the Superfortresses raiding the Japanese homeland. Born in a remote mountain village in Yamagata Prefecture in 1915, Endo enlisted in the Yokosuka Ku as a member of the 1st Yokaren Class in June 1930. After completing the Flight Training Course at Kasumigaura, he graduated in August 1933. Trained as a torpedo pilot at the Tateyama Ku, he then served aboard *Akagi* and with the Kasumigaura and Ohmura Kokutais. In December 1937 he was transferred to *Soryu*, and during the spring of 1938 was sent to central China, joining a unit being despatched to Nanking.

For about two months he was involved in supporting land operations along the banks of the Yangtze river. Then in December 1939 he was assigned to the Hyakurigahara Ku as an instructor; he later undertook similar duty with the Tateyama and Ohmura Kokutais. At this time he was promoted Wt Off in May 1940, and then to Ensign in November 1942 and the following year, again in November, to Lt(jg).

From May 1942 Endo was assigned to Kasumigaura to teach in the Yokaren programme, but in January 1943 was posted to 251 Ku where his role changed to that of piloting Type 2 land reconnaissance aircraft. He was given the task of testing the first of these aircraft to be equipped with obliquely-angled cannon, a system devised by the commanding officer, Kozono, for use of the aircraft as a night fighter. Confident of its success, Endo took the aircraft to Rabaul in May and undertook a number of night interceptions. During these engagements all the success was achieved by PO1c Kudo and Wt Off Satoru Ono, while Ens Endo was wounded in action and evacuated to Japan.

With the formation of 302 Ku in March 1944, Endo served as a buntai leader in this Gekko-equipped night fighter unit. Training was conducted at Atsugi airfield, and in June when B-29s began to undertake raids from China on the northern Kyushu areas,

Lt(jg) Endo was detached to 352 Ku at Ohmura. Here he was put in charge of six aircraft and eight Gekko flight crews. On 20 August he single-handedly pursued the departing B-29s following a raid, claiming two destroyed, one probable and two damaged, following which he made an emergency landing at Jeju Do. During a fight with B-29s raiding Ohmura on 25 October, he chased one damaged bomber for a long period, finally bringing it down.

In November he was recalled to Atsugi, from here moving with three of his crews to Hachijo Jima, but from here they soon returned to Atsugi. From mid-December onwards he was active in the night interception of B-29s attacking the Kanto and Tokai areas, where he gained further successes. On the afternoon of 14 January 1945 he shot down one aircraft and damaged a second during a daylight interception over Enshu Nada, but his own aircraft was hit and set on fire. Endo first ordered his observer, CPO Ozaki, to bale out (he was killed); he then flew the burning aircraft over Atsumi Peninsula. Here he attempted to bale out from an altitude of about 300 metres (900 ft), but he died as a result of the severe burns he had suffered. Including the aircraft he had accounted for on this date, his score stood at eight destroyed and eight damaged, all of them B-29s.

During the previous year Endo had received words of praise from the commanding general of the western area, as well as a commendation from the commander-in-chief, Sasebo Naval Station for his exploits over Kyushu. After his death a further commendation was issued by the admiral commanding the Yokosuka Naval Station and a promulgation issued by the C-in-C, Supreme Defence Command. He was also posthumously promoted two ranks to commander.

Lt Cdr Iyozo Fujita

Born in Shandong Sheng, China, Fujita attended Kitsuki High School in Oita Prefecture, then graduating with the 66th Naval Academy class in September 1938. In June 1940 he completed the 33rd Flight Training Course, and in September of the following year was ordered to serve aboard *Soryu*. His first taste of combat occurred on 7 December 1941 during the attack on Pearl Harbor, when he served as a shotai leader in the second wave fighter escort. The buntai leader, Lt Fusata Iida, was shot down and killed by anti-aircraft fire while strafing Kaneohe airfield. Lt(jg)  Fujita rounded-up the remaining fighters and headed for Point Kaena, the rendezvous for the return to the carriers. On the way, however, his group encountered a formation of P-36 fighters that included Lt Gordon H. Sterling Jr, and a dogfight ensued. After claiming one victory, Fujita's Zero was hit and he was barely able to make it back to the carrier, where a cylinder fell out of the engine.

Following his return to the homeland, he was promoted Lt, then taking part in the attacks on Darwin and on Ceylon. In June 1942 he was involved in the Battle of Midway, engaging in one of the combat air patrols. During the first interception he dived into a formation of B-26 bombers, claiming three shot down (two of them shared with other pilots). After returning to the carrier and refuelling, he took off again, this time claiming four torpedo-bombers (three shared) and three fighters (two shared). However, his own aircraft was hit in the fuselage fuel tank by anti-aircraft fire from the Japanese ships, and it caught fire. He managed to bale out into the sea from a height of only 200 metres. After four hours drifting, he was picked up by the destroyer *Nowak*, with which he returned to Japan.

He was then posted to serve as a buntai leader on board *Hiyo*, and during October undertook sorties over the Solomon Islands. Due to a mechanical failure with the carrier, the greater part of the air group was ordered to fly to Rabaul on 20th, and then to Buin. Until mid December the fighter unit from *Hiyo* was engaged in air battles around Guadalcanal where he claimed two aircraft shot down. Remaining at Rabaul for less than a month to take part in the 'I-go' operation in April 1943, he was then transferred to the Tsuiki Ku in June. In November he was nominated as hikotai leader for the newly-formed 301 Ku. During June and July 1944 this unit took part in the air defence of Iwo Jima. However, during the latter month he was transferred to become hikotai leader of S402, equipped with the new Shiden fighters. Consolidated into 341 Ku, the unit was active during air battles off Formosa in October. After hard fighting in the Philippines, the unit, having lost much of its strength, was withdrawn to Japan in January 1945, where Fujita ended the war at the Fukuchiyama base. His claims for ten shot down in one day during the Midway battle (seven of which were shared) is the highest recorded, together with those of Okumura and Okabe. After the war he returned to flying and for many years was a pilot with Japan Air Lines; he died on 1 December 2006.

Lt Cdr Sumio Fukuda

Born in Tokushima Prefecture in 1919, Fukuda graduated with the 69th class of the Naval Academy in March 1941. In February 1943 he completed the 37th Aviation Student Course and was immediately ordered aboard the carrier *Junyo*. As early as April, he had arrived at Rabaul to take part in the 'I-go' operation. The attack on Guadalcanal on 7th was his first combat sortie; however, he did not have the opportunity to engage in aerial combat. He next took part in the attack on Oro Bay on 11th and the attack on Port Moresby on 12th, following which he returned to Japan. During mid July, the *Junyo* fighter unit was sent to the base at Buin, taking part in the ferocious, almost daily, air battles there. On 1 September all pilots with this unit were transferred to 204 Ku, including Lt(jg) Fukuda. By this time nearly all the officer pilots of 204 Ku had been killed in the constant air battles and in consquence Fukuda led the flying unit for about two months until his transfer to 252 Ku on 4 November. During this period he led interceptions over Rabaul and Buin, personally claiming six aircraft shot down (two of them as probables). Following his transfer to 252 Ku he was involved in the air defence of Taroa in the Marshall Islands. On 25 November he led a force of 24 Zeros on a strafing and bombing attack on Makin Island. En route, however, his unit was intercepted by about 50 F6Fs, 11 of which were claimed shot down (four as probables). Wt Off Kojima and six aircraft were lost and the attack was abandoned, the unit returning to base. Beginning during the latter part of December, B-24s commenced continuous bombing attacks on Taroa to which the locally stationed 252 Ku pilots, led by Fukuda, responded by making close range attacks on the raiders. By using repeated vertical attacks, they were able to claim about 50 of these aircraft shot down during a period of one month. The majority of the unit's equipment and supplies had been lost by the end of January 1944, however, and the remaining pilots were evacuated to Truk by transport aircraft, eventually reaching Japan.

In March 1944 Fukuda was promoted Lt and assigned as an instructor to Konoike Ku. In July, following the conclusion of the Marianas Campaign, he was posted to command the newly-formed S163, 634 Ku. He undertook training activities at Tokushima, but due to the commencement of the 'Sho-1-go' operation during October 1944, the unit was ordered to Clark Field in the Philippines, flying there via Formosa and Okinawa. On 24 October, the day of the general offensive, Fukuda led all the serviceable fighters of 634 Ku aloft for an attack on the US carrier force which had been located east of Luzon. On the way to this target the formation was intercepted by a large force of American fighters and Lt Fukuda failed to return, being deemed killed in action. The total number of aircraft which he was credited with shooting down was 11.

Lt Yoshi-o Fukui

Fukui was born in Kagawa Prefecture in 1913 and enlisted at the Sasebo Barracks during 1931. Initially he served as a maintenance man, but subsequently graduated from the 26th Pilot Training Class in 1935, then becoming a fighter pilot. Following service with both the Ohmura and Kanoya Kokutais, he was posted on board *Ryujo* at the outbreak of the China Incident. In August 1937 the carrier arrived off Shanghai but was primarily involved in providing air support to naval surface actions. On 22nd of that month when he was flying as No. 2 in Lt(jg) Kaneko's four-aircraft flight on a patrol over Paoshan, 18 Curtiss Hawks were encountered. A fight immediately ensued, and eight of the Chinese fighters were claimed shot down, three of them by Fukui personally.

The carrier next moved to the south China coast from where, during an attack on Canton on 21 September, PO2c Fukui's acute vision spotted a group of hostile fighters; he at once dived into this group, claiming two shot down. In December 1937 he was transferred to the Yokosuka Ku, but during the following April was again assigned to carrier duty, this time joining *Kaga*. Until late in the autumn he was engaged primarily in action over southern China, and during this period he managed to surprise and despatch a further Curtiss Hawk during an attack on Liuchow on 13 September 1938. He also undertook many strafing attacks on ground targets at this time. In December 1938 he was transferred to *Akagi*, but then later served as an instructor at the Yatabe, Kanoya, Hyakurigahara and Oita Kokutais, also being promoted Wt Off during April 1941.

After a long period off operations, he was posted to *Zuiho* in June 1943, and by the start of November had reached Rabaul. On 2 November, the day after his arrival, Rabaul suffered a major air raid by Allied fighters and bombers. More than 100 Japanese fighters were launched to oppose this attack. Ens Fukui claimed a B-25 shot down at low level just above a palm grove, but he was then attacked by a number of P-38s, his aircraft being hit and set afire. He managed to bale out, but suffered burns to his right foot in doing so. Nevertheless, he was back in action three days later, and continued to take part in a number of interceptions and offensive sorties until he was posted back to Truk.

During February 1944 he joined 601 Ku, going aboard *Zuikaku*, taking part in the 'A-go' operation. On 19 June he served as a shotai leader in the escort for the first wave of attackers heading for the US task force. During this sortie he claimed one probable victory. Next day he was undertaking an evening patrol when he intercepted an American torpedo aircraft; following this engagement he made a successful night landing back on his carrier. In July 1944 he was posted to S165, 653 Ku, and with this unit took part in the sea battle off the Philippines, taking off from *Zuikaku* to take part in an attack on the US task force. Following this attack, he returned to his carrier to find it had been sunk. He force-landed in the sea from where he was rescued by a destroyer. In November he was again posted, this time to S304, moving with this unit to Bamban in the Philippines. In January 1945 he was withdrawn to the homeland where he ended the war attached to 352 Ku. His flying career had thus lasted for more than ten years; he was considered a highly-skilled master pilot with a fine character as well as being a well-seasoned combat pilot. His total of claims amounted to 11, four of which were probables.

PO1c Kurakazu Goto

Born in 1921 in Fukuoka Prefecture, he completed his training as a Ko 6th student and was posted to 202 Ku in the Netherlands East Indies. His first victory was over a B-24 on 15 February 1943, also claiming a second damaged during the same encounter, which was on his 15th operational sortie. Between March and May he took part in three escort missions to Australia, while on 15 May he claimed another B-24 over the Kai Islands. Two days later he shared with three other pilots in claiming three Australian Beaufighters probably shot down over Tual in the Kai Islands. He then took part in two more missions over Australia during late June, whilst on 7 September he claimed three Spitfires shot down and a fourth as a probable. Two days later he failed to return from a sortie over Merauke, central New Guinea. He had been credited with eight victories at the time of his death.

Lt(jg) Matsuo Hagiri

Known as 'Mustachio Hagiri', he exhibited a very good head for detail and consequently spent a long time with the Yokosuka Ku testing new aircraft and weapons. His combat time was, as a result, comparatively brief. Possessed of a powerful physique, on one occasion he underwent a special examination of his heart after recording 9.5 Gs in a diving experiment; at the time this broke all records.

Born in Shizuoka Prefecture in 1913, Hagiri commenced work as an engineer before transferring to aviation. He graduated from the 28th Pilot Training Class in August 1935, when he was posted to the Yokosuka Ku. During

August 1940 he was transferred to the 12th Ku, flying a new Zero fighter to Hankow to join this unit. On 19 August he took part in the first operation by a Zero-equipped fighter unit, with Chungking the target; on this occasion no enemy aircraft were sighted. On 4 October Chengtu was raided, and after one aerial victory had been claimed, some of the fighter pilots undertook a pre-planned stunt. Hagiri, Higashiyama, Nakase and Oh-ishi landed on Taipingsze, Chengtu airfield and set fire to whatever could be found. The four then took off again, Hagiri then single-handedly attacked three Chinese fighters which he spotted flying in formation, claiming two of these shot down. Another attack on Chengtu in March 1941 allowed him to claim three Polikarpov I-16bis biplane fighters. With the arrival of the summer, he returned to Japan where he was promoted Wt Off and posted to the Yokosuka Ku again. In July 1943 he was transferred to 204 Ku, then becoming involved in the fighting over the Solomons. Two months later, on 24 September, he claimed two aircraft shot down during an encounter over Buin, but his own aircraft was hit and he was severely wounded. Evacuated to Japan, he was then assigned to operational testing and air defence duties with the Yokosuka Ku until the end of the war. During an interception of B-29s in April 1945, he was again wounded. At the close of hostilities he had claimed 13 victories.

Lt(jg) Watari Handa

Born in Fukuoka Prefecture in 1911, Handa entered the navy in 1928, serving as a seaman for the next five years. In March 1933 he graduated from the 19th Pilot Training Class and was assigned to the carrier *Ryujo*, subsequently serving with the Ohmura and Yokosuka Kokutais. Following the outbreak of the China Incident, he was posted to *Kaga*'s air group in August 1937, operating over the front at Shanghai. By this time he was already a 26 year-old PO1c when he claimed his first victory on 7 September. Flying as No 2 to Lt Chikamasa Igarashi in a three-aircraft section, they were escorting a torpedo-bomber unit when over Tahu they were engaged by seven Curtiss Hawks, one of which Handa claimed to have shot down. During a raid on Nanking on 29 September he was able to claim a further three victories. In June 1938 he was transferred to the 15th Ku, taking part in air battles over Nanchang, and by the time he returned to Japan in November he had raised his total to six. In November 1940 he was promoted Wt Off and released to the Reserve, but was remobilized on the same day. In February 1942, following the outbreak of the Pacific War, he was posted to the Tainan Ku, putting his earlier experience to good use during operations over the Dutch East Indies, Rabaul and New Guinea. He had contracted tuberculosis by this time, however, and as this was worsening, he was evacuated to the homeland at the end of the year, his total of claims having risen to 13. After struggling against his disease for six years, he died during 1948.

Lt(jg) Kaname Harada

Born in Nagano Prefecture in 1916, Harada graduated from the 35th Pilot Training Class in February 1937. After initial service with the Saeki Ku, he joined the 12th Kokutai in October, being posted to central China. Here there were no opportunities for aerial combat, however, and he returned to Japan in January 1938. He then served as an instructor with the Saeki, Tsukuba, Hyakurigahara and Oita Kokutais until September 1931 when he was posted to the carrier *Soryu*. During the attack on Pearl Harbor he was assigned to patrol duties over the carriers, thus seeing none of the action. However, during the attack on Colombo, Ceylon, on 5 April 1942, he was able to claim three British fighters shot down and two more as probables as they were rising to intercept. On 9 April Nagumo Force was attacked by Blenheim bombers, two of which he shared in shooting down. He next saw action during the Battle of Midway, again tasked with combat air patrol. On this occasion during the course of three interceptions, he claimed three US torpedo-bombers shot down (two of them shared). However, *Soryu* was sunk and he had to ditch in the sea, being picked up by a destroyer. In July he was posted to *Hiyo*, and early in October left Japan with this unit to take part in the battle for Guadalcanal. On 17 October while escorting torpedo aircraft to attack targets at the island, his formation was attacked from above by a force of F4Fs, and while he was able to claim one probably shot down, he was hit and badly wounded. He managed to make a forced landing close to Rekata floatplane base, following which he was evacuated by hospital ship to Japan. His wounds to his left shoulder and arm were severe, and he had not fully recovered when the war ended. He was credited with nine confirmed victories. During 2002, now the principal of an infants' school, he and his wife travelled to the United Kingdom to meet Lt Cdr John Sykes, a Royal Navy Fulmar pilot who had probably been one of his victims during the Colombo raid of 5 April 1942. At the time of writing he was still alive, and had recently visited the USA as one of the survivors of the Midway Battle.

Wt Off Yoshiro Hashiguchi

Born in Fukuoka City in 1918, Hashiguchi enlisted in the navy in 1937, subsequently graduating from the 42nd Pilot Training Class in September 1938. Following service with the Saeki, Oita and Ohmura Kokutais, he was posted to the 12th Ku, joining this unit in central China in June 1939. Wounded on the ground during a Chinese bombing raid in October, he required two months of hospitalization and treatment to achieve a complete recovery. In January 1940 he returned to the Suzuka Ku to serve as an instructor until November 1941, when he was transferred to the 3rd Ku (later 202 Ku). Following the outbreak of the Pacific War, he saw action over the Philippines and Dutch East Indies. In the latter area on 9 February 1942 he and another pilot strafed a radio station at Ile Yande, which they set on fire. However he was hit in the right thigh by a bullet from the ground. Staunching the flow of blood, he nevertheless completed his assigned reconnaissance mission before returning to base. From April until mid year he took part in raids on Darwin, but in September, now a PO2c, he moved with the unit to Rabaul. For the next two months he took part in attacks on Guadalcanal. On 18 October he led a shotai

Hashiguchi's A6M with victory symbols on the fin.

of three which became involved in a fierce combat, the trio of pilots claiming five aircraft shot down plus two probables. In November the unit returned to the south-western areas to undertake further attacks on Darwin. During these he flew as wingman to succeeding hikotai leaders until June 1943, when he returned to Japan, to join the Oita Ku.

During December 1943 he became a carrier fighter pilot and conducted training as the senior NCO pilot aboard *Shokaku* with that carrier's 601 Ku. In June 1944 he took part in the 'A-go' operation, which he survived. The following month he was posted to S164, 653 Ku and in October was involved in the 'Sho-go' actions, making his way via land bases to Okinawa. Due to a technical malfunction with his aircraft, he was sent instead aboard the carrier *Chiyoda*, from which he made a number of sorties during the Battle of Cape Egano. During this action the carrier was sunk on 25 October and he was reported missing in action, presumed killed. By this time he had been credited with more than ten victories.

Ldg Sea Kazuo Hattori

Born in Aichi Prefecture during 1921, he graduated from Hei 10th and was assigned to 201 Ku for service in the Solomons area. He first engaged in combat over Buin on 12 August 1943, but was wounded in a fight in which the 15 pilots of his unit claimed 17 victories. Between 4 September-29 October he took part in 20 engagements, although details of his claims at this time are not known. On 2 November he claimed a P-38 shot-down and probably a second over Rabaul. As No.3 in Lt(jg) Yoshio Ohba's section he claimed steadily until Ohba's death on 23 December. He continued to operate over Rabaul until 2 January 1944, now flying in Wt Off Masao Taniguchi's section alongside aces PO1c Shinsaku Tanaka and SupSea Hiroshi Inagaki. Following a return to Japan, he then accompanied the unit to Palau, but on 30 March 1944 he failed to return from a fight with US carrier aircraft. His claims were recorded as eight destroyed, three probables and one shared.

Lt(jg) Hatsu-o Hidaka

Hidaka was born in Kagoshima Prefecture in 1915; after enlisting in the navy he graduated from the 24th Pilot Training Class in July 1934 to become a fighter pilot. In November 1935 he was assigned to *Kaga* on the outbreak of the China Incident, the carrier moving to operate off the coast near Shanghai. His first claim was made on 11 November 1937 when he pursued three Northrop bombers, shooting down one. During a raid on Canton on 13 April 1938 he made two further claims, but in June was transferred to the 15th Ku. With this unit he

took part in a raid on Nanchang on 18 July, claiming two more victories during the mission. He returned to Japan in December, then becoming an instructor successively with the Kasumigaura, Kure and Ohmura Kokutais. In October 1942 he was promoted Wt Off and posted to 204 Ku. Towards the end of December he moved with the unit to Buin in the Solomons. Led by Lt Miyano, he undertook numerous sorties until hospitalized due to malignant malaria. This resulted in his evacuation to the homeland where he served as an instructor at Yatabe Ku until the end of the war. He had claimed 11 victories.

Lt Ichiro Higashiyama

Born in Nagano Prefecture in 1915, Higashiyama entered the Yokosuka Kokutai as a student in the Otsu 2nd Yokaren Class. Graduating in the top place from a Flight Training Course in April 1935, he received an Imperial Award and became a fighter pilot. Following service with the Tateyama Ku, he was posted to the Yokosuka Ku where he remained for two years. Then, in December 1937, he was sent to serve aboard *Soryu*. In the following May he joined the Nanking detachment and undertook operations over central China. In July, however, he was transferred to the 15th Ku at Anking under the command of Lt Nango. Here he took part in the support of naval ground forces, in attacks on Nanchang, and in the interception of Russian-built Tupolev SB bombers.

His first combat occurred on 28 June when the shotai of which he formed a part, intercepted three SBs and shot down two of them. On 14 July, whilst engaged in the interception of nine more of these bombers, he and PO3c Yoshiharu Matsumoto claimed three shot down. Then during an attack on Nanchang on 18th of the month while flying as No.2 to Lt Nango, he became involved in a very fierce combat during which he claimed two Chinese aircraft shot down. However, Nango's aircraft collided with one of the opponents, and Higashiyama witnessed his commander's fighter crashing into a lake.

In November 1938 he returned to the Yokosuka Ku where he was involved in the testing of newly developed and more powerful aircraft, but in August 1939 he was involved in an aerial accident which resulted in him suffering serious injuries including a fractured skull. He was to make a full recovery, and in May 1940 was promoted Wt Off and posted to 12th Ku, flying out one of the first new A6Ms to this unit's base at Hankow. He took part in the first raid by these new fighters on Chungking on 19 August, although no opposition was met.

During the third attack on Chengtu on 4 October he flew as 3rd shotai leader. During the previous evening he, Hagiri, Nakase and Hideo Oh-ishi had hatched a plan to land on the enemy airfield there and try and set fire to whatever offered itself. On the day in question, Chinese aircraft on the ground at Taipingsze, Chengtu airfield had already been set aflame by strafing when the quartet landed as planned, Higashiyama and Nakase setting the control centre on fire. Having taken off again, he and PO3c Iki Arita then attacked a formation of SBs, sharing in the claimed destruction of three of these. No losses or damage was suffered by 12th Ku during these actions.

In March 1941 Higashiyama once again returned to the Yokosuka Ku to resume test flying. Referred to as 'The Master of Yokosuka Ku', he was next ordered to serve as a buntai leader in the newly-formed 261 Ku. In February 1944 he accompanied this unit to Saipan. From here on 30 March in response to a raid on Palau by a US task force, 32 261 Ku fighters led by Lt Ibusuki escorted a group of dive-bombers to a forward base on Peleliu Island. Next day, however, American carrier fighters attacked again. The Zeros which were scrambled to intercept them were still en route to their rendezvous point when attacked by F6Fs, Japanese losses amounting almost to annihilation. Lt(jg) Higashiyama fought desperately, claiming three shot down, but his own aircraft was hit and set on fire. He managed to bale out, but suffered severe burns. He was transferred to hospital in Saipan, but was still there when US forces invaded, nothing more being heard; he was deemed to have died fighting the invaders about 8 July. Considered to exhibit both an outstanding character and skills to match, he was certainly a superior pilot and an excellent example of graduates from the Yokaren system. His confirmed score was nine.

Ens Minoru Honda

Born in Kumamoto Prefecture in 1923, Honda graduated from the 5th Yokaren Course in January 1942 just after the outbreak of the Pacific War. Although assigned to the Fighter Unit attached to 22nd Air Fleet HQ, he had already missed any aerial combat in the southern area. In April however, he was transferred to the Kanoya Ku (later renamed 253 Ku) and moved to various bases in the south-west area. In September he arrived at Rabaul, later moving to Kavieng, taking part in air battles over New Guinea and the Solomons until May 1943. Transferred to S407 in April 1944, he accompanied this unit to the Philippines in October. Following withdrawal from the area to Japan, he aided his hikotai leader, Lt Hayashi, in rebuilding the unit which became part of 343 Ku. He remained active on home defence until the end of the war, by which time he had claimed 17 victories. He later became a test pilot, testing the MU-2 aircraft.

Wt Off Mitsu-o Hori

Born in Gifu Prefecture in 1921, he completed the Flight Training Course for the Otsu 10th Yokaren Class in March 1942. Posted to the Tainan Ku at Rabaul, he found the unit contained a large number of veteran pilots as a result of which there were few opportunities for him to take part in offensive sorties, although he was enabled to undertake some interceptions. In November he was transferred to 582 Ku and became engaged in south-east air operations. On 7 January 1943 whilst escorting a convoy near Lae, he was engaged by some P-38s and was shot down, baling out. Although wounded, he was successfully rescued and was evacuated to Japan. After a spell serving with the Ohmura Ku, he became an instructor in November at the Sanya Ku at

Hainan Tao. On 15 April 1944 a Zero fighter unit was formed from instructors at the Sanya and Haiko Kokutais and raided Nanning, China. The raiders were bounced by P-40s and suffered heavy casualties. After he had claimed one victory, CPO Hori's aircraft was hit and he carried out an emergency landing on Hainan Island. In June of that year he transferred to the 2nd Takao Ku, and while training new pilots, also took part in air defence battles off the coast of Formosa. As the unit's senior NCO pilot, Hori was constantly engaged in operations over western Japan as well as above the Okinawa area which he continued to undertake until the end of the war, by which time he had claimed 11 victories. After the war he joined All Nippon Airways, becoming an aircraft captain; he changed his name to Mikami during this period.

PO1c Matao Ichioka

Born in 1925 in Gifu Prefecture, Ichioka was a member of the Hei 12th Yokaren Class, graduating from the 28th Flight Training Centre in July 1943. At the end of September he was posted to 204 Ku at Rabaul where he claimed his first victory on 25 December during an interception. Despite his youth, he continued to operate almost daily, claiming 13 victories (five of which were probables or shared) during a one month period in January 1944. Following the move of the main part of 204 Ku to Truk at the end of January, he was transfered to 253 Ku at Tobera (Rabaul) from where he continued to operate until 19 April 1944 when he was killed in action.

Lt Cdr Masao Iizuka

Masao Iizuka was born in 1915 in Tochigi, graduating with the 34th Flying Students in April 1941 when he was assigned to the carrier *Shokaku*. During the Pearl Harbor attack on 7 December 1941 he was involved only in defensive CAP duties, but subsequently in April 1942 he became a buntai leader. On the morning of 5 June 1942 he led nine fighters in the first strike on Midway, sharing with the pilots of these in claiming 12 defending aircraft shot down. He moved in July 1942 to become buntai leader on *Ryujo*. From this ship he led 12 Zeros on defensive patrol on 24 August,

personally claiming one of ten torpedo-bombers shot down and a second probably so. These victories did not, however, prevent the parent ship from being sunk. He was then posted to 253 Ku, serving in the Solomons and New Guinea from November 1942 until May 1943, when he again returned to Japan. During this period he shared three victories with five other pilots off Lunga Point on 12 November, and four more on 14 November, again with five others in a battle with 33 US aircraft. Over friendly shipping sailing to Lae on 8 January 1943 he and eight others claimed seven victories, while on 11 March he led a force of 18 A6Ms to escort IJAAF heavy bombers, 13 victories being claimed during this operation. He led 26 fighters to engage raiders over Russell Island on 1 April, and during a 30-minute combat, 23 shot down plus five probables were claimed. Following his return home, he became hikotai leader of Sento 302, taking part in the air battles of October 1944. On 15th of that month he led six Zeros to escort Gingas seeking to attack the US carrier force, but these were themselves attacked by F6Fs and Iizuka was one of those who did not return. He had been credited with eight victories.

CPO Teigo Ishida

Ishida completed training with Hei 6th in November 1942, then joining 202 Ku. He undertook his first sorties in late January 1943, but did not experience his initial combat until 7 September when engaged in an escort mission over north-western Australia. Here he claimed three Spitfires shot down. He then formed part of a group led by Wt Off Yamanaka which was transferred to 204 Ku in the Solomons. He took part in attacks off Bougainville where he claimed one victory on 17 December, adding further successes on 23rd and 24th while flying as wingman to CPO Ogiya. He claimed three more on 17 January, but was himself shot down, baling out of his burning Zero. After landing, he managed to make his way to Buin, but as he did so he encountered two shot-down US pilots, both of whom he knocked unconscious. Following a return to Japan, he joined 332 Ku, later moving to Luzon with a detachment of this unit. When no aircraft remained, he was evacuated home, this time joining S407 in 343 Ku during spring 1945. On his first engagement with this unit over Kikai on 16 April 1945, he failed to return; by this time he had claimed nine victories.

Ens Susumi Ishihara

Born in 1921 in Aichi Prefecture, Ishihara entered the navy in October 1938 as a student at the Ko 3rd Yokaren Class, graduating in April 1941 from the Flight Training Course. He was posted to the newly-formed 1st Ku at Kanoya. In July he was sent to China, to Hankow, where he was involved only in patrolling sorties, having no opportunity to experience aerial combat. He returned to Kanoya during September and the following month was transferred to Tainan Ku. With the outbreak of the Pacific War he took part in operations over the Philippines and Dutch East Indies. On 8 December, the first day of the war, PO2c Ishihara flew as No.2 for the 3rd shotai of the Wakao chutai, strafing Clark Field, Luzon. Two days later he claimed his first victory during an air battle over Del Carmen. In April 1942 he returned to Japan to serve as an instructor at Tokushima Ku where he remained for some time. In June 1943 he was posted to the south-east area as a member of 582 Ku, but soon after arrival was moved to 204 Ku. Flying primarily from Buin and Rabaul, he was involved in a period of very intense fighting during which he proved to be particularly good at shooting down larger aircraft. On 18 October he claimed three B-26 medium bombers over Rabaul, while on 2 November he again claimed three bombers. In December he transferred to 202 Ku and moved to the south-west area with this unit. During the following March, however, he moved to Truk. Following participation in the Biak operation, in late May he moved forward to Yap due to the US attack on the Marianas. During 18-19 June he took part in an attack on enemy vessels off Saipan. He was particularly heavily engaged with fighters on 18th, claiming four shot down, shared with other pilots. In July he returned to the homeland and was posted to Kure Ku. From August until the end of the war he flew both Zeros and Raidens in defence of Kure and Osaka as a member of 332 Ku, frequently engaging B-29s, during which time he raised his total of victories to 16 confirmed. After the war he subsequently became a jet pilot with the Air Self-Defence Force, reaching the rank of major, but was killed in a flying accident.

CPO Isamu Ishi-Ii

Isamu Ishi-i was born in 1924 in Osaka Prefecture, graduating from Hei 10th, following which he was posted to 204 Ku in the Solomons area on 6 September 1943. He took part in his first aerial combat as No.4 man in Lt(jg) Tetsusi Ueno's element, claiming his first victory on 15 September. He continued to operate over the central Solomons, making several more claims by 24 October. He proved to be particularly effective against P-38s, claiming three of these shot down, two as probables and two more shared with other pilots. In December with Wt Off Susumu Ishihara and several other pilots, he was transferred to 202 Ku, leaving the Solomons area. In March 1944 he was engaged against US 7th Air Force B-24s raiding Truk, while during June and July he undertook interceptions of these bombers over Yap. During the battle for Okinawa, he served with S312 in 203 Ku, but was killed in action on 11 May 1945. His officially recognised total of victories was in excess of ten.

Wt Off Shizuo Ishi-i

The eighth of nine children, Ishi-i was born into a farming family in Fukushima Prefecture in 1920. He enlisted in the navy in 1937, subsequently transferring from maintenance duties to pilot training. He graduated from the 50th Pilot Training Class in June 1940, and during the following April was posted to the 12th Ku, taking part in the fighting over central China with this unit. Although aerial battles were by this time fairly rare, PO3c Ishi-i was lucky, in being in the right place at the right time. His first victory was claimed on 22 May during an attack on Chengtu when he shot down an SB bomber. By the time of his transfer to the Tainan Ku in October, he had already claimed three victories. During the opening phases of the Pacific War he was active over the Philippines and the Dutch East Indies. Posted to Ohmura Ku in April 1942, he was then assigned aboard *Junyo* in September, sailing with this ship into the south-east area. Following participation in actions over the Solomons and New Guinea fronts, he was posted to 204 Ku in September 1943. Following distinguished service in both offensive and defensive operations over Buin and Rabaul, Ishi-i was killed in action over the latter base on 24 October whilst engaged in an interception.

Considered rather unusual for a person from the north-eastern hinterlands, he had proved a swift learner and a bright pupil as a pilot. He proved to be particularly adept at attacking and shooting down larger aircraft. During January 1943 whilst patrolling over convoys making for Wewak in New Guinea, he managed to claim the destruction of two B-24s single-handed. His activities during the six weeks prior to his death had been outstanding and he had been operating at a hectic rate, claiming 17 aircraft shot down (four of them shared). Particularly notable was 23 September when 27 A6M fighters of 204 Ku, led by Lt(jg) Sumio Fukuda undertook an attack on Barakoma Beach. Here 13 intercepting US fighters were claimed shot down, five of them by Ishi-i (one of them shared). By the time of his death he had claimed 29 victories.

Lt Chitoshi Isozaki

Isozaki was a veteran pilot who achieved a record in flying more than 4,000 hours during a 13-year period – all on fighter aircraft. He showed exemplary traits not only in his combat techniques, but also as a leader. He had commenced service as an ordinary seaman, 4th Class, and rose to the rank of Lt. Born in Matsyama City in 1913, he enlisted in the navy immediately after he had graduated from Matsuyama High School. In March 1933 he completed the 19th Pilot Training Class and became a fighter pilot. Following service aboard *Ryujo* and with Kasumigaura Ku, he was posted to serve on *Kaga* in October 1937, then spending about 18 months on the Chinese battlefront. At the end of 1939 he was transferred to the 12th Ku, then taking part in attacks on Kweilin (Guilin). However, throughout this period he never had a chance

to become involved in aerial combat. In November 1940 he was promoted Wt Off, and a year later was posted to Tainan Ku, serving with this unit in the southern area operations. After one and a half years, he returned to Japan to serve as an instructor at Ohmura Ku. Promoted Ensign in April 1943, he was sent to Rabaul where he joined 251 Ku and with this unit he at last was able to claim his first aerial victory over the Russell Islands on 16 June. By this time he was already more than 30 years old. He later served as a buntai leader in 204 Ku, and then with 201 Ku, surviving six months of the desperate fighting around Buin and Rabaul. In March 1944 he was posted to 302 Ku at Atsugi, and later to 210 Ku. In May 1945 he was transferred to S301 as buntai leader, being involved in defensive duties in the Kyushu area when the war ended. By this time he had claimed 12 aircraft shot down.

Wt Off Kiyoshi Ito

Born in Niigata Prefecture in 1921, Ito graduated from Hei 2nd Yokaren Course in November 1941, immediately before the outbreak of the Pacific War, and he was posted at once to 3rd Ku. The majority of his fellow pilots in the unit at this time were veterans with 1,000 or more flight hours. As a result he was given no chance to take part in air combats during the earlier stages of the war. On 4 April 1942 he claimed his first victory during an air raid on Darwin. From that time until August he took part in many raids on this city, quickly showing a very rapid growth in his flying skills. During September, when the kokutai was temporarily split, the main body of the unit moved to the south-east area led by Lt Aioi, Ito accompanying this detachment to Rabaul. By the time he returned to the south-western area in late November, he had taken part in many testing sorties over Guadalcanal and New Guinea. In the south-western area he was then engaged in air defence sorties over the Arafura Sea and further attacks on Darwin between the early spring of 1943 and November. By that time he had taken part in 30 engagements during which he had claimed 18 aircraft shot

down and 14 damaged for which he was awarded a special commendation. Following a return to the homeland, he served as an instructor in the Oita and Tsukuba Kokutais until the end of the war. Following the conclusion of hostilities, he later changed his name to Kato. He is undertsood still to be alive at the time of writing.

Lt(jg) Tsutomu Iwai

Born in Kyoto Municipality in 1919, Iwai graduated from the Flight Training Course for the Otsu 6th Yokaren Class in August 1938. In January 1940 he was posted to the 12th Ku, joining the unit at its base at Hankow. On 13 September 13 Zero fighters led by Lt Saburo Shindo fought Chinese fighters over Chungking, claiming an outstanding 27 shot down. During this fight Iwai flew as No.3 in Lt Shirane's shotai and claimed two of the victories. During an attack on Chengtu on 26 October, he and Wt Off Koshiro Yamashita strafed the airfield, following which he intercepted and shot down an unarmed training aircraft. He returned to Japan in November 1940, then serving successively with

Tsukuba, Saeki and Ohmura Kokutais. Finally, in November 1942 he was posted aboard the carrier *Zuiho*, in January 1943 going to Rabaul and taking part in the evacuation of Guadalcanal, in the Lae convoy operations, and in the attack on Buna. During the 'I-go' operation in April he participated in attacks on Guadalcanal, Oro Bay, Port Moresby and Milne Bay. In August 1944 he was transferred to 601 Ku and underwent additional training at Matsuyama. With the launch of the 'Sho-go' operation in October, he joined *Zuikaku*, sailing to the area east of the Philippines. During the afternoon of 24 October a 56-plane attack unit was despatched to attack a US task force, Iwai forming part of the fighter escort which was led by Lt Hohei Kobayashi. Before reaching the target area, the formation was intercepted by US fighters and after the fight which followed, Iwai was obliged to break away and land at Aparri on Luzon. Here, together with pilots from other carriers who had landed there, he was placed under the command of Lt Nakagawa. This 'scratch' unit moved to Manila, from where next day it flew cover for Vice-Admiral Kurita's ships departing the area. While so engaged, he claimed the destruction of two torpedo aircraft which attempted to attack the fleet; he then landed at Batangas. For the next few days he took part in attacks on Leyte, flying from Clark Field. He was then flown out to Matsuyama in an IJAAF heavy bomber. During November he was promoted Ensign, and during the spring of 1945 moved to southern Kyushu to take part in the fighting over Okinawa. At the end of April he moved back to Hyakurigahara, ending the war at the Suzuka base; he had claimed a total of 11 victories.

PO1c Yoshio Iwaki

Born in Sahakin in 1923, following graduation with the Ko 2nd class, Iwaki was posted to *Akagi*. On the opening day of the Pacific War he flew as No. 2 in the second element of the initial wave of the attack force, attacking Hickham Field, Oahu, where his element claimed seven aircraft burned on the ground. He next participated in the attack on Ceylon, where on 4 April 1942 he claimed a British flyingboat shot down for his first victory (a Catalina of 413 Squadron, RCAF). He added a claim for a fighter probably destroyed over Colombo next day.

On 5 June during the Midway battle, he was involved in the interception of USAAF bombers, taking part in the claimed destruction of four twin-engined and three four-engined bombers. Following the battle, he was re-assigned to *Shokaku*, and from this carrier on 24 August 1942 during the Eastern Solomons battle, he shot down another twin-engined bomber. However, his own aircraft was hit by return fire and his Zero set ablaze and crashed. His officially recognised total of claims was eight.

Lt(jg) Tetsuzo Iwamoto

Born in a farmhouse in Shimane Prefecture in 1916, Iwamoto graduated from the Masuda Agricultural and Forestry School before entering Kure Naval Barracks as a seaman in 1934. He transferred to maintenance duties a year later, but in December 1936 he graduated from the 34th Pilot Training Course and became a fighter pilot. In February 1938 he became the youngest Japanese pilot ever to fight over central China with the 12th Ku, taking part in an attack on Nanchang on 25th of that month during

which he claimed Chinese aircraft shot down. After that baptism, he became the top IJN ace of the China Incident with 14 victories. During the Pacific War he served on board the carrier *Zuikaku*, taking part in operations over Hawaii, Ceylon and the Coral Sea during the opening months of fighting. In August 1942 he was posted to a training unit to become an instructor. In March 1943 he was posted to the newly-formed 281 Ku, spending the next six months or so at Paramushir Island, engaged in the air defence of the northern area. In November he was part of a unit of 15 aircraft despatched to Rabaul, the whole unit being incorporated into 204 Ku, and later to 253 Ku. For more than three months he was engaged in almost daily interceptions, but at the end of February 1944 he was withdrawn to Truk Atoll where he took part in interceptions of B-24s. Finally, in June he returned to Japan where in September he was attached to S316 of 252 Ku and promoted Ensign. In October he was engaged in air battles off Formosa, then moving to the Philippines. From here he returned to the homeland at the end of October. He was posted to 203 Ku in spring 1945, moving forward with this unit to Kyushu. Following the fighting over Okinawa, he was not to see further action before the war ended.

Iwamoyo's front line service as a fighter pilot lasted for nearly eight years; no other Japanese pilots compare with him regarding the length or variety of battle experience. With the exception of his time with 204 Ku (where he claimed a total of 20 victories and five probables during a one-month period), the kokutais with which he served during the Pacific War did not maintain records of individual claims. It is therefore difficult to assess the total number of aircraft which he did claim to have shot down. He became sick after the war and following his death his detailed diary claimed an astounding 202 victories (142 as a result of combats in the Rabaul area); many of these may actually have been attacks, rather than actual shoot-downs, however. Conservatively, therefore, it may perhaps be more appropriate to estimate the number of claims he made as around 80. He referred to himself as Kotetsu (a well-known swordsmith) and was of a chivalrous bent; his favoured technique in combat was a vertical diving single pass. It is possible that Iwamoto's total may exceed that of Nishizawa, which would make him Japan's top fighter ace.

PO2c Hideo Izumi

Born in Toyama in 1921, Izumi graduated from the Ko 3 class and was posted to Tainan Ku just before the outbreak of the Pacific War. On the first day, 8 December, he and another wingman jointly claimed a probable victory over Luzon. In the course of his first 46 sorties during this early stage of the war, he claimed two shared victories, but then began a more successful phase. On 18 April 1942 he claimed two P-40s shot down over Port Moresby, while on 24 April he added claims for a B-26, a P-39 and a P-40. A further claim on 28 April brought his total claims to nine, but on 30 April he was shot down and killed.

Lt(jg) Keishu Kamihira

Born in Yokohama City in 1920, Kamihira graduated from the city's Second High School, then responding to recruitment to join the Ko 1st Yokaren Class, newly-established during September 1937. He graduated in June 1939 and was successively assigned to the Oita, Ohmura and Yokosuka Kokutais. In August 1940 he was posted to the 12th Ku stationed at Hankow. On 14 March 1941 he flew as No.2 to PO1c Hagiri during an attack on Chengtu during which he managed to claim four victories over enemy aircraft (one of them a probable). He was transferred to the Tainan Ku

in October and with this unit claimed one victory over Luzon in the Philippines on the day on which the Pacific War began. He then saw further action over the Philippines and Dutch East Indies until April 1941 when he was posted to the newly-formed 6th Ku for service in the Aleutians. On board the carrier *Junyo* he then joined members of the unit scheduled to join the occupying forces on Midway Island. He was then sent home to join Oita Ku where he was promoted Wt Off during the following April. In October 1943 came a posting to 381 Ku where he became engaged in the air defence of Borneo and Celebes. Intercepting raids by B-24s, he became very successful in the use of Type 3 aerial bombs, and while with the unit here he claimed two P-38s and five B-24s shot down, two of the latter being probables. He was wounded during this period which occasioned his return to Japan. He ended the war as a Tokko unit instructor, being considered representative of the best of the Ko class of pilots. He died in 1970 while piloting a Maritime Safety Agency helicopter which crashed near Hakodate City.

Ens Takao Kanamaru

Kanamaru was born in 1920 in Yamanashi Prefecture. He enlisted in the navy, graduating from the 44th Pilot Training Class in January 1939. Posted in August to the 12th Ku, he saw service over central China but without becoming engaged in any aerial combat. Later he was posted aboard *Akagi*, subsequently serving with Ohminato and Oita Kokutais. In June 1943 he joined 202 Ku and was despatched to the northern Australia battlefront. Here he gained his first combat experience on 30 June flying on the wing of Lt Cdr Suzuki during an escort to Rikkos raiding

Brocks Creek. Intercepted by Spitfires, he claimed his first aerial victory over one of these. He next took part in raids carried out on 6th and 7 September, followed by a long distance attack on Merauke, New Guinea on 9th. In December, with about 20 other pilots, he was transferred to 204 Ku at Rabaul, taking part in daily interceptions there. He was severely wounded on 7 January 1944 after having claimed three aircraft shot down, and had to undertake an emergency landing. During the one-month period prior to this date he had claimed 12 aircraft shot down (two of them probables). Sent home to

Japan in a hospital ship, he was to remain under treatment for six months. Following recovery from his wounds, he joined Tsukuba Ku, ending the war with this unit at Naruo. His personal total of confirmed victories numbered 12. He has since died following illness.

PO2c Saji Kanda

Kanda was born in 1921 in Oita Prefecture, completing Hei 2 training in November 1941, when he was posted to 3rd Ku. He undertook his first sortie with this unit on 12 January 1942, but had not gained any successes in combat when he was transferred to 6th Ku. With this unit he made his first claim on 28 September over an F4F. Thereafter he claimed regularly, including three victories on 1 April 1943. However, on 16 June of that year, with his total at nine, he failed to return.

Cdr Tadashi Kaneko

Born in Tokyo the son of a lawyer, he attended the city's Municipal High School No.1, then entering the Naval Academy; from here he graduated with the 60th Class in 1932. In July 1935 he completed the 26th Aviation Student Course and became a fighter pilot. The outbreak of the China Incident found him a Lt(jg) and a shotai leader in the fighter unit aboard the carrier *Ryujo*. On 22 August 1937 he led three other pilots on a patrol near Shanghai during which a formation of 18 Curtiss Hawk biplanes were encountered. During the fierce combat which followed, Kaneko's shotai

claimed six shot down, two of them accounted for by himself. For this performance he became instantly famous nationwide. He later took part in the operations over Canton during September, then returning to Japan. During August 1938 he returned to central China now as a buntai leader in the 15th Ku. For the rest of the year he was involved mainly in base air defence and in direct support of the ground forces. Returning again to the homeland, he was then posted to the 12th Ku in April 1939, again as a buntai leader, based in Hankow. There was at this time no opportunity for further air combat, however. In May 1940 he served as a buntai leader in Ohmura Ku, and then from October in the same capacity with Bihoro Ku. In September 1941 he was posted to the newly-commissioned *Shokaku*, yet again as a buntai leader. He took part in the attack on Pearl Harbor, on Salamaua, New Guinea, and in the Indian Ocean operations over Ceylon. In May 1942 he was transferred to the 6th Ku as hikotai leader, this unit being carried on various aircraft carriers to be based on Midway Island when it was occupied. Lt Kaneko was aboard *Akagi* for this reason. On 5 June he watched repeated attacks by US carrier aircraft on the carrier, and finally, unable to restrain himself, he and four of his pilots took off from the carrier to join in the defence; they intercepted and shot down two flyingboats. However, the carrier was sunk, but he was amongst those rescued and was returned to Japan. There he was now posted to *Hiyo* with the air group aboard that vessel, as hikotai leader. In October the carrier sailed for the Solomons but due to an engine malfunction the air group was ordered to fly off to the base at Rabaul. From late October 1942 he took part in almost daily attacks over Guadalcanal. Promoted Lt Cdr on 1 November, he always led the raids personally, and on 11th of that month he claimed three F4Fs shot down himself. On 14 November he took off from Buin, leading 11 other fighters to provide cover for a convoy transporting the 38th Division to Guadalcanal. At about 1430, whilst in combat with a force of US fighters and dive-bombers, he was shot down and killed. Renowned as an excellent leader with highly developed senses and a strong will to fight, he death was mourned by many Japanese pilots. He is known to have been credited with in excess of eight personal victories.

Cdr Naoshi Kanno

Born in 1921 in Miyagi Prefecture, Kanno graduated from the Naval Academy in December 1941 with the 70th Class. He completed the 38th Aviation Student Course in September 1943 and was to become one of the most noted wartime aces. In April 1944 he went as a buntai leader in 343 Ku (the 1st) to Micronesia, but by July he had been transferred to lead S306 of 201 Ku, where he soon became hikotai leader. He was heavily involved in fighting over Yap and the Philippines area, and in particular during a series of interceptions over Yap against B-24s during July. His unit was credited with destroying or damaging more than 60 raiders. Kanno himself rammed and damaged one bomber, and his unit received a unit citation from the commander-in-chief of 1st Air Fleet. In October when the first Kamikaze Suicide Attack Unit was formed it was thought that he would be assigned to lead it. At the time, however, he had returned to Japan to procure additional aircraft, so in his place Lt Seki was nominated as commanding officer of the special unit. Even during the desperate fighting over the Philippines when Japanese forces experienced successive defeats, he continued to show an indomitable will and led his forces brilliantly. In particular, during an engagement on 27 October over Marinduque Island, he led 17 aircraft to attack 16 Grumman F6Fs, 12 of which were claimed shot down. During this period he made repeated requests to be a kamikaze pilot, but was never accepted because his superior air combat skills were considered to be too valuable to lose. With the formation of 343 Ku (the 2nd) in December 1944 he was ordered to become hikotai leader of S301, which became known as 'Shinsen-gumi'. In this new post he was active in air defence operations over the homeland as well as in the fighting over Okinawa. During a fight over Yakushima on 1 August an explosion occurred in the barrel of one of the 20mm cannon of his aircraft, rendering its handling very difficult. He refused an offer of help from his wingman, Wt Off Mitsu-o Hori, and continued to fight. When the air cleared, Hori sought to find his leader but without result. Kanno was deemed to have been killed in action and was subsequently mentioned in an all units bulletin – a considerable honour – and was posthumously promoted two ranks. Short in stature but with the fighting spirit of a bulldog, he had also exhibited a rather wild temperament. His confirmed total of victories was 25.

CPO Tomokazu Kasai

Born in Hyogo Prefecture in 1926, Kasai joined Tsuchiura Ku in 1942 as a student of the Ko 10th Yokaren Class. He completed pilot training with the 32nd Flight Training Course in November 1943 and was promoted PO2c and posted to 263 Ku. Halfway through his operational training in March 1944 he was posted to the Marianas, but whilst underway he was obliged to carry out a force-landing on Pagan Island in bad weather. Arriving late in Saipan, he became wingman to CPO Sho-ichi Sugita and in May they moved forward to Kau on Halmahera Island. When US forces landed in the Marianas, Kasai returned to Palau and took part in attacks on enemy vessels off Saipan, using Yap and Guam as advanced bases. He lost his own aircraft but escaped from Guam in a Rikko which flew to Palau, and thence to Davao. With the disbandment of 263 Ku on 10 July, he was transferred to S306 in 201 Ku. Under the leadership of Lt Kanno, this unit then moved back to Yap. For one week commencing on 16 July, the unit was engaged in intercepting raids by B-24s. On one of these sorties Kasai's aircraft was hit and he had to ditch in the sea from which he was rescued. The unit then withdrew to Cebu and concentrated on training for bombing sorties against shipping. Following the US invasion of Leyte in October 1944, Kasai flew as wingman to both Kanno and Sugita; with either of these pilots he took part in attacks and interceptions, also providing direct fighter support for Tokko units. In December he returned to Japan where he joined S301, 343 Ku. He continued to serve with this unit until the end of the war by which time he had claimed ten victories.

Ens Kan-ichi Kashimura

Kashimura was born in 1913 in Kagawa Prefecture; he enlisted in the navy following graduation from Marugame High School. He completed the 24th Pilot Training Class in July 1934, then serving successively with Ohmura, Tokosuka and Kanoya Kokutais. In October 1937 he was posted to the 13th Ku and moved to central China. His first experience of combat occurred on 22 November when he was attached to the Nango unit, taking part in a raid on Nanking. During an attack on Nanchang on 9 December he became engaged with a large formation of Curtiss Hawks, claiming one of these shot down. His aircraft then collided with another Chinese fighter during a head-on attack and began to fall – his aircraft had lost almost one third of its port wing, sheared clean off. By dint of great effort he managed to pull the A5M out of its dive just before it hit

the ground, and by carefully balancing his controls, was able to return to base. Here, however, landing proved difficult and he tried four times to get the badly damaged aircraft onto the ground. Finally he managed it, but in doing so, the aircraft somersaulted and the tail was torn off. Despite this, he escaped without a scratch ! Due to the amount of time he had been over the airfield trying to land, a photograph had been taken showing clearly the fighter minus part of one wing (see above) and in consequence the whole episode was written up in Japanese newspapers, gaining considerable fame for Kashimura. The Navy Minister, Yonai, even sent the young pilot a copy of the photograph on which had been inscribed with brush strokes the words "Shidai Shigen, Shigen Simyo" (Enormous responsibilities, enormous strength; enormous masterly techniques, masterful adroitness). Following the events, and with having claimed eight victories, Kashimura was posted back to Yokosuka Ku in March 1938. At the end of 1939 he was posted for three months to the 12th Ku, taking part in attacks on Liuchow and Kweilin (Guilin) with the new A6M Zero fighter. He then returned once more to Yokosuka Ku until December 1942. A posting to 582 Ku followed, and he took part in the fighting over the Solomons. On 6 March 1943 he failed to return from an air battle over the Russell Islands and was deemed to have been killed in action. In dogfights Kashimura had been a considerable expert, brimming with aggression; however, he was also a thoughtful theoretician in regard to aerial warfare and had a considerable influence on the pilots he had trained.

CPO Kunimichi Kato

Next to Masajiro Kawato, Kato was the youngest of all naval fighter aces. Born in 1923 in Aichi Prefecture, he graduated from the Hei 10th Flight Training Course in January 1943. After being posted to Ohmura Ku, he next moved to 254 Ku on Hainan Tao. Although inexperienced, he was a qualified marksman, and at this sport he won a cup while stationed at Sasebo Naval Station. His first combat occurred during the summer of 1943 when, without stopping to put on his flight gear, he took off and shot down a raiding P-38. At the end of the year raids began by formations of B-

24s, B-25s, P-38s and P-40s of the USAAF based in south-west China, the targets including Canton, Hong Kong and Hainan. During a series of interceptions, he became the top-scoring pilot in the unit. Before being transferred to 210 Ku in autumn 1944, he returned to Japan, having by this time claimed nine aircraft shot down. He then took part in the defence of the homeland, operating from the Meiji base. During the first US carrier aircraft attack on the home islands on 16 February 1945, he single-handedly attacked a formation of ten fighters providing air cover to the aircraft attacking Hamamatsu, claiming three shot down. He then moved forward to the Kokubu base, taking part in escorts for Okinawa-based special attack force operations. His total flight time had reached 882 hours by the end of the war and he had claimed 16 victories.

Wt Off Kiyomi Katsuki

Born in Fukuoka Prefecture in 1919, Katsuki graduated from the 54th Pilot Training Course and was posted to serve on *Chitose* as pilot of a two-seat F1M reconnaissance floatplane. On 11 January 1942 he claimed a PBY flyingboat shot down during the early days of the Pacific War, while on 3 October he claimed a US dive-bomber while escorting a convoy to Guadalcanal. During the morning of the next day he rammed the lead bomber of a formation of five B-17s, and the US aircraft and his F1M were both destroyed. However, Katsuki and his observer managed to bale out safely, and he was later awarded a commendation. Meanwhile, during the afternoon of the same day he had gone up again to claim another victory and a probable. Following conversion to A6M2N float fighters, he transferred to 452 Ku in the Kuriles where he claimed a B-24 shot down on 12 September 1943. He was then posted to 934 Ku, to which unit he flew down one of the new Kyofu float fighters in which he claimed another B-24 on 16 January 1944. Posting to 381 Ku to fly land-based fighters followed, and with this unit he claimed two more victories during interceptions over Balikpapan and Singapore. In 1945 he returned to Japan where he served in 352 Ku until the end of the war. He had by then claimed a total of eight victories.

PO1c Masajiro Kawato

Kawato became well-known for a particular record which he established – he three times undertook ramming attacks, twice was shot down, and had to bale out of his aircraft on four occasions. Born into a farming family in 1925 in Kyoto Municipality, he entered the Maizuru Naval Barracks in May 1942, passing through the Hei 12th Yokaren Class and then the 28th Flight Training Course from which he graduated in July 1943. In late September he was transported by air to Truk as part of a reinforcement team of more than 40 pilots. Here he was provided with a Zero and on 10 October he joined 253 Ku, which at that time was heavily involved in combat operations. He was only 18 when he became a Sea1c pilot

and had barely completed 300 hours of flying. His superiors made fun of him, with comments such as: *"I wonder if a young boy like you can actually fight a war !"* Probably as a result of this goading, Kawato immediately displayed a recklessness and a resolution to prove his mettle with the fighter units at Rabaul. His first ramming attack was made here on 2 November when he attacked a formation of B-25s head-on. After the initial firing pass he sought to avoid escorting US fighters by flying beneath an American aircraft which was on fire, but rammed (or collided) with this and had to bale out. He survived without a scratch, and was rescued from the sea. His second experience of baling out occurred on 11th. He had just shot down a dive-bomber which was attacking Japanese vessels, when he was attacked from behind by a fighter, his aircraft being hit and set aflame. He took to his parachute at a height of 150 metres and was rescued, but was unable to take part in operations for the next month due to burns and a wound to his left leg. Returning to action, on 17 December he attacked a P-39 head-on, but was unable to avoid this, the two fighters colliding in mid air. Both pilots baled out and fell together into the sea from where they were both picked up by the same high speed launch. Paying no attention to the injuries he had suffered on this occasion, he was back in the air next day. He then sought an opportunity to ram the tail of a B-24, achieving this on 6 February 1944. Once more he had to bale out, despite suffering six injuries this time, and was once more rescued from the sea. On 20 February almost all the remaining fighter pilots at Rabaul were evacuated to Truk, but SupSea Kawato did not go with them and remained on the island. Gathering together all the discarded spares and parts of Zeros, the remaining group of pilots rendered occasional aircraft flyable, undertaking guerilla interceptions, reconnaissance flights and strafing attacks on Green Island, The Admiralties and other locations. During this period, following the disbandment of 253 Ku, Kawato's unit became the 105th Air Base Support Squadron from July onwards. On the evening of 9 March 1945 PO2c Kawato, as he had now become, flew a two-seat Zero, with PO2c Shimizu in the back seat to intercept B-25s, but they never returned and where presumed to have been killed in action. Years later the facts were unfolded. Kawato had spotted a USN destroyer and had attacked this, but was shot down by AA fire from the ship, the Zero crashing into the sea. Despite serious wounds, he was able to swim ashore on an island where he lived alone in the jungle until found and taken prisoner by Australian forces. Having been moved to a prison camp in Leyte, Philippines, he finally returned home to Japan from here in December 1945. He had claimed 18 aircraft shot down. He later emigrated to the USA.

Wt Off Tetsuo Kikuchi

Born in Iwate Prefecture in 1916, Kikuchi enlisted in the navy in 1934, initially serving as a maintenance man. In January 1938 he graduated from the 39th Pilot Training Class and became a fighter pilot. In May he was posted to the 14th Ku, taking part in operations over southern China, but at this time there was no opposition in the air in this area. In April 1939 he returned to the Kasumigaura Ku as an instructor, and subsequently with Yatabe Ku. In September 1941 he was promoted PO1c and joined the air group on *Akagi*. The Pearl Harbor attack found him confined to CAP sorties, but in April 1942 he took part in the attacks on Ceylon in the Indian Ocean. During the attack on Colombo he

flew as wingman to the hikotai leader, Lt Cdr Itaya, and while it was his first engagement, he was able to claim three aircraft shot down and two probables against the defending RAF fighters.

Kikuchi next took part in the Midway operation in June 1942. Now flying No.2 to Lt Shirane, he was engaged in the initial attack on Midway Island, claiming two aircraft shot down and two probables. After returning to his carrier he was then engaged in intercepting attacking USN aircraft, sharing in the destruction of three of these. Since *Akagi* was sunk during these attacks, he landed on *Hiryu*. During the afternoon this carrier was also hit and set on fire, and he was trapped below decks. He managed to escape before the ship sank, and was rescued. Following his return to Japan, he was transferred to *Shokaku*. With the commencement of the offensive/defensive operations around Guadalcanal in August, Kikuchi accompanied the carrier to this area, taking part in the Battle of the Eastern Solomons. At the end of August he was detached for a week to operate from Buka Island from where he took part in a number of attacks on Guadalcanal.

Promoted CPO during November 1942, he was transferred to Tsuiki Ku as an instructor, but in September 1943 was re-assigned to carrier duty, serving on board *Ryuho*, *Hiyo* and *Junyo*. He was again detached for one week at the end of December, this time to Kavieng. On 25 January 1944 he was sent to Rabaul, seeing action there until withdrawn on 19 February. In March he joined 652 Ku on *Ryuho*, and was involved in the Battle of the Philippine Sea, during which he flew as shotai leader during the second wave launched to attack the US ships. These could not be found, however, but as the formation sought to land at Guam, it was attacked by USN fighters. The Zero pilots did their utmost to protect the dive-bombers, but fighting from a disadvantageous position, most were shot down. Kikuchi was believed to have accounted for two of the attackers shot down before he was brought down himself and killed.

Of robust build, Kikuchi weighed more than 165 lbs and required the assitance of his groundcrew to get into his fighter. Nonetheless, once in the air he became quite nimble and gained a reputation for clever aerial combat techniques. When aboard carriers, he became the self-appointed 'boss' of all the petty officers present, taking an important role in the training of others. On several occasions he was known to have ignored unofficial notification of his promotion to warrant officer due to his affection for the petty officers around him, whose company he did not wish to leave. Although the only remaining official records indicate that he had claimed 12 victories, which included seven classed as shared or as probables, it is estimated that at the time of his death this number had probably reached 20.

Ens Saburo Kitahata

Born into a farming family in Hyogo Prefecture in 1915, Kitahata enlisted in the navy in 1932. During February 1933 he was selected to attend the 21st Pilot Training Class from which he graduated in September. When he entered Ohmura Ku for further training, he was a youthful 18-year-old PO3c. Later he served on *Ryujo* and then with Kasumigaura Ku, followed by a second stint on *Ryujo* and then an assignment to Saeki Ku. In July 1937 with the outbreak of the China Incident, he was posted to the 12th Ku and in September went to Shanghai. Initially assigned to air defence and ground support duties, he did not have any opportunity to experience aerial combat. In 1938 he took part in attacks on Nanchang, Hankow, and other targets, but whether he was involved in any successful air fighting at this time is not known. In September 1938 he was posted to *Soryu*, and during the autumn participated in the Canton offensive. He transferred to Yokosuka Ku in December 1939, but in July 1940 rejoined the 12th Ku at Hankow, flying over new Zeros to this unit. During the first attack on Chungking on 19 August 1940, PO1c Kitahata flew as wingman to Lt Saburo Shindo, although on this occasion no enemy aircraft were to be seen. However, on 30 September in a big engagement with 30 Chinese fighters which ended in complete victory for the Japanese unit, Kitahata personally claimed two of the opponents shot down following which he strafed Paishih, Chungking railway station airfield before returning to base. On 26 October he led a shotai of Zeros in a formation of eight led by Lt Iida to Chengtu. Here the pilots shared in the claimed destruction of ten fighters and trainer aircraft. He was promoted Wt Off in April 1942 and in May went aboard *Junyo* from which vessel he took part in raids on Dutch Harbour in the Aleutians on both 3rd and 4 June. On the first occasion no opposition was encountered, but on the second day P-40s attacked the Japanese dive-bombers as they were rendezvousing after the attack. A hard fight ensued for the D3As, but two Zeros came to their aid, flown by Kitahata and Ozeki. In a matter of moments they changed the situation completely, Kitahata claiming three of the US fighters shot down, and Ozeki two. Pursuing the survivors closely, Kitahata then spotted, through the mist, the location of the Otter Point, Aleutians airfield from which the interceptors had come, and which was the only airfield available to the Americans at that stage. Following its return to Japan after these raids, *Junyo* next proceeded to the south-east area in October, Kitahata becoming involved in attacks on Guadalcanal, and on 17th of that month, in the Battle of Santa Cruz on 26th. The carrier then withdrew to Truk. On 17 January 1943 some of the air group's fighters were detached to Wewak to undertake interceptions of raiding B-24s. Here on 23rd Kitahata attacked a trio of these bombers, but his aircraft was hit by return fire and he was killed. He was known to have claimed in excess of ten aircraft shot down.

Lt Cdr Hohei Kobayashi

Kobayashi graduated from the Naval Academy with the 67th Class in 1939; he then completed the 36th Aviation Student Course in June 1942 and was at once posted to the fighter unit of *Shokaku*'s air group. His first taste of combat occurred on 26 October during the Battle of Santa Cruz during which he undertook CAP duties. In November he was transferred to Tsuiki Ku as a buntai leader. In February 1943 he was re-posted to *Shokaku* in a similar capacity, following which he conducted training in formation air fighting and dive-bombing tactics at Truk. He was transferred to

Hiyo in November as hikotai leader, and towards the end of January 1944 he arrived at Rabaul, where he was involved in daily air fighting until his return to Truk in mid February. He moved to 652 Ku as a buntai leader in March, and during the fighting over the Mariana Islands on 19 June, he led the fighter escort for the 2nd Carrier Division's attack force sent out to attack the US fleet. This force failed to find their target, and while landing at Guam were surprised by a big force of F6Fs. Caught in an adverse position, the Zero pilots sought valiantly to provide cover for the other aircraft, but lost many of their number. Konayashi's own aircraft was hit and he had to ditch in the sea from which he was later rescued. He then took part in the fighting around the Philippines as commander of the reformed *Zuikaku* fighter unit. On 25 October, after he had completed a defensive patrol, he was again rescued from the sea, this time by the destroyer *Hatsuzuki*. During a surface naval battle that night, however, *Hatsuzuki* was sunk and he went down with the ship. Highly regarded as a military leader with superior technical skills, a great sense of responsibility and great fighting spirit, he was known to have claimed in excess of ten aircraft shot down.

Ens Yoshinao Kodaira

Born in Nagano Prefecture in 1918, Kodaira enlisted in the paymaster's corps of the navy in 1935. In November 1938, however, he graduated from the 43rd Pilot Training Course and was assigned successively to Saeki and Oita Kokutais. In September 1939 he was posted to the 14th Ku and saw service in southern China. He saw his first combat during a raid on Liuchow on 30 December, while on 10 January 1940 he claimed two victories during an attack on Kweilin (Guilin). After engaging in air support operations to the naval ground forces, he was stationed in French Indochina, but returned to Ohmura Ku in October. During April 1942 he was posted to *Shokaku*, being engaged in CAP duties during the Coral Sea battle the following month. In August he took part in the Battle of Santa Cruz, claiming one US dive-bomber shot down over the Japanese ships, but having to land back aboard his carrier when his own aircraft was hit. In February 1944 he was transferred from Tsuiki Ku to 601 Ku (on board *Shokaku* again) and during the Battle of the Philippine Sea on 19 June he was a member of the escort for the first wave attack by the 1st Air Flotilla. Just before the attack force reached its target, a substantial formation of US fighters attacked. Kodaira claimed two shot down but was then hit and had to break away. Returning alone, he force-landed in the sea near the destroyer *Fujinami* and was rescued. Following his return to Japan, he was assigned to the reorganised carrier force, serving on board *Chitose* as a member of S164. On 24 October the ship was involved in the battle off Cape Engano, its attack force of aircraft suffering heavy losses before reaching its target. After claiming one F6F shot down, Kodaira managed to reach Aparri, Luzon, together with some other surviving aircraft. Next day they moved to Manila and flew air cover for Admiral Kurita's fleet which was retreating. After these experiences, he next took part in the operations over Leyte from Clark Field. On 8 November, however, he was injured in a take-off accident and was evacuated to Japan. There, on recovery, he served at Izumi, Huangchuan and Mobara before ending the war at Koriyama as a member of S304. After the war he joined the Air Self-Defence Force. He had claimed 11 aircraft shot down. He is believed still to be alive at the time of writing.

Ens Kiyoto Koga

Koga was known throughout the IJN as its first ace. Born in 1910 in Fukuoka Prefecture, he was first a newspaper delivery boy, but volunteered for the navy, enlisting at the Sasebo Naval Barracks in 1927. In May 1931 he graduated from the 16th Pilot Training Class and became a fighter pilot. After assignments to Ohmura and Yokosuka Kokutais, he was posted to the 13th Ku at the outbreak of the China Incident. In August 1937 the unit moved to Shanghai. His first aerial combat occurred during the first attack on Nanking on 19 September, when he claimed two Curtiss Hawks shot down. During several raids on Nanking and Nanchang, he served as a shotai leader under Lt Nango. With three victories claimed over Nanking on 5 October, his tally of victories reached seven; during the course of six air battles fought before 9 December he claimed a total of 13 aircraft shot down. On return to Japan he was given a personal citation by the C-in-C, China Area Fleet, and a special promotion to Wt Off. During night air defence manoeuvres by Yokosuka Ku on 15 September 1938, he appeared to become confused by the searchlights and crashed. Badly injured, he died next day. He was a hard working person who did not push himself forward under normal circumstances, but nonetheless possessed a strong fighting spirit and obvious courage. He was considered to be an Isamu Kondo type of warrior, Kondo being a famous swordsman of the 19th Century, who was matured by actual combat experience.

Lt Fujikazu Koizumi

Born in Fukui Prefecture in 1916, Koizumi entered Yokosuka Kokutai in June 1931 as a member of the Otsu 2nd Yokaren Class; he also graduated from a Flight Training Course in April 1935. He then served at Saeki and Kasumigaura Kokutais, and had returned to Saeki for the second time when the China Incident occurred in 1937. From there he was posted to the 12th Ku in central China. At first he was engaged in air defence and support for the naval ground forces in the Shanghai area, with few chances for aerial action. During the attack on Nanchang of 25 February 1938, however, he flew as a shotai leader, engaging some 40 I-15s and I-16s in his first combat, claiming two of these shot down. In April he returned to Ohmura Ku but in December was posted to serve aboard *Ryujo*. Further postings followed to Oita Ku and then to Hyakurigahara Ku. In September 1941 he was transferred to the newly-formed 3rd Ku, and on 8 December took part as a shotai leader in the first attack on Luzon. During the second attack two days later he and his shotai were heavily engaged over Clark Field, claiming six aircraft shot down and a seventh probable. Later, he took part in fighting over the Dutch East Indies and in the attacks on

Darwin and Wyndham, as well as other operations. During September the main part of the unit was temporarily detached to the south-east area, but Ens Koizumi remained behind and was assigned to the air defence of the area off northern Australia. During the spring of 1942, 202 Ku, which 3rd Ku had become, resumed attacks on Darwin on a number of occasions, meeting Spitfires that rose to intercept them. In May 1943 he returned to Japan for the first time in 18 months, and was posted to Tokushima Ku. Immediately upon promotion to Lt(jg) in November, he was ordered to serve as a buntai leader on *Hiyo*. On 25 January 1944, in response to the threatening situation in the south-east area, the unit was sent to Rabaul. He took off for his first sortie over this area two days later, but failed to return. He had claimed 13 victories by this time.

PO2c Take-ichi Kokubu

Born in Fukushima Prefecture in 1921, Kokubu enlisted in the navy in 1938 and graduated from the 49th Pilot Training Class in June 1940. He then served successively with Oita, Sasebo and Bihoro Kokutais. In September 1941 he joined Chitose Ku in the Marshall Islands, but in February 1942 was posted to 4th Ku, accompanying this unit to Rabaul. On 1 April he was transferred to Tainan Ku, and with this unit claimed his first victory, a P-39, over Moresby on 18 May. Despite his youth, he continued to gain victories during further attacks on Moresby, interceptions over Lae and patrols over the Buna anchorage area. Following the landings of US forces on Guadalcanal, he became involved in operations over this area as well. On 2 September 1942 nine A6Ms were led by Lt Kawai to escort Rikkos to Guadalcanal, but here he was lost in a fight with ten F4Fs, and was deemed to have been killed in action. Although he had taken part in less than six months of operations, he had claimed eight individual and three shared victories and two probables.

Wt Off Sadamu Komachi

Born in Ishikawa Prefecture in 1920, Komachi graduated from the 49th Pilot Training Class in June 1940 and was assigned to the carrier *Shokaku* during the following year. He was to be involved in fighting during the attack on Hawaii, over Ceylon, the Coral Sea, the Battle of the Eastern Solomons and the Battle of Santa Cruz. He returned to Ohmura Ku in Japan at the end of 1942, a year later being posted to 204 Ku and then to 253 Ku. Now he operated over Rabaul and Truk, receiving honours from his commanding oficer for an interception of B-24s using the Type 3 aerial bomb. On 19 June 1944 in order to take part in the fighting over the Marianas, he left Truk for

Guam, the formation in which he flew being led by Lt Cdr Harutoshi Okamoto. Minutes before landing at their destination, Komachi's aircraft was damaged by F6Fs. Although badly burned, he managed to make an emergency landing on the coast and was rescued. A few days later he was on board a Rikko which was to fly him home to Japan, where on recovery from his burns he then served with Yokosuka Ku for the rest of the war. Known as a wild character and a daredevil pilot, he had claimed 18 aircraft shot down. He is understood still to be alive at the time of writing.

Ens Masa-ichi Kondo

Born in 1917 in Ehime Prefecture, Kondo graduated from the 27th Pilot Training Class in July 1935 and was posted to Ohmura Ku. In November 1936, while he was serving on board the carrier *Ryujo*, the China Incident commenced. On 23 August 1937 the carrier was off Shanghai, and Sea1c Kondo, was flying as No.2 to Lt(jg) Minoru Suzuki when nine hostile fighters were met over Paoshan, a bitter combat ensuing. For no loss the Japanese pilots claimed ten victories, Kondo personally claiming two. Subsequently, he took part in the attack on Canton on 21 September. Transferred to *Kaga* the following year, he took part in attacks on Canton on 13 April. In June came a posting to the 15th Ku, followed in November by a return to Japan. During October 1939 he was transferred to the 12th Ku, taking part in the 30 December attack on Liuchow as well as the raid on Kweilin (Guilin) on 10 January of the following year. In all these raids he claimed further victories. After the start of the Pacific War he joined *Zuiho* in July 1942, and on 26 October was involved in the Battle of Santa Cruz, when Lt Moriyasu Hidaka led nine of the carrier's Zeros to escort the first attack wave heading for the American carriers. These met US torpedo-bombers and their escorts from USS *Enterprise*, which were heading for the Japanese fleet. Lt Hidaka left the bombers and attacked the opposing formation and in a matter of minutes the Zero pilots claimed six fighters and eight torpedo-bombers shot down, PO1c Kondo accounting for three of the victories. At the end of this fight Lt Hidaka had lost his bearings and each shotai had to find its own way back. As a result, four aircraft, including that flown by Lt(jg) Utsumi, failed to return. In November Kondo was transferred to *Junyo*, from which ship he was involved in the fighting for Guadalcanal, the support of convoys to Wewak, the evacuation of Guadalcanal and the 'I-go' operation. In May 1943 Kondo returned to Japan to take part in the relief of Attu Island. Instead, however, on 2 July *Junyo* was ordered to make for Buin for the air battle around Rendova in the central Solomons. On 15 August Kondo took off as part of the direct escort to dive-bombers heading for Vella Lavella, but after he had shot down one aircraft here, he was severely wounded in his left leg. Using his right leg only, he managed to retain control of his aircraft and to get as far as Buin, where he landed and was hospitalized. Evacuated home, he spent the next 15 months in hospital. When released, he was posted to 203 Ku, but he ended the war without taking part in any further combats. Tenacious and skillful without being showy, he had claimed 13 victories.

Lt(jg) Osamu Kudo

Born in Oita Prefecture in 1915, Kudo entered the Otsu 2nd Yokaren Class in 1931 at the age of 16. In April 1935 he completed the 2nd Flight Training Class and was posted to Ohmura Ku. In July 1937 he joined the 13th Ku, going with this unit to Shanghai, On 19 September he took part in the first attack on Nanking as a wingman of Lt(jg) Suganami. He claimed one aircraft shot down over the Chinese capital, but his own aircraft was then hit and he had to ditch in the Yangtze river, from which he was rescued. He returned to Japan in November from where the following January he was posted to the carrier *Kaga*, and for the rest of the year he was involved in operations over southern China. On 30 August 1938 during an attack on Nanxiong, he was flying as No.2 to Lt Teshima, the formation leader. Here a 40-minute-combat was fought with about 20 Gloster Gladiators, and while Kudo was able to claim two shot down, Lt Teshima was killed. Kudo later served on *Akagi*, then being posted to Hyakurigahara Ku where he was promoted Wt Off. During September 1941, just before the outbreak of the Pacific War, he joined the 3rd Kokutai. During the 10 December attack on Luzon with this unit, his shotai claimed nine aircraft shot down. However, during an attack on Broome harbour in north-western Australia, he was strafing at low altitude when his aircraft was hit by ground fire, and he crashed to his death. Subsequently, he was mentioned in an all units bulletin issued by the C-in-C of the Combined Fleet, and in addition was given the honour of a two-rank promotion to Special Service Flight Lieutenant Junior Grade. He had personally claimed seven aircraft shot down.

Ens Shigetoshi Kudo

Born to a farming family in Oita Prefecture in 1920, Kudo enlisted in the navy in 1937 and was initially assigned to a maintenance crew. In August 1940 he graduated from the 53rd Pilot Training Class and was at first assigned and trained for land-based reconnaissance aircraft. Following service with Saeki Ku, he was posted to Tainan Ku in October 1941. He took part in the Philippines and Netherlands East Indies operations flying a C5M as pathfinder to the unit's A6Ms on the longer-distance missions, and also undertaking reconnaissance flights. On 29 August 1942 he gave chase to eight B-17s over Rabaul, claiming to have brought down one and probably a second by dropping aerial bombs on them. At the time Rabaul was under constant night attack by US medium and heavy bombers, the Japanese forces there eagerly awaiting the arrival of Gekko night fighters with obliquely-angled cannons installed at the suggestion of the 251 Ku commanding officer, Kozono. In the autumn Kudo returned to Japan with the unit, but in May 1943 returned to Rabaul having undertaken the necessary training to fly the new fighter. The first of these at last became available on 21 May, and during the very early hours of the next morning he took off in this with Lt(jg) Atsushi Sugawara as observer. Taking advantage of the available moonlight and aided by searchlights, he intercepted a B-17 at about 0240 and shot this down using the angled cannons. Around 0330 he claimed a second B-17 shot down, landing at 0430 to rousing cheers from personnel at the Rabaul airfield who had witnessed both combats. Following on from these initial successes, a handful of Gekkos then swiftly regained

control of the air by night above Rabaul. In July Kudo moved south to Ballale where he took part in further nocturnal interceptions. He had claimed B-17s on 11th and 13 June, two more bombers on 26th and another on 30th. During the evening of 7 July he even managed to shoot down a Hudson over Buin using the downward-pointing oblique cannons. Having claimed eight aircraft shot down in just two months of such operations, becoming known as the 'King of the Night', he was awarded a military sword carrying the inscription *"For Conspicuous Military Valour"* from the C-in-C of the South-East Area Fleet. From the autumn of 1943 the emphasis of Allied bombing raids changed to the daytime hours, and the number of available targets by night reduced considerably. In February 1944 Kudo returned to Japan again and was posted to Yokosuka Ku where he became responsible for home defence duties. However, he was injured in an aircraft accident in May 1945 and was still convalescing when the war ended. His injuries had been very serious, however, and he finally died as a direct result of them in 1960.

Wt Off Juzo Kuramoto
Born on 1 October 1924 in Setagaya, Tokyo, Kuramoto graduated from the Otsu 15th Yokaren Course in November 1942, then completing the 29th Flight Training Course in September 1943 as a Rikko pilot. He was posted to Hsinchu Ku initially, but returned to Japan in December 1943 to convert to night fighting, being transferred to Yokosuka Ku in February 1944. He became engaged here in experiments with a radar-equipped Gekko, following which he commenced operations against B-29s with Ens Shiro Kurotori in the

rear cockpit. Over the Kanto area during the night of 15 April 1945 he claimed one shot down and one damaged, while on 25 May he claimed one bomber after another shot down until his fuel ran out, his total for the night reaching five destroyed and one damaged. For this outstanding feat he received a commendation (right) and a sword from the C-in-C, Yokosuka Naval Station, and was promoted two ranks to Wt Off. This was an extremely rare honour for a serviceman who was still alive. However, he had no further chance to add to his total before the war ended.

Wt Off Toshio Kuro-iwa

Together with Lt Nokiji Ikuta, Kuro-iwa was famous in the IJN as one of the victors of the first Japanese aerial victory – namely that achieved over the Boeing demonstration pilot, Robert Short, in the export version of the Boeing P-12. Kuro-iwa was born in Fukuoka Prefecture in 1908 and enlisted in the navy in 1926. He was one of the first fighter pilots to graduate in December 1928 from the 13th Pilot Training Class. At the outbreak of the first Shanghai Incident in January 1932, he went aboard the carrier *Kaga* and sailed for Shanghai where he was to fly as No.2 to Lt Ikuta. On 22 February they flew in support of three torpedo-bombers led by Lt Kotani and over the target area engaged Short who had taken off alone and attacked the bombers. Ikuta then attacked the Boeing from behind while Kuro-iwa came up from behind and below, both opening fire and sending the fighter down. Admiral Nomura of the Third Fleet issued commendations arising from this victory. In the spring of 1938 during the subsequent China Incident, Kuro-iwa was attached to the 12th Ku, exhibiting great piloting skill during the next three months over central China which allowed him to claim victories over 13 Chinese aircraft. He left the service during the following year, entering  the reserves and joining Great Japan Airlines, engaged in air transport operations. On 26 August 1944 he was lost off the Malayan Peninsula and was deemed to have died in the line of duty.

PO2c Sei-ichi Kurosawa

Born in Saitama Prefecture in 1922, he entered the Hei 3rd Course and graduated from the 17th Flying Training Course in July 1942. He was then assigned to 204 Ku in the Solomons area where he engaged in his first combat on 8 November 1942 over Buin. On 11 January 1943 over Guadalcanal, he claimed three destroyed and a probable when in combat with F4Fs. Over Port Moresby on 12 April he claimed a further victory while flying as No.3 in Wt Off Hidaka's section, following which as wingman to Hidaka he claimed four more destroyed and two probables. He then became No.4 man to CPO Hideo Watanabe, who led a number of missions following the death of Lt Miyano. Subsequently Lt(jg) Koshida and Lt(jg) Shimada became leaders, Kurosawa acting as leader's wingman and protecting whichever was undertaking that role on a dozen occasions. On 5 August 1943 he was shot down and killed over Buin. At this time his personal total was believed to be eight destroyed and three probables, plus one shared with Wt Off Hidaka, but during the period when he was flying in the lead element, no personal scores were noted in the unit records.

Lt(jg) Hideo Maeda

Born in Mie Prefecture in 1920, Maeda responded to a call for students for the Ko 1st Yokaren Class in September 1937, and was accepted. After training at Yokosuka and Kasumigaura Kokutais, he completed his flight training in June 1939 and was promoted PO3c. After assignments to Oita and Ohmura Kokutais, he joined the 12th Ku in January 1940. Although engaged in operational flying over central China, his duties were restricted primarily to air defence and support of the naval ground forces. In November he returned to Japan to become an instructor at Yatabe Ku, but following the commencement of the Pacific War he was posted to the *Kasuga Maru*.

A posting to Ohmura Ku followed, and in April 1943 he was promted Wt Off. During October he transferred to 204 Ku and moved to Rabaul. From then until the end of January 1944 he was involved in almost daily interception sorties, until withdrawn to Truk for the unit to be replenished with aircraft and pilots; he also took part in attacks on Cape Marcus and Cape Torokina. When US carrier air groups raided Truk on 17 February, 204 Ku put up 31 fighters to challenge them. These, however, were overwhelmed by the continuous attacks and superior numbers employed by the Americans, losing 18 A6Ms. Maeda himself was reported to have shot down two of the attackers, but failed to return. Although his period of active engagement in the war lasted for less than six months, his mature personality, combatative spirit and technical skills had made a considerable impression on his superiors. Consequently, he was mentioned in an all units bulletin and given a special two-grade promotion by the C-in-C, Combined Fleet; this confirmed his total of confirmed victories as 13.

PO1c Koichi Magara

Born in 1914 in Mie Prefecture, Magara completed training with the 28th Pilot Training Class, and was later posted to 2nd Ku on its formation. Just after his arrival at Rabaul, he shared in the claimed destruction of a B-17 over the area on 7 August 1942. Due to the shorter range of their A6M3 version of the Zero, the kokutai operated mainly over New Guinea where Magara claimed two P-39s shot down on 24 August. In September the unit moved to Buka Island to take part in operations over Guadalcanal where he claimed two more victories on 12th of that month, but two days later, with his score at eight, he failed to return from a further mission to that area.

Wt Off Masao Masuyama

Masuyama was born in Nagasaki Prefecture in 1921 and graduated with the 49th Pilot Training Class in June 1940. Following service with Sasebo and 14th Kokutais, he joined 3rd Ku at the beginning of the Pacific War. From the first attack on Luzon on 8 December 1941, he took part in all the operations over the Philippines and Netherlands East Indies. At the end of the initial series of advances, he was based at Timor from where he took part in the early raids on Darwin. During this period, from September-November 1942, he was despatched to Rabaul with a detachment under the command of Lt Cdr Kiyoji Sakakibara and Lt Takahide Aioi; here he also saw action over Guadalcanal.

In April 1943 he returned to Japan to join the Naval Air Technology Arsenal Test Department as a test pilot. In total, he amassed 1,540 hours of flight time, whilst during his front line service he claimed 17 aircraft shot down.

Lt(jg) Akio Matsuba

Born in Mie Prefecture in 1914, Matsuba graduated with the 26th Pilot Training Class in March 1935. He was to spend the next ten years as a fighter pilot, becoming a true veteran, with highly developed skills and techniques. In November 1936 he went aboard *Kaga*, where he was serving when the China Incident erupted during the following year. During the course of an air battle over Shanghai on 16 August 1937 he experienced his first combat, sharing in the destruction of a Douglas O-38 reconnaissance-bomber. At the end of that year he was posted home to Kasumigaura Ku, following which he again became carrier-based on *Ryujo*. Later he served at Iwakuni, Genzan and Oita Kokutais. He was posted to 301 Ku in November 1943, and with this unit moved to Iwo Jima at the end of June 1944. During interceptions on 3rd and 4 July he claimed six F6Fs shot down, following which he transferred to S701 to fly Shiden fighters. As a member of T Force he took part in operations off Formosa and then over Leyte. Subsequently transferred to 343 Ku, he was twice wounded during interceptions in defence of the home islands, but continued to fly until the end of the war by which time he had claimed 18 aircraft shot down during his long career. He has since died following illness.

Lt(jg) Jiro Matsuda

Matsuda was born in Nagasaki during 1918, completing his flying training at the Ko 1st Class. Posted to the 12th Ku in China, he claimed two victories there. By the time of the outbreak of the Pacific War, he was serving on board *Shokaku*, taking part in the first wave attack on Hawaii on 7 December, strafing Kaneohe flyingboat base. On 5 April he shared in shooting down a 'shadowing' Catalina flyingboat near the Nagumo Force carriers during the approach to attack Ceylon in the Indian Ocean. Over the US fleet on 8 May during the Coral Sea engagement, he claimed to have shot down three F4F fighters and to have shared in the destruction of a USN bomber. Subsequently, he took part in the battles over the Solomons, but late in 1943 he left the carrier to return to Japan, where he remained until the end of the war.

Lt(jg) Kagemitsu Matsu-o (Reserve)

Born in Fukuoka Prefecture in 1920, Matsu-o was a graduate of the Miyazaki Agriculture and Forestry College in December 1941. He was selected as a reserve student for the navy in February 1942. After receiving flying training at Kasumigaura, Iwakuni and Oita Kokutais, he was commissioned as a Naval Reserve Ensign in February 1943. In August he was posted to 253 Ku and immediately departed for Rabaul. From the beginning of September he was engaged on an almost daily basis in attacks on Ant Atoll, Cape Marcus, Torokina and Buna, also taking part in many interceptions over Rabaul itself. On 23 December, whilst intercepting a large force of US aircraft numbering about 80, he was shot down and killed. Although he had been in action only three months, he had claimed ten aircraft shot down during that time. His younger brother (left in this family photo) had also become a pilot, graduating from the Military Academy; he too, would be killed in action, being lost over southern Formosa a year later.

PO3c Susumu Matsuki

Born in Niigata Prefecture in 1922, Matsuki completed the Ko 4th Class and was then posted to 3rd Ku. When the Pacific War commenced, he was initially given no chance to take part in aerial combats due to the preponderance of veteran fighter pilots in the kokutai. Only after undertaking more than 20 sorties did he finally get a chance, this occurring on 3 March 1942 during a raid on Broome harbour in northern Australia. He shared in claiming three large aircraft shot down, then taking part in a strafing attack on flyingboats in the harbour and other aircraft on the nearby airfield. Claims for 24 aircraft destroyed being submitted, although Matsuki's leader, Wt Off Kudo, was shot down and killed. Thereafter raids were made on the Darwin area during which he was engaged in combats on three occasions, although details of his claims at this time are not available. He was then transferred to Tainan Ku at Rabaul and in northern New Guinea where he commenced operations on 21 July, flying often as wingman to both Wt Off Takatsuka and PO1c Ohta, claiming regularly. On 7 August he claimed two destroyed, one probable and one shared. On 13 September while escorting a reconnaissance aircraft, three Zeros failed to return, including both Takatsuka's and Matsuki's. In the two months he had been with the kokutai he had claimed six destroyed, two probables and three shared, his official total being recorded as nine victories.

Ens Momoto Matsumura

Matsumura was born in Yamaguchi Prefecture in 1915, enlisting in the navy as a maintenance man during 1934. Selected for pilot training, he graduated from the 29th Pilot Training Class in November 1935 and became a fighter pilot. On the outbreak of the China Incident, he was posted to the 12th Ku and moved to Shanghai. In December 1937 he was transferred to the 13th Ku, but by March the following year had rejoined 12th Ku. He was involved in a number of engagements with Chinese aircraft over the mainland of China. During his first such involvement on 25 February 1938 he was flying as No.3 in PO1c Tomokichi Arai's shotai, returning to claim four destroyed and three probables personally. Again, on 31 May over Hankow while flying with the Yoshitomi chutai, he took part in a wild dogfight with a large force of opposing fighters, claiming three of these shot down. In January 1939 he was posted for carrier duty aboard *Akagi*, while later he served as an instructor with Iwakuni and Oita Kokutais. In July 1942 he was attached to 6 Ku and during the latter part of August flew to Rabaul, staging through various islands to reach this base. Here he remained until April 1943 when he was promoted Wt Off and sent home to Japan. During his period in the Solomons, he had taken part in raids on Guadalcanal and in interceptions over Buin and Munda. On arrival in Japan, he became an instructor again, serving at Iwakuni, Suzuka and Konoike Kokutais. In August 1944 he was posted to S161 in 601 Ku, and with this unit took part in the 'Sho-go' operation during October, departing from the western end of the Inland Sea. On 24th and 25th he was involved in the Battle of the Philippines. When his carrier, *Zuikaku*, was sunk he had to ditch in the sea, having been in the air at the time. After nightfall on 25th, he was picked up by the destroyer *Hatsuzuki*, but this vessel was then surrounded by US warships and was sunk. Lt Hohei Kobayashi, the hikotai leader, Matsumara himself, and others aboard were all lost and were regarded as having been killed in action. His officially-recognised total of aircraft shot down was 13.

Lt Yoshimi Minami

Minami was nicknamed 'Kashimura the second'; and was to die as a member of a Tokko unit in the Philippines. Born in Kagawa Prefecture in 1915, he entered the navy in 1933, volunteering for flying duty. In November 1935 he graduated from the 30th Pilot Training Class and in July 1937 was posted to the 13th Ku (and subsequently the 12th Ku) via Ohmura Ku. Initially engaged in the fighting in the Shanghai area, he took part in actions over Nanking from 20

September, and then over Nanchang and Hankow. During this period he claimed nine aircraft shot down and was for a time the top-scoring pilot of the 12th Ku. During 31 May 1938 an attack was made on Hankow when nine A5Ms of the Yoshitomi chutai engaged about 50 opposing fighters, Minami claiming one of these shot down. His aircraft was then hit in the fuel tank and he found himself surrounded by a dozen Chinese fighters at a time when he had used up all his ammunition. He therefore rammed one of his opponents, then, despite the fact that his left wingtip had been sheared off from the red ball roundel outwards, he regained control of his aircraft and started the flight back towards base. Ultimately he made an emergency landing on the banks of the Yangtze river and set his aircraft on fire. Fortunately for him, he was spotted by a search aircraft and was rescued by a patrol boat and returned to his unit.

In September 1938 he returned to Japan, being assigned to Saeki and Oita Kokutais, followed by carrier duty aboard *Hiryu* and *Zuiho*. In October 1941 he transferred to *Shokaku*, taking part in the Hawaiian attacks, the Indian Ocean raid and the Coral Sea battle. Transferred again to Ohmura Ku in June 1942, he was promoted Wt Off and transferred to 601 Ku in February 1944. He was assigned to serve aboard the newly-completed aircraft carrier *Taiho*. On 19 June he took part in an attack on the US task force during the Battle of the Philippine Sea. After several dogfights, he was wounded and returned to his carrier only to find that it had been sunk. Surviving this, he was again returned to Japan and transferred to 653 Ku. On 24-25 October he took part in the battles off the Philippines, then becoming involved in the attacks on Leyte after the American invasions had commenced. On 25 November 1944, now a member of the Kasagi unit of the Tokko Force, he plunged his aircraft towards the task force and was killed. He was one of only a few experienced and leading fighter pilots to be lost in this way. Following his death he was given a special two-grade promotion. Prior to his demise, he had claimed 15 victories.

Cdr Zenjiro Miyano

Born in 1917 in Osaka Municipality, Miyano graduated from the Naval Academy with the 65th Class in 1938. Following this, he completed the 32nd Aviation Student Course and was attached to the 12th Ku the following year.Although involved in operations over China, no opportunity to engage in aerial combat occurred. In October 1941 he was promoted Lt and became a buntai leader in the 3rd Ku. He claimed his first victory on 8 December, the day on which the Pacific War commenced, whilst engaged in the initial attack on Luzon. He continued to operate over the Philippines and the Netherlands East Indies until April 1942 when he was transferred to the 6th Ku as a buntai leader, intended to serve on Midway Island. Going aboard *Junyo* for transport there during June, he took part in the attack on Dutch Harbour, but was returned to Japan when the operation was cancelled as a result of the Midway battle. He was then

posted to Rabaul to join 204 Ku as a buntai leader, and later as hikotai leader. There he remained for almost a year, respected and popular as a fighter leader not only with his subordinates, but with his superiors as well. He was also the first commander in the IJN to introduce the four aircraft tactical formation in place of the traditional vic of three. He also sought to develop the difficult task of providing effective support to dive-bombers at the bottom of their dives when they were both low and slow. He volunteered to employ such tactics during an attack on a convoy off Lunga, Guadalcanal, on 16 June 1943, but while engaged in a dogfight with opposing fighters he failed to return. Following his death he was given the honour of a mention in an all units bulletin and a posthumous two-rank promotion to the rank of Cdr. The bulletin recorded that Miyano himself had accounted for 16 enemy aircraft, and that the unit under his leadership had reached a total of 228.

Lt(jg) Gitaro Miyazaki

Born in 1917 in Kochi Prefecture, Miyazaki entered the Otsu 4th Yokaren Class in 1933. He subsequently completed a flight training course in May 1937, then being posted to Saeki Ku, followed by Takao Ku. In September 1938 he joined the 12th Ku and took part in operations over central China. Nothing was to be seen of opposition in the air at first, but on 5 October during a raid on Hankow he was able to catch and shoot down one of the few remaining I-16s. In June 1939 he was transferred to Yokosuka Ku, subsequently rejoining Takao Ku. In April 1941 came a second posting to the 12th Ku which he joined at Hankow. Again, little was to be seen of Chinese aircraft at this time, although once again he did manage to claim a single victory during an attack on Chengtu on 11 August. Promoted Wt Off in October 1941, he was then attached to Tainan Ku. On the first day of the Pacific War he was able to claim a victory over Luzon, following which he was promoted to lead a shotai, seeing action throughout the conquests of the Philippines and Netherlands East Indies. In April 1942 he accompanied the unit to Rabaul and Lae, seeing much action in these areas. Although ill, he took part in the attack on Port Moresby on 1 June, but here his Zero was set on fire when opposing fighters made a surprise attack on the Rikkos that the Tainan Ku pilots were escorting. A mid air explosion followed, and Miyazaki was killed. Ever since his earlier service with 12th Ku, he had enjoyed a close relationship with Saburo Sakai. Their first victories were recorded on the same dates, and Sakai was present and a direct witness when Miyazaki was killed. Although of slim build, Miyazaki held a high black belt rating in judo, and was the strongest arm-wrestler in the entire Tainan Kokutai. Following his death, he was mentioned in an all units bulletin and promoted posthumously by two grades, becoming a Special Service Flight Lt(jg). He had taken part in 37 air battles, his shotai confirming that he had claimed 44 aircraft shot down, six destroyed on the ground and 30 damaged.

Ens Isamu Miyazaki

Isamu Miyazaki was born in Kagawa Prefecture in 1919, enlsting in the navy in 1936. After serving four and a half years as a seaman with the fleet, he volunteered to become a pilot, graduating from the Hei 2nd Toyaren Course in November 1941. He was then assigned to Yokosuka Ku until October the following year, when he was posted to 252 Ku, going to Rabaul on 9 November. Flying an A6M in the flight led by Lt Suganami to escort Rikkos on a special daytime torpedo attack on a convoy off Guadalcanal on 12 November, he experienced his first engagement with enemy aircraft, using up all his ammunition in claiming a single F4F shot down. During the next four months until the unit moved to the Marshall Islands to provide air defence there, Miyazaki was involved in the frequent hard-fought engagements over both the Solomons and New Guinea. When Wake Island was attacked by US carrier aircraft on 6 October 1943, he took part in escorting G4M Rikkos which were destined to replenish the forces on this island. Taking off from Maloelap Atoll en route to this destination, the formation became involved in a fight with a force of F6Fs. When the battle ended, he landed on Wake Island – alone! He returned to base next day, hitching a lift in a Rikko. From late November until the end of the following January he was involved in the interception of B-24s raiding Maloelap, but in February he returned to Japan. He was posted to the newly-formed S301, 343 Ku, in January 1945, seeing action during air defence missions over western Japan until the end of the war, by which time he had claimed 13 victories.

Lt Isamu Mochizuki

Born in Saga Prefecture in 1906, Mochizuki enlisted in the navy in 1925, completing the 9th Pilot Training Class in November 1926. After service on board *Hosho* and *Kaga*, he was posted to Ohmura Ku. In November 1932 he commenced a four-year period at Yokosuka Ku, renowned at the time as the very best posting for fighter pilots – the fighter units' Mecca. He was promoted Wt Off in November 1936 and returned to Ohmura Ku, but on the outbreak of the China Incident, he joined the 13th Ku and moved to Kunda base in Shanghai. Even for that early conflict he was already a veteran of 31. However, during the next six months until he was posted back to Japan, he claimed nine victories, employing

outstanding air combat techniques and an abundant fighting spirit. Later he served as an instructor in a number of home-based kokutais; he was promoted to Special Flight Ensign in October 1941. In March 1943 he was ordered to serve as a buntai leader in the newly-formed 281 Ku in the northern Kuriles. When the situation in the Marshall Islands became critical, the unit was moved to Roi, where on 29 December he claimed a B-24 shot down from a formation of six, plus a second damaged; this brought his official total of victories to ten. By this time, however, US carrier aircraft had already destroyed the majority of aircraft on the ground and it was no longer possible successfully to rescue pilots either by submarine or by air. When American forces landed on the island on 6 February 1944, Mochizuki is presumed to have met his death there. His flying experience had extended over a period of 18 years, and during his time at Yokosuka Ku he had developed a manoeuvre called Hinerikomi (half loop, roll and dive technique), widely used by Japanese fighters throughout the war. Mochizuki was considered to be the greatest expert in the one-on-one dogfight.

Lt(jg) Mitsugu Mori

Mori was born in 1908 in Shizuoka Prefecture. Enlisting in the navy in 1927, he graduated from the 14th Pilot Training Class in May 1929, becoming one of the old-timers of the fighter pilot corps. During the First Shanghai Incident in 1932, he was serving aboard *Kaga*, his unit being sent ashore at Shanghai. Here he took part in ground support sorties and patrols, but a truce soon developed and the unit was withdrawn from the area. He was then posted to Ohmura Ku, subsequently seeing service on board *Akagi* and then at Yokosuka and Tateyama Kokutais. In January 1938 he was posted to the 13th Ku, where on 18 February, one week after his arrival in China, he took part in the first air raid on Hankow, flying as a shotai leader. In this capacity his shotai joined the 12th Ku, led by Lt Kaneko, escorting Rikkos to the target area. More than 30 Chinese fighters rose to challenge the Japanese formation. Taking full advantage of his years of practice, he claimed four aircraft brought down, two of these colliding whilst attempting to engage him. Following reorganisation during March, he was transferred to the 12th Ku. He was to remain with this unit for about 18 months, taking part in many raids on Hankow, Nanchang and other targets. He was given the rare honour for an NCO pilot of being presented with the Order of the Golden Kite, 5th Class, in 1940, indicating that he almost certainly claimed numerous victories during this period. Meanwhile, at the beginning of January 1939 he became an instructor with Suzuka Ku. The following May he was promoted Wt Off and placed on the reserve.

Nevertheless, he was re-mobilized on the same day, and in July 1942 was posted aboard *Hiyo*, returning to the combat area. Although by now 34 years old, his air combat skills were finely honed, rather than in any way dulled. In October he arrived at Rabaul, and from there, Buin. For about a month he took part in attacks on Guadalcanal, where on 11 October he claimed an F4F and on 14th, a dive-bomber. On the latter occasion, his own aircraft was hit and he had to ditch in the sea, from which he was rescued. The following April in order to participate in the 'I-go' operation, he was again sent to Rabaul, flying sorties over Guadalcanal, Oro Bay, Moresby and Milne Bay. During this period he claimed two further victories. In May 1943 he was promoted Ensign and returned to Japan where he became a test pilot with the Air Technology Arsenal. Promoted Lt(jg) in August 1944, he was then retired. He became ill some years later and died in 1960 in his birthplace. During the China Incident he is known to have claimed more than four victories to which he had added five more during the Pacific War.

Lt(jg) Hideo Morinio

Morinio was born in Hiroshima City during 1917. On completion of the 41st Pilot Training Class, he was posted to China where he claimed his first victory. Returning to Japan, he then became a basic flying training instructor for a period. In 1942 he was on board *Hiyo*, accompanying this vessel to the Solomons. On 15 October he shared in the destruction of a flyingboat trailing the fleet, on the same day claiming to have shot down a dive-bomber over a Japanese convoy. He then claimed an F4F over Guadalcanal on 25 October, adding two more of these fighters on 11 November, and two more on 14th. In April 1943 *Hiyo* took part in the 'I-go' operation, during which on 12 April he claimed two P-39s and a P-38. He was not to make any further claims before the war ended, his total remaining at nine and one shared.

PO3c Toyoo Moriura

Born in Kumamoto Prefecture in 1922, Moriura completed the Otsu 9th Class and was posted to Tainan Ku. It was only when many of the unit's veteran pilots had been killed that he was able to engage in combat for the first time on 5 September 1942. On 28th of that month he claimed two F4Fs and an SBD over Guadalcanal. On 3 October he claimed a further F4F, while on 15th he claimed two more SBDs and a shared probable. However, he failed to return from a sortie over Guadalcanal on 25 October 1942.

Ens Kazu-o Muranaka

Born in Fukuoka Prefecture in 1919, Muranaka enlisted in the Otsu 6th Yokaren Class in June 1935, graduating from the Flight Training Course in August 1938. He was posted successively to Saeki, Oita and Ohmura Kokutais, but on 14 August 1939 he joined the 14th Ku for service in southern China. In December the A5M unit of the kokutai moved from Haihou to Nanning, which had been encircled by Chinese troops. Muranaka caught up with the fighter unit here on 27 December. That same afternoon he and Sea1c Nojima immediately took off to undertake a patrol over the city. They spotted a formation of single-engined aircraft which they pursued, claiming between them one shot down and one probably so. These were not only Muranaka's first victories but also the first to be achieved by the 14th Ku. He took part in an attack on Kweilin (Guilin) on 10 January 1940, and remained with the unit until August when he was posted home to Suzuka Ku. In November 1941, immediately before the outbreak of the Pacific War, he was posted to serve on board *Hiryu*, and during the attack on Pearl Harbor he formed part of the escort to the first wave. Over the harbour he saw no sign of any enemy aircraft, so strafed Ewa Field. Subsequently he took part in attacks on Darwin and on Ceylon, then being involved in the Midway battle. During the latter, he took off with the first wave fighter escort, engaging in a fight with US F4Fs which rose to intercept. During this mission his was the only A6M to stay with the torpedo-bombers throughout the whole operation. After returning to his carrier, he took off again on combat air patrol, intercepting incoming raiders of which he claimed two shot down and a third as a probable. However, his own aircraft was hit and wounded, had to ditch, being rescued by the crew of the destroyer *Nowake*. On recovery, he was first assigned to *Shokaku* and then to *Junyo*. With the latter carrier he saw action both in the Battle of the Eastern Solomons in August, and then in the Battle of Santa Cruz in October. During November he was sent to Rabaul, flying sorties during the Guadalcanal fighting. He returned to Japan in April 1943 to serve as an instructor with Tokushima Ku. A month later he moved to the 11th Ku (an advanced fighter training unit at Singapore) where he continued to instruct. He was promoted Ensign in May 1945 and posted to S701, 343 Ku, returning to Ohmura. On 24 July he claimed one US aircraft shot down over Bungo Strait, but although he took part in several other combats prior to the end of the war, this was his last victory, bringing his total to six destroyed and three probables. He later served in the Air Self Defence Force, retiring with the rank of major.

Lt(jg) Kaneyoshi Muto

Muto was born in Aichi Prefecture in 1916. He enlisted in 1935 as a fireman at the Kure Naval Barracks, but six months later he was accepted for flying training, passing out of the 32nd Pilot Training Class and being posted to Ohmura Ku. On the outbreak of the China Incident he was posted as a Sea1c to the 13th Ku, taking part in the fighting over central China. In his first combat on 4 December 1937 during an attack on Nanking, he was engaged by I-16 fighters, one of which he claimed to have shot down. He was then transferred to the 12th Ku, seeing action over Nanking, Nanchang, Hankow and other areas. In September 1941, after serving with the Oita, Suzuka and Genzan Kokutais, he was posted to the 3rd Ku. Following the outbreak of the Pacific War, he took part in operations over the Philippines and Netherlands East Indies, flying as No.2 to the hikotai leader, Lt Yokoyama. In April 1942 he was transferred to Genzan Ku (from November 1942 it became 252 Ku), moving with this unit to Rabaul in November. He remained active here in operations over the Solomons and eastern New Guinea until March 1943. In November 1943 he was promoted Wt Off, having been transferred home to Yokosuka Ku. During June and July 1944 he accompanied the unit to Iwo Jima, undertaking defensive interceptions and an attack on the US task forces. With the arrival of 1945 he took part in the air defence of the Kanto area until the end of June when at the urgent request of Capt Genda, he was transferred to 343 Ku for escort duties led by Lt Kanno, taking the place of Ens Sugita who had been killed in action. Soon after his arrival, however, he was killed in an air battle over Bungo Strait on 24 July. He was recalled as a person whose personality as well as his flying skills had fully matured. He is especially remembered for an occasion during February 1945 when, flying a Shiden-kai, he single-handed took on 12 F6Fs over Atsugi, claiming to have shot down four, each with a single short burst of fire. This episode was considered to be reminiscent of the actions of Musashi Miyamato (a famed swordsman of the early

17th Century) at the battle of Sagari-matsu. Although the final total of victories claimed by him is uncertain, it is estimated to be about 28. Short in stature, he was reportedly a man whose entire body exuded vitality; he was friendly and cheerful, and was liked by all who came into contact with him.

CPO Yoshikazu Nagahama

Born in Fukuoka Prefecture in 1921, he entered Ko 2nd Course in 1938, graduating at the end of 1939 when he was assigned to *Kaga*. He took part in the second wave attack at Hawaii, flying No.2 to PO1c Kiyonobu Suzuki. He remained with this carrier's air group until she was sunk at Midway, following which he was transferred to *Zukaku*. In August 1942 he moved to the Solomons area, on 26 August sharing in the destruction of a flyingboat over his parent vessel. During the Battle of Santa Cruz he flew as 'Tail-end Charlie' to the first wave attack force, becoming engaged in a fight with an estimated more than 30 F4Fs during which he shared in the shooting down of a claimed 14 US aircraft. He was subsequently posted home to serve with Tsuiki Ku, where he was killed in a flying accident on 6 September 1943. His claims during his time aboard the aircraft carriers were recognised as amounting to ten in number.

Wt Off Ki-ichi Nagano

Born in Shizuoka Prefecture in 1922, Nagano entered the navy in June 1939. In July 1941, aged still only 18, he graduated with the 56th Pilot Training Class. Posted to Chitose Ku in October, he was assigned to air defence of Micronesia, based on Taroa Island, when the Pacific War broke out. In June 1942 he was transferred to the 2nd Ku (later 582 Ku) and returned to Japan. In August the unit was posted to Rabaul where he served until July the following year. During this period he saw much action over the south-east area, official records indicating claims by him for 16 aircraft shot down and three probables, leading to him becoming 582 Ku's leading ace. From March 1943 onwards, this kokutai ceased recording individual claims and for this reason successes gained by Nagano after this date until his departure from Rabaul are not known. Based on his earlier actions it would seem that a figure of 20 or more by July 1943 would not be unreasonable. During a raid on Guadalcanal on 25 October 1942 he flew one of eight Zeros which were being led by Lt(jg) Futagami. After circling and observing the US airfield there from a height of 200 metres, the Zero pilots were just about to commence a strafing attack when they found themselves under attack by F4Fs from above. In the desperate struggle which followed, four of the Japanese fighters, including that flown by the leader, were shot down. PO1c Nagano's aircraft was hit ten or more times, but he managed to escape, returning to claim three shot down and one probable. He later wrote: *"It would not be an exaggeration to say that this was the day on which I was*

reborn. It is one day that I will never forget." In July 1943 he was posted to Atsugi Ku, and then to S304, 203 Ku. In April the following year he was transferred to the Kurile Islands and undertook air defence duties there. Six months later the unit moved to southern Kyushu, and following the start of the 'Sho-go' operation in October, he took part in the fighting off Formosa. On 22nd he moved to Bamban, Luzon, taking part in both interceptions and attacks on targets in Leyte. Whilst trying to intercept US carrier aircraft over Bamban on 6 November 1944, he was shot down and killed. While his final total of claims may well have been substantially higher for the reasons mentioned, a figure of 19 was confirmed by existing records.

PO Yoshimitsu Naka

Graduated from Hei 4th Yokaren Course, Naka was posted to 936 Ku as a two-seat floatplane pilot. In this role he flew 56 sorties between 15 April-30 September 1943, twice being engaged in combats by day, claiming damage to a US dive-bomber; he also claimed to have sunk two torpedo boats at night. He then converted to fighters and was posted to 302 Ku to fly the night fighter conversion of the Suisei dive-bomber. Operating with Ens Hisao Kanazawa in the back seat, he flew in defence of the Kanto area against B-29s. On 20 February 1945 he gave chase to one bomber which was beginning its return flight, making dive and zoom attacks on this until it went down – the first victory for the Suisei buntai of the kokutai. Using similar tactics, he was to claim five B-29s shot down during five encounters, including one during the night of 13 April and another in the early hours of 24 May.

CPO Yoshimasa Nakagawa

Born on 22 February 1924 in Fukuoka Prefecture, he graduated from Otsu 15th Yokaren Class in May 1943, then completing the 32nd Flight Training Class during the following October, when he was posted to 321 Ku. In February 1944 he moved with the unit to Tinian, engaging in search and patrol sorties in a Gekko. In May he was detached to Yap where he claimed to have sunk a submarine whilst engaged in patrols against these vessels. The unit withdrew to Davao in June where it was disbanded the following month; he was then posted instead to S901, 153 Ku. He then became involved in night interceptions over Davao, but it was by day that he claimed a P-38 shot down on 2 September. Two nights later, while pursuing a B-24, his guns malfunctioned, leading him to ram his opponent with the propeller of one engine, which he claimed caused it to crash. He then returned safely to base on his other engine, having achieved one of the few night victories in this area. Next came a move to Cebu, from where, despite an injury suffered in a recent flying accident, he took part in attacks on Tacloban. Promoted CPO in November, he then returned with his unit to Japan. There he trained for single-engined night fighting at Fujieda in Shizuoka Prefecture with 131 Ku, in April 1945 rejoining S901 at Kanoya as the invasion of Okinawa began. He took part in attacks both on the airfields on Okinawa captured by the Americans, and on shipping offshore, flying a Suisei. During the early morning of 10 June he engaged an F6F-5N night fighter, which he claimed to have shot down. Then during early August, just before the end of the war, with Ens Takeaki Shimizu in the back seat, he claimed three B-24s destroyed with Type 3-go phosphorus bombs. He ended the war based at Iwakuni.

Wt Off Bunkichi Nakajima

Born in 1918 in Toyama Prefecture, Nakajima entered the navy after having worked as a waiter in a police station. He graduated in September 1936 from the 33rd Pilot Training Class, and at the age of 17 became the youngest PO3c ever to serve as a fighter pilot. He was immediately assigned to Kanoya Ku, and with the outbreak of the China Incident was posted to Formosa for air defence duties. In March 1938 he was transferred to the 13th Ku and in July, to the 15th Ku. However, there was no opportunity at this time to meet opponents in the air. In September 1941 he was posted to the 3rd Ku, consequently being involved in the early attacks on the Philippines and Netherlands East Indies following the commencement of the Pacific War. In November 1942 he was transferred to 252 Ku at Rabaul, taking part in many sorties over eastern New Guinea and Guadalcanal for the next three months until the unit was sent to the Marshall Islands in February 1943. On 6 October 1943 very strong forces of US carrier aircraft attacked Wake Island. In consequence seven Zeros, including that flown by Nakajima and led by Lt Yuzo Tsukamoto, took off to escort a similar number of Rikkos to the island as reinforcements. As they approached the island they were attacked by American fighters and two of the Zeros were shot down, in one of which CPO Nakajima was lost; up to this time he had claimed 16 aircraft shot down.

Ens Kunimori Nakakariya

Born in 1920 in Kagoshima Prefecture, Nakakariya entered Yokosuka Ku in June 1937 as a student of the Otsu 8th Yokaren Class. He graduated from the Flight Training Course in March 1940, then serving with Oita, Ohmura and Kanoya Kokutais. In April 1941 he joined the 12th Ku at Hankow. With this unit on 26 May he was pilot of one of 11 Zeros led by Lt Minoru Suzuki which undertook a long-range mission to attack Tianshui and Nanzheng. Engaging intercepting Chinese fighters in this, his first combat, PO3c Nakakariya was able to claim two shot down. During September he was transferred to the 3rd Ku, then engaging in operations over the Philippines and Netherlands East Indies during the opening months of the Pacific War. He then served at Koepang on Timor Island, defending the Arafura Sea, also taking part in the opening attacks on Darwin by the kokutai. During autumn 1942 the main part of 3rd Ku moved to Rabaul, but Nakakariya remained with a detachment at Koepang. In May 1943 for the first time in two and a half years, he returned to Japan to become an instructor at Ohmura Ku. Upon promotion to Wt Off in May 1944, he was posted to 653 Ku, taking part in the June Battle of the Philippine Sea. Later as the 'Sho-go' operation commenced in October, he was ordered to the Philippines. While the main part of the kokutai were carried to the area on an aircraft carrier, he went by air to Bamban, Luzon, via Okinawa and Formosa. On rejoining the rest of the unit, he moved with it to Cebu. Interception sorties and attacks on Leyte targets followed until mid November when the greatly depleted unit transferred its surviving pilots back to Japan where they joined 601 Ku. With this unit Nakakariya continued to take part in home defence sorties and operations over Okinawa. He survived the war with a total of 16 claims to his credit.

Wt Off Wataru Nakamichi

Nakamichi was born in Osaka City in 1922 and worked in a tin can factory prior to enlisting at Kure Naval Barracks in 1940. In May the following year, wishing to become a pilot, he entered Tsuchiura Ku for training, graduating with the Hei 4th Yokaren Class, then receiving pilot training with Kasumigaura and Oita Kokutais. On 21 July 1942 he graduated from the 21st Flight Training Course and in December was posted aboard *Junyo*. However, in summer 1943 he was transferred to 204 Ku, moving to the base at Buin in the Solomons, where he served for about nine months of hectic operations until sent home to Japan in March 1944. During this period he was credited with 19 successful combats, nine of which related to shared or probable claims; his personal total was subsequently confirmed as being 15. Later, and until the end of the war, he was assigned to duties in the home islands. Particularly, after November 1944 he joined a fighter hikotai, 721 Ku, which acted as direct escort to the Jinrai Tokko unit, which operated the piloted Ohka rocket bombs.

Wt Off Yoshi-o Nakamura

Born in a farming family in Hokkaido in 1923, he enlisted in the navy in 1940 as an aircraft engineer. In February 1941 he was posted to Tsuchiura Ku as a pilot training member of the Hei 3rd Class, and in July moved to the 18th Flight Training Course in January 1942, then receiving further training at Ohmura Ku. In July 1942 he was posted to 6th Ku as a PO1c and in October arrived at Rabaul. For the next year he survived the many furious air battles in the area, using both Rabaul and Buin as bases. On 7 June 1943 during a combat over the Russell Islands, he claimed three victories (one of them a probable). On 16th, however, while flying as No.4 in Lt Miyano's shotai, he was hit and wounded, making an emergency landing on Kolombangara Island. During January 1944 he was transferred to 302 Ku and returned to Japan. After serving in Yokosuka Ku, he was later transferred to S701, 343 Ku in January 1945, remaining with this unit until the end of the war. His confirmed total of claims was nine.

Ens Masayuki Nakase

Born in 1918 in Tokushima Prefecture, Nakase joined Yokosuka Ku in 1934 on volunteering for the Otsu 5th Yokaren Class. In March 1938 he completed the Flight Training Course and was promoted PO3c. Following service with Ohmura Ku, he was posted to the 14th Ku during the latter part of the year, where he served for six months, but without seeing any aerial combat. Returning to Yokosuka Ku, he was then assigned to the 12th Ku to fly one of the first A6M Zeros which that unit took into action in July 1940. During the 4 October attack on Chengtu, he, Wt Off Ichiro Higashiyama, PO1C Hagiri and PO1c Hideo Oh-ishi undertook the daring feat of landing on Chungking airfield and setting fire to the command post there. On 14 March 1941 during another attack on Chengtu, Nakase was able to claim six Soviet-made I-15s shot down, one of these classed as a probable. He claimed three further victories over Nanzheng on 26 May. In September 1941 he was transferred to the 3rd Ku. From the start of the Pacific War three months later, he proved to be one of a core of experienced NCO pilots who carried the fighting through the Philippines and Netherlands East Indies. On 9 February 1942, however, he was hit by ground fire while strafing armoured cars near Makassar on the island of Celebes, and he was killed. His name was then published in an all units bulletin and he was given an honour two-grade promotion posthumously to the rank of Special Service Flight Ensign. He was remembered as a gentle character, but a fine all-round athlete. Always the first to take action, there was considered to be a touch of genius in his air fighting technique. At the time of his death he had claimed 18 victories.

Wt Off Yoshi-ichi Nakaya

Nakaya was born in 1921 in Nagano Prefecture into a farming family. Initially he worked in a bicycle shop before enlisting in the navy as a maintenance man, but was transferred to the Hei 2nd Flight Training Class, from which he graduated in November 1941, when he was immediately posted to Chitose Ku. With this unit he took part in the air defence of Micronesia, seeing action on 1 February 1942 when a US task force entered the area; taking off from Roi Island, he claimed a dive-bomber shot down. In August he was detached to Rabaul as part of a reinforcement group for Tainan Ku, and until October he was involved in action over the Solomons. In December he returned to 201 Ku (which Chitose Ku had been amalgamated into) and became involved in the defence of the Marshall Islands until spring 1943, when he returned to Japan. In July he returned to the Solomons to be based at Buin. Here on one occasion he sought to force a P-38 to land on a Japanese-occupied airfield so that it might be captured. However, as he did so other members of his unit, unaware of what he was attempting, shot the American fighter down. On another occasion he attacked a pair of P-38s, reporting that he had thereby forced them to crash into each other in mid air. During December 1943 he was transferred to 331 Ku, moving to Sabang, Sumatra. He was then transferred to 202 Ku in March 1944, and then to 221 Ku, finally returning to the homeland. He ended the war as an instructor with Tsukuba and Yatabe Kokutais. A steady, reliable but unspectacular pilot, he ended the war with claims of 16 aircraft shot down.

Lt Cdr Mochifumi Nango

Nango was to make his name during the early phase of the China Incident as the leading commander of fighter units. After his death he became known as Gunshin Nango Shosa (Lt Cdr Nango, War God). His grandfather was a high official in the Navy Ministry during its early days, as well as being one of the elder statesmen who were members of the National Diet. Nango's father was a rear admiral and had served as head of the Kodokan (Judo Centre in Tokyo). Born into such a naval-oriented family, it was not surprising that he should enter the Naval Academy after attending the Gakushuin High School. He graduated from the Academy in 1927 with an excellent scholastic record and was ordered to sea. In November 1932 he completed the 22nd Aviation Student Course and became a fighter pilot. After service on board *Akagi*, and with Yokosuka Ku, he was sent to England to serve as assistant naval attache in the Japanese Embassy. After two years' residence in London, he returned home. Following the outbreak of the China Incident, he was transferred in October 1937 to the 13th Ku and moved to central China as a buntai leader. On 2 December he led six A5Ms over Nanking, engaging some 30 Chinese fighters which attempted to intercept. 13 victories were claimed, two of them by Nango himself. For achieving such victories, the C-in-C of the China Area Fleet immediately sent Nango a letter of commendation.

In December he was posted to *Soryu* as a hikotai leader, but during summer 1938 he was transferred to the newly-formed 15th Ku as hikotai leader and went again to China, this time to Anking to assist in the Wuhan operation. During the course of an air fight over Nanchang on 18 July, despite his aircraft still having an overload drop tank attached, he shot down a Gladiator. At that point his A5M was struck by a Chinese aircraft which was falling out of control, and he crashed into a lake, being killed. At the time of his death he was 30 years old, unmarried and was mourned very widely in Japan. It was reported that the vice minister of the navy, Isoroku Yamamoto, whilst attending the wake held by the Nango family, wept in front of the altar. Nango was described as being a gentle, well-trained, youthful model officer who enjoyed the trust and affection of those both above and below him. However, inside there burned a strong competitive urge. One episode which is felt to define him related to an attack on Anking during which his aircraft was hit by enemy fire. He held in place for two and a half hours a piece of ruptured oil piping until he was able to reach his home airfield. He claimed eight aircraft shot down over China before his death. It is interesting to note that his younger brother, Lt Col Shigeo Nango, later became one of the Japanese Army Air Force's leading fighter aces before his own death in action.

Lt(jg) Hiroyoshi Nishizawa

Lt(jg) Nishizawa was considered to have been the top fighter ace of the Pacific War. Born in 1920 in a mountain village in Nagano Prefecture, the fifth son of Shuzoji Nishizawa, his father, and Miyoshi, his mother. Following graduation from higher elementary school, he worked for a time in a textile factory, but then responded to a recruiting poster to join the Yokaren programme. He was accepted, and in June 1936 qualified as a student in the Otsu 7th Class. In March 1939 he completed the flight training course, being placed 16th in a class of 71. Following service with Oita, Ohmura and Suzuka Kokutais, he was assigned to Chitose Ku in October 1941, just before the outbreak of the Pacific War, being posted to the Marshall Islands on air defence duties. On 3 February 1942 he claimed his first victory when he shot down a flyingboat at night over Rabaul. During the month the detachment with which he was serving was transferred to Tainan Ku and was sent to eastern New Guinea. Following the US invasion of Guadalcanal on 7 August he also took part in the fighting over the Solomons. Before withdrawal to Toyohashi in November, he had recorded 30 (confirmed) victories. In combat on 7 August he had claimed six Grumman F4Fs shot down. During this fight his own aircraft was hit and he decided to commit suicide by ramming an opposing aircraft. However, when he turned to do so, there were none to be seen; instead, he just managed to undertake the long flight back to base in his damaged Zero.

With the move of 251 Ku to Rabaul in May 1943 for the second time, he took part in further fighting over the south-east area until September when he was transferred to 253 Ku, but returned to Japan in October. During this period Nishizawa's abilities had increased greatly, and he was honoured with a gift from Admiral Kusaka, C-in-C, South-East Area Fleet of a military sword inscribed Buko Batsugun (For Conspicuous Military Valour). He was by now held in complete trust by both superiors and subordinates. It was recorded that following one combat after which Nishizawa's return was delayed, Admiral Kusaka himself stood by for many hours at the airfield awaiting his arrival, showing a quite extraordinary degree of deep personal concern over the fate of what was, when all is said and done, a mere noncommissioned officer. According to the official records Nishizawa had claimed six victories since his return to Rabaul by mid June. However the kokutais with which he served then discontinued the practice of recording individual claims, and Nishizawa's own record is not clear for the rest of this period.

In November he was promotoed Wt Off and posted to 203 Ku which moved from Atsugi to the northern Kuriles for air defence. With the activation in October 1944 of the 'Sho-go' operation, a move was made on 24th to Mabalcat, Luzon, to become part of the northern command there. Next day Nishizawa took off to lead a flight of three fighters serving as escort for the first kamikaze suicide mission. Brushing aside interference from F6Fs, he shared in shooting down two US aircraft, completed the mission and landed at Cebu. Next day, however, while returning to base in a transport aircraft after leaving his own fighter behind, two F6Fs intercepted in the skies above Calapan, Mindoro Island, and shot the aircraft down, Nishizawa dying with the others aboard.

He had remained convinced that he led a charmed life and would never be shot down in aerial combat. However, as a passenger in a transport aircraft, there was nothing he could do in his own defence. Following his death, the commander of the Combined Fleet honoured him with an all units bulletin and a special double promotion. However, in the confusion surrounding the closing days of the war, publication of the bulletin was delayed. Funeral services were held on 2 December 1947. He was given the posthumous name Bukai-in Kohan Gida*. Various figures have been given for Nishizawa's tally of victory claims; these range from 147 (reported to his family); more than 150 (newspaper articles at the time of his death); 102, and other figures. It is believed, however, that the figure of 86 aircraft mentioned to the hikotai leader of 253 Ku, Lt Cdr Okamoto, at the time of the evacuation from Rabaul is the most trustworthy count. If the results of his final air battle are added, then perhaps 87 may be near the truth. Officially, the estimate given is around 60 to 70. In any event, it is the highest total ever recorded in the history of either the Army or the Navy fighter units. As is the case with many abnormally outstanding characters, a number of legends have arisen surrounding Nishizawa; he certainly left a number of quite different impressions on people. Some recorded that he was lively and social; others, that he had been taciturn, aloof, stubborn and obstinate. Although accomplished in the arts of judo and sumo, he had a pale face and complained frequently of having problems with his internal organs. Tall for a Japanese, he measured around 177 cm in height.

Wt Off Yoshinao Norichi
Born in Miyazaki in 1919, he graduated from Otsu 9th Class and was posted initially to 12th Ku, and then to 3rd Ku. With this latter unit he took part in fighting over Luzon on 8th and 10 December 1941 as wingman to Lt Takeo Kurosawa. On 3 February 1942 over Surabaya, Java, he and two other pilots claimed to have jointly shot down four aircraft, while two days later he participated with his shotai in claiming five P-40s shot down and two probables. Then, on 20 February, he and Sea 1c Masao Masuyama, shared five victories over Bali Island. On conclusion of the first phase of the fighting, he returned to Japan. In June 1944 he was with 253 Ku on Truk, where he was again involved in combat. However, while flying to Guam with 12 others on 16 June, their formation was bounced by F6Fs just as they were preparing to land, and he was among five who were shot down and killed. He had claimed a total of eight victories at this time.

* The Buddhist phrase contains several elements: reference to the deceased person in a form reserved for members of the nobility or other high personages "Bukai-in"; the posthumous name "Kohan Gida"; and a term including information that the person is deceased and a member of the Zen sect of Buddhism. The phrase could be translated: In the ocean of the military, reflective of all distinguished pilots. an honoured Buddhist person.

Ens Ki-ichi Oda

Born in Niigata Prefecture in 1913, Oda enlisted in the navy in 1931 and served briefly in the paymaster corps. In November of the following year, however, he graduated from the 18th Pilot Training Class and became a fighter pilot. Following service with Ohmura and Yokosuka Kokutais, he served on *Hosho* for a time before returning to Yokosuka Ku. In August 1937, immediately following the outbreak of the China Incident, he was posted to *Kaga* and moved to Shanghai. During the first attack on Nanking on 19 September, the main attacking force included 12 Type 96 (A5M) fighters from the 13th Ku. These were joined by three fighters from *Kaga*, led by Lt Igarashi. However both the leader's aircraft and that of his No.2 malfunctioned and had to turn back, only PO2c Oda continuing with the 13th Ku formation. In combat over the Chinese capital, he then claimed three Curtiss Hawks shot down (one of them being classed as a probable). He was to add a further victory on 7 October during a raid on Shaoguan. He returned to Japan in December, to Kasumigaura Ku, but in March 1938 joined the 13th Ku, moving in June to the 15th Ku. He remained active over China until November, when he was again posted home. Subsequently, at a time when the 1st Air Fleet was being reorganised, he was posted to the carrier *Soryu*. During the attack on Pearl Harbor on 7 December 1941, he flew as leader of the 3rd shotai in the second attack wave's fighter escort, strafing Kaneohe Naval Air Station. During April 1942 he took part in the Indian Ocean operation when attacks were made on the British naval bases on Ceylon, claiming three victories during fierce fighting over Colombo on 5th. Four days later he was flying Combat Air Patrol over the carriers when a formation of Blenheim bombers sought to attack. Oda's shotai shot four of these down. During the Battle of Midway two months later, he again undertook CAP duty. After *Soryu* was sunk, he was rescued, and was returned to Japan. In May 1943 he was promoted Wt Off and placed on the reserve, but was immediately mobilized again, then serving as an instructor at Iwakuni and Kure Kokutais. In April 1944 he was posted to 261 Ku and moved to the Marianas from where he took part in a number of battles in the Micronesia area. He was transferred to S306 in July, remaining on Truk when no aircraft were left available. He managed to obtain a ride home in submarine I-365, but on 10 December the vessel was sunk in the Ogasawara area, Oda sharing her fate. Promoted Ensign posthumously, he had claimed nine victories.

Ens Nobu-o Ogiya

The name of Ogiya was not well-known even amongst his colleagues since his air combat activities covered only a short period of three months, from the end of 1943 until his death in action in February 1944. This was a period which saw the tide turning against the Japanese forces. The speed with which he claimed 18 aircraft shot down over Rabaul in a period of just 13 days is the highest for such a

concentration of victories on record in Japan; it exceeded even that of Wt Off Shinohara, the Army's top ace during the Nomonhan air battles. Ogiya was born in 1918 in Ibaraki Prefecture into a family descended from a line of swordsmen. After graduating from Minato Commercial School, he initially wished to be a swordsman himself, attaining the third rank in kendo (Japanese fencing). Ultimately, he chose the path of being a warrior in the skies instead, and enlisted in the navy in 1938. After graduating with the 48th Pilot Training Class in January 1940, he was posted to the fighter unit of Chitose Ku. At the time of the outbreak of the Pacific War, he had been assigned to the air defence of the Micronesia area, but here he saw no air combat. He was then posted to 281 Ku, going to the northern Chishima (Kuriles). Transferred to 204 Ku in November 1943, he then went to serve in the Solomons, becoming engaged in the furious fighting over Rabaul, by which time he was nearly 26 years old. He claimed his first aerial victory on 16 December during an attack on Cape Marcus, while by the end of January 1944 when he transferred to 253 Ku, he had been credited with shooting down 24 aircraft. Particularly, on 20 January he had single-handedly claimed five American aircraft – two F4Us, two SBDs and a P-38. Following an interception on 13 February, his aircraft, decorated with 32 cherry blossom victory marks on its fuselage, failed to return.

Wt Off Ryoji Oh-hara

Born in Miyagi Prefecture in 1921, Oh-hara was a wartime pilot who graduated with the Hei 4th Flight Training Course in July 1942. Immediately upon graduation he was assigned to the 6th Ku, in October of that year moving to Buin as a replacement. After achieving his first victory during an attack on Guadalcanal on 23 October, he became involved in a series of fierce battles over the Solomons, Rabaul and New Guinea until his return to Japan later in 1943. On 13 May, flying as No.2 to Lt Miyano, he claimed an F4U Corsair shot down over the Russell Islands, but immediately his own aircraft was hit 38 times and the horizontal stabilizer was badly damaged. He managed to break off, but was pursued by two more Corsairs; turning on these, he claimed one of them shot down, then carrying out an emergency landing on Kolombangara Island. As a result of this particular combat, he received a good conduct commendation from the commander of 204 Ku. Following his return home, he was assigned to Yokosuka Ku. After taking part in the air defence of the Kanto area during spring 1945, he survived the war, having claimed 16 aircraft shot down. Subsequently, he served with the Maritime Self-Defence Force, and was believed still to be alive at the time of writing.

Lt(jg) Yoshio Oh-ishi

Born in 1923 in Shizukoka Prefecture, Oh-ishi qualified for the Otsu 9th Yokaren Class in June 1938 and enlisted at Yokosuka Ku. In October 1941 he graduated from the Flight Training Course and was posted to Ohmura Ku. At the end of July 1942 he was posted to serve aboard *Zuikaku*, immediately going with the vessel to the Solomons area. His first air combat occurred during the Battle of Santa Cruz on 26 October when four Zeros from the carrier led by Wt Off Katsuma Shigemi escorted torpedo-bombers of the

second wave attack on the American carriers. A fight with defending fighters ensued during which Oh-ishi shared in the claimed destruction of nine aircraft (including two probables). Towards the end of January 1943, the *Zuikaku* fighter unit was detached for two weeks to Rabaul to assist in covering the evacuation from Guadalcanal. This island was raided on 1st, 4th and 7 February, heavy air battles taking place on each occasion. In order to take part in the 'I-go' operation in April, the unit returned to Rabaul for a second time, attacks following on Guadalcanal, Oro Bay, Moresby and Milne Bay, Oh-ishi claiming two shot down during these missions. Following this, the *Zuikaku* unit then underwent further training in Japan and at Truk Atoll. On 1 November the unit again moved to Rabaul and took part in repeated attacks on Torokina and Bougainville, while also being involved in several interceptions until 11th, when it withdrew to Truk. At this time the situation in the Marshall Islands was becoming ominous, and without any pause for rest, the unit was moved to Roi. Here during a big raid on 1 December, the unit was virtually destroyed; CPO Oh-ishi survived, however, and was returned to Japan. Here he became an instructor at Oita and Tsukuba Kokutais. In June 1944 he was posted to S302, 252 Ku, and with the activation of the 'Sho-go' operation, he moved to the Philippines via Okinawa and Formosa. On the day of the general offensive against the US shipping, he took off for an attack on a task force off the eastern coast of Luzon, claiming one aircraft shot down during this mission. He then continued to take part in a number of battles over the Philippines until the end of the year. In January 1945 he withdrew to Formosa and was transferred to 205 Ku. With the coming of the Okinawa battle, he was posted to a Tokko unit. On 4 May 21 Zeros of the 17th Taigi-tai ('Noble Cause Unit') carrying bombs and escorted by seven Zero fighters, attacked enemy vessels off Okinawa. That same evening Wt Off Oh-ishi took off by himself from Ilan (Giran) on a battle verification flight. He failed to return and was deemed to have been killed in action, being given a two-grade promotion posthumously, to Lt(jg). Due to inadequate records, the number of his victories can only be verified as about 11. However, he was known to be the top-scoring pilot of the *Zuikaku* fighter unit, and it is estimated that the toal of his claims should actually be at least 15.

Ens Yoshio Ohki

Born in Ibaraki Prefecture in 1916, Ohki entered the navy as an engineer in 1933, but later changed to being a maintenance man. His desire, however, was to be a pilot, and in July 1937 he graduated with the 37th Pilot Training Course, becoming a fighter pilot. In July 1940 he was posted from Yokosuka Ku to the 12th Ku in China. During an air raid on Chungking on 13 September in an A6M, he experienced his first combat, claiming four Chinese aircraft shot down. He was posted to Tainan Ku in July 1942, seeing action over New Guinea and the Solomons until November, when he returned to Japan. In

the following May he was again posted to Rabaul as a member of 251 Ku, but on 16 June during an air battle over the Russell Islands, he was shot down and killed. He had claimed 17 victories.

Ens Shigetaka Ohmori

Born in 1916 in Yamanashi Prefecture, Ohmori enlisted in the navy in May 1933; he graduated from the 33rd Pilot Training Class in September 1936 and became a fighter pilot. In February 1938 he was posted to the 13th Ku in central China where he gained his first victory over Nanchang on 25th of that month. In March he transferred to the 12th Ku and remained active over the same area until December. Subsequently he was posted to *Akagi*, and then to Tsukuba and Ohminato Kokutais. At the outbreak of the Pacific War he was on *Hosho*, but in May 1942 transferred to *Akagi*. In June, during the Battle of Midway, Ohmori served as a shotai leader in the *Akagi* fighter unit, which was led by Lt Shirane. While escorting the first wave which was heading for Midway Island, US fighters engaged and Ohmori claimed two F4Fs shot down. Following his return to the carrier, he and his fellow pilots took off again almost immediately on Combat Air Patrol, intercepting torpedo-bombers which were seeking to attack the carrier, the Japanese pilots claiming six shot down, However, Ohmori's Zero was hit 14 times and as *Akagi* was on fire, he landed instead on *Hiryu*. During the afternoon he took off from this carrier, which was the last surviving, and patrolled overhead. Despite this, *Hiryu* too was hit and started to burn. He continued flying until 1900 hours that evening and then, with fuel exhausted, he and Lt Shirane ditched near the light cruiser *Nagara*, both pilots being rescued. Following the return to Japan, he was transferred to *Shokaku*. During the Battle of Santa Cruz on 26 October, he was again ordered up on CAP, leading his shotai to intercept dive-bombers, five of which he claimed to have shot down. Seeing that one was about to release its bomb on *Shokaku*, he rammed this aircraft to protect the carrier, and was killed. Because of this sacrificial act, he was mentioned in an all units bulletin and was promoted two ranks posthumously. Ens Ohmori was taciturn by nature, seldom speaking, but was considered to have a kind and conservative character; he was generally well-liked by all.

Lt Takeyoshi Ohno

Born in Ishikawa Prefecture in 1921, he graduated with the 68th Class of the Naval Academy in 1940. In June 1942 he completed the 36th Aviation Student Course and was at once posted to Tainan Ku, which was at that time based at Rabaul. His first experience of combat occurred on 27 August when he was patrolling over Buna with two other Zeros. Spotting a P-39, he attacked this alone and claimed to have shot it down. By the time he returned to Japan in November he had claimed five victories, showing a rapid advance in his shooting ability. At the time it was hoped that he would become a successor to Lt(jg) Sasai, who had

been killed in action. In May 1943 he returned to Rabaul as a buntai leader in 251 Ku, and at once took part in an attack on Oro Bay on 14 May. He followed this with sorties over the Russell Islands on 7th and 12 June, and in attacks on shipping off Lunga Point on 16th. During the two latter missions, despite his youth, he acted as overall 251 Ku leader, also claiming two aircraft shot down and two shared. He led a buntai to Rendova on 30 June 1943, but in combat here he was killed in action. He had claimed eight victories by this time.

CPO Shinkichi Ohshima

Graduating from the floatplane course of Hei 4th in September 1942, he was assigned to the newly-formed float fighter unit of the 14th Ku, reaching Shortland on 12 October. The float fighters were engaged in combat from the very next day, using Rekata floatplane base to operate over Guadalcanal. On 7 November the unit lost five aircraft and pilots, including Lt Hidero Goto, and from then only CPO Eiji Matsuyama and LdgSea Ohshima were left to fly sorties until more aircraft and crews arrived. On 10 December they intercepted a mixed formation of B-17s and P-38s over Shortland, claiming one destroyed and one probable. Following the arrival of the new 802 Ku, Ohshima claimed a probable over a P-38 on 15 January 1943, claimed one shot down on 13 February, and next day shared a B-24 destroyed. In March the unit moved to Jaluit in the Marshall Islands, and in July Ohshima was repatriated to Japan. Here he was transferred to land-based fighters, joining 204 Ku. He intercepted US carrier aircraft over Truk on 17 February 1944, claiming five F6Fs shot down. However, his own aircraft was hit and he was badly wounded. On recovery, he was posted to S316, 201 Ku as the unit was about to fly to the Philippines. Here, on 15 June, he was obliged to force-land in the sea while on patrol, and was killed. He had claimed seven and one shared shot down and two probables.

Wt Off Toshio Ohta

Born to a farming family in Nagasaki Prefecture in 1919, following graduation from higher elementary school, Ohta enlisted at the Sasebo Naval Barracks in 1935. In September 1939 he graduated from the 46th Pilot Training Class, becoming a fighter pilot. Following service with Ohmura and Yatabe Kokutais, he was posted to the 12th Ku in June 1941. Although he reached Hankow, he was to find no opportunties for aerial action over China at this time. In October of that year he was transferred to Tainan Ku, then seeing action over the Philippines and Netherlands East Indies during the early months of the Pacific War, claiming his first victory on 8 December, the first day. During a fight with B-17s over Balikpapan, Borneo, on 29 January, he was wounded by return fire from the bombers, and was kept off flying for a time. In April 1942 he commenced a period of great success as a member of the Sasai Chutai, becoming involved in a spectacular race for victories with Sakai, Nishizawa, and others, flying mainly from Rabaul during this period. From August onwards many of the unit's missions were flown over Guadalcanal. On 21 October 1942, he was seen to shoot down one opposing aircraft, but failed to return and was presumed killed in action. Reportedly, he was a gentle person with a smile always on his face, and was liked by all. At the same time his thirst for combat was strong. On one occasion he pursued a B-17 for over an hour until finally shooting it down. He was officially credited with 34 victories, despite having been killed quite early in the war. At the time of his death his rate of success was the highest in the Sasai Chutai, and it was believed that, had he lived, he might well have matched the totals of his two great rivals, Nishizawa and Sakai.

Ens Kenji Okabe

Okabe hold the official record for having claimed eight aircraft shot down (including three probables) in a single encounter. Born in Fukuoka City in 1915, he enlisted in the navy after graduating from Shuyukan High School. In November 1937 he graduated from the 37th Pilot Training Class and was then posted to Saeki and Ohmura Kokutais. In July 1937 he joined the 12th Ku, serving on the battlefront in China; however, no opportunity presented itself for him to obtain any aerial victories here. Just before the outbreak of the Pacific War, he was transferred to the carrier *Shokaku*, but during the Pearl Harbor attack was retained for defensive Combat Air Patrol over the Nagumo Force carriers. His first chance for success occurred during the 9 April 1942 attack on the British naval base at Trincomalee, Ceylon during which he claimed two Hurricanes shot down. At the Coral Sea battle on 8 May he was again assigned CAP duty, He waited at a higher level than that employed by USN dive-bombers as these attacked the Japanese vessels, diving on one after another, climbing back above after each attack, and in this way he was able to make eight claims. However, his own carrier had been hit and he was unable to land aboard, having to ditch in the sea from where he was rescued.

In July 1943 he was again assigned to *Shokaku* after she had been repaired, and on 1 November went with the ship's air group to Rabaul. Including the major air raid next day on 2nd, he took part in a number of interceptions whilst there. He was later posted back to Japan to Ohmura Ku, but in July 1944 was assigned to the newly-formed 634 Ku. By this time there were no longer any aircraft carriers left to serve on, and with the launching of the 'Sho-go' operation in October, he was despached via Okinawa and Formosa, eventually arriving in the Philippines. After taking part in the fighting over Leyte, he returned to Japan towards the end of the war, ending this period with 601 Ku. Something of a theoretician, but with a strong character, he openly expressed opposition to the concept of Tokko (suicide attack). His final total of claims was 15.

Wt Off Juzo Okamoto

Born in Tokushima Prefecture in 1916, he enlisted in the navy in 1933, subsequently graduating from the 31st Pilot Training Course in March 1936. During July 1937 he was posted to the 13th Ku in central China. Here he claimed his first victory on the evening of 19 September, when flying as No.3 in Lt Shichiro Yamashita's shotai in the second wave of an attack on Nanking. During the sixth attack on this city on 22nd, he made a second claim, while on 22 December over Nanchang, he was able to claim two, although one was classed as a probable. He was then sent back to Japan, serving as an instructor at Saeki, Oita and Suzuka Kokutais. He then undertook a second tour of duty in China, this time with the 14th Ku in the southern part of the country. In September 1941 he was transferred to the 3rd Ku, in the opening attacks of the Pacific War on Luzon on 8th and 10 December, acting as a shotai leader. During these two raids he personally claimed six aircraft shot down. In April 1942 he returned again to Japan, and on this occasion was then posted to the newly-formed 6th Ku. Transported aboard the carrier *Junyo* to Midway Island, he took part in the initial attacks on Dutch Harbour on 3rd and 4 June, but the unit was then returned to the home islands. In the autumn of 1942 he was sent to the Solomons, where he saw considerable action. On 11 October he took off from Buka to patrol off the Guadalcanal anchorage, but on the way back five Zeros led by Lt(jg) Kubo disappeared in very bad weather, and all were subsequently assumed to have died or been killed, the veteran PO1c Okamoto amongst them. He had claimed nine victories.

Wt Off Hiroshi Okano

Born in Ibaraki Prefecture in 1921, Okano enlisted at the Yokosuka Naval Barracks in June 1938. In May 1941 he graduated from the 54th Pilot Training Class and was assigned to Yokosuka Ku, and then in September to Chitose Ku. When the Pacific War broke out he was assigned to the air defence of the Marshall Islands, flying a Type 96 carrier fighter (A5M). In April 1943 he was transferred to the 1st Ku, but late in May was sent as part of a reinforcement group led by Lt Joji Yamashita to strengthen Tainan Ku at Rabaul. For about six months, until he returned to his parent unit in November, he took part in many furious air battles in the south-east area. During this

period, he claimed his first victory over Moresby on 25 June, subsequently increasing his total to six, of which one was a probable. In December 1942 he was posted to 201 Ku, engaged in the air defence of the Marshall Islands, but returned to the home islands with this unit in March the following year. Following additional training at Matsushima, the kokutai returned south to the Solomons in July. Travelling on the carrier *Junyo*, he then moved further south to Buin. In November he was promoted CPO and was transferred to 331 Ku. During almost daily sorties until he was moved to the south-west area, his score kept on rising. In March 1944 he left 331 Ku for S603, 202 Ku, taking part in operations over Biak until September, when he returned to Ohmura Ku in Japan. He ended the war serving with S701, 343 Ku, having claimed 19 victories. In later years he remained an active pilot, and was believed still to be alive at the time of writing.

Wt Off Takeo Okumura

Okumura was born in 1920 in Fukui Prefecture. He entered the Kure Naval Barracks in 1935 and in September 1938 graduated from the 42nd Pilot Training Class. In March 1940 he was assigned to the 14th Ku in southern China. Here on 7 October as a member of a force led by Lt Mitsugu Kofukuda to attack Kunming, he had his first experience of combat during which he was able to claim four opposing fighter aircraft shot down. During July 1942 he was posted aboard *Ryujo*, accompanying this carrier and its air group to the Solomons. Here, on 24 August, he took part in an attack on Guadalcanal; although initially reported to have been lost, he managed to return later. Since the carrier had been sunk in the meantime, he was then transferred to Tainan Ku at Rabaul. From early September until late October he took part in raids on Guadalcanal, claiming 14 aircraft shot down. In December he returned to Japan where in May of the following year he joined 201 Ku. In early July 1943 he returned to the Solomons, to Buin, from here taking part in very heavy fighting. On 14 September he flew five sorties during which he claimed ten aircraft shot down; these included eight fighters, one dive-bomber and a share in a B-24. Although these claims remained unofficial, this is believed to have been the highest number of victories claimed in a single day throughout the entire Pacific War. In recognition of this feat, Admiral Kusaka, C-in-C, the South-East Area Fleet, presented Okumura with a white sheathed military sword inscribed Buko Batsugun ('For Conspicuous Military Valour'). A week after this citation was announced, but before he had received the sword, he took part in an attack on an enemy convoy off Cape Cretin on 22 September, led by Lt Shiro Kawai. A strong defending force estimated as being 50-strong was encountered in poor weather conditions, and from this mission Okumura failed to return, being assumed to have been killed in combat. His commanding officer recommended a posthumous two-rank promotion, but this did not materialize. During his life the bright and cheerful Okumura had demonstrated glimpses of genius. His total of victories has been estimated to include four in China and 50 in the Solomons. The latter figure is felt to be somewhat exaggerated, perhaps a figure of 30 being more appropriate.

Lt(jg) Satoru Ono

Born in Oita Prefecture in 1915, Ono enlisted in the navy in 1932. While serving as a fireman at Ohmura Ku, he volunteered to become a pilot and joined the 23rd Pilot Training Class in September 1933. Graduating in April 1934, he was posted to Ohmura Ku as a Sea2c, trained to fly dive-bombers.

After serving aboard *Ryujo* and *Kaga*, he was then posted to Kasumigaura Ku. Immediately upon the outbreak of the China Incident he returned for further service on *Kaga*, and with this vessel's air group, took part in bombing attacks on various targets in central and southern China. During an attack on Canton on 13 April 1938, although flying a dive-bomber, he engaged Chinese aircraft, claiming one of these shot down. During the offensive against Nanchang on 18 July he was posted to the 15th Ku and flew as No.2 to the commanding officer, Lt Cdr Matsumoto. Together with Lt(jg) Ogawa, Ono landed on an enemy airfield, setting fire to aircraft on the ground there. During this period he also claimed two further aircraft shot down in aerial combat, as a result becoming well-known at home almost overnight. In November of that year he returned to Japan, and after serving as an instructor with Ohmura and Usa Kokutais, was promoted PO1c. In February 1942 he was transferred to Yokosuka Ku and conducted operational tests with the Gekko aircraft. During April he was transferred to Tainan Ku and moved to Rabaul from where he flew reconnaissance sorties over a wide area. Following the US landings on Guadalcanal on 7 August, he flew frequent reconnaissance flights over the Solomons with good results.

Although sent home to Japan in November, Ono returned to Rabaul in May 1943. Now using the night fighter version of the Gekko, modified and equipped with obliquely angled cannons, he served with Lt(jg) Kisaku Hamano as a team, the latter in the rear cockpit. The result was that Ono became known as 'King of the Night' together with CPO Kudo. Ono's first combat experience nocturnally was during the night of 5 June 1943, when he claimed two B-24s shot down; five nights later he made two more claims, one for a B-17 and the second as a probable. In July he moved to Ballale airfield, undertaking both night interceptions and bombing sorties. In January 1944 he returned to Japan, to Atsugi Ku, but in April was posted to S804, 322 Ku, being promoted Ensign the following month. In October 1944 he joined 141 Ku and was despatched to the Philippines. Using Nichols Field as his base, he undertook night bombing sorties against targets on Leyte, still using a Gekko. During this period his own aircraft was attacked by enemy fighters, and on one occasion was riddled with 147 bullet holes. At the end of the year he was pulled back to Japan and transferred to 352 Ku, spending the rest of the war attempting to intercept B-29s, although he was only to manage to shoot down one of these, raising his total of victories to eight.

Ens Yukiharu Ozeki

Born in Aichi Prefecture in 1918, Ozeki enlisted at the Kure Naval Barracks in 1935. During January of the following year he was selected for the 32nd Pilot Training Course from which he graduated in July; he was then posted to Ohmura Kokutai. In December 1937 he was transferred to the 12th Ku, seeing action over central China. Here for a period of almost a year, ending in October 1938, he was active as one of the youngest fighter pilots in the IJN. He experienced his first air combat on 25 March 1938 during a raid on Nanchang when as a PO3c he became engaged in a long fight with a group of I-15s and I-16s, claiming three of the hostile fighters shot down.

Later he was successively posted to Saeki, Ohmura and Genzan Kokutais, but finally, in September 1941, just before the start of the Pacific War, he joined the 3rd Ku. With this unit he took part in the operations over the Philippines and Netherlands East Indies. He was involved in a particularly ferocious battle over Luzon on 10 December when as wingman to Lt Ichiro Mukai, he and Mukai jointly claimed five aircraft shot down. In April 1942, newly-promoted to PO1c, he was posted to the newly-formed 6th Ku. Thus in June he was on board *Junyo*, heading for Midway Island to become part of the island's garrison after its occupation, and to take part in raids on Dutch Harbour. Upon cancellation of this operation, he returned to Japan. Towards the end of the same year he was posted to Buin, Solomons Islands, as a member of 204 Ku, taking part in the daily air battles until May of the following year. Returning home again, he then served with Atsugi Ku. He was transferred to S304, 230 Ku in February 1944, and after defensive duties in the Kurile Islands, moved to Kyushu in October on activation of the 'Sho-go' operation. On 21st he was put in charge of the unit's rear guard and moved with this to Formosa. Rejoining the main part of the unit, he then moved to Bamban in Luzon. Here, during the general offensive against the US fleet approaching, he headed out for an attack but was intercepted by defending fighters. From the dogfight which followed, he failed to return. By this time his personal tally had exceeded 14 victories.

Ens Saburo Saito

Born in Yamagata Prefecture in 1917, Saito enlisted in the navy in 1939 and was assigned to sea duty. In January 1939 he graduated with the 44th Pilot Training Class, then receiving further training at both Oita and Ohmura Kokutais. In October he was posted to the 12th Ku at Hankow, but was soon transferred to Weizhoudao in southern China. Assigned to Lt Aioi's buntai, he saw his first combats on 30 December and 10 January of the following year during attacks on Kweilin (Guilin). In January 1940 he was transferred to *Akagi*, subsequently being posted to Oita, Tokushima and Tsuiki

Kokutais. Upon being promoted CPO in November 1942, he was posted to *Zuikaku*. In January 1943 he moved to Rabaul to assist in the withdrawal from Guadalcanal. On 1 February the *Zuikaku* fighter unit, in conjunction with land-based fighter units at Rabaul, escorted dive-bombers attacking enemy vessels off Tulagi. During this mission, Saito claimed his first victory, sharing in shooting down an F4F. On 4th and 7 February, he formed part of patrols of A6Ms over the evacuation convoy, claiming two aircraft shot down on each of these two days. He then took part in the 'I-go' operation in April, and joined in attacks undertaken off Guadalcanal, and over Oro Bay, Moresby and Milne Bay. Although *Zuikaku* returned to Japan for a while, she then sailed again to Truk in July. On 1 November the carrier's air group was again sent to Rabaul to take part in the 'Ro-go' operation. On 2nd, 5th, 8th and 11 November, Wt Off Saito served as the shotai leader in interceptions. He also participated in attacks carried out off Cape Torokina on 2nd and 3rd. During these operations Saito claimed eight aircraft shot down (including one shared), these including P-38s, F4Us and B-25s. However, during this one week of ferocious battles, more than 70% of the *Zuikaku* fighter pilots were lost. Returning to Truk on 13 November, Saito later led seven others to Roi Island on 26th, then moving to Taroa Island on 3 December to engage in air defence sorties here. On 5th, as he returned to land at Roi, he became engaged in a fight with raiding US carrier aircraft, claiming four shot down. However, four of the Zeros he was leading, including that flown by Ens Ohyama, were lost. The following day he withdrew to Truk aboard a Rikko and was then transferred to Tokushima Ku on return to Japan. In July he was posted to S317, 252 Ku at Mobara. In order to take part in the 'Sho-go' operation, the unit moved to the Philippines on 13 October via Kasanbara, Okinawa and Formosa. He took part in the general offensive against the US task force on 24 October, engaging in combat with US fighters over the waters east of Luzon. After shooting down one, his aircraft was hit and

Zero-Sens of 204 Ku at Lakunai, Rabaul.

A6Ms of 582 Ku at Buin.

he had to make an emergency landing on the shore of Lamon Bay, and was injured in doing so. Rescued by an army guard unit, he was again returned to Japan, but remained bedridden until after the war had ended. He had completed 2,188 hours of flight time and had claimed 18 aircraft shot down plus another six shared or probables, according to his flight log.

Lt(jg) Saburo Sakai

The top-scoring ace to survive the war, Lt(jg) Sakai wrote *Samurai in the Big Sky: The Record of the Air Combats of Saburo Sakai*; this book became a best-seller which was translated into several languages (in English simply as *Samurai*). Born on a farm in Saga Prefecture in 1916, Sakai enlisted as a seaman at Sasebo Naval Barracks in May 1933. After graduating at the top of the 38th Pilot Training Class in November 1937, he was posted in September 1938 to the 12th Ku in central China. At this time aerial opposition had almost disappeared, and Sakai's only success here occurred during an attack on Hankow on 5 October when he was able to claim a single victory. In April 1941 he was posted for the second time to the 12th Ku, this time at Hankow. He took part in attacks on Chengtu and Lanchow. During June he was promoted to PO1c and was transferred to Tainan Ku during October. Immediately on the outbreak of the Pacific War he took part in the battles over the Philippines and Netherlands East Indies as a shotai leader. During April 1942 he accompanied the unit both to Rabual and to Lae, and was active as the senior NCO pilot in the Sasai Chutai. On the first day of the air battle for Guadalcanal, 7 August, he was hit by fire from an SBD rear guner, receiving severe head and eye wounds. Despite these injuries, he managed to make the long return flight to Rabaul from where he was transported back to Japan. Official records show that during this period the number of aircraft shot down by Sakai had reached 28. Later, following long hospital treatment, he was posted to Ohmura Ku and then to Yokosuka Ku. During June 1944 he took part in interceptions over Iwo Jima, but was then forced to give up aerial combat because of failing vision; he therefore ended the war as an instructor at Yokosuka Ku.

It has been widely claimed that the total number of aircraft he had shot down numbered 64; the majority were claimed during the early stages of the war and in New Guinea. Sakai was a typical hard-working fighter pilot. His aerial combat techniques were finely crafted and logical in nature. It is noteworthy that he never lost a single wingman during some 200 or more sorties. He is also credited with being the pilot who, on 10 December 1941, shot down over Luzon the B-17 bomber flown by Colin Kelly, who became an American air hero early in the war*. Also, shortly thereafter, he, together with fellow pilots Nishizawa and Ohta, reputedly carried out an ostentatious display of a formation loop over the Port Moresby airfield. Sakai was certainly a man with many colourful episodes in his life; he died in 2002.

---

* Saburo Sakai, who has often been credited with destroying this aircraft, was indeed a shotai leader engaged in this fight with the bomber, but he and his two wingmen do not appear to have been given official credit for its despatch, which was shared by five other Tainan Ku pilots. Lt Colin Kelly Jr was posthumously awarded a Distinguished Service Order, becoming the first American air hero of the war; an address in his honour was broadcast to the nation by President Franklin D. Roosevelt personally.

Wt Off Moriji Sako

Born in 1923 in Kumano City, Moriji Sako entered the navy in the Ko 6th Flight Reserve Enlisted Trainee Class. He completed flight training during 1942 and was assigned to 331 Ku. During September 1943 he was posted to Car Nicobar in the Nicobar Islands, Indian Ocean. Here on 22 September, shortly after arrival, he joined with eight other pilots in shooting down a Liberator bomber which they believed to be a US aircraft; it was in fact from the RAF 160 Squadron (FL939 'M'). He achieved his first individual victory on 5 December 1943 whilst engaged in the only joint IJN/IJAAF raid on Calcutta, India. Flying as part of the rear element of a force of 27 A6Ms, he engaged eight RAF Hurricanes, claiming one shot down.

In early April 1943 he moved with Sento 603 to Truk, and here on 30th of that month he claimed three F6Fs shot down over the Atoll. He made two attacks on the landing beaches on Biak Island during early June, then returning to Japan when S603 was disbanded. He subsequently saw action over the Philippines and the homeland, ending the war with Sento 308, and with a total of nine victories. His claims totalled 19, (including probables). After the conclusion of hostilities, he joined Japan Airlines, flying all over the world until his retirement in 1986. He remained in good health at the time of writing.

Lt Cdr Jun-ichi Sasai

Known as the 'Richthofen of Rabaul', Sasai was the highest scoring ace amongst graduates of the Naval Academy. The exploits of the Sasai Chutai, in which such experts as Sakai, Nishizawa, Ohta, and others flew, are vividly described in Saburo Sakai's popular book, *Samurai*. Born in 1918 in Tokyo, the eldest son of naval engineer Capt Kenji Sasai, during his early years young Jun-ichi was not in good health. From around the time he entered Tokyo Municipal High School No.1, his health improved and he came to gain a grade in judo. In 1939 he graduated with the 67th Class of the

Naval Academy, and shortly before the outbreak of the Pacific War, he completed the 35th Aviation Student Course and was immediately posted to Tainan Ku. As early as 10 December 1941 he took part in one of the first attacks on Luzon, but was forced to return early due to engine trouble. His first claim was made on 3 February 1942 during a big fight over Java. After April, when the kokutai moved to Rabaul, his skill in aerial combat improved rapidly and his score increased dramatically. However, he lost his most valuable subordinate, Saburo Sakai, during the first attack on the US forces landing on Guadalcanal during August. Not long after this event, Sasai led eight other pilots over

Guadalcanal on 26 August where 15 F4Fs were encountered, and from this flight he failed to return. Although Sasai's fate has not been verified, it is assumed that he may have been shot down by the US Marine Corps ace, Capt Marion Carl of VMF 223 – but certainly by a pilot of that unit. In a letter despatched shortly before his death, he reported that he had shot down 54 aircraft and implied that he would soon exceed the record of 80 set by Richthofen. However, the official figure stands at 27. Following his death, Sasai's exploits were announced to all units and he was given the honour of being promoted two ranks to Lt Cdr. During his days at the Academy his classmates had referred to him as Shamo (Gamecock) because of his dislike for losing and his vigorously combatative personality.

Pilots of 5 Ku float fighter unit at Kiska in August 1942. Back row, extreme left, Sea2c Hachiro Norita; 2nd from left, PO2c Giichi Sasaki.

CPO Gi-ichi Sasaki

Born in Miyagi Prefecture, he entered the navy in June 1937, finishing the 44th Pilot Training Course to become a two-seat floatplane pilot. He took part in operations over China, while during the early stages of the Pacific War he served on the floatplane tender *Mizuho*, operating over the Philippines and Netherlands East Indies. After *Mizuho* was sunk he was transferred to a newly-formed float fighter unit in Toko Ku which moved to Kiska Island in the Aleutians, equipped with A6M2N aircraft. The waters of Kiska Bay never froze over, and from here they were able to intercept USAAF fighters and bombers operating from Alaska. As the area was always cloaked in dense cloud, the American aircraft were forced to operate at low levels where the performance of the float fighters allowed the Japanese pilots to challenge them on equal terms. On 16 July 1942 six A6M2N pilots shot down a B-24 for the unit's first victory. In August the float fighter element of Toko Ku was formed into a separate unit, the 5th Ku, and on 8th of that month Sasaki and Lt Kushichiro Yamada, the unit leader, shared in shooting down a

USN reconnaissance floatplane. Sasaki's greatest day occurred on 15 September when at the head of four A6M2Ns, he led an attack on a force estimated as 28 fighters and 12 B-25s. In the fight which followed at an altitude of only 800 metres, he claimed four of the fighters shot down. However, his own aircraft was hit and he force-landed on the sea where the aircraft overturned.

As float fighters were lost either in combat or due to inclement weather, supply of new aircraft was sporadic, the 5th Ku pilots operating as and when was possible. Sasaki claimed a share with a fellow pilot in bringing down a B-24 on 3 October, plus another as a probable. On 14 January 1943 he claimed a half share in a P-39 destroyed, bringing his total of claims to four individual and five shared shot down and one shared probable. However, on 24 January US forces landed on Amchitka Island, 130 km from Kiska, building a runway there. Sasaki and another pilot undertook strafing runs on three consecutive days, but on 19 February when the pair took off on a reconnaissance over the area, they failed to return. USAAF records show that they were shot down over Amchitka by patrolling P-40 fighters.

Ens Masao Sasakibara

Born in Aomori Prefecture in 1921, Sasakibara graduated from the Ko 4th Yokaren Course in September 1941 and was posted to the aircraft carrier *Shokaku*. He took part in the Hawaiian and Indian Ocean raids, joining the fighter escort for the attack on a US carrier task force at the Battle of the Coral Sea on 8 May 1942, during which he claimed four aircraft shot down as his first victories. For the Aleutians campaign of the following month, he was temporarily transferred to *Junyo*, but returned to *Shokaku* at the end of that operation. In August the vessel arrived in the south Pacific where on 24th during the Battle of the Eastern Solomons, he flew Combat Air Patrol. On 28 August he was part of a 15-strong detachment led by Lt Shingo to Buka Island, from where attacks were made on Guadalcanal on 29th and 30th, Sasakibara claiming victories on both dates. He took part in an attack on the US task force during the Battle of Santa Cruz on 26 October, as one of Lt Shingo's wingmen. He was then detached to Rabaul to assist in the withdrawal from Guadalcanal, where on 4 February 1943 he was part of an attack by 15 carrier fighters led by Lt Kenjiro Notomi.

After claiming to have shot down two dive-bombers and two fighters, Sasakibara attacked another fighter head-on, but his Zero was hit and he was forced to ditch in the sea. He was rescued, but had received a severe wound to his forehead, which caused him to be hospitalized and evacuated to Japan. In June 1945 he was sufficiently recovered to join 343 Ku where he remained until the end of the war; he had claimed 12 victories during his earlier service. He was believed still to be living at the time of writing.

Wt Off Kiyoshi Sekiya

Sekiya was born in Tochigi Prefecture in 1921, and enlisted in the navy in 1939. In November of the following year he was selected for the Hei 2nd Yokaren Class, subsequently graduating from the 12th Flight Training Course in November 1941. He was posted to the 3rd Ku in spring 1942, but had little chance to undertake operational sorties due to the number of experienced pilots already serving in the unit. He finally experienced his first air combat during a raid on Darwin on 16 June. During the autumn of that year he was posted to 582 Ku at Rabaul. Here he took part in

fighting over the Solomons and eastern New Guinea, making regular claims at this time. He was transferred to 204 Ku in July 1943, serving with this unit until November when he returned to Japan, where he joined Yokosuka Ku. In order to take part in the 'A-go' operation in June 1944, he moved to Iwo Jima, but from here on 24th he failed to return from the first interception over that island and was presumed to have been killed in action, his final total of claims being 11.

PO3c Hiroshi Shibagaki

Born in Niigata Prefecture in 1924, he enlisted in the navy in May 1942 at the age of 18. As a member of the Hei 12th Yokaren Class, he entered Iwakuni Ku in August. Completing the 28th Flight Training Course, he graduated in July 1943 and in the autumn of that year joined 201 Ku at Rabaul. He achieved his first victory during an interception on 7 November, and in January 1944 was transferred to 204 Ku, continuing to engage in daily interceptions over the Rabaul area. Here he was killed in action on 22 January, having shown outstanding skill which led to 13 confirmed victories.

Wt Off Sekizen Shibayama

Born in Saitama Prefecture in 1922, Shibayama completed Otsu 13th Class, then arriving at Rabaul in September 1943. Here he joined 201 Ku, engaging in his first combat on 21 October during which he claimed a P-39 shot down. In January 1944 he was transferred to 253 Ku, and he continued to operate over Rabaul until he was hit and wounded on 15 February. A few days later the main party of the kokutai withdrew to Truk, but he remained at Rabaul. Following his recovery, he continued to fly on those occasions when an aircraft could be rendered serviceable, and here he remained until the end of the war. He had claimed ten victories.

Wt Off Shigeru Shibukawa

Shibukawa was born in Osaka Municipality in 1923, enlisting in the navy in 1940. He graduated from the 23rd Flight Training Course in September 1942 and was assigned to 253 Ku in December of that year. Early in 1943 he arrived at Kavieng from where he took part in many operations over the Solomons and New Guinea until May, when he was posted to Saipan with the unit for rest and recuperation before returning to Rabaul in early September. Here he was again heavily involved in operations until the end of October. On 1 November he took part in an escort to units attacking shipping off Cape Torokina, Bougainville, where he claimed one F6F shot down. At that point his aircraft was hit from behind and his left hand was wounded sufficiently badly for him to be evacuated to

Japan in a hospital ship. After recovery, he was posted to Tsukuba Ku as an instructor, remaining with that unit until the end of the war. His flight log listed 767 hours of flight time and 15 victories, all claimed during his service in the Solomons. He has died since the war following an illness.

Ens Masami Shiga

Born to a farming family in Ibaraki Prefecture in 1919, Shiga enlisted in the navy in 1937 and was assigned to maintenance duties. He was later accepted for aircrew training, graduating with the 50th Pilot Training Class in June 1940, becoming a fighter pilot. In September 1941 he was posted to Chitose Ku and at the outbreak of the Pacific War was engaged in the air defence of the Micronesia area, based on Taroa Island. He claimed his first victory on 1 February 1942 when he intercepted US carrier aircraft in this area. After that, due to being stationed in the Marshall Islands, he had no further opportunity to engage in aerial combat until July 1943 when he was posted to Buin with 201 Ku. For the next six months until he was sent home to join Yokosuka Ku, he saw much heavy fighting over the Solomons,

Rabaul and New Guinea. In June 1944 he was sent to Iwo Jima, taking part in fighting here as wingman to Saburo Sakai. During the 24 June attack on the US task force he was able to fly safely through the night and return to base. He was later transferred to 203 Ku in February 1945, becoming engaged in the air battles over Okinawa; he remained with this unit until the end of the war, having raised his total of claims to 16. After the war he changed his name to Ohtomo, and served for some years in the Air Self-Defence Force.

Cdr Yasuhiro Shigematsu

Born in Tokyo in 1916, Shigematsu attended Tokyo Municipal High School No.8, then graduating from the Naval Academy in September 1938 as a member of the 66th Class. In April 1941 he completed the 34th Aviation Student Course and in September arrived on board *Hiryu*. At the time he was the youngest officer pilot to take part in the attack on Pearl Harbor on 7 December, and in January 1942 was promoted to become a buntai leader, then seeing action over the Netherlands East Indies and Indian Ocean. During the Battle of Midway in June, he was initially assigned to be part of the escort covering the first wave attack. Having returned to his carrier, he then led the escort to the Kobayashi dive-bomber unit that sought the US carrier force. He was rescued following the sinking of *Hiryu* and was transferred to *Junyo*, again as a buntai leader. During the Battle of Santa Cruz on 26 October he was again part of the escort for the first wave of attack aircraft, but immediateely after take-off his aircraft malfunctioned. He landed on again, leaping into a replacement fighter, taking off and racing after the rest of the formation, which he joined in time to take part in the assault on the American carriers. During the spring of 1943, after taking part in the 'I-go' operation, he returned to Japan to serve at Tokosuka Ku. He was nominated to lead the newly-formed 263 Ku in October and the following February led an advance party to the Marianas. However, in a fight with a force of US carrier aircraft on 22nd, the kokutai suffered heavy losses. After receiving reinforcements, the unit took part in the battles in the Marianas and over Palau. On 8 July he led the unit's five remaining aircraft, including one flown by CPO Sho-ichi Sugita, to Palau, but during a combat over the Yap area Shigematsu was killed in action. He received mention in an all units bulletin and a two-grade promotion posthumously. Although of small stature, he had been a good all-round athlete, referred to as 'Undo shinkei no katamari' (a bundle of athletic vigour); his total of victory claims was estimated to exceed ten.

Wt Off Katsuma Shigemi

Born in 1914 in Shimane Prefecture, he completed the 28th Pilot Training Class. He took part in early operations over China aboard *Kaga* in 1937, claiming three victories here. Following the outbreak of the Pacific War he served on *Ryujo*, on 24 August 1942 sharing with two other pilots in claiming four US aircraft shot down and two probables. Subsequently he took part in the Santa Cruz battle on board *Zuikaku*, leading four fighters as part of the escort to the second attack wave on the US carriers. Here ten or more F4Fs were encountered, of which six were claimed shot down plus one probable by his flight. On 1 February 1943 he took part in a fight over Tulagi during which 11 A6M pilots claimed 11 victories and one probable for a single loss. Three days later he engaged a formation of F4Fs and SBDs over a Japanese convoy, but failed to return. Officially it was confirmed that he had claimed eight victories.

PO1c Kiyoshi Shimizu

Born in Kyoto City in 1919, Shimizu enlisted in the navy in 1940 and served as a maintenance man in China. He then commenced aircrew training, graduating from the 24th Flight Training Course and then being posted to 253 Ku in the south-east area. He was transferred to 204 Ku in November 1943, but when this unit withdrew to Truk on 25 January 1944, he returned to 253 Ku. He was killed in action the very next day, whilst involved in an interception. Until that moment he had appeared to be an almost irresistible force, for following his first victory on 4 January, he had added 11 more within the next 20 days.

Wt Off Yoshijiro Shirahama

Born in Tokyo in 1921, he volunteered at the Yokosuka Naval Barracks in 1938, completing the 56th Pilot Training Class as a floatplane pilot during 1941. He was initially posted successively to Kashima, Hakata and Kure Kokutais, but then to the 16th Ku. Immmediately upon the outbreak of the Pacific War he took part in the invasion of Mindanao Island in the Philippines. He was then posted to the 32nd Ku, and then to Ohtsu Ku. Later, now at Tsuiki Ku, he was reassigned to become a carrier fighter pilot. He was attached to the 1st Air Flotilla command in December 1943, moving to Singapore with the reorganised task force fighter unit. Shortly after this move, he went aboard *Shokaku* at Lingga anchorage where he continued to undergo intensive training. During the Battle of the Philippine Sea on 19 June he flew as No.2 to Lt Fumio Yamagata, who was commander of the 3rd Daitai, as part of the escort for the first attack wave sent out by 601 Ku. En route to their target, the formation was attacked by a large force of US fighters and a wild melee ensued from which Shirahama returned to the carrier alone, claiming one victory. During the evening of the next day he was again heavily engaged with US carrier aircraft, three of which he claimed destroyed on this occasion. Far from base, he ditched in the pitch darkness, where he was lucky to be picked up by a destroyer, allowing him to return home to Japan. Here he became a key figure in the rebuilding of the carrier striking force. As a senior NCO pilot of S166, 653 Ku, he concentrated his energies on training new pilots at Oita. When the 'Sho-go' operation required most of the unit to go aboard the carriers for transport to the Philippines during October 1944, he remained behind. During the latter part of the month, however, he was ordered to undertake air transportation duties, moving to Bamban, Luzon, via Formosa. From here he then moved to Cebu where he rejoined the main part of the unit, taking part in interceptions and attacks on targets on Leyte. During these actions he lost his aircraft and two weeks later returned to Oita. When 601 Ku was reorganised in December, he was transferred to S310, instructing younger pilots at Matsuyama and Iwakuni. In February 1945 he moved to the Kanto area where, on 16th and 17th he took part in interceptions of US carrier aircraft. Towards the end of March, in order to participate in the final battle for Okinawa, he moved to Kokubu in southern Kyushu. During an air battle over Kikaishima (or Kikai-go-shima), he claimed three aircraft shot down. During May he returned to the Kanto area, ending the war stationed at Suzuka. He had claimed 11 victories.

PO1c Toshihisa Shirakawa

Born in 1924, Shirakawa, a native of Kagawa Prefecture, completed Ko 5th Class and was posted to 204 Ku. He experienced his first combat on 24 March 1943 over Oro Bay, New Guinea, but it was on 7 April that he claimed his first victory over Russell Island. On 13 May over Buin he claimed one destroyed and one shared over F4U Corsairs, while from late in June he flew as the No.3 man in CPO Watanabe's element. From early July until the end of August he was involved in severe fighting over the central Solomons as an element leader on at least 16 occasions. However, the fighting reports of 204 Ku during this period made no record of personal claims, so it is uncertain how many he made at this time. On 6 September he claimed a P-38 over Hopoi, adding a shared F4U over Vella Lavella on 13th. On 14th he shared in claims for six aircraft destroyed, including a B-24 over Buin. He flew as wingman in the leading element of Zeros over Finschhafen on 22 September, where he reportedly shot down a P-40 and an F4F, but he did not return from this mission. Officially, his total of claims was recorded as nine.

Cdr Aya-o Shirane

Born in 1916 as the fourth son of Takesuka Shirane, who was later to assume the post of Cabinet Secretary, the younger Shirane attended Tokyo Municipal High School No.4, then graduating from the Naval Academy. In March 1939 he completed the 31st Aviation Student Course and became a fighter pilot. In September of the same year he was posted to the 12th Ku at Hankow. From here on 19 August 1940 he took part in the first attack on Chungking using the new and advanced A6M Zero fighter, During a subsequent attack on this target on 13 September, he was a member of the 13 aircraft unit commanded by Lt Saburo Shindo, which claimed 27 enemy fighters shot down on this date. Even Lt(jg) Shirane in his first combat was able to claim one of these. He was promoted Lt in May 1941, and during the Pacific War he took part in the first attack on Darwin by the carrier force, followed by the Indian Ocean operations and then the Midway battle, by which time he was flying as a buntai leader on board *Akagi*. In July 1942 he was transferred to *Zuikaku*, seeing considerable action during the Battles of the Eastern Solomons in August, and Santa Cruz in October. In November he was posted home to Yokosuka Ku. He was then selected as hikotai leader of 341 Ku in November 1943, this unit beginning to equip with the powerful new Shiden fighter. Towards the end of October 1944 he led S701, which he had personally trained on Shidens, to Mabalacat on Luzon. Here the unit took part in the fighting over Leyte, where on 24 November while engaged with P-38s of the USAAF's 433rd Fighter Squadron, 475th Fighter Group, Shirane was shot down and killed. He had been considered an excellent fighter commander, well liked and trusted by all; at the time of his death he had claimed nine victories.

Ens Toshiyuki Sueda

Born in Fukuoka Prefecture in 1913, Sueda enlisted in the navy in 1933 and graduated from the 32nd Pilot Training Class in July 1936. He then served at Ohmura Ku, but immediately upon the outbreak of the China Incident, he was posted to the 13th Ku, moving to Kunda airfield, Shanghai. He took part in the first air raid on Nanking on 19 September 1937, claiming his first two victories here. On 6 October, although the machine guns in his aircraft were not working properly, he pursued one Chinese aircraft so vigorously that it crashed into the ground while trying to evade him. This resulted in him being identified as the proponent of the 'Mutekatsu-ryu gekitsui' method of air combat ('knock'em down without shooting'em'). During this combat his A5M had been hit, the compass being damaged. Unable to find his bearings to return to base, he followed the banks of the Yangtze river, finally getting back as evening fell. In November, having returned to Japan, he was assigned to the 12th Ku, and in May 1940 joined this unit at Hankow. He then participated in the attack on Chungking on 13 September, but immediately thereafter he was posted to the 14th Ku in southern China, being involved in the attack on Kunming on 7 October; in November he returned home again, this time to Oita Ku. In October 1942 he was promoted Wt Off and joined 252 Ku. Towards the end of December came a move to the south-east area, where on 27 December he claimed two aircraft shot down over Munda, although his A6M was damaged and he had to undertake an emergency landing. He remained heavily engaged in the actions here until March 1943, when he was assigned to the air defence of Wake Island. On 6 October 1943, after intercepting a large group of US carrier aircraft here, he failed to return, being deemed to have been killed in action. He had claimed six victories in China plus a further three during the Pacific War.

CPO Masao Sugawara

Sugawara was born in 1924 in Akita Prefecture, entering the navy with the Otsu 15th Yokaren; he subsequently completed the 29th Flying Cadet programme in September 1943, and was assigned to Sento 302 Hikotai, 252 Ku, in January 1944. After undertaking training at Misawa, he arrived on Iwo Jima in June of that year as wingman of the hikotai leader. Here on the morning of 24 June he was about to take off, his aircraft carrying a 250 kg bomb beneath the fuselage, when US carrier aircraft were seen approaching. He took off at once, but the majority of the unit's aircraft were shot down, including that flown by the hikotai leader. Sugawara flew away southwards, then turned after gaining some height, and dived on the F6Fs, claiming three shot down before having to bale out of his own stricken Zero. He was rescued from the sea after swimming for ten hours. He again intercepted F6Fs on 3 July, claiming one shot down although his own aircraft was hit 34 times. He landed on one wheel, but suffered no wounds or injury himself. In the air again next day, he claimed two further victories, one over a floatplane spotting for the fire of US battleships. Thereafter he was repatriated to Japan in a transport on 6 July, where he joined 601 Ku. He next saw action during the Okinawa campaign, then moving to Yamato with the unit where he remained until the end of the war. He claimed a total of seven victories.

Wt Off Kazuo Sugino

Born in 1921 in Yamaguchi Prefecture, Sugino worked in a cement factory until he volunteered for the navy, enlisting in June 1939. In February 1941 he commenced training with the Hei 3rd Yokaren Class, completing the 17th Flight Training Course in March 1942. In April of the same year he was posted to the 6th Ku, and during the Battle of Midway served on board the carrier *Akagi* which was carrying his unit to the island to be left there as air garrison following its capture. However, the events that took place precluded any possibility of his being involved in the actions over the carriers, while the sinking of *Akagi* resulted in him being rescued from the sea and returned to Japan. He was then posted to the converted carrier *Kasuga Maru* for shipping convoy escort and air transport duties. In October he became an instructor at Ohmura Ku until February 1943, when he went aboard *Zuikaku*. The ship sailed to Truk Atoll, where under buntai leader Lt Hohei Kobayashi, the unit was intensively trained in both air combat and dive-bombing techniques. In November Sugino accompanied the carrier's air group to Rabaul where he was to stay even when the group withdrew, being posted to 253 Ku at that point. Until the time of his return to the homeland in March 1944, he was engaged in almost daily interceptions. His first combat occurred on 2 November 1943 when he claimed three aircraft shot down. He then served as an instructor at Tsukuba Ku and other units until August 1944, when he joined 634 Ku. After being involved in the fighting over Formosa and the Philippines, he withdrew to Formosa in February 1945. His next assignment was as an instructor of Tokko pilots at Hakata Ku, where he remained until the end of the war. Since his experience of air fighting started only late in 1943, he was not well-known. However, according to his logbook his flight time amounted to 1,994 hours and he was engaged in 495 operational sorties, about 100 of these resulting in combat as a result of which he claimed 32 victories. After the war he served with the Maritime Self-Defence Force.

Lt(jg) Shigeo Sugi-o

Born in Miyazaki Prefecture, Sugi-o was inspired by the activities of a civilian aviator, Yukichi Goto, who was also a native of his home town of Nobeoka. From his youth his ambition was to be a pilot, and in 1934 he passed the examinations for the Otsu 5th Yokaren Class. He completed his flight training in March 1938 and in September was posted to the 12th Ku in central China. At this time there was little opposition to be met in the air, and while he did take part in escorts to Rikko units raiding Nanyang and other areas, he met nothing. After a year with this unit he returned to Japan, but in April 1941 returned to China for a second tour of duty with the same unit, now based at Hankow. Again the skies were quiet when he was flying, although he did manage to shoot down a transport aircraft during a raid on Kwangyuan. In October he was transferred to the newly-formed 3rd Ku and as a shotai leader he took part in the air battles over the Philippines and Netherlands East Indies following the start of the Pacific War. On the very first day, 8 December, while strafing Iba airfield, Luzon, his Zero was hit by ground fire, Although the fuel gauge then indicated that his tanks were empty, he nonetheless managed to fly back to Takao. On another occasion on 12th over Batangas, he and the other pilots of the shotai attacked eight Filipino Boeing P-26s fighters, claiming all shot down, though it is not clear whether Sugi-o claimed four of these personally, or two shared; actual losses on this occasion were three shot down and two badly damaged – roughly half the number claimed. On 3 February over the Surabaya area he and his shotai of three were able to claim nine victories, eight of them over Dutch fighters, whilst undertaking a surprise attack as these were trying to intercept Japanese bombers, In April he moved to Timor, undertaking a number of attacks on Darwin until August, where heavy losses were claimed to have been inflicted on defending USAAF P-40s. From September-November he was part of a detachment sent to Rabaul which became engaged in heavy fighting over Guadalcanal, Sugi-o now operating in place of the chutai leader. Following a return to the south-west area, he was promoted Wt Off, taking part in further attacks on Darwin. He returned home for the first time in two years during April 1943, where six months later he joined Haikow Ku to help train Otsu class pilots. He also took part in the 1944 spring offensive on Nanning where he fought China-based USAAF P-40s. In May 1944 he again returned to Japan, transferring to Tsukuba Ku, ending the war as an instructor on Shiden fighters. It is estimated that he had claimed more than 20 aircraft shot down.

Ens Sho-ichi Sugita

Born in 1924 in a mountain village in Niigata Prefecture, Sugita withdrew from agricultural school to volunteer for the navy in 1940 at the age of 15. In March 1942 he graduated from the Hei 3rd Yokaren Class and was assigned to the 6th Ku (later 204 Ku). Although he was involved in the Midway battle, there were no chances for air combat for him, but during the autumn he arrived in the Solomons as one of the youngest pilots at the front. Based at Buin, he saw considerable heavy fighting during almost a year in the area. His first victory was claimed on 1 December 1942 when he single-

handedly rammed a B-17 during an interception over Buin. The collision sheared the right wing off the bomber, which then fell, this episode indicating the uniquely strong fighting spirit which Sugita exhibited. On 18 April 1943 his was one of six Zeros escorting the G4M Rikkos carrying Admiral Yamamato and his staff on an inspection trip around Buin. Although Sugita claimed to have shot down two of the P-38s which attacked, his efforts were to no avail because the Admiral's aircraft was shot down and he was killed. On 26 August 1943 Sugita himself was shot down, baling out with burns all over his body. This put an end to his flying for the time being, and he was evacuated to Japan.

In March 1944 he was posted to 263 Ku, then taking part in the air battles over the Marianas and Carolines before transferring to 201 Ku in July with which he served in the Philippines. In January 1945 he was posted to the newly-formed S301, 343 Ku. Here, serving under Lt Kanno, he flew Shiden-kai fighters on both home defence and during the Okinawa fighting. On 15 April, due to a delayed warning, he was caught while still taking off from Kanoya airfield, he and his wingman's aircraft both crashing in flames; both pilots were killed. Earlier in the spring of 1945 he and Ens Saburo Sakai had been decorated for gaining large numbers of victories. Immediately upon Sugita's death he was posthumously honoured by the publication of his name in an all units bulletin, and he was given a double promotion. His claims for the destruction of 70 enemy aircraft and shares in the shooting down of a further 40 were confirmed in his personal commendation. It may, however, be more appropriate to view the total number of aircraft which he had shot down as being in the thirties.

Ens Teruo Sugiyama

Born in Yamaguchi Prefecture in 1920, Sugiyama was selected for the Otsu 7th Yokaren Class in June 1936, subsequently graduating from the Flight Training Course in March 1939. Following assignments at Suzuka, Ohmura and Genzan Kokutais, he was posted to *Ryujo* after the Pacific War had started. He took part in the June 1942 Aleutians campaign, following which he was sent to the Solomons battlefront. During the attack on Guadalcanal on 24 August, PO1c Sugiyama undertook Combat Air Patrol over the carrier, claiming the destruction of two dive-bombers. However, *Ryujo* was sunk and he had to ditch in the sea, from where he was picked up by a destroyer. Following his return to Japan in autumn 1943, he was promoted Wt Off and was posted to 201 Ku at Rabaul. The following January he joined 253 Ku, continuing to see action over the area on a daily basis. In a period of a little over four months, he claimed more than ten aircraft shot down – three of them probables. During an interception on 4 February 1944 he was shot down and killed, with a confirmed total of ten victories.

Lt Cdr Motonari Suho

Born in Tottori Prefecture in 1912, Suho graduated with the 62nd Naval Academy Class in 1934. After serving with the fleet, he completed the 28th Aviation Student Course in September 1937. Advanced training followed at Saeki and Yokosuka Kokutais after which as a brand new Lt(jg) he was posted to the 12th Ku

Suho's favourite A5M with 14th Ku.

in central China in February 1938. Soon after arrival he was injured in a flying accident, but was only retained in hospital for two weeks. On 29 April he first saw action in the air, claiming one aircraft shot down during an attack on Hankow. However, his own aircraft was hit as a result of which it ran out of fuel while he was on his way back to base and he had to undertake an emergecy landing at Anking. Undaunted, he claimed two further victories on 4 July during an air battle over Nanchang. He was then transferred to the 15th Ku, but returned to Japan within ten days. Following service as a training instructor at Kasumigaura and Oita Kokutais, he was promoted Lt in October 1939, He was then ordered to serve as a buntai leader in the 14th Ku, returning to the China front. On 30 December 1939 and 10 January 1940 he took part in attacks on Liuchow and Kweilin (Guilin), flying via Nanning. Undertaking some inspired flying, he was able to claim two victories on each of these dates. During autumn 1940 at Hankow he was provided with special training on the new A6M Zero-Sen fighter. On 7 October he flew to Hanoi from where as shotai leader in a second unit of six Zeros led by Lt Kofukuda, they flew on to Kunming. Here he personally claimed four aircraft shot down, one of which he chased down a valley until it crashed without Suho himself having to fire a shot. In November he was transferred to Genzan Ku, while during the following April he was posted to the Naval Air Technology Arsenal as a test pilot. For the next 18 months he was engaged in the operational testing of Zero fighters and Raiden interceptors. In December 1942, however, he was posted to 252 Ku as hikotai leader in place of Lt Suganami, who had been killed in action. He moved to Munda at New Georgia in the Solomons, where for about two months he took part in various missions including the defence of Munda and Ballale Island, attacks on Buna, patrols over convoys withdrawing from Guadalcanal, and other convoys heading for Lae. In February 1943 he moved to the Micronesia area, later being assigned to lead a detachment on Wake Island. This unit, however, was totally destroyed during a major US air attack on 6 October, the surviving pilots being flown out in Rikkos to Roi Island. Re-equipped, Lt Suho led a group of 19 Zeros on a strafing and bombing attack on Makin Island immediately following its capture by US forces, but while on the way there the formation was suddenly attacked by F6Fs. In their first pass, fire from one of these hit Suho's aircraft in the fuel tank and he had to break away and land on Deboyne Island; as a result, the planned attack by the Japanese unit came to an immediate end without achieving any success. In March 1944 Suho returned to Japan where in May he was promoted Lt Cdr and assigned to Genzan Ku, and later to Tsuiki Ku as a hikocho, and in this post he ended the war. A quiet and self-possessed man and an expert at his job, he was considered by some to have been the best officer pilot in terms of flying skills. His claims for 11 victories over China amounted to the highest figure to be claimed by an officer pilot in this theatre, his subsequent service raising his total to 15. After the war he served as a jet pilot with the Air Self-Defence Force, rising to the rank of general before his retirement; he died on 16 April 1983.

CPO Hiroshi Suzuki

Hiroshi Suzuki was born in Chiba Prefecture in 1922 and completed Hei 2nd Class, then being posted to 204 Ku. He took part in the fighting over the central Solomons from 12 April-18 October 1943. During this period he claimed one and one probable on 13 May, two on 7 June, one shared destroyed on 12 June, and also one F4U and one P-39, both shared. During July and August no individual claims were recorded in the kokutai's report, so details of his activities during these two months are not available. It is, however, known that he took part in 22 engagements during these weeks. On 14 September over Buin he claimed one more victory and one probable, while next day he claimed a P-39  plus an F4U as a probable. After returning to Japan, he was posted to 201 Ku. While flying with this unit to the Philippines on 13 October 1944, he was engaged in an air battle off Formosa during which he was shot down and killed. He had claimed eight victories.

Wt Off Kiyonobu Suzuki

Suzuki was born in 1914 in Fukuoka Prefecture; he enlisted in the navy in May 1933 as a fireman. Selected for flying training, he graduated in August 1935 from the 28th Pilot Training Class, becoming a fighter pilot. Following initial service with Ohmura Ku, he was posted to Suzuka Ku where he was serving at the outbreak of the China Incident. He moved to Taipei where he was engaged in air defence. Towards the end of October 1937 came a posting to the 13th Ku and a move to the base at Shanghai. His first encounter with

Kiyonobu Suzuki's A5M with 12th Ku in China.

hostile aircraft occurred during a raid on Nanking on 2 December. On this mission he flew as No.3 in the 2nd shotai of six A5Ms led by Lt Nango. During a fierce fight with a number of I-16s, he claimed one shot down and a second as a probable, and was given a unit commendation. He took part in an attack on Nanchang on 25 February, flying as No.2 to Lt Takuma, claiming another victory on this date. However, his own aircraft was hit and he was wounded, while his leader's aircraft failed to return. In March 1938 he was transferred to the 12th Ku with which unit he participated in a number of further missions before returning to Japan in September. Whether he made further claims while serving with 12th Ku is not certain, but he did receive an award of the rare Order of the Golden Kite, 5th Class, for his service at that time. Following time as an instructor with Saeki and Oita Kokutais, he was ordered aboard *Kaga* in October 1941. During the attack on Pearl Harbor he took part as a shotai leader, attached to the second wave fighter escort. No US aircraft were spotted, and because of the amount of smoke generated, it was not possible to verify the results of the strafing undertaken by the first wave. Following the raid on Darwin in February 1942, Suzuki was next involved in the

Battle of Midway in June. During the morning of 5th he formed part of the fighter escort for the attack on Midway Island where his shotai claimed 12 aircraft shot down. After returning to the carrier, he took off twice more in defence of the Japanese carriers. During these, he and his shotai claimed 14 US torpedo-bombers and dive-bombers shot down between them. During the same evening, he ditched in the sea and was rescued by the destroyer *Hagikaze*. Following his return to Japan, he was transferred to serve on *Junyo*, being with this carrier during the Battle of Santa Cruz in October 1942. On 26th he joined the escort accompanying the second attack wave heading for the US carriers. However, he and a second pilot who was flying with him, both failed to return and were believed to have been killed in action. His total of victories was believed to have exceeded nine.

Wt Off Ken-ichi Takahashi

Born in Nagano City in 1924, Takahashi entered the Otsu 13th Yokaren Class in 1940. Subsequently, he completed the 26th Flight Training Course in March 1943, then serving successively with Nagoya and Atsugi Kokutais before being posted to 204 Ku towards the end of September. Thus he was thrust into the furious battles taking place over Rabaul. He claimed his first victory here on 24 October, continuing to fly many sorties here, until 204 Ku was withdrawn to Truk at the end of January 1944. He did not accompany the unit, remaining at Rabaul with 253 Ku. Following further actions here, he withdrew with his new unit to Truk in mid February, just after the US carrier air groups had attacked that island. During the 'A-go' operation on 19 June 1944, he left Truk under the command of Lt Cdr Okamoto, heading for Guam. Minutes before landing, he was attacked by enemy fighters and barely managed to get his Zero down onto the ground. A few days later he was flown out of Guam in a Rikko. The rest of the journey back to Japan was then made in a submarine. On arrival, in July he was posted to S308, 221 Ku and in mid October moved with this unit to Clark Field, Luzon, and thence to Formosa to take part in the 'Sho-go' operation. He was then involved in attacks on targets on Leyte, and in interceptions. During this period he also flew to Jolo Island to take part in dawn attacks against Morotai Island. He survived to return to Japan in January 1945, ending the war as an instructor at Tsukuba Ku; he had claimed 14 victories.

Ens Shigeru Takahashi

Born in Miyagi in 1922, he completed Ko 6th Class in January 1942. Posted to Tainan Ku, he undertook his first combat sortie on 6 September of that year. On 11 October he claimed an F4F over Guadalcanal – the first IJN victory over the island. After returning to Japan, he served with 201 Ku, seeing action between 21 July and 25 October 1943. In 1945 he operated over Kyushu with S316 until the end of the war by which time his total of victories was believed to have been about ten.

CPO Kaoru Takaiwa

Born in Nagano Prefecture in 1923, Takaiwa entered the Otsu 13th Yokaren Class at the age of 17, completing this at Tsuchiura, followed by the 26th Flight Training Course at Nagoya and Oita Kokutais. He then underwent advanced training at Suzuka and Atsugi Kokutais. He was posted to 201 Ku in December 1943, and sent to Rabaul where heavy fighting was underway. He claimed his first victory on 17th, immediately after his arrival, then demonstrating his personal prowess by claiming ten victories (three of them probables) during the next two weeks. With the withdrawal of 201 Ku to Truk on 3 January 1944, he was transferred to 253 Ku,

with which unit he continued to operate. His further claims are not known, but are believed as a minimum to have exceeded those claimed during his short service with 201 Ku. However, on 10 February 1944 he failed to return from a sortie and was deemed to have been killed in action.

Ens Tora-ichi Takatsuka

Born in Shizuoka Prefecture in 1914, Takatsuka was a veteran pilot who graduated with the 22nd Pilot Training Class in November 1933. During the China Incident he took part in the 13 September 1940 attack on Chungking by Zero fighters attached to the 12th Ku, personally claiming three victories during this mission. In October 1941 he was promoted Wt Off and retired from active duty, but was at once re-mobilized and assigned to Tainan Ku. In June 1943 he moved to Rabaul where, using unspectacular but well-planned tactics, he quickly built up a personal total of 16 victories. On 13 September 1942, however, he failed to return from a raid over Guadalcanal, and was assumed to have been killed in combat.

CPO Jiro Tanaka

Born in Saitama Prefecture in 1919, he entered the navy where he was selected for pilot training. He graduated from the 39th Pilot Course as top of the class. Serving aboard *Soryu* during the Hawaiian attacks of 7 December 1941, he flew in the second wave, claiming one US aircraft shot down. Subsequently when taking part in the attack on Colombo, Ceylon, on 5 April 1942 he claimed two victories and a third shared against intercepting RAF Hurricanes. Four days later in defence of the fleet, he shared with two others in shooting down three attacking Blenheim bombers; one of the other

pilots involved was PO1c Ki-ichi Oda. Following the loss of *Soryu*, he was assigned instead to *Hiyo*, accompanying this ship's air group to the Solomons area. Here on 17 October 1942 he was part of the escort for a raid on Guadalcanal as wingman to Lt Tadashi Kaneko, the nine Zero pilots involved engaging 15 F4Fs and claiming seven shot down. On 25 October he claimed another F4F as a probable, while on 11 November he claimed one more shot down. Three days later during a very fierce battle above a Japanese convoy during which Lt Kaneko was killed, Tanaka attacked about 18 dive-bombers, claiming four of these shot down, but then had to force-land his own aircraft, being rescued. While patrolling with five other pilots on 10 December, he became engaged with 11 B-17s and five P-38s, and was shot down and killed. By this time he had claimed eight and four shared destroyed, and one probable.

Ens Kuniyoshi Tanaka

Tanaka was one of the leading aces of the China Incident. Born in Saga Prefecture in 1917, he enlisted in the navy in 1934, graduating from the 31st Pilot Training Class in March 1936. Posted to Ohmura Ku initially, he was transferred to the 13th Ku in October 1937 on the outbreak of fighting in China, being sent to Shanghai. His first experience of combat occurred on 9 December, the day of the attack on Nanchang. He expended all his ammunition in shooting down a Curtiss Hawk and a Vought Corsair, both of them biplanes. At this time the number of pilots in fighter units greatly exceeded the number of available aircraft, making it quite difficult for many to gain battle experience. The youngest pilot in the unit, Sea1c Tanaka seemed to be favoured by the gods of war, however, and he took part in six engagements, the largest number achieved by any pilot at this time, allowing him to amass a total of claims for 12 aircraft shot down. In July 1938 he returned to Ohmura Ku, later serving aboard *Ryujo* as well as with Suzuka and Kanoya Kokutais. In October 1941 he was promoted PO1c and posted to Tainan Ku with which he was serving when the Pacific War broke out. On 8 December during the first attack on Luzon, he flew as No.2 to the hikotai leader, Lt Shingo, the two sharing in the destruction of a P-40. He then saw action over the Netherlands East Indies, demonstrating a particular adeptness in the special techniques found necessary for attacking the high-flying and heavily-armed B-17s. While involved in a patrol over ships in the Balikpapan area on the morning of 24 January 1942, he attacked a formation of seven of these bombers. Although hit several times and wounded himself, he made a number of firing passes, shooting down one B-17 which then collided with one of the others; he stayed around long enough to verify that both had crashed into the sea. On 8 February he joined nine other pilots led by Lt Shingo on a sortie over the Java Sea. Here a formation of nine B-17s led by Capt J.L.Dufrane Jr were spotted flying eastwards, the Zeros splitting into two flights in order to attack from both sides simultaneously. Each of three of the bombers were then attacked head-on resulting in two of them, including that flown by Dufrane, being shot down. During April 1942 Tanaka was posted home to Oita Ku as an instructor. He was unable to return to operational flying, however, when it was discovered that he was suffering from a valvular heart condition, and he remained as an instructor with Tsukuba and Kasumigaura Kokutais until the end of the war. He had claimed 17 victories. He was believed still to be alive at the time of writing.

Wt Off Minpo Tanaka

Born in 1923 in Nagasaki Prefecture, Tanaka was to become one of the youngest of the Japanese aces. He entered the Otsu 11th Yokaren Class during June 1939, later graduating from the 23rd Flight Training Course in September 1942 from where he was posted to 261 Ku in June the following year. When the Marianas became threatened towards the end of February 1944, he was sent to Saipan as part of a buntai commanded by Ens Ichiro Higashiyama. During the evening of 30 March Tanaka moved to Peleliu under the command of Lt Ibusuki when US carrier aircraft raided there. Consequently, the next day proved to be Tanaka's introduction to combat, although 261 Ku lost 20 of its 28 aircraft. Tanaka later moved to Meleyon Island (Woleai Atoll), beginning to intercept PBY flyingboats from here. Attacking from in front and below – a dead angle for defence for the PBY – he shot down two. Early in June he moved to Halmahera Island, but from here transferred back to Yap in preparation for the main defence of the Marianas. On 18th he escorted a Ginga unit for an attack on American ships off Saipan, but here he became involved in a fierce battle with interceptions F6Fs, claiming two of these shot down before landing on Guam. He continued to fly interceptions from here over the next few days, but the unit's number of operable aircraft steadily decreased until most of the key personnel had to be evacuated. Only *"that wild Tanaka and his gang of four"* remained on the island. They continued to operate almost constantly, attacking shipping convoys off Saipan, strafing airfields and participating in the air defence until mid-July. On the day that US forces landed on the island, Tanaka took off carrying a 60 kg bomb, dropping this on the invasion beach before making good his escape to Meleyon Island. From there he flew on to Palau, and thence to Cebu. He was then transferred to 201 Ku, and from that unit to 203 Ku. He ended the war engaged in air defence sorties in the Kanto and Kyushu areas, providing escorts to Okinawa Tokko units. During the course of the war he had claimed 15 aircraft shot down, but had been shot down and baled out on three occasions himself. During the postwar period, he served for many years as a pilot with All Nippon Airways, subsequently dying following an illness.

CPO Shinsaku Tanaka

Born in Kumamoto Prefecture in 1924, he entered the navy in November 1939 as Otsu 12th, graduating with the 25th Flying Course in January 1943, when he was posted to 204 Ku. Arriving in the Solomons area in July 1943, he first became engaged in aerial combat on 18 October when flying as No.3 to CPO Kazuo Umeki. After returning from a force-landing during a fight over Torokina, Bougainville, on 1 November, he made his first claims over Rabaul on 24 December and 1 January 1944. During the opening weeks of this year he claimed five fighters destroyed and one probable, shared in shooting down a B-24 and claimed another probably shot down; during this period he flew mainly as wingman to Wt Off Masao Taniguchi. Departing Rabaul with his unit for recoupment in Japan, he was then engaged over Palau on 30 March where he claimed three F6Fs shot down plus one probable. On 12 September when US carrier forces attacked Cebu, he took off with three wingmen to intercept, but failed to return. By this time he had claimed nine destroyed, two probables and one shared shot down.

Ens Masao Taniguchi

Taniguchi was born in 1919 in Fukuoka Prefecture. He enlisted in the navy in 1936, subsequently graduating from the 51st Pilot Training Class in July 1940. During the following April he was posted aboard A*kagi*, where he was still serving at the outbreak of the Pacific War. He saw action as part of the fighter escort during the attacks on Hawaii, Darwin and Colombo. During the attack on Trincomalee, Ceylon, on 9 April 1942, he claimed his first victory. During the Midway battle which followed, as part of the Combat Air Patrol over the Japanese carriers, he was able to claim three incoming torpedo-bombers shot down. However, *Akagi* was sunk, and he had to ditch in the sea, being rescued by a destroyer. Following his return to Japan, he was then posted to *Shokaku*, taking part in the Battle of the Eastern Solomons on 24 August and the Battle of Santa Cruz on 26 October 1942. In November he left the carrier to serve at Ohmura Ku, where he remained until July 1943. Posted then to 331 Ku, he was sent to Sabang, Sumatra. From here he took part in the combined IJN/IJAAF attack on Calcutta on 5 December. During that same month he was transferred to 201 Ku at Rabaul, where he was at once involved in the very heavy air fighting over the Solomons. Whilst there only for about one month, he claimed five victories plus three probables before the unit was withdrawn to Truk. Here he was engaged with US carrier aircraft on 17 February, and then again over Peleliu on 30 March. After withdrawing to the Philippines, and while attempting to regroup on 23 October after an interception over Manila, he was hit and badly wounded, making an emergency landing. He was at once evacuated to Japan, where he had still not recovered when the war ended. He had claimed 14 victories.

Wt Off Takeo Tanimizu

Graduating from the 17th Flight Training Course in March 1942, Tanimizu joined 6th Ku in April. In June he was assigned to the training carrier *Kasuga Maru*, from where he transferred to *Shokaku* in February 1943; in November the air group arrived at Rabaul, where during his first engagement on 2 November, he claimed two P-38s shot down during an interception. On 2 March he was transferred to Tainan Ku, taking part in regular interceptions over a period of four months. Subsequently, at Tainan Ku he served not only as an instructor, but also in making a number of interceptions off

Formosa. During November 1944 while patrolling over a convoy of ships off Amoy, his aircraft was attacked by a P-51 which took him by surprise, and he had to bale out as it fell in flames. He landed in the sea and was rescued, but had suffered burns for which he was hospitalized. Towards the end of 1944 he returned to Japan and was assigned to 203 Ku where he remained until the end of the war. With this unit he participated in the defence of the Kyushu area and also in the fighting over Okinawa. When the war came to an end, he had completed 1,425 hours of flight time and claimed 18 aircraft shot down. Tanimizu died in 2009.

Wt Off Yoshinao Tokuji

Tokuji was born in Miyazaki Prefecture in 1919; on completion of flight training with 6th Otsu, he was posted to 12 Ku, but was later transferred to 3 Ku. On 8 December 1941 he flew as wingman to Lt Kurosawa over the Philippines where on 10th he shared one aircraft destroyed with two others, suffering slight wounds during the combat. Still flying with Kurosawa, he shared four more victories over Surabaya on 3 February 1942, also claiming three aircraft in flames on the ground. Two days later, now leading the last element, he and two others claimed four P-40s shot down

and two more as probables. He and Sea 1c Masuyama claimed a further five between them over Bali on 20 February, but as this initial stage of the war drew to a close, he was returned home to Japan.. He was assigned to 253 Ku in June 1944, operating over Truk against B-24s initially, and then moving to Guam on 19th. Here, as the Zeros arrived, they were pounced upon by F6Fs and he was amongst those shot down and killed. He had claimed eight victories.

Ens Tadashi Torakuma

During the early days of the China Incident, Tadashi Torakuma was renowned amongst all Japan's fighter pilots for his daring. Born in Oita Prefecture in 1911, he enlisted in the navy in 1929 as a maintenance man. Selected for pilot training, he graduated from the 20th Pilot Training Class in July 1933 and was posted to Ohmura Ku. He later served on *Ryujo* before again joining Ohmura Ku. Following the outbreak of the China Incident, he was transferred to the 13th Ku, and on 17 September 1937 arrived at Kunda airfield,

Shanghai. Two days later, on 19th, PO1c Torakuma flew as a shotai leader during the first attack on Nanking, claiming two Chinese aircraft shot down as his first victories on his first operational sortie. During the sixth attack on this city on 22 September, he was returning to base alone when he was surprised by a Curtiss Hawk fighter near Changchow. His aircraft was peppered with 21 13mm and 7.7mm bullets, although none penetrated the fuselage, being deflected off. Turning on his attacker, he claimed to have shot it down, then returning safely. This incident proved to the Japanese the superiority of the all-metal monocoque fuselage of the A5M fighter, but also served as an important lesson of the need to develop guns of larger calibre. Subsequently, during an attack on Nanking on 12 October, one formation of 11 carrier fighters led by Lt Cdr Nakano, found themselves surrounded and under attack by a force of Breda fighters, and three A5Ms were lost; the Torakuma shotai lost Sea1c Ino, but its remaining pilots were able to claim five of the attackers shot down. Torakuma's own aircraft was hit several times, but he managed to undertake an emergency landing and survived to be rescued. This was the first occasion on which any A5M fighters had been lost in combat.

With a personal score of seven aircraft shot down to his credit, Torakuma returned to Japan in November, but in December the following year (1938) he left to join the 12th Ku at Hankow. This time there were no opportunities to take part in further aerial combat, while during October 1939 Torakuma received severe wounds during a Chinese bombing raid and was evacuated home. In April 1941 he was promoted Wt Off and retired to inactive duty. On the same day, however, he was re-mobilized and assigned to the Oh-ita Ku as a training instructor. In June 1942 he was posted to Kanoya Ku and moved to Kavieng. In December he was sent home again to Ohmura Ku. On 16 April 1943 while engaged in a training flight, he was involved in an accident and was killed. Torakuma referred to himself as Torakuma Hyozo ('Tiger-Bear Panther-Elephant').

Lt(jg) Kazu-o Tsunoda

Born in Chiba Prefecture in 1918, Tsunoda qualified for the Otsu Yokaren Class as a trainee and entered Yokosuka Ku. In March 1938 he completed the 5th Flight Training Course and was promoted PO3c. Following initial service with Saeki and Ohmura Kokutais, he served on board *Soryu*. In February 1940 he was posted to the 12th Ku, joining this unit at Hankow. On 26 October eight A6M fighters led by Lt Fusata Iida made their third attack on Chengtu where Tsunoda gained his first experience of air combat and his first success, claiming an unarmed trainer aircraft shot down whilst flying as No.2 to Wt Off Yamashita. He was posted to Tsukuba Ku in November, returning to Japan. At the end of May 1942 he joined the newly-formed 2nd Ku (later 582 Ku) and in August arrived at Rabaul to begin air operations in the south-east area. For the next ten months until his return to Japan in May 1943, he undertook many sorties over Guadalcanal and Buna, as well as being involved in interceptions over Rabaul and Buin. In particular, on 14 November he led a group of eight A6Ms to provide cover for a convoy of transport ships heading for Guadalcanal. Engaging US dive-bombers, he claimed one shot down but then, when his own aircraft was hit, had to ditch near the Russell Islands; he was rescued by the destroyer *Amagiri*. From May 1943 he served as an instructor at Atsugi Ku, but in March 1944 he was posted to S302, 252 Ku. For a period of one week from 30 June, he flew from Iwo Jima which came under attack

Kazu-o Tsunoda flying a 12th Ku A5M over China during 1940.

by strong forces of US carrier aircraft on 3rd and 4 July. During these attacks he claimed one F6F. Having returned home, he was again ordered to Iwo Jima on 14 July, this time remaining there for a month, intercepting B-24s. In October, in order to take part in the air battles then occurring off Formosa, he departed Kyushu yet again, this time for Clark Field in the Philippines, from there proceeding to Okinawa and Formosa. Here he became involved in providing escorts to Tokko units until November when he was himself posted to one such unit, the Baika-tai. During the early part of December he and three of his fellow pilots moved to Cebu to seek an opportune moment to attack a shipping convoy. However, no such opportunity arose, and towards the end of the month he withdrew to Clark Field. During February he walked all the way to Tuguegarao in the northern Philippines from where he managed to escape by sea to Formosa. Having arrived there, he was posted to S317, 205 Ku, and joined another Tokko unit, Taigi-tai. After the start of the Okinawa campaign, Tsunoda flew a number of sorties against the British Pacific Fleet carriers, but again found no good opportunities to attack, and therefore was still alive at Ilan when hostilities ceased in August 1945. He had claimed nine aerial victories.

PO2c Kazushi Uto

Uto was a gentle, white-faced and boyish-looking pilot, known affectionately to his colleagues in Tainan Ku as 'Poppo' (in Japanese Uto's name can also be read as 'Hato', which meant dove; so his nickname meant dove's twitter). Despite this, he flew as wingman to Saburo Sakai, quietly racking up his own score of victories. Born in Ehime Prefecture in 1922, he graduated with the Otsu 9th Yokaren Class in October 1941. He joined Tainan Ku during the following February, accompanying this unit to Rabaul in April, and then to Lae. He claimed his first victory over Port Moresby on 10 April 1942, then seeing action over New Guinea and the Solomons. During a period of less than six months he claimed 19 victories, but he was lost in action over Guadalcanal on 13 September 1942.

Lt Yoshio Wajima

Born in Hokkaido in 1911, Wajima graduated from the 18th Pilot Training Class in November 1932. He became a true veteran and old-timer amongst the fighter pilots. Following the outbreak of the China Incident, he was posted to the 12th Ku and moved to central China. In April 1938 he was transferred to the 14th Ku, then taking part in operations over south China. His duties, however, were confined to patrol and direct support sorties to the army, with no opportunities for aerial combat. He was transferred to Kasumigaura Ku in April 1939, and in May the following year was promoted Wt Off. He joined Chitose Ku in September 1940, and when the Pacific War began he was posted to the air defence of Taroa Island in Micronesia, flying a Type 96 carrier fighter (A5M). On 1 February 1942 US carrier aircraft launched a surprise attack on the Marshall Islands during which the fighter unit on Taroa undertook a number of interceptions. Wajima, now aged 30, claimed three aircraft shot down as his first victories during these raids. In May he was posted to the newly-formed 2nd Ku (later 582 Ku) and returned to Japan. In August he was sent to Rabaul for about four months, where he took full advantage of his well-honed skills to claim ten aircraft shot down (three of them probables). He then returned to Japan to the Yokosuka Ku where in December he was promoted Ensign. When 263 Ku was formed in October 1943, he joined this unit as a buntai leader. On 21 February 1944 Lt(jg) Wajima moved to Tinian in the Marshalls as part of the advanced unit to reach there, but during the course of a fight with US carrier aircraft which raided the island two days later, he was killed in action. His confirmed total of claims was 11 by this time.

Wt Off Hideo Watanabe

Born in 1920 in a farming family in Fukushima Prefecture, he enlisted in the navy in 1937 and graduated from the Hei 2nd Yokaren Class in November 1941, immediately before the start of the Pacific War. He was posted to Chitose Ku in March 1942, assigned to the air defence of Micronesia. During March 1943 he transferred to 204 Ku, serving at Rabaul and then at Buin. Here he took part in the ferocious battles over the Solomons. On 16 June 1943 the hikotai leader, Miyano, was killed, leaving the unit with no officer or warrant officer pilots remaining. Consequently, the youthful 23-year -old Watanabe had to lead the chutai, and even on

some occasions, the whole unit. During the afternoon of 26 August he took off alone to intercept an escorted bomber formation. He had just claimed a B-24 shot down over Buin, followed by an F4F, when he was hit from behind, one bullet passing out through his right eye and fracturing his forehead. Despite this serious wound, he regained control of his A6M just above the surface of the sea and managed to land before losing consciousness. For his gallantry, while still in naval hospital at Rabaul, he was awarded a military sword by Admiral Kusaka, C-in-C, South-East Area Fleet. The sword was inscribed with the words: Buko Batsugun ('For Conspicuous Military Valour'). He was then evacuated home to Japan, but was not able to return to duty before the war ended. He had claimed 16 victories.

Lt Cdr Sada-o Yamaguchi

Born in Hiroshima Prefecture in 1919, Yamaguchi graduated with the 67th Naval Academy Class in July 1939. He then completed the 35th Aviation Student Course in November 1941, and was at once posted to the 3rd Kokutai in Formosa. At the start of the Pacific War 3rd Ku had a large number of veteran pilots and Yamaguchi was scarcely given the opportunity to take part in operational missions at first, undertaking instead additional training coupled with defensive patrol flights over the home base. He experienced his first combat on 3 February 1942 when flying as leader of the second shotai during an attack on Surabaya, Java; he claimed two victories on this date. He was then promoted to lead a chutai, and during the spring and summer took part in a number of raids on Darwin from the island of Timor. From September until December he was based at Rabaul, joining in attacks on Guadalcanal. Here on 3 October his shotai claimed three victories, but his aircraft was hit and he had to make an emergency landing on the north coast of Guadalcanal. He was rescued by friendly forces and was returned to his base. In mid November he returned to Kendari, Timor, and in May the following year was promoted Lt. From here he was again involved in raids on Darwin, making a number of claims against the RAAF Spitfire units based there. In July he was posted home to Yokosuka Ku as a senior student, and on graduating from this course in October he was posted as a buntai leader to 204 Ku, returning at once to Rabaul to take part in the heavy fighting there. By the time he was withdrawn to Truk towards the end of January 1944 he had claimed five aircraft shot down. In May 1944 he returned again to Japan, serving as a buntai leader in the fighter unit of Yokosuka Ku. The following month he was moved to the Hachiman Force, going to Iwo Jima to take part in the 'A-go' operation. On three occasions he intercepted raids by US carrier aircraft, while on 24 June he accompanied the escort force for an attack on the US task force. However, on 4 July, while involved in an interception sortie, he was shot down and killed. His final total of victories was 12.

Lt(jg) Akira Yamamoto

Born in 1913 in Shizuoka Prefecture, Yamamoto graduated from the 24th Pilot Training Class in July 1934 and became a fighter pilot. Following service with Tateyama and Ohminato Kokutais, he was posted to serve aboard *Hosho*. At the start of the China Incident he was sent to central and southern China. On 27 September 1937 while flying as wingman to Lt Hanamoto during a raid on Canton, he spotted two Curtiss Hawks. Challenging them alone, he shot down the aircraft on the leader's wing, thereby achieving his first victory. In December he returned to Japan, joining Kasumigaura Ku where he served until October 1939 when he was posted to the 12th Ku. Here he was again engaged in the aerial fighting over China until July of the following year. He then served as an instructor at Oita Ku. Immediately before the outbreak of the Pacific War he was posted aboard *Kaga* and during the attack on Pearl Harbor he flew as leader of the 4th shotai in the first wave of fighter escort. Immediately following arrival over the target area, PO1c Yamamoto spotted a civilian aircraft apparently sightseeing over the island in a leisurely manner. He shot this down with one quick burst of gunfire, thereby achieving the first victory of the new war. He then strafed Hickam Field, claiming to have destroyed six aircraft on the ground here.

During the Battle of Midway he was assigned to combat air patrol over his carrier, intercepting incoming dive-bombers and torpedo aircraft, five of which he claimed to have shot down. Since *Kaga* had been hit by bombs, he landed instead on *Hiryu*. From this carrier he then took part in the escort to the Tomonaga torpedo-bomber unit which took off to attack the USS *Yorktown*. During a furious engagement with the USN CAP he claimed four of the American aircraft shot down. In July 1942 he was transferred to *Zuiho*, taking part in the Battle of Santa Cruz on 26 October. Promoted Wt Off, he was detached to Rabaul in March and April 1943. Here, while providing air cover for naval convoys and during attacks on Port Moresby, Buna, Oro Bay and Guadalcanal, he added four more victories to his total. In May he was posted home to Yokosuka Ku. A year later in June 1944 he took part in the air battles over Iwo Jima during which he was wounded by fire from a naval vessel; he made a swift recovery and soon returned to operational flying. When the first raids on the homeland by B-29s occurred on 24 November, he was amongst those sent up to intercept. However, his aircraft was hit by return fire and he baled out. His parachute failed to deploy and he fell to his death. Recalled as being warm and sincere, he was a veteran who was modest enough always to volunteer to act as rear guard in any major air battle. Available records show that he claimed 13 victories.

Ens Ichiro Yamamoto
Born in a fishing village in Ehime Prefecture in 1918, Ichiro Yamamoto enlisted in the navy when he graduated from higher elementary school. He was selected for the 50th Pilot Training Class in December 1939, graduating in June of the following year. After initial postings to Oita and Sasebo Kokutais, he then served aboard *Zuiho* for a short time, but was transferred to *Shokaku* in time for the outbreak of the Pacific War. During the attack on Pearl Harbor he was assigned to CAP over the carrier whilst the attack took place. He then took part in the fighting over Ceylon in the Indian Ocean, and then in the Coral Sea. On 8 May 1942 he was a part of the fighter escort led by Lt Takumi Hoashi which protected the unit sent to attack the US carriers. During the fighting over the American fleet he was able to claim four defending aircraft shot down. During the Battle of Santa Cruz on 26 October he was once more assigned to CAP duties, intercepting US torpedo aircraft and dive-bombers seeking to attack the Japanese carriers. On this occasion he and two other pilots shared in claiming nine aircraft shot down. During the following month he was promoted CPO and was posted home to Oita

Ku as an instructor. At the end of 1943 he was again assigned to carrier duty, and in May 1944 was promoted Wt Off. Upon the launching of the 'A-go' operation, he flew sorties from *Zuikaku* as part of 601 Ku. On 19 June, flying as a shotai leader in the *Zuikaku* fighter unit, he led a section of three as part of the first wave attack force against the US task force. En route to the target, however, the Japanese formation was intercepted by a strong force of F6Fs. Counter-attacking from an inferior position, he was reported to have shot down two of the interceptors before his own aircraft was shot down and he was killed. He had claimed 11 victories by the time of his death.

Wt Off Tomezo Yamamoto

Born in Kitami, Hokkaido, in 1922, he enlisted in the navy in 1939, later being selected for the Hei 2nd Yokaren Class in November 1940. Posted to Tsuchiura Ku, he subsequently completed the 12th Flight Training Course at Oita Ku in November 1941, immediately being posted to Chitose Ku for the air defence of Micronesia. In June 1942 he was transferred to 2nd Ku (later 582 Ku) and returned to Japan. In August he moved to Rabaul, claiming his first victories during an attack on Guadalcanal on 12 September when he claimed two aircraft shot down. From then until July 1943 he was active in the Solomons and over New Guinea, despite his youth claiming more than 14 victories (including six classed as probables or shared). Notably, during an interception over Shortland on 5 June 1943 he shared with others in the claimed destruction of five dive-bombers, then adding an individual claim for an F4U Corsair. He was posted back to Japan, joining the Atsugi Ku in August, and in November was promoted PO1c. During February 1944 he was posted to 203 Ku, and following the spring thaw, went with this unit to Shumshu Island in the northern Kuriles. Here he was assigned to intercepting US heavy bombers that were raiding the area from bases in the Aleutians. On 24 June 1944, now a CPO, he took off to undertake a patrol mission, but as his aircraft left the ground, it suddenly lost speed and crashed from an altitude of only 50 metres and he was killed. By this time he had claimed 11 victories, officially confirmed, all with 582 Ku.

Ens Tadao Yamanaka

Born in Kochi Prefecture in 1918, Yamanaka completed the 44th Pilot Training Class. Details of his early career are not known, but when he joined 202 Ku and undertook his first sorties in April 1943, he was already a CPO and an element leader. On 10 May he gained his first experience of combat over Bilimbimbi, when his aircraft was hit and he had to force-land in the sea; he was rescued safely. He subsequently took part in at least five more sorties over Australia. In December 1943 he led a group of pilots to join 204 Ku in exchange for a group from this unit, led by Wt Off Dudumu Ishihara. From mid December Wt Off Yamanaka took part in actions over

Rabaul, Bougainville and New Guinea as a buntai leader. On 24 December he claimed two F4Us shot down and a third as a probable over Rabaul, while on 30th he claimed one aircraft shot down and a B-26 shared. He then claimed victories on 4th and 23 January 1944. Following withdrawal to Truk the following month, he led fighters to intercept a raid by US carrier aircraft on 17 February. He continued to serve at various front line locations before seeing action over Okinawa with 210 Ku. He then ended the war based in the home islands. Officially by that time he had claimed nine victories.

Lt(jg) Koshiro Yamashita

Born in 1910 in a farmhouse in Kochi Prefecture, he enlisted at the Yokosuka Naval Barracks in 1927 as a fireman. In May 1931 he transferred to Kasumigaura Ku, joining the 17th Pilot Training Class, which he completed in March of the following year, then becoming a fighter pilot. He then served successively with Ohmura Ku, aboard *Akagi*, and then the Yokosuka and Kanoya Kokutais. When the China Incident broke out he was serving aboard *Ryujo*, this carrier operating off central and southern China from August to November. On 21 September, now a PO1c and a shotai leader, he took part in an attack on Canton where he made his first claims for one shot down and a second shared. Following a number of attacks in direct support of the naval ground forces, he returned to Japan. In May 1940, following service as an instructor at Suzuka Ku, he was posted to 12th Ku as a Wt Off. Although now 30 years old, he continued to show great skill during air battles over mainland China. During a raid on Chungking by Zero fighters on 13 September when he was leading the 2nd shotai of the 1st chutai, the unit was intercepted by some 30 opposing fighters. 27 of these were claimed shot down, Yamashita personally accounting for five of these; the last of these he chased to within 50 metres of the ground, forcing it to crash into a rice paddy. He and PO Kitahata then undertook a spectacular loop just 50 metres above the Baishi railway station. On 14 March 1941 during an attack on Chengtu, a dogfight took place at low altitude in conditions of dense mist. On this occasion Yamashita was able to claim victories over three aircraft (one a probable). Other engagements also took place, and after one year on this front, he had claimed ten victories, making him the 12th Ku's top-scorer. Subsequently, he served as an instructor with the Kasumigaura, Tainan and Tsuiki Kokutais. In January 1944, however, he was posted to 201 Ku, going to Palau in Micronesia. Here in an interception of an attack by US carrier aircraft on 30 March, he was shot down and killed during his first operational sortie in the Pacific War.

Ens Sahei Yamashita

Born in Shizuoka Prefecture in 1918, Yamashita entered the navy in 1934 as a member of the Otsu 5th Yokaren Class, completing a flight training course. Together with classmate PO1c Yoshino, Yamashita was posted to the Chitose Ku in October 1940. When the Pacific War commenced, he was in the Marshall Islands, assigned to air defence in the Micronesia area. Promoted Wt Off in April 1942, he was then detached to Rabaul. Towards the end of May he was formally transferred to the Tainan Ku and was

sent to Lae. He gained his first experience of combat on 27 May during a raid on Port Moresby, where his shotai claimed a P-39 shot down. He later increased his victories over eastern New Guinea and during raids on Guadalcanal following the American invasion of that island in August. When the Tainan Ku was withdrawn to the homeland early in November, he was transferred to 201 Ku and once again assigned to the Micronesia area on air defence. On 9 February 1943 he and four fellow pilots intercepted a B-17 that was seeking to reconnoitre Nauru Island, and this they shot down. However, Yamashita then lost his way when about 150 miles north of the island and never returned; he was deemed to have died in action. By this time his list of personal claims was confirmed at 13.

PO1c Ichirobei Yamazaki

Yamazaki was born in 1920 in the mountains near the village of Hinohara, in the western part of Tokyo. After entering the Yokosuka Naval Barracks in 1937, he graduated from the 54th Pilot Training Class in May 1941. During February 1942, having served with the Oita Ku he was posted to the 4th Ku at Rabaul. During April he was transferred to the Tainan Ku with which unit he was active over eastern New Guinea and the Solomons, using Rabaul and Lae as bases. On one occasion he pursued a Lockheed Hudson bomber alone, but his Zero was hit and he had to make an emergency landing deep in the

mountains. Aided by local people, he built a log raft and floated downstream, ultimately arriving safely back at his base. During May 1942 he was wounded during a combat over Port Moresby, whilst during an attack on Buna on 16 August he was again hit and had to make another emergency landing. On this occasion he was seriously injured and was evacuated to Japan. On recovery he was posted to 251 Ku in May 1943, returning with this unit to Rabaul. During an attack on Rendova Island on 4 July 1943 he was shot down and killed in action. He had claimed 14 aircraft shot down (confirmed).

PO1c Shigeru Yano

Yano was born in 1917 in Tochigi, enlisting in the navy and completing the 44th Pilot Training Class. He was posted to 3rd Ku before the opening of the Pacific War, and on 8 December 1941, as wingman to Wt Off Akamatsu, he took part in a strafing attack on airfields on Luzon during which Akamatsu's element claimed six aircraft destroyed on the ground. Two days later Akamatsu's element claimed seven victories, Yano taking part in this success, On 31 December Yano personally claimed a P-40 over Davao, while on 5 February the element claimed four P-40s jointly over Bali. Flying No.2 to Lt(jg) Sada-o

Yamaguchi over Darwin on 30 March, he took part in another shared success when the element claimed three destroyed and a probable. Having brought his personal total to eight, he was then killed in an accident on 17 April 1942.

Ens Kozaburo Yasui

Yasui was born in 1916 in Kyoto Municipality. He enlisted in the navy and graduated from the 40th Pilot Training Class in February 1938. At the end of the following year, 1939, he was posted to the 14th Ku, seeing action over the south China front. He returned to Japan in November 1940 to serve as an instructor at Hyakurigahara Ku. In November 1941, immediately before the outbreak of the Pacific War, he was posted to the fighter unit attached to the 22nd Air Flotilla HQ, accompanying this unit to French Indochina. During January and February 1942 he took part in attacks on Singapore, in convoy escort patrols and on other duties. He is known to have made some claims during this period, but details are not known. In March he was promoted PO1c and was transferred to the Kanoya Ku fighter unit. Another move came in August when he joined the battle-hardened Tainan Ku at Rabaul. He saw his first action in this new area during an attack on Rabi, New Guinea, taking part in a fight with P-39s. He claimed two of these shot down and a third as a probable. During the two and a half months he was with the unit he claimed about ten victories before being posted home early in November. In March 1943 he joined the Oita Ku as an instructor, but after a year moved to 652 Ku with which he took part in the Battle of the Philippine Sea. During the morning of 19 March 1943 he escorted carrier bombers from the 2nd Carrier Division to attack a US task force west of the Marianas. Unable to find the target, his formation then flew to Guam. As he went in to land, US fighters attacked and both he and the other pilots with him were all lost. He had claimed 11 victories (confirmed) by the time of his death.

Wt Off Katsuyoshi Yoshida

Born in 1923 in Hyogo Prefecture, Katsuyoshi Yoshida completed Ko 6th Class and was posted to 202 Ku. Undertaking his first operational flight on 30 January 1943, he experienced his first aerial engagement on 15 March during an escort mission to Darwin. Following this, he was to fly over Australia on at least three more occasions. A year later when S301 was formed from the kokutai's flying unit, he saw further action over Truk, Biak, western New Guinea and Guam, following which he returned to Japan. As a member of S304 he then took part in the fighting over Okinawa, remaining with this hikotai until the end of the war, by which time he had claimed ten victories. At the time of writing he was alive and possessed of a very good memory

Wt Off Mototsuna Yoshida

Born in Okayama Prefecture during 1918, Yoshida enlisted in the Kure Naval Barracks in June 1935. He served as a fireman aboard *Chogei* initially, but the following year transferred to maintenance work. In January 1939 he graduated from the 44th Pilot Training Class and became a fighter pilot. Following postings to Oita Ku and Ohmura Ku, he joined the 12th Ku in September 1939 and was sent to China. During Chinese bombing of Hankow on 14 October, he was wounded. On recovery he moved to southern China, taking part in attacks on Liuchow and Kweilin (Guilin), but in July 1940 he became an instructor with the Yokosuka Ku. In February 1942 he was posted to the 4th Ku fighter unit, stationed at Rabaul. Soon after his arrival, on 23 February he intercepted and shot down a B-17 over the unit's base. He was wounded during March and was not able to fly operationally for some time. However, during April he was promoted PO1c and was posted to the Tainan Ku. From May onwards he flew as number two to Lt Shiro Kawai, participating in a number of attacks on Port Moresby, claiming victories on almost every occasion. During the attack on Tulagi on 7 August 1942 he failed to return and was believed to have been killed in action. It is recorded that during his time at Rabaul he had personally claimed 12 destroyed, one probable and three shared.

PO3c Keisaku Yoshimura

Born in Niigata Prefecture in 1922, Yoshimura enlisted in the navy in 1939 at the age of 17. In July 1941 he graduated from the 56th Pilot Training Class and was posted to the 1st Ku. In October he was transferred to the Tainan Ku to undergo further training on Formosa. As of 1 December he was detached to southern French Indochina to join the fighter unit attached to the 22nd Air Flotilla HQ. He then took part in the Malayan and Netherlands East Indies invasions until the end of May 1942. At that point he returned to the Tainan Ku at Rabaul. Sea1c Yoshimura then moved to Lae where he was assigned the number three position in the 3rd shotai, Kawai Chutai. On 16 June he was involved in a raid on Port Moresby during which he claimed two P-39s whilst engaged in a fight with 20 or more of these fighters. Following the return flight to base, he gave chase to another P-39 that had been raiding Lae, and claimed this shot down as well. During the attack on Tulagi on 7 August he flew as wingman to the hikotai leader, Lt Cdr Nakajima, becoming engaged in a pitched battle with F4Fs. Single-handed, he claimed to have shot down five of the US fighters (two as probables). Newly-promoted PO3c, he took part in the attack on Guadalcanal on 25 October, but here he was involved in a battle with at least ten F4Fs, as a result of which his aircraft was hit and fell in flames. Only 20 years old at the time of his death, during less than six months at Rabaul he had claimed nine destroyed, four probables and one shared – all of them fighters.

Ens Satoshi Yoshino

Born in Chiba Prefecture in 1918, Yoshino was accepted in June 1934 at the age of 16 as a student of the Otsu 5th Yokaren Class. Graduating in August 1937, he then undertook the Flight Training Course which he completed in March 1938. Proceeding via the Ohmura Ku and duty aboard *Soryu*, he was posted to Chitose Ku's fighter unit, newly formed in October 1940. In October 1941, immediately before the outbreak of war, he was posted to the Marshall Islands and was engaged in the defence of the Micronesia area. In February of the following year he was transferred to the 4th Ku and moved to Rabaul. On 11 February he led four Type 96 (A5M) fighters on patrol over Gasmata (Surumi), New Britain, which encountered three Hudson bombers of the Royal Australian Air Force, and at once two of these were claimed shot down. Two days later Yoshino claimed another aircraft shot down, again over Gasmata. The unit then re-equipped with Zero fighters and on 11 March, the first day of the landings at Lae, he led seven aircraft which between them shot down one enemy aircraft. Until April, when the Tainan Ku moved into the area, Yoshino was one of the core pilots of the 4th Ku fighter unit. During attacks on Port Moresby, Horn Island and during interceptions he claimed four more aircraft shot down.

Even after his promotion to Wt Off and his transfer to the Tainan Ku, Yoshino continued to participate in the attacks on Moresby and in interceptions over Lae. On 9 June 1942 he intercepted a flight of B-26s over Lae, pursuing them to Cape Ward Hunt, where he was pounced upon by an 11-plane formation of P-39s. He failed to return and was presumed to have died in action. He had claimed 15 aircraft shot down (confirmed) by this date.

Wt Off Tokushige Yoshizawa

Born in Akita Prefecture in 1923, Yoshizawa enlisted at the Yokosuka Naval Barracks as a seaman in June 1940. In October of the following year, wishing to become a pilot, he was posted to Tsuchiura Ku as a student in the Hei 7th Yokaren Class. In March 1943 he completed the Flight Training Course, and in the autumn of the same year was posted to 201 Ku at Buin in the Solomon Islands. At the end of the year he moved to Rabaul where he took part in almost daily interceptions, during this period personally claiming more than nine aircraft shot down. Following transfer to 204 Ku, he remained at Rabaul even after this unit was withdrawn to Truk, when he became part of the Air Base 105 unit. On 9 January 1945 he took off on a mission to reconnoitre the Admiralty Islands, flying a two-seat Zero which had been assembled at Rabaul from parts of wrecked A6Ms. He never returned and was deemed to have been killed in action. Twenty-seven years later, in August 1972, a Zero fighter and a human skeleton were dragged up by Australians from waters at a depth of eight metres just off the mouth of the river at Cape Lambert. These relics were returned to Japan in February 1975, and are believed to have been the remains of Yoshizawa and his aircraft.

A-Z of Fighter Aces' Planes

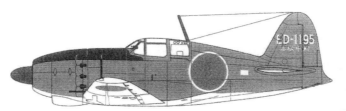

Lt (JG) Akamatsu, Sadaaki / 302 Ku, late 1944 – early 1945 at Atsugi.

Lt Endo, Sachio / 302 Ku.
January 1945 at Atsugi.

WO Hagiri, Matsuo / 204 Ku summer 1943 at Buin.

PO2/c Ishihara, Susumu / Tainan Ku in February 1942 at Bali.

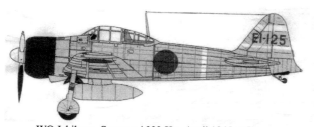

WO Ishihara, Susumu / 332 Ku, April 1945 at Kanoya.

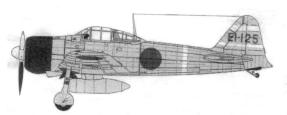

Lt (JG) Iizuka, Masao / *Shokaku*.

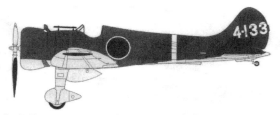

Sea1c Iwamoto, Tetsuzo / 13 Ku, February 1938 at Nanking.

PO1/c Iwamoto, Tetsuzo / *Zuikaku*, December 1941 in Hawaii Operation.

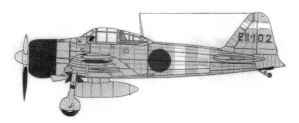

Ens Iwamoto, Tetsuzo / 252 Ku.

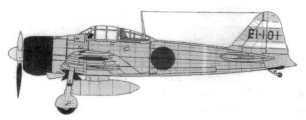

Lt Cmdr Kaneko, Tadashi / *Shokaku*.

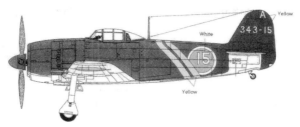

Lt. Kanno, Naoshi / 343 Ku, S301, March 1945 at Matsuyama.

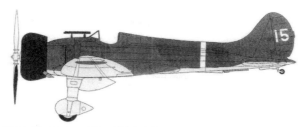

PO3/c Kashimura, Kan-ichi / 13 Ku, December 1937 at Shanghai.

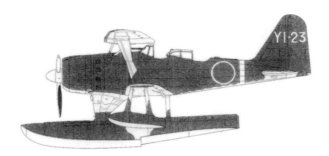

Sea 1/c Katsuki, Katsumi / *Chitose* on October 4 1942
when he shot down a B-17 by ramming.

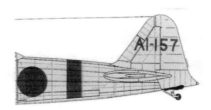

PO1/c Kikuchi, Tetsuo / *Akagi*. December 1941.

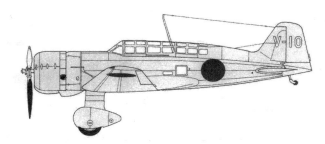

PO2/c Kudo, Shigetoshi / Tainan Ku, February 1942 at Bali.

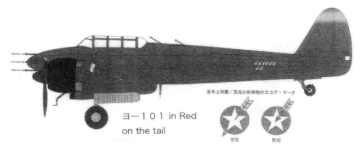

ヨ－１０１ in Red
on the tail

CPO Kuramoto, Jozo / Yokosuka Ku, May 1945 at Yokosuka.

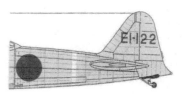

PO1/c Matsuda, Jiro / Shokaku.

WO Miyazaki, Isamu / 343 Ku, S407 April 1945 at Matsuyama.

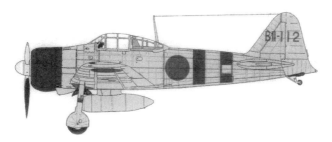

PO1/c Muranaka, Kazuo / *Hiryu.*

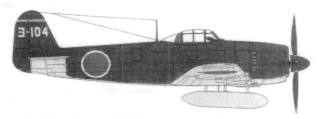

Ens Muto, Kaneyoshi / Yokosuka Ku, April 1945 at Yokosuka.

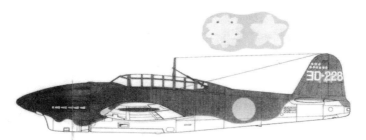

CPO Naka, Yoshimitsu / 302 Ku, June 1945 at Atsugi.

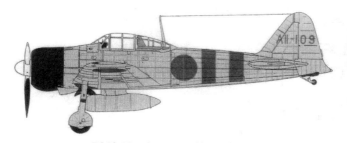

PO1/c Nagahama, Yoshikazu / *Kaga.*

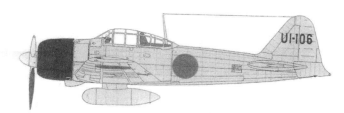

CPO Nishizawa, Hiroyoshi /251 Ku, spring 1943 at Toyohashi.

PO1/c Nishizawa, Hiroyoshi / 4 Ku,
March 1942 at Rabaul.

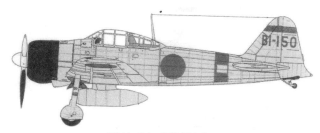

PO1/c Oda, Kiichi / *Soryu*.

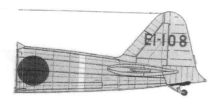

PO1/c Okabe, Kenji / *Shokaku*,
December 1941.

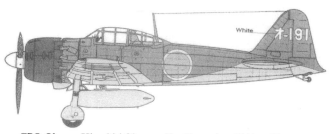

CPO Okano, Hiroshi / Ohmura Ku, December 1944 at Ohmura.

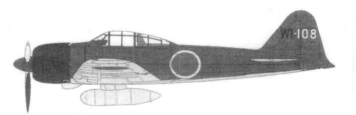

CPO Okumura, Takeo / 201 Ku, September 1943 at Buin.

CPO Saito, Saburo / *Zuikaku*, January 1943.

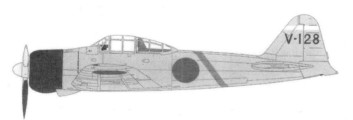

PO1/c Sakai, Saburo / Tainan Ku, August 7, 1942 at Rabaul.

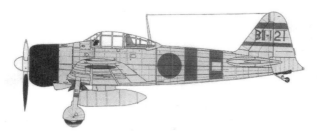

Lt (JG) Shigematsu, Yasuhiro / *Hiryu*.

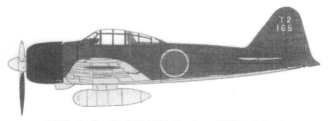

PO2/c Sugita, Shoichi / 204 Ku, June 1943 at Rabaul.

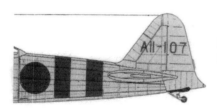

PO1/c Suzuki, Kiyonobu / *Kaga*
December 1941.

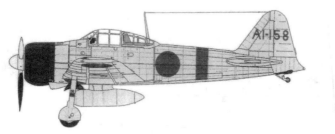

PO2/c Taniguchi, Masao / *Akagi*, December 1941.

CPO Tanimizu, Takeo / 2nd Tainan Ku. September 1944 at Tainan.

CPO Tanimizu, Takeo / 203 Ku, June 1945 at Kagoshima.

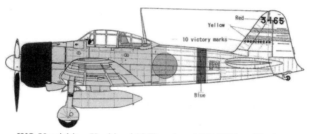

WO Yamishita, Koshiro / 12 Ku, circa 1940-1941 at Hankow.

SECTION FIVE

Maps

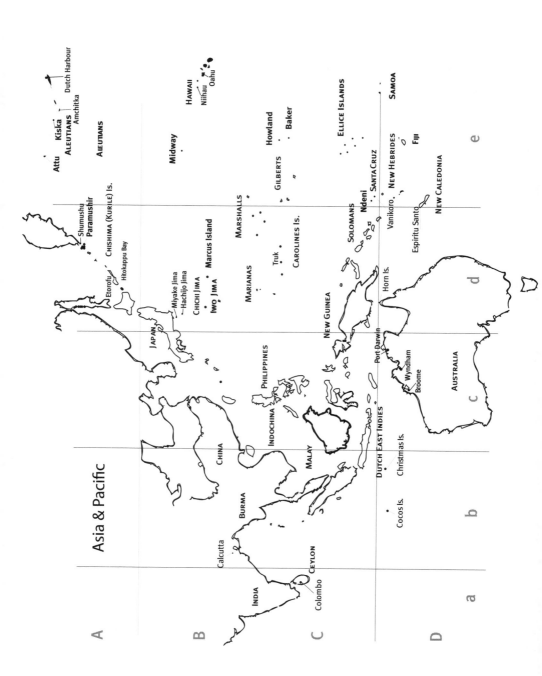

Asia & Pacific

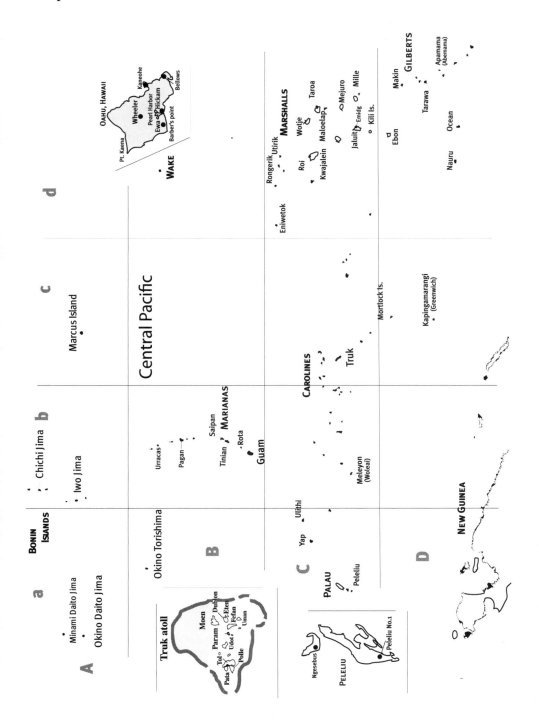

Central Pacific

FORMOSA

Taoyuan

Hsinchu
Houlung · · **Taipei**

D

Taichung
·Changhua

E

·Tainan

Takao · ·Pindong
(Gaoxiong)
·Toko
(Donggang)

f

KOREA

Pyongyang

Genzan
(Wonzan)

A

Seoul

B

Kwangju

JEJU DO

k

JAPAN, KOREA & FORMOSA

HOKKAIDO

Bihoro

·Chitose

·Hakodate

·Ohminato

Misawa

A

HONSHU

Matsushima

Koriyama

Kasumigaura

Kounoike

B

Komatsu

KOREA

CHUGOKU

Miho

Fukuchiyama

Taisha

Hiroshima

Kure

Tsuiki

Usa

TOKAI

Suzuka **NAGOYA**

HANSHIN
Naruo
KOBE Itami

Hamamatsu
Meiji

Nara

Toyohashi
ATSUMIPENINSULA

Yamato

MT.FUJI

TOKYO

Katori
Yokosuka

Tateyama

KANTO AREA
Miyake Jima

Fujieda

ENSHU NADA

Fukushima

Matsuyama

SHIKOKU

INLAND SEA

Sukumo
BUNGO STRAIT

Sasebo
Ohmura

Oita

KYUSHU

NAGASAKI

Izumi

Kokubu Miyazaki

Kagoshima

Kanoya

YAKU SHIMA   TANEGA SHIMA

Hachijo Jima

a   b   c   d   e

C

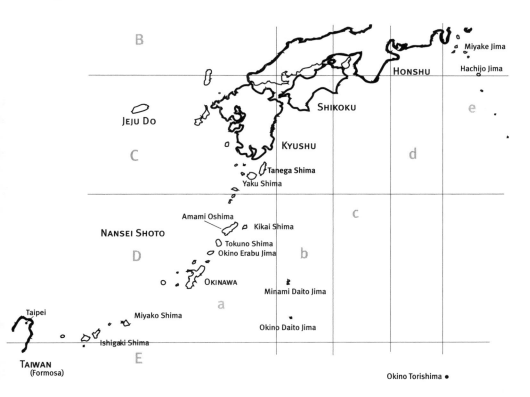

B

Miyake Jima

Hachijo Jima

HONSHU

e

JEJU DO

SHIKOKU

KYUSHU

C

d

Tanega Shima

Yaku Shima

c

Amami Oshima

Kikai Shima

NANSEI SHOTO

Tokuno Shima

Okino Erabu Jima

D

b

OKINAWA

Minami Daito Jima

a

Taipei

Miyako Shima

Okino Daito Jima

Ishigaki Shima

E

TAIWAN
(Formosa)

Okino Torishima ●

OKINAWA & NANSEI SHOTO

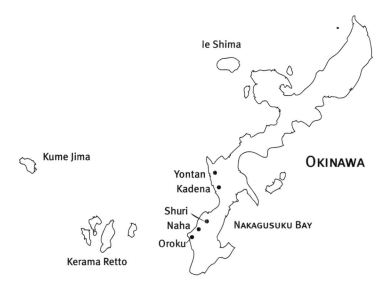

Ie Shima

Kume Jima

OKINAWA

Yontan

Kadena

Shuri

Naha

NAKAGUSUKU BAY

Oroku

Kerama Retto

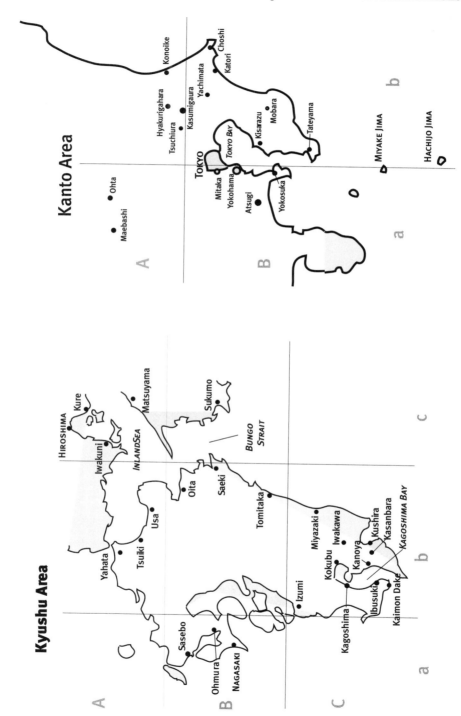

Kanto Area

Kyushu Area

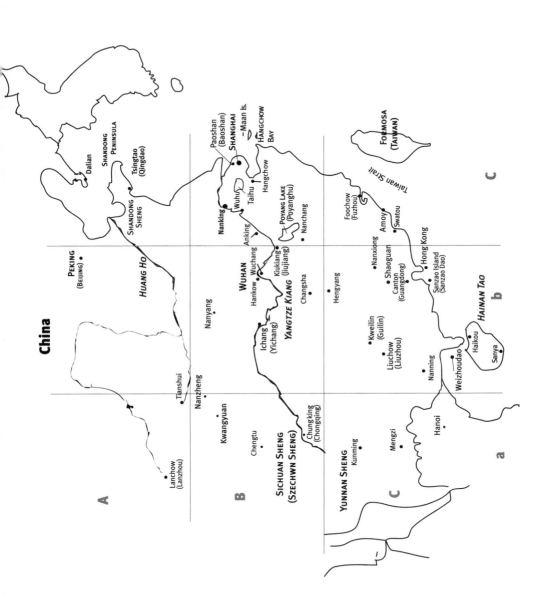

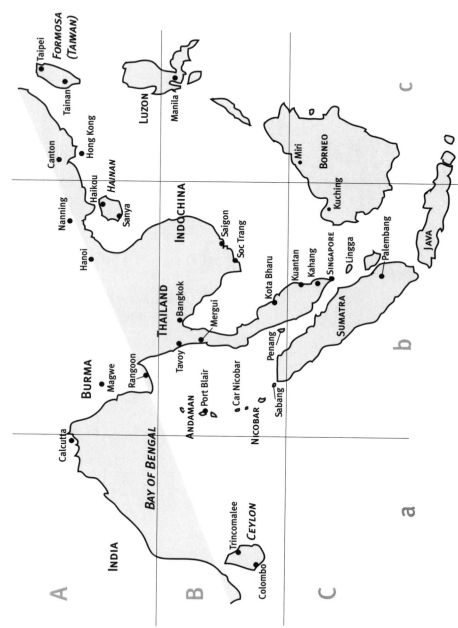

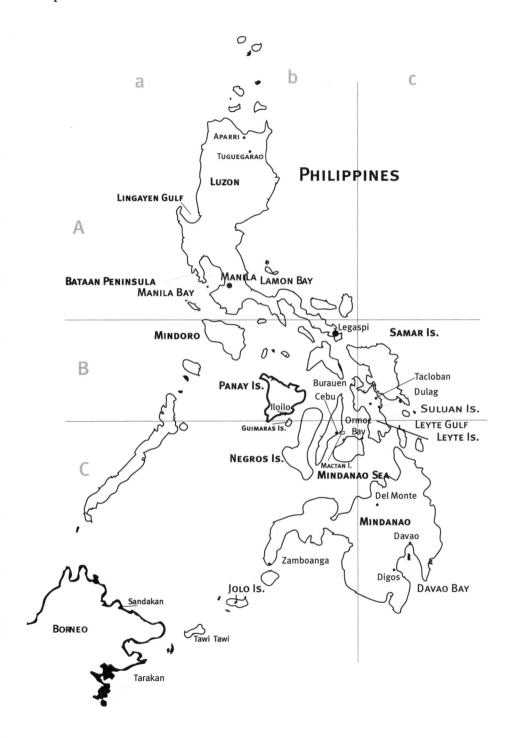

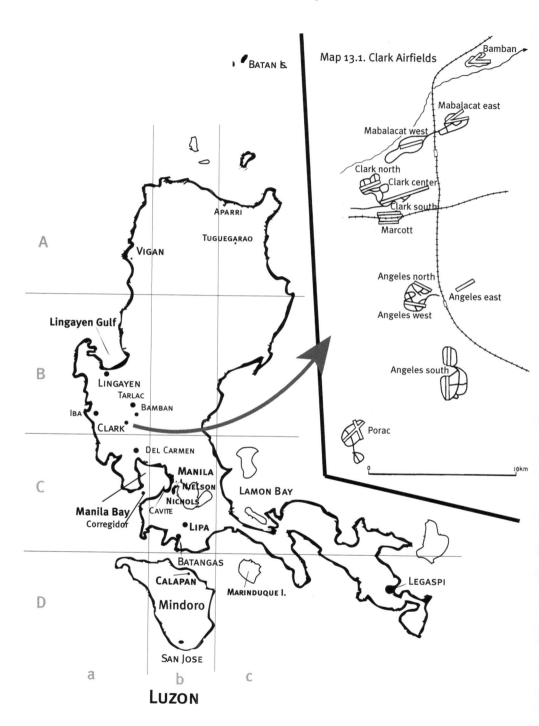

Map 13.1. Clark Airfields

LUZON

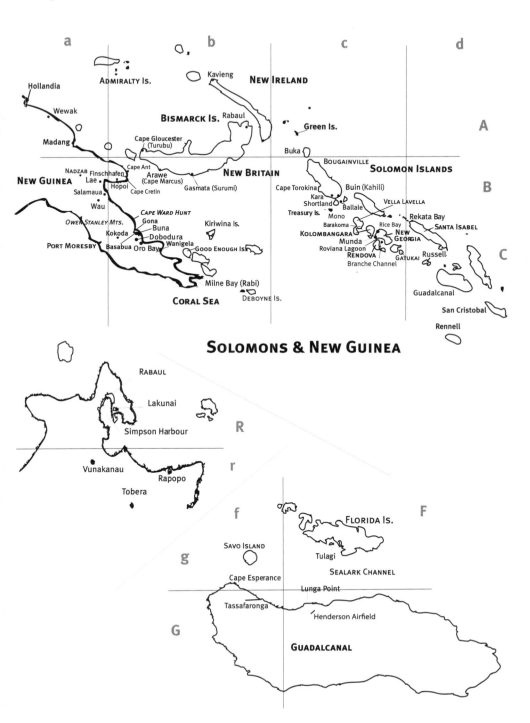

SOLOMONS & NEW GUINEA

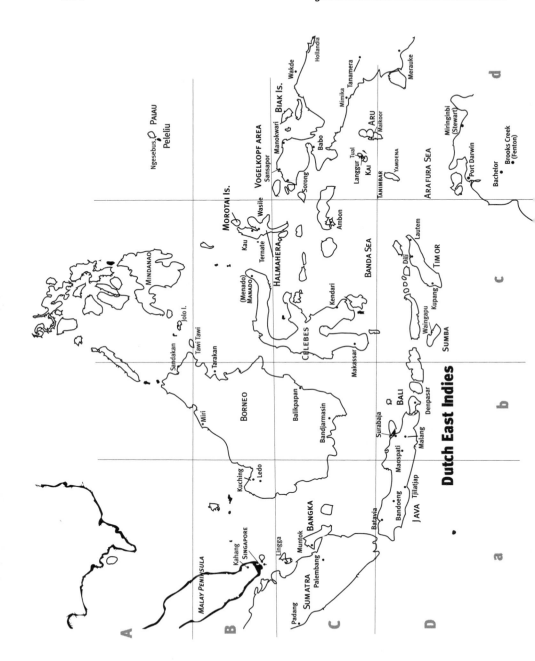

APPENDIX A

List of Naval Fighter Aces

1. The term ace is normally considered to be applicable to any person who has shot down five or more aircraft; however, in the following table the listing is restricted in most cases to those responsible for claiming the downing of eight or more aircraft.

2. In the 'Class' column, Western numerals indicate the Naval Academy Class Number (for example, 57 means the 57th class at the Academy); 'Pilot' indicates Pilot Training Class (Soju renshusei) numbers. 'Ko', 'Otsu' and 'Hei' indicate the respective classes within the Flight Reserve Enlisted Trainee System for aviation students (Hiko yoka renshusei). 'Res' (Yo) indicates classes held for Reserve Flight Officers (Yobi hiko shikan). Further details regarding these entry and training categories are provided at the commencement of the Appendix C section and reading of these will give the reader a somewhat more detailed understanding of this rather complex system.

3. In the 'Status' column, the date shown is that of the pilot's death. If blank, the pilot survived the war. Deaths caused by accidents during the war years are included.

4. In the 'Victories' column (off) means that the figure has been officially recognised, i.e. a detailed battle report (Sento Shoto), a letter of appreciation (Kanno), a letter of commendation (Hyosho) and/or flight logbooks (Koku Kiroku).

5. An asterisk by the name indicates mention in an all units bulletin (Zengun Fukoku), and the person was given a two-rank promotion (exclusive of Tokko personnel).

Name Family First	Final Rank	Class	Native Prefecture	Birth Date	Status	Total Victories	China Incident
*NISHIZAWA, HIROYOSHI	Wt Off	Otsu 7	NAGANO	27 Jan 1920	26 Oct 1940	87	
IWAMOTO, TETSUZO	Lt(jg)	Pilot 34	SHIMANE	14 Jun 1916		c.80	14
*SUGITA, SHO-ICHI	CPO	Hei 3	NIIGATA	1 Jul 1924	4 Apr 45	c.70	
SAKAI, SABURO	Lt(jg)	Pilot 38	SAGA	26 Aug 1916		64	2
OKUMURA, TAKEO	CPO	Pilot 42	FUKUI	27 Feb 1920	22 Sep 43	54	4(off)
OHTA, TOSHIO	PO1c	Pilot 46	NAGASAKI	20 Mar 1919	21 Oct 42	34(off)	
SUGINO, KAZUO	Wt Off	Hei 3	YAMAGUCHI	5 Aug 1921		32	
ISHII, SHIZUO	CPO	Pilot 50	FUKUSHIMA	18 Nov 1920	24 Oct 43	29(off)	3(off)
MUTO, KANEYOSHI	Ens	Pilot 32	AICHI	18 Jun 1916	24 Jul 45	28	5
*SASAI, JUN-ICHI	Lt(jg)	67	TOKYO	13 Feb 1918	26 Aug 42	27(off)	
AKAMATSU, SADAAKI	Lt(jg)	Pilot 17	KOCHI	30 Jul 1910		27	11
*KANNO, NAOSHI	Lt	70	MIYAGI	13 Oct 1921	1 Aug 45	25	
OGIYA, NOBUO	Wt Off	Pilot 48	IBARAGI	20 Feb 1918	13 Feb 44	24(off)	
SUGIO, SHIGEO	Lt(jg)	Otsu 5	MIYAZAKI			20+	

Name Family First	Final Rank	Class	Native Prefecture	Birth Date	Status	Total Victories	China Incident
UTO, KAZUSHI	PO3c	Otsu 9	EHIME	18 Aug 1922	13 Sep 42	19(off)	
NAGANO, KI-ICHI	CPO	Pilot 56	SHIZUOKA	26 Oct 1922	6 Nov 44	19(off)	
OKANO, HIROSHI	Wt Off	Pilot 54	IBARAGI	27 May 1921		19	
*NAKASE, MASAYUKI	PO1c	Otsu 5	TOKUSHIMA	1 Jul 1918	9 Feb 42	18	9
MATSUBA, AKIO	Lt(jg)	Pilot 26	MIE	25 Oct 1914		18	2
KOMACHI, SADAMU	Wt Off	Pilot 49	ISHIKAWA	18 Apr 1920		18	
TANIMIZU, TAKEO	Wt Off	Hei 3	MIE	14 Apr 1919		18	
KAWATO, MASAJIRO	CPO	Hei 12	KYOTO	19 Sep 1925		18	
SAITO, SABURO	Ens	Pilot 44	YAMAGATA	1 Oct 1917		18	
ITO, KIYOSHI	Wt Off	Hei 2	NIIGATA	13 Nov 1921		18	
OHKI, YOSHIO	CPO	Pilot 37	IBARAGI	1 Feb 1916	16 Jun 43	17(off)	12(off)
TANAKA, KUNIYOSHI	Ens	Pilot 31	SAGA	2 Mar 1917		17(off)	
MASUYAMA, MASAO	CPO	Pilot 49	NAGASAKI	2 Mar 1921		17	
KAMIHIRA, KEISHU	Lt(jg)	Ko 1	KANAGAWA	3 Dec 1920		17	4
HONDA, MINORU	Ens	Ko 5	KUMAMOTO	5 Jan 1923		17	
ISHIHARA, SUSUMU	Ens	Ko 3	AICHI	9 Jan 1921		16(off)	
TAKATSUKA, TORA-ICHI	Wt Off	Pilot 22	SHIZUOKA	22 Feb 1914	13 Sep 42	16(off)	3(off)
*MIYANO, ZENJIRO	Lt	63	OSAKA	2 Mar 1917	16 Jun 43	16	
OH-HARA, RYOJI	Wt Off	Hei 4	MIYAGI	25 Feb 1921		16	
NAKAJIMA, BUNKICHI	CPO	Pilot 33	TOYAMA	20 Oct 1918	6 Oct 43	16	
NAKAYA, YOSHI-ICHI	Wt Off	Hei 2	NAGANO		15 Sep 1921	16	
KATO, KUNIMICHI	CPO	Hei 10	AICHI	Jun 1923		16	
SHIGA, MASAMI	Ens	Pilot 50	IBARAGI	15 Feb 1919		16	
WATANABE, HIDEO	Wt Off	Hei 2	FUKUSHIMA	22 Jun 1920		16	
NAKAKARIYA, KUNIMORI	Ens	Otsu 8	KAGOSHIMA	13 Mar 1920		16	2
MINAMI, YOSHIMI	Ens	Pilot 30	KAGAWA	15 Dec 1915	25 Nov 44	15(off)	9
YOSHINO, SATOSHI	Wt Off	Otsu 5	CHIBA	21 Feb 1918	9 Jun 42	15(off)	
NAKAMICHI, WATARU	Wt Off	Hei 4	OSAKA	22 Aug 1922		15(off)	
SHIBUKAWA, SHIGERU	Wt Off	Hei 6	OSAKA	12 Aug 1923		15(off)	
SUHO, MOTONARI	Lt Cdr	62	TOTTORI	14 Dec 1912		15	11
TANAKA, MINPO	Wt Off	Otsu 11	NAGASAKI	3 Oct 1923		15	
OKABE, KENJI	Ens	Pilot 38	FUKUOKA	7 May 1915		15	
ENDO, MASUAKI	PO1c	Otsu 9	FUKUSHIMA	20 Dec 1920	7 Jul 43	14(off)	
YAMAZAKI, ICHIROBEI	PO1c	Pilot 54	TOKYO	5 May 1920	4 Jul 43	14(off)	
YOSHIDA, MOROTSUNA	PO2c	Pilot 44	OKAYAMA	1 Jan 1918	7 Aug 42	14(off)	
TANIGUCHI, MASAO	Ens	Pilot 51	FUKUOKA	7 Jan 1919		14(off)	
OZEKI, YUKIHARU	Wt Off	Pilot 32	AICHI	2 Feb 1918	24 Oct 44	14+(off)	3(off)
TAKAHASHI, KEN-ICHI	Wt Off	Otsu 13	NAGANO	5 May 1924		14	
KOGA, KIYOTO	Wt Off	Pilot 16	FUKUOKA	30 Jun 1910	16 Sep 38	13(off)	13(off)
HANDA, WATARI	Lt(jg)	Pilot 19	FUKUOKA	22 Aug 1911		13(off)	6(off)
YAMAMOTO, AKIRA	Ens	Pilot 24	SHIZUOKA	13 Jun 1913	24 Nov 44	13(off)	1(off)
YAMASHITA, SAHEI	Wt Off	Otsu 5	SHIZUOKA	22 May 1918	9 Feb 43	13(off)	
MATSUMURA, MOMOTO	Wt Off	Pilot 29	YAMAGUCHI	21 Sep 1915	25 Oct 44	13(off)	10(off)
KUROIWA, TOSHIO	Wt Off	Pilot 13	FUKUOKA	5 Dec 1908	25 Aug 44	13(off)	13(off)
*MAEDA, HIDEO	Wt Off	Ko 1	MIE	11 Jul 1920	17 Feb 44	13(off)	
*MIYAZAKI, GITARO	Wt Off	Otsu 4	KOCHI	19 Jun 1917	1 Jun 42	13(off)	2
SHIBAGAKI, HIROSHI	SupSea	Hei 12	NIIGATA	9 Dec 1924	22 Jan 44	13(off)	
KONDO, MASA-ICHI	Ens	Pilot 27	EHIME	5 Nov 1917		13	6(off)
*OHMORI, SHIGETAKA	PO1c	Pilot 33	YAMANASHI	15 Jan 1916	26 Oct 42	13	
HAGIRI, MATSUO	Lt(jg)	Pilot 28	SHIZUOKA	10 Nov 1913		13	7
MIYAZAKI, ISAMU	Ens	Hei 2	KAGAWA	5 Oct 1919		13	
KOIZUMI, FUJIKAZU	Lt(jg)	Otsu 2	FUKUI	1 Mar 1916	27 Jan 44	13	2
FUJITA, IYOZO	Lt Cdr	66	OITA	2 Nov 1916		13	
KASHIMURA, KAN-ICHI	Wt Off	Pilot 24	KAGAWA	5 Jul 1913	6 Mar 43	12(off)	10(off)

Name Family First	Final Rank	Class	Native Prefecture	Birth Date	Status	Total Victories	China Incident
YOSHIMURA, KEISAKU	PO1c	Pilot 56	NIIGATA	16 Feb 1922	25 Oct 42	12(off)	
KANAMARU, TAKEO	Wt Off	Pilot 44	YAMANASHI	30 Nov 1920		12	
KIKUCHI, TETSUO	CPO	Pilot 39	IWATE	13 Aug 1916	19 Jun 44	12	
SHIMIZU, KIYOSHI	PO1c	Hei 7	KYOTO	3 Feb 1919	26 Jan 44	12(off)	
ISOZAKI, CHITOSHI`	Lt	Pilot 19	EHIME	12 Jan 1913		12	
YAMAGUCHI, SADA-O	Lt	67	HIROSHIMA	13 Jan 1919	4 Jul 44	12	
SASAKIBARA, MASAO	Ens	Ko 4	AOMORI	17 Nov 1921		12	
YAMASHITA, KOSHIRO	Ens	Pilot 17	KOCHI	8 Apr 1910	30 Mar 44	11(off)	11(off)
SEKIYA, KIYOSHI	CPO	Hei 2	TOCHIGI	5 Feb 1921	24 Jun 44	11(off)	
YASUI, KOZABURO	Wt Off	Pilot 40	KYOTO	11 Aug 1916	19 Jun 44	11(off)	
YAMAMOTO, TOMEZO	CPO	Hei 2	HOKKAIDO	30 Sep 1922	24 Jun 44	11(off)	
WAJIMA, YOSHIO	Lt(jg)	Pilot 18	HOKKAIDO	7 May 1911	23 Feb 44	11(off)	
ICHIOKA, MATAO	PO2c	Hei 12	GIFU	1 Mar 1925	9 Apr 44	11(off)	
KOKUBU, TAKEICHI	PO3c	Pilot 49	FUKUSHIMA	1 May 1921	2 Sep 42	11(off)	
HIDAKA, HATSUO	Lt(jg)	Pilot 24	KAGOSHIMA	7 May 1915		11(off)	5(off)
OH-ISHI, YOSHIO	Wt Off	Otsu 9	SHIZUOKA	25 May 1923	4 May 45	11	
IWAI, TSUTOMU	Lt(jg)	Otsu 6	KYOTO	20 Jul 1919		11	3
SHIRAHAMA, YOSHIJIRO	CPO	Pilot 56	TOKYO	15 May 1921		11	
HORI, MITSUO	Wt Off	Otsu 10	GIFU	4 Mar 1921		11	
FUKUDA, SUMIO	Lt	69	TOKUSHIMA	20 Nov 1919	14 Oct 44	11	
KODAIRA, YOSHIMA-O	Ens	Pilot 43	NAGANO	18 Apr 1918		11	2
YAMAMOTO, ICHIRO	Wt Off	Pilot 50	EHIME	19 Jun 1918	19 Jun 44	11	
KITAHATA, SABURO	Wt Off	Pilot 21	HYOGO	12 Jan 1915	23 Jan 43	10+(off)	4(off)
SUGIYAMA, TERUO	Wt Off	Otsu 7	YAMAGUCHI	4 Jan 1920	4 Feb 44	10(off)	
TANAKA, JIRO	CPO	Pilot 39	SAITAMA	8 Sep 1919	10 Dec 42	10(off)	
TANAKA, SHINSAKU	CPO	Otsu 12	KUMAMOTO	17 Apr 1924	12 Sep 44	10(off)	
BANNO, TAKAO	PO2c	Pilot 53	AICHI	2 Dec 1921	7 Oct 43	10(off)	
ISHI-I, ISAMU	CPO	Hei 10	OSAKA	15 Jan 1924	11 May 45	10(off)	
HATTORI, KAZUO	LdgSea	Hei 10	AICHI	4 Dec 1921	30 Mar 44	10+(off)	
NAGAHAMA, YOSHIKAZU	CPO	Ko 2	FUKUOKA	26 Aug 1921	6 Sep 43	10+(off)	
KUROSAWA, SEI-ICHI	PO2c	Hei 4	SAITAMA	20 Dec 1922	6 Aug 43	10(off)	
HASHIGUCHI, YOSHIRO	CPO	Pilot 42	FUKUOKA	10 Sep 1918	25 Oct 44	10+	
SHIGEMATSU, YASUHIRO	Lt	66	TOKYO	23 Jun 1916	8 Jul 44	10+	
KOBAYASHI, HOHEI	Lt	67	GUMMA	1 Mar 1918	25 Oct 44	10+	
MATSUO, KAGEMITSU	Res Ens	Res. 10	FUKUOKA	27 Aug 1920	23 Dec 43	10+	
KUDO, SHIGETOSHI	Ens	Pilot 53	OITA	14 Feb 1920		10(off)	
MOCHIZUKI, ISAMU	Lt(jg)	Pilot 9	SAGA	15 Sep 1906	6 Feb 44	10(off)	9(off)
TAKAHASHI, SHIGERU	Ens	Ko 5	MIYAGI	5 Oct 1922		c.10	
KASAI, TOMOKAZU	CPO	Ko 10	HYOGO	8 Mar 1926		10	
ATAKE, TOMITA	Wt Off	Pilot 47	CHIBA			c.10	
YOSHIDA, KATSUYOSHI	Wt Off	Ko 6	HYOGO	2 Jul 1923		10	
AIOI, TAKAHIDE	Cdr	59	HIROSHIMA	4 Jan 1912		10	5
ABE, KEN-ICHI	CPO	Otsu 9	OITA	1 Mar 1923		10	
SHIBAYAMA, SEKIZEN	Wt Off	Otsu 13	SAITAMA	15 Dec 1922		c.10	
SHIRANE, AYA-O	Lt Cdr	64	TOKYO	7 Aug 1916	24 Nov 44	9(off)	1(off)
SUEDA, TOSHIYUKI	Wt Off	Pilot 32	FUKUOKA	27 Jan 1914	6 Oct 43	9(off)	6(off)
SUZUKI, KIYONOBU	PO1c	Pilot 28	FUKUOKA	5 Feb 1914	26 Oct 42	9(off)	3(off)
OKAMOTO, JUZO	PO1c	Pilot 31	TOKUSHIMA	18 Sep 1916	11 Oct 42	9(off)	4(off)
ODA, KI-ICHI	Wt Off	Pilot 18	NIIGATA	5 Apr 1913	10 Dec 44	9(off)	4(off)
MORI, MITSUGU	Lt(jg)	Pilot 14	SHIZUOKA	5 May 1908		9+(off)	4+(off)
TSUNODA, KAZUO	Lt(jg)	Otsu 5	CHIBA	11 Oct 1918		9(off)	1(off)
MORINIO, HIDEO	Lt(jg)	Pilot 41	HIROSHIMA	3 Oct 1917		9(off)	1
MATSUDA, JIRO	Lt(jg)	Ko 1	NAGASAKI	26 Apr 1918		9(off)	2(off)
HARADA, KANAME	Lt(jg)	Pilot 35	NAGANO	11 Aug 1916		9(off)	

Name Family First	Final Rank	Class	Native Prefecture	Birth Date	Status	Total Victories	China Incident
IZUMI, HIDEO	PO2c	Ko 3	TOYAMA	19 Jun 1921	30 Apr 42	9(off)	
MATSUKI, SUSUMU	PO3c	Ko 4	NIIGATA	15 Feb 1922	13 Sep 42	9(off)	
NAKAMURA, YOSHIO	Wt Off	Hei 3	HOKKAIDO	20 Jan 1923		9(off)	
KANDA, SAJI	PO2c	Hei 2	OITA	22 Sep 1921	10 Jun 43	9(off)	
YAMANAKA, TADAO	Ens	Pilot 44	KOCHI	23 Jan 1918		9(off)	
SAKO, MORIJI	Wt Off	Ko 6	KUMAMOTO	Oct 1923		9	
SHIRAKAWA, TOSHIHISA	PO1c	Ko 6	KAGAWA	23 Jul 1924	22 Sep 43	9(off)	
YOSHIZAWA, TOKUSHIGE	CPO	Hei 7	AKITA	18 Jun 1923	6 Jan 45	9(off)	
ISHIDA, TEIGO	CPO	Hei 6		28 Jun 1920	16 Apr 45	9(off)	
OHSHIMA, SHINKICHI	CPO.	Hei 4			15 Jun 44	9(off)	
HIGASHIYAMA, ICHIRO	Lt(jg)	Otsu 2	NAGANO	25 Aug 1915	8 Jul 44	9(off)	
FUKUI, YOSHIO	Lt	Pilot 26	KAGAWA	10 Apr 1913		9	6(off)
KANEKO, TADASHI	Lt Cdr	60	YAMAGATA	14 Nov 42		8(off)	3(off)
IIZUKA, MASAO	Lt	66	TOCHIGI	6 Jul 1916	15 Oct 44	8(off)	
OHNO, TAKEYOSHI	Lt(jg)	68	ISHIKAWA	18 Jan 1921	30 Jun 43	8(off)	
*ENDO, SACHIO	Lt	Otsu 1	YAMAGATA	9 Sep 1915	14 Jan 45	8(off)	
MORIURA, TOYO-O	PO3c	Otsu 9	KUMAMOTO	20 Nov 1922	25 Oct 42	8(off)	
SUZUKI, HIROSHI	CPO	Ko 5	CHIBA	30 Nov 1922	13 Oct 44	8(off)	
TAKAIWA, KAORU	PO1c	Otsu 13	NAGASAKI	27 May 1923	10 Feb 44	8(off)	
IWAKI, YOSHIO	PO1c	Ko 2	KARAFUTO	Feb 1923	24 Aug 42	8(off)	
EMA, TOMOKAZU	Wt Off	Pilot 22	NARA	8 Nov 1912	29 Oct 44	8(off)	2(off)
MAGARA, KO-ICHI	PO1c	Pilot 28	MIE	8 Jun 1914	14 Sep 42	8(off)	
SHIGEMI, KATSUMA	Wt Off	Pilot 20	SHIMANE	27 Jan 1924	4 Feb 43	8(off)	3(off)
GOTO, KURAKAZU	PO1c	Ko 6	FUKUOKA	19 Nov 1921	9 Sep 43	8(off)	
YANO, SHIGERU	PO1c	Pilot 44	TOCHIGI	20 Feb 1917	17 Apr 42	8(off)	
TOKUJI, YOSHIHISA	Wt Off	Otsu 6	MIYAZAKI	29 Sep 1919	19 Jun 44	8(off)	
ONO, SATORU	Lt(jg)	Pilot 23	OITA	5 Mar 1915		8(off)	3
OHSHIMA, SHINKICHI	CPO	Hei 4			5 Jun 44	8(off)	
MURANAKA, KAZUO	Ens	Otsu 6	FUKUOKA	27 Feb 1919		8	1
NANGO, MOCHIFUMI	Lt	55	TOKYO	18 Jul 1938		8	8
KATSUKI, KIYOMI	Wt Off	Pilot 54	FUKUOKA	1919		8	
SASAKI, GI-ICHI	PO1c	Pilot 44	MIYAGI		19 Feb 43	7	
CHONO, JIRO	Wt Off	Pilot 15	EHIME	1907	21 Feb 41	7	7(off)
KUDO, OSAMU	Wt Off	Otsu 2	OITA	1915	3 Mar 42	7	3(off)
SUGAWARA, MASAO	CPO	Otsu 15	AKITA	1924		7	
TORAKUMA, TADASHI	Wt Off	Pilot 20	OITA	1911	16 Apr 43	7	7
KURAMOTO, JUZO	Wt Off	Otsu 15	TOKYO	1 Oct 1924		6	
NAKAGAWA, YOSHIMASA	CPO	Otsu 15	FUKUOKA	22 Feb 1924		6	
NAKA, YOSHIMITSU	CPO	Hei 4				5	

APPENDIX B

Key Fighter Pilots Killed in Action

(by date)

1. Rank indicated is at the time of death in action.
2. In the 'Class' column, Western numerals indicate Naval Academy class number (for example, 57 means 57th class). 'Pilot' (So) indicates Pilot Training (Soju renshusei) class number. 'Ko', 'Otsu', and 'Hei' represent the respective Aviation Student (Hiko yoka renshusei) Course number, and 'Res' (Yo) indicates Reserve Flight Officer (Yobi hiko shikan) class number.
3. Deaths by accident are included.
4. An asterisk means that an all units bulletin was issued and a double rank promotion given out.

Name (First name first)	Rank	Class	Unit	Date of Death (mm/dd/yy)	Place	Major Unit History and Other Data
Shichiro Yamashita	LT	57	13th Ku Buntai Ldr.	Sep/26/37	Central China	POW; then died
Nagaharu Umeda	PO3c	Pilot 32	13th Ku	Oct/12/37	Nanking	
Also PO3c Torata Takiguchi (Pilot 25) and POlc Masazumi Ino (Pilot30)						
Hei-ichiro Mase	WO	Pilot 8	12th Ku	Nov/8/37	Shoko [near Shanghai]	by AA fire
Rizo Harada	PO1c	Pilot 29	12th Ku	Nov/15/37	Shanghai	by AA fire
Koji Miyazaki	PO3c	Pilot 29	13th Ku	Nov/22/37	Nanking	
Naoshi Teramatsu	PO3c	Pilot 31	13th Ku	Dec/9/37	Nanking	
Norito Obayashi	LT	55	13th Ku Buntai Ldr.	Dec/22/37	Nanchang	
Ryohei Ushioda	LT	57	2th Ku Buntai Ldr.	Jan/7/38	Nanchang	
Takashi Kaneko	LT	59	12th Ku Buntai Ldr.	Feb/18/38	Hankow	
Also PO1c Shigeo Miyamato (Otsu 1) and Sealc Hiroji Hayakawa (Pilot 29)						
Inao Hamada	PO1c	Pilot 34	13th Ku	Feb/18/38	Hankow	
Shigeo Takuma	LT	58	13th Ku Buntai Ldr.	Feb/25/38	Nanchang	
Also PO1c Hisao Ochi (Pilot 31)						
Naoshi Eitoku	PO1c	Pilot 14	*Kaga*	Apr/13/38	Canton	
Also PO3c Yukio Miyasato and PO3c Yuji Mori (both Otsu 3)						
Ken-ichi Takahashi	PO2c	Pilot 19	12th Ku	Apr/29/38	Hankow	
Also PO3c Kinji Fujiwara (Pilot 29)						
Hiromitsu Takahara	Sea1c	Pilot 36	12th Ku	May/31/38	Hankow	
Sakae Kato	WO	Pilot 11	15th Ku	Jun/28/38	Anking	
Kyushichi Kobayashi	PO1c	Pilot 18	12th Ku	Ju1/4/38	Nanchang	

Mochifumi Nango Hikotai Ldr.	LT	55	15th Ku	Jul/18/38	Nanchang	
Naohisa Shinjo	LTJG	62	15th Ku	Aug/3/38	Hankow	
Also PO2c Hitoshi Fukazawa (Pilot 32)						
Namitaro Matsushima	PO3c	Pilot 30	15th Ku	Aug/3/38	Hankow	POW, returned
Masumi Tsutsumi	PO1c	Pilot 24	*Kaga*	Aug/16/38	Canton	
Hideo Teshima Buntai Ldr.	LT	58	*Kaga*	Aug/30/38	Nanchang	
Also PO2c Seizaburo Sugino(Otsu 3)						
Kiyoto Koga	WO	Pilot 16	Yoko Ku	Sep/16/38	Yokosuka	Accident
Hironori Shimomura	PO2c	Pilot 33	13th Ku	Oct/23/38	Central China	AA fire
Sukesada Senda	PO1c	Otsu 2	*Ryujo*	Mar/16/39	Homeland	Accident
Shiro Ayukawa	PO1c	Pilot 28	*Akagi*	May/9/39	Homeland	Accident
Genkichi Ogawa	PO3c	Pilot 40	14th Ku	Jun/5/39	South China	
Mitsuo Kaneko	PO1c	Pilot 22	Tsukuba Ku	Aug/9/39	Homeland	Accident
Kanetake Okazaki	LTJG	62	12th Ku	Oct/3/39	China	
Isamu Ochi	PO3c	Pilot 41	12th Ku	Oct/14/39	Hankow	Killed by bombing
Hiroshi Fujita	PO2c	Pilot 42	14th Ku	Dec/30/39	Liuchow	
Hiro-o Natori	PO1c	Otsu 2	12th Ku	Mar/8/40	Hankow	Accident
Jiro Chono	WO	Pilot 15	14th Ku	Feb/21/41	Kunming	by AA fire
Manbei Shimokawa Buntai Ldr.	LT	58	Yoko Ku	Apr/17/41	Yokosuka	Accident
Ei-ichi Kimura	PO1c	Otsu 5	12th Ku	May/20/41	Central China	by AA fire
Otoshiro Kobayashi	PO1c	Pilot 48	12th Ku	Jun/23/41	Lanchow	by AA fire
Masayuki Mitsumasu	PO1c	Otsu 5	3rd Ku	Nov/8/41	China	Accident
Also PO1c Sakae Nakasawa (Pilot 48)						
Osamu Hatanaka	Sea1c	Pilot 54	Tainan Ku	Nov/24/41	Formosa	Accident
Shime Inoue	PO1c	Pilot 35	22nd AF	Nov/26/41	South China Sea	Accident
Also PO2c Takaaki Shimotsuka (Ko 3)						
*Fusata Iida Buntai Ldr.	LT	62	*Soryu*	Dec/7/41	Hawaii	AA fire, 12th Ku
*Saburo Ishii	PO2c	Pilot 41	*Soryu*	Dec/7/41	Hawaii	Lost on return; 14th Ku
Also PO1c Shun-ichi Atsumi (Ko 2)						
*Ippei Goto	WO	Pilot 19	*Kaga*	Dec/7/41	Hawaii	13th Ku
*Tomio Inenaga	PO1c	Otsu 7	*Kaga*	Dec/7/41	Hawaii	13th Ku
Also PO2c Seinoshin Sano (Pilot 41), PO2c Toru Hada (Pilot 35)						
*Takashi Hirano	PO1c	Ko 1	*Akagi*	Dec/7/41	Hawaii	12th Ku
*Shigenori Nishikaichi	PO1c	Ko 2	*Hiryu*	Dec/7/41	Hawaii	Force land, killed
Ryoichi Nakamizo	WO	Otsu 3	Tainan Ku	Dec/8/41	Luzon	*Kaga*
Also PO1c Yasuhisa Sato (Otsu 6), PO1c Yasujiro Kawano (Otsu 7), and PO3c Yoshio Aoki (Pilot 56)						
Yoshio Hirose	PO3c	Pilot 40	Tainan Ku	Dec/8/41	Luzon	12th Ku
Saburo Yoshii	PO3c	Pilot 45	3rd Ku	Dec/8/41	Luzon	12th Ku
Hiroshi Kawanishi	PO2c	Pilot 38	*Ryujo*	Dec/8/41	Davao	Force land, suicide
Masaharu Higa	Sea1c	Pilot 40	Tainan Ku	Dec/10/41	Luzon	
Tamotsu Kojima	PO2c	Pilot 41	3rd Ku	Dec/10/41	Luzon	12th Ku
Also PO1c Kiyoharu Tezuka (Otsu 7)						
Hiroshi Kuratomi	PO3c	Pilot 44	Tainan Ku	Dec/13/41	Luzon	12th Ku
Kaneo Suzuki	PO2c	Pilot 46	3rd Ku	Dec/13/41	Luzon	
Toshio Kikuchi	PO1c	Otsu 7	Tainan Ku	Dec/24/41	Legaspi	14th Ku
Hiroshi Suyama	PO1c	Pilot 54	22nd AF	Jan/15/42	Singapore	
Yoshihiro Sakuraba	PO2c	Pilot 44	22nd AF	Jan/18/42	Malay	2th Ku
Isao Hiraishi	PO2c	Pilot 27	*Kaga*	Jan/23/42	Rabaul	
Yoshimitsu Harada	WO	Otsu 1	Tainan Ku	Jan/24/42	Meraku	AA fire

Akira Wakao	LT	65	Tainan Ku	Jan/25/42	Balikpapan	
Buntai Ldr.						
Also PO3c Akimizu Seki (Otsu 9)						
Yoshikane Sasaki	PO1c	Ko 1	3rd Ku	Jan/26/42	Ambon	
*Toshiyuki Sakai	PO1c	Pilot 25	Tainan Ku	Jan/29/42	Balikpapan	12th Ku
Hatsumasa Yamaya	PO2c	Pilot 40	3rd Ku	Feb/3/42	Surabaya	
Also PO3c Sho-ichi Shoji (Pilot 48)						
Masaru Morita	PO3c	Otsu 9	3rd Ku	Feb/3/42	Surabaya	
Kyoji Kobayashi	Sea1c	Pilot 55	Tainan Ku	Feb/3/42	Malang	
*Masayuki Nakase	PO1c	Otsu 5	3rd Ku	Feb/9/42	Celebes	12th Ku; AA fire
*Masao Asai	LT	63	Tainan Ku	Feb/19/42	Surabaya	12th Ku
Buntai Ldr.						
Hajime Toyoshima	Sea1c	Pilot 56	*Hiryu*	Feb/19/42	Darwin	
Tomekichi Otsuki	Sea1c	Pilot 56	3rd Ku	Feb/20/42	Bali	
Isaburo Yawata	PO3c	Pilot 44	3rd Ku	Feb/22/42	Bali	Killed by bombing
Makoto Ueda	PO3c	Otsu 9	Tainan Ku	Feb/24/42	Borneo	
Sueharu Ide	PO1c	Otsu 6	22 AF	Feb/25/42	Batavia	AA fire;12th Ku
*Toyoo Sakai	PO1c	Otsu 6	Tainan Ku	Feb/27/42	Indian Ocean	Attack on *Langley*
Katsuaki Nagatomo	Sea1c	Pilot 49	4th Ku	Feb/28/42	Moresby	
*Osamu Kudo	WO	Otsu 2	3rd Ku	Mar/3/42	Broome	AA fire, 13th Ku, *Kaga*
Tsutomu Kobayashi	PO1c	Otsu 5	Oita Ku	Mar/11/42	Homeland	Accident; 12th Ku
Nobuhiro Iwasaki	LTJG	67	4th Ku	Mar/14/42	Horn Island	
Also PO1c Genkichi Oishi (Pilot 54)						
Keiji Kikuchi	PO3c	Pilot 47	4th Ku	Mar/22/42	Lae	
Kyo-ichi Yoshii	PO2c	Pilot 34	4th Ku	Mar/23/42	Moresby	
Sachio Higashi	Sea1c	Pilot 56	*Soryu*	Apr/5/42	Colombo	
Yukihisa Tan	PO2c	Ko 4	Tainan Ku	Apr/7/42	Lae	
Sumio Nono	LT	61	*Hiryu*	Apr/9/42	off Ceylon	Intercepting Blenheims
Buntai Ldr.						
Masatoshi Makino	LT	65	*Zuikaku*	Apr/9/42	Trincomalee	
Buntai Ldr.						
Also Sea1c Toru Matsumoto (Pilot 50)						
Fujio Hayashi	PO1c	Otsu 7	*Shokaku*	Apr/9/42	Trincomalee	
Toshio Makinoda	PO1c	Ko 1	*Hiryu*	Apr/9/42	Trincomalee	
Masayoshi Sonoyama	PO3c	Pilot 47	3rd Ku	Apr/13/42	Kupang	
Yoshimi Sakai	PO2c	Ko 4	Tainan Ku	Apr/17/42	Port Moresby	
Shigeru Yano	PO1c	Pilot 44	3rd Ku	Apr/17/42	Kendari	12th Ku, accident
Shiro Murakami	PO1c	Pilot 54	3rd Ku	Apr/25/42	Darwin	
Yoshimitsu Maeda	PO3c	Hei 2	Tainan Ku	Apr/28/42	Lae	
Hideo Izumi	PO2c	Ko 3	Tainan Ku	Apr/30/42	Lae	
*Gisuke Arita	PO1c	Ko 3	Tainan Ku	May/1/42	Port Moresby	12th Ku
*Haruo Kawanishi	Sea1c	Pilot 56	Tainan Ku	May/2/42	Port Moresby	
Toshikazu Tamura	PO2c	Ko 4	*Shoho*	May/3/42	Tulagi	Accident
Shigeshi Imamura	WO	Pilot 29	*Shoho*	May/7/42	Coral Sea	12th Ku
Tadao Aoki	PO2c	Ko 4	*Shoho*	May/7/42	Coral Sea	
Also PO2c Takeo Inoue (Ko 4), PO2c Yukio Hayakawa, and PO2c Hachiro Kuwabara (both Ko 2)						
Takeo Miyazawa	PO1c	Ko 3	*Shokaku*	May/8/42	Coral Sea	Ramming
Hisashi Ichinose	PO2c	Ko 4	*Zuikaku*	May/8/42	Coral Sea	
Yasushi Nikaido	LT	64	*Kaga*	May/8/42	off Izu	Accident
Buntai Ldr.						
*Toshiaki Honda	PO3c	Pilot 49	Tainan Ku	May/13/42	Port Moresby	12th Ku
Toru Oshima	PO1c	Ko 1	Tainan Ku	May/14/42	Port Moresby	12th Ku, 4th Ku
Tadao Fujiwara	PO2c	Ko 5	Tainan Ku	May/16/42	Lae	

Kaoru Yamaguchi	LTJG	67	Tainan Ku	May/17/42	Port Moresby	
Also PO2c Tsutomu Ito (Ko 4)						
Masao Watanabe	Sea1c	Hei 2	Tainan Ku	May/25/42	Lae	
Hisao Komori	PO2c	Pilot 45	Tainan Ku	May/29/42	Port Moresby	
*Gitaro Miyazaki	WO	Otsu 4	Tainan Ku	Jun/1/42	Port Moresby	
Yoshihiro Kobayashi	WO	Otsu 5	Kanoya Ku	Jun/2/42	Port Blair	Emergency landing (Nicobar Islands)
Shigeru Mori	LT	64	*Hiryu*	Jun/4/42	Midway	12th Ku
Buntai Ldr.						
Yoshimi Kodama	WO	Otsu 2	*Hiryu*	Jun/4/42	Midway	12th Ku, *Zuikaku*
Masato Hino	PO1c	Pilot 27	*Hiryu*	Jun/4/42	Midway	13th Ku, 15th Ku, *Kaga*
Noboru Todaka	PO2c	Otsu 8	*Hiryu*	Jun/4/42	Midway	
Toru Yamamoto	PO2c	Pilot 41	*Hiryu*	Jun/4/42	Midway	
Michisuke Tokuda	PO1c	Pilot 40	*Hiryu*	Jun/4/42	Midway	
Also PO3c Yutaka Chiyojima (Pilot 50), PO2c Haruo Nitta (Pilot 48),						
PO2c Ichiro Sakai (Otsu 8), and Sea1c Suekichi Yoshimoto (Pilot 54)						
Hiroyuku Yamaguchi	Sp.ENS	Otsu 1	*Kaga*	Jun/4/42	Midway	
Iwao Hirayama	PO1c	Pilot 38	*Kaga*	Jun/4/42	Midway	
Yukuo Tanaka	PO1c	Otsu 6	*Kaga*	Jun/4/42	Midway	
Also PO1c Hiromi Ito (Pilot 47), PO2c Shigeto Sawano (Pilot 46) and Sea1c Ei-ichi Takahashi (Pilot 53)						
Shinaji Iwama	PO1c	Ko 2	*Akagi*	Jun/4/42	Midway	
Toichiro Hanyu	PO3c	Pilot 43	*Akagi*	Jun/4/42	Midway	
Also PO2c Yozo Kawada (Ko 4), and Sea1c Shinpei Sano (Pilot 49)						
Takeo Takashima	PO2c	Pilot 44	*Soryu*	Jun/4/42	Midway	
Also PO3c Teruo Kawamata (Pilot 54) and Sea1c Genzo Nagasawa (Pilot 50)						
Tadayoshi Koga	PO1c	Ko 3	*Ryujo*	Jun/4/42	Aleutians	12th Ku, Force land, KIA
Satoshi Yoshino	WO	Otsu 5	Tainan Ku	Jun/9/42	Lae	4th Ku
Also PO1c Sakyo Kikuchi (Ko 3)						
Katsuji Matsushima	WO	Pilot 15	3rd Ku	Jun/13/42	Darwin	15th Ku
Also Sea1c Mikio Tanikawa (Pilot 56)						
Takeichiro Hidaka	Sea1c	Hei 2	Tainan Ku	Jun/16/42	Lae	
Mitsuo Suizu	Sea1c	Pilot 54	Tainan Ku	Jul/4/42	Lae	Rammed to B-25
Tatsuo Hori	PO1c		Hama Ku	Jul/17/42	Tulagi	
Katsumi Kurihara	LTJG	67	Chitose Ku	Jul/20/42	Port Moresby	
Also PO1c Katsumi Kobayashi (Ko 3) and PO3c Yoshimi Onishi (Otsu 9)						
Saburo Matsui	Sea1c	Pilot 54	Hama Ku	Jul/23/42	Tulagi	
Tadashi Sato	LTJG	68	Kanoya Ku	Jul/30/42	Andamans	
Yoshio Motoyoshi	Sea1c	Pilot 53	Tainan Ku	Aug/2/42	Buna	
Shigeto Kobayashi	Sea1c	Pilot 51	Hama Ku	Aug/4/42	Tulagi	Rammed to B-17
Mototsuna Yoshida	PO2c	Pilot 44	Tainan Ku	Aug/7/42	Guadalcanal	14th Ku
Also PO2c Kunimatsu Nishiura (Ko 4)						
Ri-ichiro Sato	LT	66	Hama Ku	Aug/7/42	Tulagi	Presumed KIA
Yomichiyo Hirahashi	WO	Otsu 3	Hama Ku	Aug/7/42	Tulagi	POW, returned
Tadashi Hayashidani	LTJG	67	Tainan Ku	Aug/8/42	Guadalcanal	
Also PO3c Yutaka Kimura (Otsu 9)						
Shunkichi Tashiro	LTJG	68	3rd Ku	Aug/8/42	Kendari	Accident
Isao Murata	LTJG	68	Tainan Ku	Aug/13/42	Lae	
Masami Arai	PO3c	Otsu 9	Tainan Ku	Aug/14/42	Lae	
Norio Tokushige	PO2c	Pilot 42	Tainan Ku	Aug/17/42	Port Moresby	12th Ku
Tadatsune Tokaji	LT	64	3rd Ku	Aug/23/42	Darwin	
Buntai Ldr.						
Nobutoshi Furukawa	PO2c	Pilot 45	3rd Ku	Aug/23/42	Darwin	
Also PO2c Itsuzo Shimizu (Ko 5) and PO3c Yoshiyuki Hirata (Otsu 10)						

Yoshio Iwaki	PO1c	Ko 2	*Shokaku*	Aug/24/42	Solomons	*Akagi*
Shigeru Makino	PO1c	Pilot 27	*Zuikaku*	Aug/24/42	Solomons	

Also PO2c Goro Sakaida (Pilot 43) and PO2c Toshiharu Okubo (Ko 5)

Matsutaro Takaoka	Sea1c	Pilot 54	*Zuikaku*	Aug/24/42	Solomons	*Kaga*: sank with the carrier
Jinsaku Nojima	PO2c	Pilot 42	*Ryujo*	Aug/24/42	Guadalcanal	

Also Sea1c Shoji Ishihara (Pilot 55)

*Jun-ichi Sasai	LTJG	67	Tainan Ku	Aug/26/42	Guadalcanal	
Kunisuke Yuki	LTJG	68	Tainan Ku	Aug/26/42	Guadalcanal	

Also PO3c Ken-ichi Kumagaya (Otsu 9)

Ki-ichi Iwase	PO1c	Pilot 34	2nd Ku	Aug/26/42	Buna	12th Ku

Also PO3c Daizo Ihara and PO3c Kiyoshi Nakano (both Otsu 9)

Joji Yamashita Buntai Ldr.	LT	66	Tainan Ku	Aug/27/42	Rabi	
Sadao Yamashita	PO1c	Pilot 34	Tainan Ku	Aug/27/42	Rabi	

Also PO2c Enji Kakimoto (Pilot 47), PO3c Takeo Matsuda (Pilot 56), and Sea1c Kihachi Ninomiya (Pilot 56)

Mitsuyoshi Takasuga	PO3c	Pilot 5	*Shokaku*	Aug/28/42	Buka	Accident, *Akagi*
Seiji I-ishi	PO2c	Pilot 50	*Shokaku*	Aug/29/42	Guadalcanal	
Tsuguo Ogihara	PO1c	Pilot 30	*Shokaku*	Aug/30/42	Guadalcanal	13th Ku
Ko Nakamoto	PO1c	Pilot 31	*Shokaku*	Aug/30/42	Guadalcanal	12th Ku

Also PO2c Jin-ichiro Kawanishi (Ko 3) and PO3c Yoshizo Tanaka (Pilot 46)

Tsuyoshi Sumita	WO	Pilot 26	*Zuikaku*	Aug/30/42	Guadalcanal	*Shokaku*
Terusada Chuman	PO1c	Ko 1	*Zuikaku*	Aug/30/42	Guadalcanal	

Also PO3c Minoru Awao (Otsu 10)

Takeichi Kokubun	PO3c	Pilot 49	Tainan Ku	Sep/2/42	Guadalcanal	4th Ku

Also Sea1c Ken-ichiro Yamamoto (Pilot 54)

Shigejiro Murakami	Sea1c	Hei 2	6th Ku	Sep/11/42	Guadalcanal	
Torakichi Okazaki	PO2c	Pilot 44	2nd Ku	Sep/12/42	Guadalcanal	12th Ku, *Shoho*
Tora-ichi Takatsuka	WO	Pilot 22	Tainan Ku	Sep/13/42	Guadalcanal	12th Ku
Kazushi Uto	PO3c	Otsu 9	Tainan Ku	Sep/13/42	Guadalcanal	4th Ku

Also PO2c Susumu Matsuki (Ko 4), PO3c Noboru Sato (Otsu 9)

Ko-ichi Magara	PO1c	Pilot 28	2nd Ku	Sep/13/42	Guadalcanal	Chitose Ku
Masa Kawashima	LTJG	67	*Kamikawa Maru*	Sep/14/42	Guadalcanal	

Also WO Makio Kawamura, PO2c Toshio Ohyama (Ko 4)

Matsutaro Ohmura	PO2c		*Kamikawa Maru*	Sep/14/42	Guadalcanal	
Kiyomi Saito	ENS	Pilot 16	5th Ku	Sep/15/42	Kiska	

Also PO3c Katsutaro Uchiyama (Pilot 54)

Torao Mirikawa	PO2c	Ko 5	5th Ku	Sep/26/42	Kiska	MIA
Gi-ichi Yamanouchi	WO	Otsu 2	3rd Ku	Sep/27/42	Guadalcanal	13th Ku, 14th Ku
Tadashi Sasaki	Sea1c	Pilot 56	5th Ku	Sep/29/42	Kiska	
Kozo Miyazawa	LTJG	68	5th Ku	Sep/29/42	Kiska	
Yu-ichi Kobayashi	Sea1c	Hei 2	6th Ku	Oct/2/42	Guadalcanal	
Takeru Imahashi	PO1c	Ko 1	Kanoya Ku	Oct/3/42	Guadalcanal	22 AF
Toshikazu Iwata	PO2c	Pilot 40	Kanoya Ku	Oct/3/42	Guadalcanal	

Also Sea1c Tsutau Suematsu (Pilot 56)

Joji Taniguchi	PO2c	Ko 5	3rd Ku	Oct/3/42	Guadalcanal	

Also PO2c Masashi Tomita (Ko 5)

Hiroshi Sato	PO3c	Ko 6	5th Ku	Oct/4/42	Kiska	
Shigeo Morimoto	Res.ENS	Res 7	3rd Ku	Oct/5/42	Celebes	Accident
Masaharu Hosono	PO3c	Hei 3	6th Ku	Oct/7/42	New Ireland	Accident while moving

Also LdgSea Yoshiro Shoji and Sea1c Soroku Kawakami

Jiro Kawai	PO1c	Pilot 37	*Kamikawa Maru*	Oct/10/42	over convoy to Guadal.	

Also Sea1c Maruyama

Kazuto Kuho	LTJG	68	6th Ku	Oct/11/42	Solomons	Ditched
Ya-ichi Sagane	WO	Otsu 5	6th Ku	Oct/11/42	Solomons	Ditched

Juzo Okamoto	PO1c	Pilot 31	6th Ku	Oct/11/42	Solomons	13th Ku and
Also Sea1c Shigeto Kawakami (Hei 3) and Sea2c Shigeo Hirano (Hei3)						3rd Ku
Mitsuma Hirai	WO	Otsu 2	6th Ku	Oct/13/42	Buka	Killed by
						bombing,
						Kaga, Ryujo,
						12th Ku
Mari Ito	Sea1c	Hei 2	3rd Ku	Oct/13/42	Guadalcanal	
Toshio Igarashi	LTJG		14th Ku	Oct/13/42	Shortland	Died next day
Chuji Sakurai	PO2c	Ko 5	Tainan Ku	Oct/15/42	Guadalcanal	
Also Sea1c Yoshifusa Iwasaka (Hei 3)						
Tomoji Sawada	PO2c	Ko 4	Kanoya Ku	Oct/18/42	Guadalcanal	
Also Sea2c Mitsuo Goto (Hei 3)						
Tamotsu Fujita	PO3c	Otsu 10	Kanoya Ku	Oct/18/42	Guadalcanal	POW, returned
Takeo Inaba	WO	Pilot 19	*Hiyo*	Oct/19/42	Solomons	*Kaga*
Sadamu Tamai	Sea1c	Hei 2	6th Ku	Oct/20/42	Guadalcanal	
Toshio Ohta	PO1c	Pilot 46	Tainan Ku	Oct/21/42	Guadalcanal	12th Ku
Masayoshi Baba	LTJG	67	Kanoya Ku	Oct/23/42	Guadalcanal	22nd AF
Also Sea2c Hachiro Mitsunaga (Hei 3)						
Mukumitsu Kanemitsu	LTJG	68	6th Ku	Oct/23/42	Guadalcanal	
Gunji Suzuki	PO1c	Otsu 7	6th Ku	Oct/23/42	Guadalcanal	
Also PO3c Shimpei Takagaki and PO3c Hiroshi Fukuda (both Ko 6)						
Koji Ikeda	PO3c	Otsu 10	3rd Ku	Oct/24/42	Goodenough	Bad weather
Rokuzo Iwamoto	PO1c	Pilot 42	3rd Ku	Oct/25/42	Guadalcanal	AA fire
Also PO2c Naoichi Maeda (Ko 5)						
Keisaku Yoshimura	Sea1c	Pilot 56	Tainan Ku	Oct/25/42	Guadalcanal	
Also PO3c Ryusuke Goto and PO3c Toyoo Moriura, (both Otsu 9)						
Tokitane Futagami	LTJG	68	2nd Ku	Oct 25/42	Guadalcanal	
Also PO3c Toyoo Morita (Otsu 9) and Sea1c Naoichi Ubukata (Hei 2)						
Shiro Ishikawa	PO2c	Ko 5	2nd Ku	Oct/25/42	Guadalcanal	POW, returned
Kyoichiro Ogino	PO2c	Pilot 44	*Hiyo*	Oct/25/42	Guadalcanal	12th Ku
Yukuo Hanzawa	WO	Otsu 5	*Shokaku*	Oct/26/42	Battle of Santa Cruz	12th Ku
Also PO2c Katsuo Kanno (Hei 3)						
Suekichi Osanai	WO	Otsu 2	*Zuikaku*	Oct/26/42	Battle of Santa Cruz	13th Ku,
						Akagi
Tomio Kamei	PO1c	Ko 2	*Zuikaku*	Oct/26/42	Battle of Santa Cruz	
Kasuke Hoshiya	PO2c	Pilot 39	*Zuikaku*	Oct/26/42	Battle of Santa Cruz	
Also Sea1c Takashi Nakagami (Pilot 53) and PO3c Kozo Takayama (Hei 3)						
*Shigetaka Ohmori	PO1c	Pilot 33	*Shokaku*	Oct/26/42	Battle of Santa Cruz	Ramming,
						12th Ku, *Akagi*
Kiyonobu Suzuki	PO1c	Pilot 26	*Junyo*	Oct/26/42	Battle of Santa Cruz	13th Ku,
Also Sea2c Kiyoshi Nakamoto (Hei 3)						12th Ku, *Kaga*
Shigeru Okamoto	WO	Otsu 5	*Zuiho*	Oct/26/42	Battle of Santa Cruz	
Zenpei Matsumoto	PO3c	Pilot 48	*Zuiho*	Oct/26/42	Battle of Santa Cruz	
Also LTJG Shu-ichi Utsumi (68), PO1c Kazuo Seki (Otsu 7),						
PO1c Masao Kawasaki (Otsu 6), and PO3c Shizuta Takagi (Hei 2)						
Akira Takita	Sea1c	Otsu 11	6th Ku	Oct/30/42	Buka	
Teruo Watanabe	Sea1c	Pilot 56	*Kamikawa Maru*	Oct/30/42	Rekata Bay	
Toshio Kaneko	PO1c	Pilot 29	251 Ku	Nov/1/42	Lae	12th Ku
Katsutoshi Kawamata	LTJG	67	202 Ku	Nov/3/42	Guadalcanal	Tainan Ku
Also LdgSea Wataru Takeda (Hei 2)						
AH Matsumoto	LdgSea		*Kamikawa Maru*	Nov/3/42	Mono Island	
Hidero Goto	LT	66	802 Ku	Nov/7/42	over convoy	
Sadao Jito	PO2c	Pilot 49	802 Ku	Nov/7/42	over convoy	
Also PO1c Sadanori Hagiwara(Ko 5), PO2c Jiro Kondo (Ko 6),						
LdgSea Hajime Mizoguchi						

Name	Rank	Class	Unit	Date	Location	Notes
Isamu Yoshiwara	PO2c	Otsu 10	*Hiyo*	Nov/11/42	Guadalcanal	
Also PO2c Toshio Morita (Otsu 10)						
Isao Ito	PO2c	Otsu10	582 Ku	Nov/12/42	Guadalcanal	
Zenji Ono	WO	Otsu 2	*Junyo*	Nov/13/42	Guadalcanal	*Kaga*
Also CPO Tatsuzo Hasegawa (Otsu 8) and CPO Tasuke Mukai (Pilot 29)						
Tadashi Kaneko Hikotai Ldr.	LCDR	60	*Hiyo*	Nov/14/42	Guadalcanal	*Ryujo*,15th Ku,12th Ku *Shokaku*, 6th Ku
Masaji Suganami Hikotai Ldr.	LT	61	252 Ku	Nov/14/42	Guadalcanal	13th Ku, *Soryu*
Tsumoru Ohkura	CPO	Pilot 36	253 Ku	Nov/14/42	Guadalcanal	Kanoya Ku
Also PO1c Akeji Hikuma (Ko 5) and PO1c Minoru Tanaka (Pilot 54)						
Sanae Matsumoto	PO1c	Pilot 42	204 Ku	Nov/14/42	Guadalcanal	
Also PO2c Koichi Hoshino (Ko 6) and PO2c Toshio Nagata (Otsu 11)						
Masahiro Ueno	LdgSea	Hei 3	*Hiyo*	Nov/15/42	Guadalcanal	
Yoshikazu Ohara	PO1c	Otsu 9	582 Ku	Nov/18/42	Solomons	3rd Ku
Iwao Gono	PO2c	Hei 3	252 Ku	Nov/20/42	Lae	
Also LdgSea Yasutaka Kanoya (Hei 3)						
Toru Oda	PO3c	Pilot 56	252 Ku	Nov/22/42	Lae	
Also LdgSea Tomio Maeda (Hei 3)						
Tomoyuku Sakai	LTJG	66	582 Ku	Nov/30/42	Buna	Attack vessels, *Hiryu*
Also PO1c Takashi Yokoyama (Otsu 9)						
Shigeru Yoshihashi	CPO	Ko 1	252 Ku	Dec/3/42	Rabaul	Force land
Naoshi Uematsu	PO2c	Otsu 10	582 Ku	Dec/7/42	Buna	
Nobumichi Takebe	LdgSea	Hei 3	204 Ku	Dec/10/42	Solomons	
Jiro Tanaka	CPO	Pilot 39	Hiyo	Dec/10/42	Munda	Soryu
Also PO2c Shigeo Motegi (Otsu 10)						
Toshimi Sato	LdgSea	Hei 3	582 Ku	Dec/14/42	Buna	
Yaichiro Fukunishi	Res ENS	Res. 8	252 Ku	Dec/24/42	Munda	
Also LdgSea Kazufusa Harano (Hei 3)						
Chuichiro Hata	LdgSea	Hei 2	252 Ku	Dec/27/42	Munda	
Susumu Ohtsuki	PO2c	Hei 2	582 Ku	Jan/7/43	Lae	
Hikoji Kawada	PO2c	Hei 2	204 Ku	Jan/9/43	Buin	
Sueji Itsukaichi	CPO	Pilot 32	204 Ku	Jan/11/43	Munda	12th Ku
Tatenoshin Tanoue	LTJG	68	204 Ku	Jan/15/43	Munda	
Also LdgSea Sho-ichi Fujisada (Hei 3) and LdgSea Yutaka Kimoto (Hei 4)						
Mitsuoki Asano	Res.ENS	Res.8	582 Ku	Jan/17/43	Rabi	
Minoru Minazawa	LdgSea	Hei 3	802 Ku	Jan/18/43	Shortland	
Saburo Kitahata	WO	Pilot 21	*Junyo*	Jan/23/43	Wewak	Intercepting B-24s, 12th Ku
Ryusuke Sato	CPO	Pilot 43	*Junyo*	Jan/23/43	Wewak	Ramming, *Hiryu*
Kiyoharu Shibuya	LTJG	67	204 Ku	Jan/23/43	Guadalcanal	
Also PO2c Yoshio Imamura (Otsu 11)						
Mitsunori Nakajima	WO	Pilot 29	253 Ku	Jan/24/43	Guadalcanal	POW, returned, 12th Ku, 13th Ku
Makoto Iwamoto	PO1c	Ko 5	253 Ku	Jan/25/43	Guadalcanal	
Mitsuyoshi Inoue	PO1c	Hei 2	202 Ku	Jan/28/43	Ambon	
Korenobu Nishide	CPO	Ko 1	252 Ku	Feb/1/43	Guadalcanal	*Shokaku*
Saburo Hotta	PO2c	Hei 2	582 Ku	Feb/1/43	Guadalcanal	
Also PO2c Tatsuo Morioka (Otsu 10)						
Saishi Ohkawa	CPO	Otsu 8	452 Ku	Feb/2/43	Amchitka Island	
Also PO2c Hitoshi Naito (Otsu 11)						
Katsuma Shigemi	WO	Pilot 20	*Zuikaku*	Feb/4/43	Guadalcanal	*Kaga*, 14th Ku, *Ryujo*
Also CPO Soji Chiba (Pilot 40)						
Masami Takemoto	CPO	Pilot 37	582 Ku	Feb/7/43	Guadalcanal	12th Ku, 3rd Ku

Sahei Yamashita	WO	Otsu 5	201 Ku	Feb/9/43	Nauru	Intercepted B-17s, Chitose Ku, Tainan Ku
Hifumi Yamamoto	LdgSea	Hei 6	204 Ku	Feb/13/43	Buin	
Kotaro Takano	PO2c	Otsu 11	252 Ku	Feb/13/43	Buin	
Yoshio Yoshida	PO2c	Otsu 11	252 Ku	Feb/14/43	Ballale	
Yo-ichi Kenmochi	LTJG	68	253 Ku	Feb/19/43	Surumi [New Britain Island]	
Kunitsugu Nakamachi	WO	Otsu 5	452 Ku	Feb/19/43	Amchitka Island	
Gi-ichi Sasaki	PO1c	Pilot 44	452 Ku	Feb/19/43	Amchitka Island	
*Masanao Maki	LdgSea	Hei 3	*Zuiho*	Mar/3/43	off Lae	Ramming
Takio Dannoue	CPO	Ko 1	*Zuiho*	Mar/3/43	off Lae	12th Ku
Shizuki Nishiyama	PO2c	Pilot 54	204 Ku	Mar/3/43	off Lae	Tainan Ku
Kan-ichi Kashimura	WO	Pilot 24	582 Ku	Mar/6/43	Russell	13th Ku, 12th Ku
Sei-ichi Kitaoka	PO1c	Pilot 45	*Zuiho*	Mar/11/43	Buna	
Also LdgSea Hiroshi Koyama (Hei 3)						
Seiji Tajiri	PO2c	Pilot 50	202 Ku	Mar/15/43	Darwin	
Ko-ichi Yoshida Buntai Ldr.	LT	Res.4	252 Ku	Mar/26/43	Nauru	Killed by bombing, 14th Ku
Tomo-o Inoue	WO	Otsu 5	253 Ku	Mar/28/43	Oro Bay	
Also PO2c Kaneyuki Kamikatahira (Ko 6)						
Shigeto Kawahara	LTJG	68	204 Ku	Apr/1/43	Russell	
Also PO2c Ei-ichi Sugiyama (Ko 6)						
Kiyoshi Ono	PO2c	Hei 2	253 Ku	Apr/1/43	Russell	
Hideo Shimizu	PO1c	Otsu 9	253 Ku	Apr/1/43	Russell	
Also PO2c Ichiro Kawahata and PO2c Izumi Tanaka (both Ko 6), LdgSea Yoshiharu Izumi (Hei 4), and SupSea Sueo Mizuno (Hei 6)						
Makoto Murata	LdgSea	Hei 3	204 Ku	Apr/7/43	Guadalcanal	
Tsuguo Matsuyama	WO	Otsu 5	*Hiyo*	Apr/7/43	Guadalcanal	13th Ku, *Hiryu*
Shozo Katayama	WO	Pilot 21	*Junyo*	Apr/7/43	Guadalcanal	12th Ku
Chiune Yotsumoto	PO1c	Ko 5	*Junyo*	Apr/7/43	Guadalcanal	
Also LTJG Shiro Itesono (69), CPO Matsutaro Kobayashi (Ko 3), PO2c Yuji Ando (Pilot 48), and LdgSea Ippei Ninomiya (Hei 3)						
Yasuo Kanemitsu	LdgSea	Hei 3	253 Ku	Apr/7/43	Guadalcanal	
Taizo Okamoto	LTJG	Pilot 16	*Zuikaku*	Apr/11/43	Oro Bay	12th Ku
Also PO1c Shuryu Uenuma (Otsu 9)						
Sadakazu Iwaki	PO2c	Hei 3	253 Ku	Apr/11/43	OroBay	
Jiro Mitsumoto	WO	Otsu 6	*Zuikaku*	Apr/14/43	Milne Bay	*Zuiho*
Tadashi Torakuma	WO	Pilot 20	Omura Ku	Apr/16/43	Kyushu	Accident, 13th Ku 12th Ku, 253 Ku
Tomekichi Hinako	LT	Pilot 1	Sasebo Ku	May/2/43	Kyushu	Accident
Hayato Noda	WO	Otsu 4	204 Ku	May/13/43	Russell	
Also PO2c Yuki Kariya (Hei 2)						
Tamari Tanigaki	PO2c	Otsu 11	253 Ku	May/13/43	Buin	
Shogo Sasaki	PO2c	Hei 4	582 Ku	May/13/43	Russell	
Akira Kimura Buntai Ldr.	LT	65	251 Ku	May/15/43	New Britain	
Also PO2c Gi-ichi Nakayama (Hei 2)						
Kuratoshi Yasuda	PO1c	Hei 2	202 Ku	May/15/43	Dutch East Indies	
Jun-ichi Takahashi	PO1c	Otsu 10	253 Ku	May/17/43	Kavieng	Accident
Also PO1c Katsutoshi Maetsuji (Ko 5)						
Satoru Ogawa	PO1c	Hei 2	582 Ku	Jun/5/43	Buin	3rd Ku
Also PO2c Shigehiko Ito (Hei 3)						
Yoshimi Hidaka	CPO	Pilot 48	204 Ku	Jun/7/43	Russell	
Also PO1c Yasushi Okazaki (Ko 6) and PO2c Kameji Yamane (Hei 3)						
Masuaki Endo	PO1c	Otsu 9	251 Ku	Jun/7/43	Russell	

Toshitaro Sekiguchi	PO2c	Ko 7	251 Ku	Jun/7/43	Russell	
Also PO2c Kan-ichi Masuda (Otsu 11), PO2c Misao Matsuyoshi (Hei 3), and PO2c Yutaka Fukano (Hei 7)						
Gi-ichi Noguchi	LTJG	68	582 Ku	Jun/12/43	Russell	
Also PO2c So-ichi Fujioka (Hei 6)						
Shigenobu Kozuki	PO2c	Hei 7	251 Ku	Jun/12/43	Russell	
Also PO2c Katsujiro Matsumoto (Ko 7)						
*Zenjiro Miyano	LT	65	204 Ku	Jun/16/43	Lunga Point	12th Ku, 3rd Ku
Hikotai Ldr. Also PO2c Yamato Tamura (Hei3)						
Takeshi Morizaki	Res.LTJG	Res.7	204 Ku	Jun/16/43	Lunga Point	
Also PO2c Saji Kanda (Hei 2)						
Daizo Fukumori	PO1c	Otsu 10	582 Ku	Jun/16/43	Lunga Point	
Also PO1c Katsumi Furumoto (Otsu 10) and PO2c Moto-omi Ishibashi (Hei 4)						
Ken-ichi Shinozuba	PO2c	Hei 2	582 Ku	Jun/16/43	LungaPoint	
Shuhei Ohya	LTJG	Pilot 14	251 Ku	Jun/16/43	LungaPoint	12th Ku
Yoshio Oki	WO	Pilot 37	251 Ku	Jun/16/43	LungaPoint	12th Ku
Also PO2c Suehiro Yamamoto (Hei 1), PO2c Hiroshi Kanda (Hei 7), and PO2c Ikuzu Shimizu (Ko 7)						
Takashi Koshita	LTJG	69	251 Ku	Jun/16/43	Lunga Point	
Takumi Hoashi	LT	63	*Kujisho*	Jun/16/43	Yokosuka	Accident of Raiden; 14th Ku, *Shokaku,*
Ichiro Mukai	LT	63	251 Ku	Jun/30/43	Rendova	12th Ku
Hikotai Ldr.						
Takeyoshi Ohno	LTJG	68	251 Ku	Jun/30/43	Rendova	
Mitsuteru Hashimoto	LTJG	69	251 Ku	Jun/30/43	Rendova	
Also PO2c Hiroshi Iwano (Ko 7), PO2c Nobuo Konishi (Ko 7), PO2c Uichiro Ando (Hei 3),						
PO2c Kazuo Fukui (Hei 7) and PO2c Shun-ichi Hiromori						
Nobutaka Yanami	PO1c	Hei 5	582 Ku	Jun/30/43	Rendova	
Also PO2c Takamichi Sasamoto (Hei 3)						
Toyomitsu Tsujinokami	CPO	Ko 5	204 Ku	Jul/1/43	Rendova	
Yosaburo Shinomiya	PO1c	Otsu 9	251 Ku	Jul/1/43	Rendova	
Also PO2c Hideo Tsukamoto (Hei 4)						
Ichirobei Yamazaki	PO1c	Pilot 54	251 Ku	Jul/4/43	Rendova	4th Ku
Mitsugu Ohtani	CPO	Pilot 41	*Ryuho*	Jul/7/43	Rendova	*Hiyo*
Takeo Ohtsu	PO2c	Otsu 12	253 Ku	Jul/7/43	Rendova	
Saburo Nozawa	CPO	Pilot 45	253 Ku	Jul/9/43	Rendova	12th Ku
Jisuke Iwase	PO2c	Hei 3	*Ryuho*	Jul/11/43	Rendova	
Yoshio Ohsawa	PO2c	Hei 3	582 Ku	Jul/11/43	Rendova	POW, returned
Tsukijiro Fujii	PO2c	Otsu 11	253 Ku	Jul/11/43	Rendova	
Fusayoshi Murasaki	CPO	Pilot 41	582 Ku	Jul/12/43	Rendova	
Also PO2c Kaoru Ohmiyaji (Otsu 11)						
Tamotsu Tsujioka	PO2c	Hei 7	251 Ku	Jul/12/43	Rendova	POW, returned
Also PO1c Hiroji Ishizaki (Ko 6)						
Saburo Saito	LTJG	68	253 Ku	Jul/15/43	RendoVa	
Also PO2c Shichiro Nagai and PO2c Fukumitsu Shimada, (both Hei 3)						
Taiji Suzuki	PO2c	Hei 3	*Ryuho*	Jul/15/43	Rendova	*Shokaku*
Sei-ichi Nakazawa	PO2c	Hei 4	204 Ku	Jul/16/43	Buin	
Kenji Moriyama	CPO	Pilot 35	*Junyo*	Jul/17/43	Buin	
Hisa-aki Fujimaki	LTJG	68	*Junyo*	Jul/17/43	Buin	
Also PO1c Kiyoshi Kojima (Ko 6) and PO1c Hideya Takezawa (Otsu 10)						
Masaharu Hiramoto	CPO	Pilot 38	*Ryuho*	Jul/17/43	Buin	
Also PO2c Yoshio Nakazono (Hei 3)						
Kisaku Koshida	LTJG	69	204 Ku	Jul/17/43	Buin	
Minoru Kuranaga	PO2c	Hei 2	201 Ku	Jul/25/43	Buin	
Also LdgSea Takeshi Oh-ura (Hei 7)						
Tetsuro Nihei	PO1c	Ko 6	204 Ku	Jul/25/43	Rendova	
Also PO2c Kenkichi Nemoto (Hei 4)						

Sachio Hayama	PO2c	Hei 2	252 Ku	Jul/25/43	Wake	
Also PO2c Kenji Wada (Hei 3)						
Kiyoshi Ohsuga	PO2c	Hei 2	252 Ku	Jul/27/43	Wake	
Chuji Sakagami	CPO	Pilot 45	251 Ku	Aug/4/43	Rendova	
Matsuhimaru Kashibo	PO2c	Pilot 54	201 Ku	Aug/4/43	Buin	
Also PO2c Fukuyoshi Morino (Hei 3)						
Sei-ichi Kurosawa	PO2c	Hei 3	204 Ku	Aug/6/43	Buin	
Yasuhiro Nakamura	PO2c	Hei 3	*Junyo*	Aug/13/43	Buin	
Keigo Fujiwara	LTJG	68	*Junyo*	Aug/15/43	Ballale	
Also PO2c Munenori Shimizu (Ko 7)						
Masao Shimada	LTJG	69	204 Ku	Aug/15/43	Ballale	
Seizaburo Watanabe	PO2c	Hei 3	204 Ku	Aug/15/43	Ballale	
Also PO1c Norimasa Narahara (Ko 5), PO2c Motoharu Imazeki (Otsu 11), and PO2c Tetsuo Hidaka (Hei 3)						
Takeshi Takahashi	CPO	Pilot 42	202 Ku	Aug/16/43	Balikpapan	14th Ku
Osamu Yoza	SupSea	Hei 8	934 Ku	Aug/17/43	Maikoor	
Hideo Inohana	PO2c	Ko 7	934 Ku	Aug/21/43	Maikoor	Died 2 days later
Jisaku Kaneko	PO2c	Hei 1	201 Ku	Aug/30/43	Ballale	
Iki Arita	CPO	Pilot 41	251 Ku	Aug/30/43	Buin	12th Ku
Also PO2c Shigeo Hayashi (Otsu 12)						
Michitaka Kashihara	PO1c	Otsu 10	251 Ku	Sep/4/43	Buin	
Yoshikazu Nagahama	CPO	Ko 2	Tsuiki Ku	Sep/6/43	Homeland	Accident, *Kaga*, *Zuikaku*
Yoshio Terai	PO1c	Ko 6	202 Ku	Sep/7/43	Darwin	
Tsukasa Kondo	WO	Otsu 6	201 Ku	Sep/9/43	Buin	12th Ku
Morio Miyaguchi	ENS	Otsu 1	202 Ku	Sep/9/43	Merauke	*Ryujo*, 12th Ku, 22nd AF
Kuraichi Goto	PO1c	Ko 6	202 Ku	Sep/9/43	Merauke	
Also PO2c Koshiro Agawa (Ko 7)						
Wataru Kubota	WO	Pilot 36	Yoko Ku	Sep/12/43	Homeland	Accident, *Soryu*
Iro-ichi Yokota	CPO	Pilot 43	Yoko Ku	Sep/12/43	Homeland	12th Ku, *Zuikaku*
Sato-o Yoshizawa	CPO	Otsu 6	Yoko Ku	Sep/12/43	Homeland	Accident
Makoto Terao	PO2c	Otsu 11	204 Ku	Sep/14/43	Buin	
Also PO2c Tokuji Yoshizaki (Hei 4)						
Eiji Nishida	PO2c	Hei 3	201 Ku	Sep/14/43	Buin	
Also PO2c Hiroshi Mure (Hei 3)						
Tetsuji Ueno	LTJG	69	204 Ku	Sep/16/43	Buin	
Also LTJG Sue-ichi Ohshima (69)						
Ryo-itsu Ohkubo	WO	Pilot 29	204 Ku	Sep/18/43	Buin	
Yukuo Nanao	PO2c	Hei 6	331 Ku	Sep/22/43	Car Nicobar	
Takeo Okumura	CPO	Pilot 42	201 Ku	Sep/22/43	Cape Cretin	14th Ku, *Ryujo*, Tainan Ku
Shunzo Hongo	WO	Pilot 30	201 Ku	Sep/22/43	Cape Cretin	13th Ku, 12th Ku
Also Res.ENS Minoru Tanaka (Res.10)						
Toshihisa Shirakawa	PO1c	Ko 6	204 Ku	Sep/22/43	Cape Cretin	
Shigeru Tanaka	SupSea	Hei 11	204 Ku	Sep/23/43	Viroa	
Masa-ichi Enomoto	CPO	Hei 1	204 Ku	Sep/25/43	Viroa	3rd Ku, 582 Ku
Hachiro Tsuboya	PO1c	Ko 6	204 Ku	Oct/3/43	Munda	
Also LdgSea Yutaka Shimizu (Hei 10)						
Toshiyuki Sueda	WO	Pilot 32	252 Ku	Oct/6/43	Wake	13th Ku, 12th Ku
Bunkichi Nakajima	CPO	Pilot 33	252 Ku	Oct/6/43	Wake	13th Ku, 15th Ku, 3rd Ku
Yukuo Miyauchi	CPO	Ko 3	252 Ku	Oct/6/43	Wake	
Hisashi Hide	CPO	Otsu 7	252 Ku	Oct/6/43	Wake	

Yoshio Shiode	PO2c	Otsu 11	252 Ku	Oct/6/43	Wake	

Also CPO Kazuo Tobita (Hei 3), PO1c Tamotsu Okabayashi (Hei 3), PO1c Motosuke Shibata (Ko 6), PO2c Saburo Fujiuma (Hei 4), PO2c Katsunobu Shiba (Hei 6), PO2c Kazuyoshi Tokuhara (Hei 6), and PO2c Kiyoshi Takei (Otsu 12)

Takao Banno	PO2c	Pilot 53	204 Ku	Oct/7/43	Vella Lavella	*Kaga, Shoho, Junyo*
Koshio Enomoto	SupSea	Hei 10	201 Ku	Oct/10/43	Buin	

Also SupSea Hisashi Akahori (Hei 10)

Makoto Endo	WO	Pilot 25	253 Ku	Oct/12/43	Rabaul	12th Ku, *Ryujo*
Tomokichi Arai Buntai Ldr.	LTJG	Pilot 15	201 Ku	Oct/15/43	Buin	13th Ku
Shigenobu Adachi	PO1c	Otsu 9	204 Ku	Oct/15/43	Buin	

Also PO2c Kisaku Tanaka (Hei 3)

Tsunehiro Yamagami	CPO	Ko 3	253 Ku	Oct/15/43	Buna	12th Ku

Also PO2c Eiji Sekiguchi (Otsu 12)

Hisayoshi Mori	PO2c	Hei 3	253 Ku	Oct/15/43	Buna	
Haruo Sagara	LTJG	69	253 Ku	Oct/17/43	Buna	

Also SupSea Kazunori Mishima (Hei 10)

Yoshiharu Matsumoto	WO	Pilot 28	253 Ku	Oct/17/43	Buna	15th Ku
Yoshimi Yamauchi	PO2c	Hei 3	204 Ku	Oct/17/43	Buin	

Also PO2c Shunji Ito (Hei 4)

Yuji Yokoyama	CPO	Pilot 38	204 Ku	Oct/18/43	Rabaul	
Yoshio Aizawa	Res.ENS	Res.9	201 Ku	Oct/18/43	Buin	

Also PO1c Yoshio Mogi (Otsu 9)

Hidemichi Moriyama	Res.ENS	Res.10	204 Ku	Oct/23/43	Rabaul	
Shizu-o Ishi-i	CPO	Pilot 50	204Ku	Oct/24/43	Rabaul	12th Ku, Tainan Ku, *Junyo*
Tsuyoshi Tanaka	PO1c	Hei 6	204 Ku	Oct/24/43	Rabaul	12th Ku,

Also PO2c Masaru Kubo (Otsu 13) and PO2c Masakazu Kobayashi (Hei 2)   Tainan Ku, *Junyo*

Yoshinori Noguchi	PO1c	Otsu 9	201 Ku	Oct/29/43	Rabaul	
Shigetsune Ohgane	CPO	Pilot 45	201 Ku	Nov/1/43	Bougainville	
Gen-ichi Seki	CPO	Ko 5	201 Ku	Nov/1/43	Bougainville	

Also PO1c Ichiro Yamashita, PO1c Manabu Yoshi-i, and PO1c Yonehachiro Ishikura (all Ko 7)

Yutaka Ohtani	WO	Ko 3	204 Ku	Nov/1/43	Bougainville	22nd AF, Kanoya
Masatsugu Kawamura	CPO	Ko 6	*Shokaku*	Nov/2/43	Rabaul	

Also PO1c Genshichi Sato (Hei 3) and PO1c Take-o Yamamoto (Hei 6)

Shigeru Ohkura	CPO	Pilot 41	*Zuikaku*	Nov/2/43	Rabaul	
Kazuo Komaba	PO1c	Hei 3	*Zuikaku*	Nov/2/43	Rabaul	

Also PO1c Saburo Yoshida (Ko 7) and PO2c Masayoshi Miyakawa (Hei 7)

Fusa-ichi Kaneko	PO1c	Hei 3	204 Ku	Nov/2/43	Rabaul	

Also CPO Katsumi Shimamoto (Ko 6)

Kazunori Miyabe	LTJG	Otsu 2	*Shokaku*	Nov/3/43	Rabaul	12th Ku, injured the previous day
Kosoku Minato	CPO	Pilot 44	*Zuiho*	Nov/5/43	Rabaul	Tainan Ku
Hiroshi Nishimura	PO1c	Otsu 11	*Zuikaku*	Nov/5/43	Rabaul	
*Kenjiro Notomi Hikotai Ldr.	LT	62	*Zuikaku*	Nov/8/43	Bougainville	*Shoho, Ryujo*
Nobuyuki Muraoka	PO1c	Hei 3	*Zuiho*	Nov/8/43	Bougainville	
Shigeru Araki Buntai Ldr.	LT	67	*Zuikaku*	Nov/11/43	Rabaul	
Masao Sato Hikotai Ldr.	LT	63	*Zuiho*	Nov/11/43	Bougainville	12th Ku, *Zuikaku, Kaga*

Also Res.ENS Shoichiro Yamada (Res. 9)

Hitoshi Sato	WO	Pilot 6	*Shokaku*	Nov/11/43	Bougainville	15th Ku

Also PO1c Kazuo Tachizumi (Otsu 11) and PO1c Ryuzo Isobe (Hei 7)

Yukio Aiso	ENS	Otsu 3	201 Ku	Nov/17/43	Torokina	
Also PO1c Jisuke Yoshino (Ko 8)						
Masateru Tomoishi	WO	Pilot 44	253 Ku	Nov/21/43	Torokina	14th Ku, *Ryujo*
Also PO1c Hidemichi Matsumura and PO1c Isao Fukuda (both, Hei 3)						
? Kubo	LdgSea		934 Ku	Nov/21/43	Maikoor	
Takeshi Kawaguchi	WO	Otsu 4	934 Ku	Nov/21/43	Maikoor	
Shotoku Yamaguchi	PO1c	Hei 3	204 Ku	Nov/22/43	Cape Torokina	
Also PO1c Yorihisa Kobayashi (Otsu 11)						
Masahiro Chikanami	Res. LTJG	Res.7	252 Ku	Nov/24/43	Makin	
Tsutomu Kawai	LTJG	69	252 Ku	Nov/24/43	Makin	
Masao Kuramoto	LTJG	Res. 8	252 Ku	Nov/24/43	Makin	
(Seinan Gakuin)						
Also WO Kumaji Tsugame (Otsu 4), PO1c Hiroshi Ueda (Otso 11), and PO1c Kiyoshi Tokunaga (Hei 3)						
Shin-ichi Yamawaki	PO1c	Hei 4	252 Ku	Nov/24/43	Makin	
Also PO1c Morimasa Hirai (Hei 6)						
Shizuo Kojima	WO	Pilot 32	252 Ku	Nov/25/43	Makin	
Kumaichi Kato	PO1c	Otsu 11	252 Ku	Nov/25/43	Makin	13th Ku
Also PO1c Shigekazu Suzuki (Hei 4), PO1c Koichi Ishikawa (Otsu 12), and PO2c Seiji Sakamaki (Hei 11)						
Genshichiro Ohyama	Res.ENS	Res.9	*Zuikaku*	Dec/5/43	Roi	
Hisashi Aoyama	WO	Pilot 39	*Zuikaku*	Dec/5/43	Roi	
Also PO1c Yasukichi Yamakawa (Hei 6) and PO1c Osamu Ichikawa (Ko 7)						
Yoshio Yoshida	WO	Pilot 46	*Zuikaku*	Dec/5/43	Roi	
Ichiro Imamura	LTJG	69	281 Ku	Dec/5/43	Roi	
Shigeru Tagami	PO1c	Hei 3	281 Ku	Dec/5/43	Roi	
Also PO1c Sadamu Imazono and PO1c Itsuro Kubo (both Ko 8)						
Ten-ichi Tagami	PO1c	Hei 3	201 Ku	Dec/16/43	Cape Marcus	
Kazuo Umeki	CPO	Ko 5	201 Ku	Dec/19/43	Rabaul	
Kaneji Shimizu	Res.ENS	Res.10	253 Ku	Dec/19/43	Rabaul	
Also PO1c Tokio Yokoi (Otsu 12)						
Ko-ichiro Takishita	LdgSea	Hei 10	281 Ku	Dec/19/43	Mili	
Nihei Iwano	PO1c	Ko 8	281 Ku	Dec/20/43	Mili	
*Yoshio Ohba	LTJG	69	201 Ku	Dec/23/43	Rabaul	
Also PO1c Masanobu Kijiya (Otsu 12)						
Kagemitsu Matsuo	Res.ENS	Res.10	253 Ku	Dec/23/43	Rabaul	
Takeshi Fujii	CPO	Ko 6	204 Ku	Dec/23/43	Rabaul	
Also PO1c Jiro Uchida (Otsu 12)						
Jinkichiro Kudo	WO	Pilot 31	281 Ku	Dec/24/43	Mili	14th Ku
Tadashi Hirai	PO1c	Otsu 11	253 Ku	Dec/26/43	Turubu	
Also PO1c Chikara Kitaguchi (Hei 6)					Cape Gloucester	
Yasujiro Ohno	CPO	Pilot 43	253 K	Dec/27/43	Rabaul	12th Ku, *Zuiho*
Also PO1c Matsukichi Matsui (Hei 3)						
Manzo Iwaki	LTJG	Pilot 13	*Hiyo*	Jan/1/44	Kavieng	
Also PO1c Hitoshi Nagano (Ko 7)						
Michio Takeshita	PO1c	Ko 7	201 Ku	Jan/1/44	Rabaul	
Also PO1c Tsuneo Suzuki (Otsu 13)						
Den Katayama	Res.ENS	Res.9	204 Ku	Jan/2/44	Rabaul	
Also PO1c Tokio Iseki (Hei 6)						
Itsuyoshi Yamazaki	PO1c	Otsu 11	251 Ku	Jan/2/44	Rabaul	
(night fighters)						
Toshio Fujita	PO2c	Otsu 15	204 Ku	Jan/7/44	Rabaul	
Sadayoshi Yokota	WO	Ko 1	934 Ku	Jan/19/44	Ambon	
Hiroshi Shibagaki	SupSea	Hei 12	204 Ku	Jan/22/44	Rabaul	
Hiroshi Oh-iwa	CPO	Pilot 40	204 Ku	Jan/23/44	Rabaul	
Also PO1c Susumu Kanenobu (Ko 8) and PO1c Namio Hashimasa (Otsu 11)						
Takeshige Senuma	PO1c	Otsu 11	202 Ku	Jan/23/44	Rangoor	

Taira Kubota	PO1c	Otsu 13	253 Ku	Jan/23/44	Rabaul	
Also PO1c Takeshi Katayama (Otsu 12)						
Kiyoshi Shimizu	PO1c	Hei	253 Ku	Jan/26/44	Rabaul	204 Ku
Sanemori Yoshioka	PO1c	Otsu 14	*Ryuho*	Jan/26/44	Rabaul	
Fujikazu Koizumi	LTJG	Otsu 2	*Hiyo*	Jan/27/44	Rabaul	12th Ku, 3rd Ku
Naoshi Mae	LTJG	69	*Junyo*	Jan/27/44	Rabaul	
Also CPO Sukeo Kawasaki (Otsu 9)						
Yoshiaki Hatakeyama	WO	Ko 1	252 Ku	Jan/27/44	Maloelap	3rd Ku
Also PO1c Shiro Tsukahara (Hei 3)						
Toshio Komiya	PO1c	Hei 3	252 Ku	Jan/27/44	Maloelap	
Masashi Ishida	PO1c	Pilot 56	253 Ku	Jan/28/44	Rabaul	*Akagi, Shokaku*
Shigeru Tsuda	PO1c	Otsu 14	*Junyo*	Jan/28/44	Rabaul	
Yoshihide Ishizawa	CPO	Otsu 10	*Ryuho*	Jan/28/44	Rabaul	*Shokaku*
Minoru Kitazaki	PO1c	Otsu 12	253 Ku	Jan/29/44	Rabaul	
Also PO2c Shozo Hara (Otsu 15)						
Hiroshi Arai	PO1c	Otsu 14	*Ryuho*	Jan/30/44	Rabaul	
Keizaburo Yamazaki	CPO	Ko 6	252 Ku	Jan/30/44	Roi	
Teruo Sugiyama	WO	Otsu 7	253 Ku	Feb/4/44	Rabaul	*Ryujo*, 201 Ku
Takaichi Hasuo Hikotai Ldr.	LT	65	281 Ku	Feb/6/44	Roi	Presumed killed, 12th Ku, 3rd Ku
Also LTJG Masayuki Goto (Res.10)						
Isamu Mochizuki Buntai Ldr.	LTJG	Pilot 9	281 Ku	Feb/6/44	Roi	Presumed killed, 13th Ku
Yoshijiro Minegishi	ENS	Otsu 2	281 Ku	Feb/6/44	Roi	Presumed killed, 15th Ku, *Hiryu*
Also CPO Shunji Nakazawa and CPO Shigeichi Sasa (both Otsu 10)						
Akira Kikuchi	WO	Pilot 31	281 Ku	Feb/6/44	Roi	Presumed killed, 12th Ku
Also PO1c Toshikazu Yahiro (Hei 7) and PO1c Sakae Okamura (Otsu 12)						
Taketoshi I-io	PO1c	Otsu 11	281 Ku	Feb/6/44	Roi	
Also PO1c Fumio Kato (Hei 3)						
Ginji Kiyosue	WO	Ko 2	*Ryuho*	Feb/7/44	Rabaul	*Shokaku, Zuikaku*
Toshio Ito	Res. LTJG	Res.10	253 Ku	Feb/9/44	Rabaul	
Kaoru Takaiwa	PO1c	Otsu 13	253 Ku	Feb/10/44	Rabaul	201 Ku
Shichijiro Mae	PO1c	Pilot 54	*Junyo*	Feb/11/44	Rabaul	*Zuikaku*
Sanenori Kuroki	CPO	Pilot 42	*Hiyo*	Feb/12/44	Rabaul	
Nobuo Ogiya	WO	Pilot 48	253 Ku	Feb/13/44	Rabaul	Chitose Ku, 281 Ku, 204 Ku
Takeshi Kobayashi	PO1c	Hei 2	254 Ku	Feb/14/44	Nanning	
Sukeichi Yamashita	Res.LTJG	Res.10	252 Ku	Feb/16/44	Micronesia	On transport
Also CPO Juji Kido (Ko 6)						
*Hideo Maeda	WO	Ko 1	204 Ku	Feb/17/44	Truk	12th Ku
Hisao Harada	WO	Otsu 7	204 Ku	Feb/17/44	Truk	
Also LTJG Tadao Satoh (70), PO1c Iwao Taneda (Hei 6), and POle Yasumasa Tsutsumi (Ko 8)						
Haruo Tomita	CPO	Ko 6	201 Ku	Feb/17/44	Truk	
Hiroshi Sonokawa (night fighters)	LT	62	251 Ku	Feb/17/44	Truk	
Michiharu Kawano	LTJG		902 Ku	Feb/17/44	Truk	
Also CPO Yoshio Suzuki (Pilot 49),PO1c Yoshinori Mitani (Ko 8),						
LdgSea Yoshimune Horinouchi, SupSea Shigeyoshi Haruna						
Yoshihiro Yoshikawa	LTJG	70	263 Ku	Feb/23/44	Marianas	
Also PO1c Tomimasa Ohkubo and PO1c Ryozo Okada (both Ko 7)						
Yoshio Wajima	LTJG	Pilot 18	263 Ku	Feb/23/44	Marianas	12th Ku, 14th Ku, Chitose Ku, 582 Ku
Kazuo Ikumi	CPO	Otsu 10	203 Ku	Mar/15/44	Atsugi	Accident, 582 Ku
Masaru Tsukiji	PO1c	Otsu 13	253 Ku	Mar/24/44	Cape Torokina	

Yasuto Abe	CPO	Ko 6	202 Ku	Mar/27/44	Ponape	
Kaoru Miyazaki	PO1c	Hei 3	Yoko Ku	Mar/27/44	Homeland	
Mitsu-omi Noda	WO	Ko 6	202 Ku	Mar/29/44	Truk	
Also PO1c Furuo Yoshimitsu (Hei 4)						
Koshiro Yamashita	ENS	Pilot 17	201 Ku	Mar/30/44	Peleliu	*Ryujo*, 12th Ku
Kazuo Hattori	LdgSea	Hei 10	201 Ku	Mar/30/44	Peleliu	
Shin-ichi Suzuki	CPO	Pilot 45	201 Ku	Mar/30/44	Peleliu	
Also LTJG Shigeo Hayashi (70), CPO Haruo Kunihiro (Ko 6), and PO1c Masao Wada (Ko 9)						
Tomojiro Yamaguchi	LT	69	S351	Mar/30/44	Peleliu	
Buntai Ldr.						
Masato Okasako	LTJG	70	261 Ku	Mar/31/44	Peleliu	
Masao Shibuya	Res.ENS	Res.11	261 Ku	Mar/31/44	Peleliu	
Also PO2c Kenji Nihonmori (Pilot 45), PO1c Shigeru Hayashi (Pilot 55),						
PO1c Tashiro Koga (Otsu 11), and PO2c Ken-ichi Iryu (Otsu 15)						
Nobuhiko Muto	LT	70	263 Ku	Mar/31/44	Peleliu	
Takeichi Kikuchi	WO	Ko 1	263 Ku	Mar/31/44	Peleliu	
Also Res. ENS Yoshito Shimoda (Res. 11), WO Masaro Nagase (Ko 1),						
CPO Shoji Kurita (Ko 6), and CPO Shokichi Nishimoto (Otsu 10)						
Takeshi Okui	CPO	Ko 5	253 Ku	Apr/5/44	Truk	
Tsuneo Nakahara	LTJG	Pilot 12	Sanya Ku	Apr/5/44	Nanning	Tainan Ku
Takeo Kume	CPO	Otsu 9	Sanya Ku	Apr/5/44	Nanning	Tainan Ku
Also PO1c Tasuke Okabe, PO1c Tomio Kitaoka (both Otsu 13), and PO1c Asagoro Ishioka (Otsu 12)						
Kaname Yoshimatsu	CPO	Pilot 41	Kaiko Ku	Apr/5/44	Nanning	*Soryu*, *Hiryu*
Also CPO Sakae Mori (Pilot 50)						
Hiroshi Maeda	LT	69	254 Ku	Apr/8/44	Hainan Tao	
Isao Tahara	CPO	Pilot 45	Sanya Ku	Apr/9/44	Hainan Tao	*Hiryu*
Kishichiro Ayukawa	LTJG	70	261 Ku	Apr/16/44	Meleyon [Woleai]	
Matao Ichioka	PO2c	Hei 12	253 Ku	Apr/19/44	Solomons	
Shiro Kawakubo	LT	69	202 Ku	Apr/30/44	Truk	
Hiroshi Suzuki	Res.LTJG	Res.9	202 Ku	Apr/30/44	Truk	
Also PO1c Noboru Nakaoka (Ko 9),PO1c Hiroshi Nishida (Otsu 12), and PO1c Kikuo Ikeda (Ko 7)						
Usao Nishimura	PO1c	Ko 9	253 Ku	Apr/30/44	Truk	
Also PO2c Masaru Moriyama (Hei 6)						
Shunji Horiguchi	CPO	Pilot 51	265 Ku	May/5/44	Saipan	*Shokaku*,
Also PO1c Norio Nakajima (Ko 10)						*Junyo*, accident
Ken-ichi Honda	LT	69	263 Ku	May/7/44	Guam	
Also Res.ENS Shozo Kajikawa (Res.11)						
San-ichiro Tanaka	WO	Pilot 43	263 Ku	May/7/44	Guam	12th Ku,
						Tainan Ku
Sukemasa Kato	WO	Otsu 7	S311	May/7/44	Biak	
Also CPO Tsutomu Shibata (Otsu 9)						
Sanenobu Maehara	WO	Ko 2	203 Ku	May/13/44	Northern Kuriles	
(night fighters)						
Shinji Ishida	WO	Ko 1	S311	May/29/44	Biak	
Also CPO Tadashi Goto (Ko 8)						
Yuji Sato	WO	Ko 2	202 Ku	Jun/2/44	Biak	331 Ku
Also CPO Kenji Yamashita (Ko 9)						
Hiro-o Takao	LT	69	S603	Jun/3/44	Babo	
Buntai Ldr.						
Also CPO Michiaki Ichikawa (Ko 9)						
Toshio Ushikubo	Res.LTJG	Res.10	S301	Jun/3/44	Babo	
(Kiryu HighTech) Also CPO Koshichi Izumida (Otsu 12)						
Otojiro Sakaguchi	WO	Ko 1	S301	Jun/3/44	Biak	POW, returned
Ryutaro Masuda	LT	70	253 Ku	Jun/9/44	Truk	Intercepted B-24s
Kunishige Hasegawa	WO	Otsu 6	S901	Jun/10/44	Truk	
(night fighters)						

Juji Torimoto	LT	69	S316	Jun/11/44	Saipan	
Buntai Ldr.						
Tetsuji Koga	WO	Otsu 7	S316	Jun/11/44	Saipan	
Also CPO Ryoji Saito, CPO Tsuneaki Shimada (both Ko 9), and PO1c Takemi Kasuda (Ko 10)						
Nobuyuki Ikura	LT	70	265 Ku	Jun/11/44	Saipan	*Akagi*
Katsushi Tanaka	WO	Ko 1	265 Ku	Jun/11/44	Saipan	
Also ENS Masaji Yoshifuku (Res. 11), CPO Ryo-ichi Sugiura (Ko 5),						
PO1c Shinya Sugita (KO 10), and PO2c Gunji Kawachi (Hei 10)						
Ryuzo Yamamoto	CPO	Ko 8	261 Ku	Jun/11/44	Saipan	
Zenji Saito	CPO	Ko 8	261 Ku	Jun/11/44	Saipan	
Also CPO Kiyoshi Fushimi (Otsu 14) and CPO Yotaro Ishikawa (Ko 9)						
Takashi Jiromaru	PO1c	Ko 10	263 Ku	Jun/11/44	Marianas	
Koji Shimizu	CPO	Otsu 11	343 Ku	Jun/12/44	Marianas	
Also CPO Kiyohira Kaneyama (Otsu 13)						
Motoi Kaneko	LT	69	S401	Jun/15/44	Iwo Jima	
Buntai Ldr. Also PO1c Toshiro Kanazawa and PO1c Rinzo Toya (both Ko 10)						
Yoshio Torishima	CPO	Hei 2	S401	Jun/15/44	Iwo Jima	3rd Ku
Mamoru Shimura	LTJG	Pilot 18	265 Ku	Jun/15/44	Iwo Jima	
Shigeo Juni	LT	67	S316	Jun/15/44	Iwo Jima	
Hikotai Ldr.						
Suminori Kawahata	CPO	Pilot 43	S316	Jun/15/44	Iwo Jima	
Also Res.LTJG Chobei Morira (Res. 9) and CPO Kazuo Oh-ishi (Otsu 14)						
Shiro Sakamoto	Res.ENS	Res.11	S301	Jun/15/44	Saipan	
Aiji Sato	PO1c	Ko 10	265 Ku	Jun/15/44	Saipan	
Also PO1c Takashi Tomioka (Ko 10)						
Masanori Sato	CPO	Otsu 12	201 Ku	Jun/15/44	Naha	
Also CPO Shinkichi Ohshima (Hei 3)						
Tetsuo Kadomatsu	PO1c	Ko 10	343 Ku	Jun/17/44	Marianas	
Hiroshi Kurihara	WO	Ko 1	201 Ku	Jun/18/44	Marianas	12th Ku, *Ryujo*
Kiji Kaburagi	CPO	Hei 3	263 Ku	Jun/18/44	Marianas	
Ikuro Sakami	LT	69	601 Ku	Jun/19/44	Battle of	
Buntai Ldr. Also LT Kiyoshi Fukagawa (70)					Philippine Sea	
Toshitada Kawazoe	LT	67	601 Ku	Jun/19/44	Battle of	3rd Ku
Buntai Ldr. Also CPO Mamoru Morita (Otsu 9)					Philippine Sea	
Akira Maruyama	ENS	Otsu 4	601 Ku	Jun/19/44	Battle of	12th Ku, 14th Ku
					Philippine Sea	*Ryujo, Shokaku*
Ichiro Yamamoto	WO	Pilot 50	601 Ku	Jun/19/44	Battle of	*Shokaku*
Also CPO Masayuki Hanamura (Ko 7)					Philippine Sea	
Saburo Sugai	WO	Pilot 26	601 Ku	Jun/19/44	Battle of	*Ryuho*
Also CPO Shun-ichi Koyanagi (Hei 3)					Philippine Sea	
Hiroshi Yoshimura	LT	68	652 Ku	Jun/19/44	Battle of	*Ryuho*
Also LT Kenkichi Takasawa (69)					Philippine Sea	
Yoshihiko Takenaka	WO	Ko 1	652 Ku	Jun/19/44	Battle of	3rd Ku
					Philippine Sea	
Kenta Komiyama	WO	Otsu 7	652 Ku	Jun/19/44	Battle of	*Zuikaku*
					Philippine Sea	
Ko-ichi Imamura	CPO	Pilot 56	652 Ku	Jun/19/44	Battle of	
					Philippine Sea	
Hiroshi Shiozaka	LT	70	653 Ku	Jun/19/44	Battle of	
Also PO1c Kiyotaka Sawazaki (Hei 3)					Philippine Sea	
Isao Kondo	CPO	Ko 6	653 Ku	Jun/19/44	Battle of	
					Philippine Sea	
Kozaburo Yasui	WO	Pilot 40	652 Ku	Jun/19/44	Battle of	14th Ku,
Also CPO Masami Komaru (Hei 7)					Philippine Sea	22 AF,
						Tainan Ku

Tetsuo Kikuchi	CPO	Pilot 39	652 Ku	Jun/19/44	Battle of Philippine Sea	14th Ku, *Akagi, Shokaku, Ryuho*
Fumio Yamagata	LT	70	601 Ku	Jun/19/44	Battle of Philippine Sea	
Also LT Yutaka Yagi (70)						
Toshio Fukushima	LT	70	601 Ku	Jun/19/44	Battle of Philippine Sea	
Also CPO Fumio Ito (Otsu 12)						
Mitsunobu Kaga	LTJG	Otsu 2	601 Ku	Jun/19/44	Battle of Philippine Sea	
Also CPO Iwao Yamamoto (Otsu 11)						
Susumu Horio	LT	70	601 Ku	Jun/19/44	Battle of Philippine Sea	
Also CPO Takeo Nagai (Pilot 44)						
Shinya Ozaki Hikotai Ldr.	LT	68	343 Ku	Jun/19/44	Guam	
Isao Doikawa	CPO	Pilot 47	343 Ku	Jun/19/44	Guam	*Soryu*
Also PO1c Kiyoshi Yoshioka (Ko 10)						
Tatsuo Hirano Hikotai Ldr.	LT	Eng 47	S309	Jun/19/44	Guam	
Yoshinao Tokuji	WO	Otsu 6	253 Ku	Jun/19/44	Guam	12th Ku, 3rd Ku
Also CPO Noboru Kayahara (Ko 9)						
Ko-ichi Yamauchi	CPO	Otsu 12	201 Ku	Jun/19/44	off Saipan	
Naoto Sato	CPO	Ko 8	261 Ku	Jun/19/44	Guam	
Also CPO Shigenori Hayashi (Otsu 14)						
Takumi Kai	WO	Otsu 8	652 Ku	Jun/20/44	Battle of Philippine Sea	*Kaga Zuikaku*
Jiro Imura	WO	Otsu 7	652 Ku	Jun/20/44	Battle of Philippine Sea	*Hosho*
Also CPO Masahiro Motoki (Ko 8)						
Masamichi Minokata	LT	69	253 Ku	Jun/23/44	Guam	
Toshiharu Ikeda Hikotai Ldr.	LT	67	S603	Jun/23/44	off Saipan	
Saneo Imamura	WO	Pilot 23	S603	Jun/23/44	off Saipan	331 Ku
Tadashi Nakamoto	CPO	Ko 5	201 Ku	Jun/23/44	off Saipan	Tainan Ku
Also CPO Toshio Tanaka (Ko 9)						
Kakuro Kawamura Buntai Ldr.	LT	69	343 Ku	Jun/24/44	Iwo Jima	
Also LTJG Naoyoshi Yonemasu (71)						
Katsumi Koda Buntai Ldr.	LT	69	S601	Jun/24/44	Iwo Jima	
Nobuo Awa Hikotai Ldr.	LT	69	252 Ku	Jun/24/44	Iwo Jima	
Sadayoshi Masumoto	LT	70	252 Ku	Jun/24/44	Iwo Jima	
Also WO Mitsuzo Hashimoto (Ko 2), CPO Makoto Nagura (Ko 9), and Res.ENS Masao Katsuta (Res. 11)						
Tadao Shiratori	WO	Ko 3	Yoko Ku	Jun/24/44	Iwo Jima	
Also CPO Takao Banno (Hei 3)						
Kiyoshi Sekiya	CPO	Hei 3	Yoko Ku	Jun/24/44	Iwo Jima	3rd Ku, 582 Ku, 204 Ku
Kazuo Oh-hata	LT	70	Yoko Ku	Jun/24/44	Iwo Jima	Attack task forces
Toshitsugu Nisugi	CPO	Pilot 54	Yoko Ku	Jun/24/44	Iwo Jima	
Also PO1c Kinji Koike (Otsu 16)						
Tomezo Yamamoto	CPO	Hei 2	203 Ku	Jun/24/44	Shumushu	Accident, 582 Ku
Masateru Kurokawa	Res.ENS	Res.11	261 Ku	Jun/24/44	Marianas	
Mitsuho Tanaka	LT	70	253 Ku	Jun/26/44	Truk	Intercepted B-24s
Susumu Ishida	CPO	Otsu 8	252 Ku	Jul/3/44	Iwo Jima	Chitose Ku
Also CPO Akio Maeda (Hei 4)						
Junzo Okutani	WO	Ko 4	S601	Jul/3/44	Iwo Jima	
Korekiyo Kawakita	LT	70	S601	Jul/3/44	Iwo Jima	

Mibuichi Shimada	CPO	Pilot 54	S601	Jul/3/44	Iwo Jima	
Also CPO Yoshio Goto (Otsu 11), CPO Kikuichi Ishikawa (Otsu 14), and POlc Teruo Doi (Otsu 16)						
Sadao Yamaguchi	LT	67	Yoko Ku	Jul/4/44	Iwo Jima	3rd Ku
Buntai Ldr.						
Bangoro Myokei	CPO	Hei 3	Yoko Ku	Jul/4/44	Iwo Jima	582 Ku
Also CPO Sadao Kubo (Hei 3)						
Masami Iwatsubo	CPO	Hei 3	301 Ku	Jul/4/44	Iwo Jima	
Iwao Mita	WO	Ko 2	301 Ku	Jul/4/44	Iwo Jima	*Soryu, Hiyo*
Also PO2c Takayoshi Morita (Hei 10)						
Tomotoshi Ishikawa	WO	Ko 1	Omura Ku	Jul/6/44		Homeland Accident
Ryogo Nakahachi	PO1c	Ko 10	265 Ku	Jul/7/44	Saipan	Presumed killed
Also PO1c Sakuro Takahashi and PO1c Yoshimaro Miwa (Ko 10)						
Hitoshi Ishibashi	PO1c	Ko 10	261 Ku	Jul/7/44	Saipan	Presumed killed
Also PO1c Takeshi Suzuki, PO1c Tamotsu Harada (both Ko 10)						
*Yasuhiro Shigematsu	LT	66	263 Ku	Jul/8/44	Marianas	*Hiryu, Junyo*
Hikotai Ldr.						
Also LTJG Yoshishi Busujima (Res. 11)						
Jizo Nishiyama	CPO	Ko 5	201 Ku	Jul/8/44	Marianas	
Also CPO Shigehisa Aoki (Otsu 13)						
Ichiro Higashiyama	LTJG	Otsu 2	261 Ku	Jul/8/44	Saipan	Presumed killed, 15th Ku, 12th Ku
Buntai Ldr.						
Tetsutaro Kumagaya	WO	Pilot 28	253 Ku	Jul/8/44	Saipan	Presumed killed, 13th Ku, 12th Ku 253 Ku, 204 Ku
Kenzo Asatsu	CPO	Otsu 11	301 Ku	Jul/8/44	Saipan	Presumed killed
Teiji Kagami	Res.LT	Res.7	S603	Jul/11/44	Micronesia	
(Nagoya Pharma.Col.) Buntai Ldr.						
Tetsuo Matsuo	PO1c	Ko 10	201 Ku	Jul/21/44	Yap	Rammed into a B-24
Also PO1c Ryuji Tomita (Ko 10)						
Shigeru Itaya	LCDR	57	51st AF	Jul/24/44	Kuriles	Accident; *Ryujo*, 15th Ku, 12th Ku, *Akagi*
Yasuhiko Ukimura	LT	70	S301	Jul/25/44	Micronesia	265 Ku
Buntai Ldr.						
Jiro Iwai	CPO	Otsu 10	331Ku	Jul/25/44	Sabang	*Hiyo*
Also PO1c Hiroshi Kataoka (Ko 10)						
Rikio Aizawa	CPO	Ko 8	254 Ku	Jul/29/44	Sanya	
Ichiro Shimoda	LT	66	321 Ku	Aug/2/44	Tinian	Presumed killed
Hikotai Ldr. (night fighters) Also LT Hideo Kume and LT Tametsugu Yonishi (both 70)						
Ken-ichi Ban	LT	69	S306	Aug/10/44	Guam	Presumed killed
Hikotai Ldr.						
Yoshio Iwabuchi	CPO	Pilot 56	601 Ku	Aug/10/44	Guam	*Soryu*, presumed killed
Mizuho Tanaka	LT	70	S301	Aug/17/44	Truk	Intercepted B-24s
Buntai Ldr.						
Shin Yamauchi	LT	69	S311	Aug/20/44	Borneo	Accident
Buntai Ldr.						
Rikio Tomioka	LT	70	S315	Aug/31/44	Iwo Jima	
Kunio Mori	LT	69	S902	Sep/10/44	off Leyte	
Buntai Ldr.						
Also LTJG Nobuichi Takahata (14th Recon)						
Hiroshi Mori-i	LT	69	S306	Sep/12/44	Cebu	
Hikotai Ldr.						
Also Res.LTJG Hayao Ishihashi (Res. 11)						

Hideo Oh-ishi	WO	Pilot 26	201 Ku	Sep/12/44	Cebu	*Hosho, Soryu,* 12th Ku
Tadahiro Sakai	WO	Ko 4	201 Ku	Sep/12/44	Cebu	3rd Ku
Hyakuro Makiyama	CPO	Hei 3	201 Ku	Sep/12/44	Cebu	582 Ku
Shinsaku Tanaka	CPO	Otsu 12	201 Ku	Sep/12/44	Cebu	

Also CPO Masami Futaki (Otsu 10), CPO Tokuharu Noda (Otsu 11), CPO Tsuneji Mitani (Ko 9), and CPO Hiroshi Yasumatsu (Hei 4)

Kishio Kadora	CPO	Otsu10	201 Ku	Sep/12/44	Legaspi	
Shitau Sato	Res.ENS	Res.13	201 Ku	Sep/13/44	Bacolod	

Also PO1c Sumio Sasamoto (Ko 10)

Shigenobu Takahara	WO	Otsu 8	S311	Sep/21/44	Manila	*Akagi*
Nobuo Takano	LT	70	S311	Sep/21/44	Manila	

Also CPO Kikuo Nishimori (Otsu 10) and PO1c Tadashi Ikeda (Otsu 16)

Hisateru Tabuchi	LTJG	Pilot 14	S901	Sep/21/44	East of Philippines	

Buntai Ldr. (night fighters)

Kinya Yanogawa	PO1c	Ko 10	201 Ku	Sep/22/44	East of Philippines	
Yukio Maki	LT	65	Tsukuba Ku	Sep/28/44	Homeland	Accident, 14th Ku, Tainan Ku, 6th Ku

Hikotai Ldr.

Keizaburo Uchiyama Res.LTJG			S602	Oct/3/44	Balikpapan	

Also CPO Kitami Kikuchi (Otsu 14)

Eizo Ohta	CPO	Otsu 12	S602	Oct/10/44	Balikpapan	
Sadao Ozaki	LT	Res.7	S902	Oct/10/44	Balikpapan	

(Kansai Univ.) (night fighters)

Akira Tanaka	LT	Res.8	S309	Oct/10/44	Balikpapan	

(Kanto Gakuin) Hikotai Ldr.

Also CPO Tsumiro Tanaka (Otsu 13) and PO1c Akio Fukui (Ko 10)

Yoshio Murata	Res.LT	Res.4	Takao Ku	Oct/12/44	Taiwan	

(Nippon Univ.) Hikotai Ldr.

Also LT Toshikazu Taguchi (Engineer 50) and CPO Toshihiro Matsunaga (Otsu 13)

Ryo-ichi Iwakawa	CPO	Pilot 56	Takao Ku	Oct/12/44	Taiwan	3rd Ku, 253 Ku
Noboru Okugawa	CPO	Pilot 29	Tainan Ku	Oct/12/44	Taiwan	*Zuikaku*

Also LTJG Takeshi Kihara (72)

Junjiro Ito	WO	Ko 1	221 Ku	Oct/12/44	Taiwan	12th Ku, *Zuikaku*
Katsuhiko Kawasaki	CPO	Hei 2	221 Ku	Oct/12/44	Taiwan	253 Ku

Also LTJG Yozaburo Ohtsuki (71), CPO Kiyoshi Yamazaki (Otsu 11), and Res.LTJG Takeo Kawaguchi (Res. 11)

Fumio Shigeta	LT	69	S304	Oct/12/44	Taiwan	

Buntai Ldr.

Shigemi Wakabayashi	LTJG	71	S401	Oct/12/44	Taiwan	

Also LTJG Katsumi Yamaguchi (72)

Usaburo Suzuki	LT	68	S301	Oct/13/44	off Taiwan	202 Ku, 582 Ku, 204 Ku

Hikotai Ldr.

Tadashi Sakai	CPO	Hei 2	341 Ku	Oct/13/44	Taiwan	
Hiroshi Suzuki	CPO	Ko 5	201 Ku	Oct/13/44	off Philippines	
Kunio Kimura	LT	70	S302	Oct/14/44	Taiwan	
Masao Iizuka	LT	66	S302	Oct /15/44	off Taiwan	*Kaga*

Hikotai Ldr. Also LTJG Mitsuo Tanabe (72)

Osamu Takahashi	LT	70	S308	Oct/15/44	off Taiwan	

Also PO1c Jinpei Isozaki (Otsu 15)

Hiromichi Hojo	CPO	Pilot 43	S315	Oct/15/44	off Taiwan	3rd Ku, *Hiyo*
Hiroshi Nagakura	CPO	Ko 7	653 Ku	Oct/15/44	off Taiwan	
Satoshi Kano	WO	Otsu 6	254 Ku	Oct/16/44	off Taiwan	*Zuikaku*
*Hidehiro Nakama	LT	70	S317	Oct/21/44	Iwo Jima	Rammed into B-24

Buntai Ldr.

Yoshiyasu Kuno (Hosei Univ.)	Res.LTJG	Res.11	S301	Oct/21/44	off Leyte	Kamikaze attack
*Masahisa Uemura (Rikkyo Univ.)	Res.ENS	Res.13	201 Ku	Oct/21/44	off Leyte	Kamikaze attack
Minoru Tsukahara	CPO	Ko 9	221 Ku	Oct/21/44	Morotai	
*Kaoru Sato	CPO	Hei 4	201 Ku	Oct/23/44	off Leyte	Kamikaze attack, 253 Ku
Minoru Kobayashi Hikotai Ldr.	LCDR	64	S317	Oct/24/44	East of Philippines	*Ryuho*, 3rd Ku
Iwao Akiyama	LT	Eng 51	S316	Oct/24/44	East of Philippines	
Buntai Ldr. Also CPO Bun-ichi Fujise (Otsu 11) and CPO Takeru Wada (Ko 7)						
Sumio Fukuda	LT	69	S163	Oct/24/44	East of Philippines	*Junyo*, 204 Ku
Hikotai Ldr. Also WO Mamoru Ishi-i (Pilot 44)						
Yukiharu Ozeki	WO	Pilot 32	203 Ku	Oct/24/44	East of Philippines	13th Ku, 12th Ku, 3rd Ku, 6th Ku
Minoru Shibamura	WO	Pilot 48	S402	Oct/24/44	East of Philippines	
Also CPO Rokusaburo Shinohara (Otsu 11)						
Hiroshi Tanaka	LT	70	S407	Oct/24/44	East of Philippines	
Saneo Miyauchi	LT	70	S313	Oct/24/44	East of Philippines	
Also CPO Yoshito Azuhata (Otsu 12)						
Hisaya Hirusawa	LT	Eng 50	S308	Oct/24/44	East of Philippines	
Hikotai Ldr. Also LT Hisao Kawanishi (70)						
Kazunari Koyama	CPO	Ko 9	653 Ku	Oct/24/44	East of Philippines	
Also PO1c Mitsuo Senda (Otsu 16)						
Hohei Kobayashi Hikotai Ldr.	LT	67	S161	Oct/25/44	East of Philippines	*Shokaku*
Yoshiteru Mine Buntai Ldr.	LT	70	S161	Oct/25/44	East of Philippines	
Also PO1c Takeo Ohkawa (Ko 10), CPO Takeo Yamashiro (Otsu 10), and CPO Tsuguomi Kataoka (Otsu 10)						
Momoto Matsumura	WO	Pilot 29	S161	Oct/25/44	East of Philippines	12th Ku, 13th Ku, 204 Ku
Also CPO Yasuo Iguchi (Ko 7)						
Mitsuo Ohfuji Buntai Ldr.	LT	70	S165	Oct/25/44	East of Philippines	
Seikichi Kubota	WO	Otsu 7	S165	Oct/25/44	East of Philippines	*Shokaku*, 601 Ku,
Also CPO Oto Kataoka (Ko 8)						
Yoshiro Hashiguchi	CPO	Pilot 42	S164	Oct/25/44	East of Philippines	12th Ku, 3rd Ku, 601 Ku
Also CPO Masakazu Suzuki (Ko 6)						
Eiji Sanada	CPO	Pilot 56	634 Ku	Oct/25/44	East of Philippines	*Junyo*
Yasuhide Aoki	Res.LTJG	Res.11	S304	Oct/25/44	East of Philippines	
Also PO1c Hisao Fujii (Otsu 15)						
*Yukio Seki	LT	70	S301	Oct/25/44	off Leyte	Tokko
Buntai Ldr. Also PO1c Nobuo Tani and PO1c Iwao Nakano (both Ko 10)						
Misao Sugawa	LdgSea	Hei 15	201 Ku	Oct/25/44	off Leyte	Escort for Tokko
Masaru Sometani	LTJG	71	S401	Oct/25/44	Philippines	
Tai Nakajima Buntai Ldr.	LT	69	S161	Oct/26/44	Sibuyan Sea	
*Tomisaku Katsumata	PO1c	Ko 10	S301	Oct/26/44	off Leyte	Tokko
*Hiroyoshi Nishizawa	WO	Otsu 7	S303	Oct/26/44	Mindoro	On transport Chitose Ku, 251 Ku, 203 Ku, 253 Ku
Shingo Honda	CPO	Hei 4	S303	Oct/26/44	Mindoro	
Katsumasa Matsumoto	CPO	Otsu 9	762 Ku	Oct/26/44	Cebu	
Tomokazu Ema	WO	Pilot 22	254 Ku	Oct/29/44	Manila	14th Ku, 6th Ku,
Also LTJG Gi-ichi Minami (Otsu 2)						

Name	Rank		Unit	Date	Place	Carriers/Units
Kyoji Handa	CPO	Hei 3	341 Ku	Oct/29/44	Manila	
Tsune-ishi Nakamura	CPO	Ko 6	653 Ku	Oct/29/44	Bamban	
Tsunesaku Sakai	CPO	Hei 4	203 Ku	Nov/1/44	Cebu	
Shoji Kato	CPO	Ko 9	203 Ku	Nov/2/44	Leyte	
Chozo Nakaya	CPO	Otsu 15	653 Ku	Nov/2/44	Leyte	
Kenji Nakagawa	LT	67	S165	Nov/3/44	Leyte	*Zuiho, Zuikaku,*
Hikotai Ldr. Also LTJG Satoru Iguchi (72)						653 Ku
Tei-ichi Kato	Res.LTJG	Res.10	S303	Nov/4/44	Cebu	
(Tokyo Univ.)						
Torajiro Haruta	LT	69	S316	Nov/5/44	Mabalacat	253 Ku
Hikotai Ldr. Also CPO Soji Wakatsu (Otsu 16)						
Makoto Inoue	WO	Pilot 46	203 Ku	Nov/5/44	Bamban	12th Ku, 3rd Ku,
						253 Ku
Yoshinori Nakada	LdgSea Spe.	Otsu 2	S407	Nov/5/44	Bamban	
Ki-ichi Nagano	CPO	Pilot 56	203 Ku	Nov/6/44	Bamban	582 Ku
Shoku Kimura	LTJG	72	254 Ku	Nov/11/44	Leyte	
*Korekiyo Otsuji	LTJG	71	S163	Nov/12/44	Leyte	Tokko
Buntai Ldr. Also CPO Yaozo Wada (Otsu 13)						
Sozaburo Takahashi	ENS	Pilot 30	341 Ku	Nov/18/44	Tacloban	13th Ku, 12th Ku,
						Soryu, Hiyo
*Yoshita Toda	LTJG	72	221 Ku	Nov/19/44	Leyte	Tokko
*Mikihiko Sakamoto	LTJG	71	352 Ku	Nov/21/44	Omura	Rammed into
						B-29
Masa-aki Kawahara	ENS	Pilot 26	Omura Ku	Nov/21/44	Omura	Intercepted B-29s
Ayao Shirane	LCDR	64	S701	Nov/24/44	Ormoc Bay	12th Ku, *Akagi,*
Hikotai Ldr. Also CPO Sadao Koike (Otsu 10)						*Zuikaku*
Akira Yamamoto	ENS	Pilot 24	Yoko Ku	Nov/24/44	Yachimata	Intercepted B-29s,
						Hosho, Ryujo,
						12th Ku, *Kaga,*
						Zuiho
*Yoshimi Minami	ENS	Pilot 30	302 Ku	Nov/25/44	Philippines	Tokko, 12th Ku,
						Shokaku
Tatsu Nagato	Res.LTJG	Res.11	S305	Nov/26/44	Philippines	Tokko
(Hikone Com.Col.)						
Mune-ichi Oshoya	CPO	Otsu 9	Yoko Ku	Nov/26/44	Atsugi	Accident, 204 Ku
Susumu Takeda	LTJG	71	S303	Nov/27/44	Philippines	
Buntai Ldr. Also WO Kyoichi Inuzuka (Otsu 8)						
*Yasunori Ono	WO	Ko 2	252 Ku	Nov/27/44	Saipan	Tokko
Also LTJG Kenji Ohmura (72)						
Keiji Kataki	LT	70	S312	Nov/29/44	Philippines	
Buntai Ldr. Also PO1c Tsunehiro Yoshii (Otsu 16)						
Masuzo Seto	LCDR	64	S315	Dec/4/44	off Taiwan	Tainan Ku,
Hikotai Ldr.						*Shokaku,* ditched
*Tetsuro Yano	Res.LTJG	Res.11	S316	Dec/7/44	Philippines	Tokko
Ushi-o Nishimura	CPO	Ko 9	S302	Dec/8/44	Iwo Jima	
Ki-ichi Oda	WO	Pilot 18	S306	Dec/10/44	off Ogasawara:	*Kaga,*
					killed on board	13th Ku,
					submarine	15th Ku, *Soryu*
Yoshiharu Kagami	CPO	Otsu 12	S701	Dec/14/44	Philippines	
Also LT Sumio Arikawa (71)						
Seiya Nakajima	LT	71	S402	Dec/15/44	Philippines	
Buntai Ldr.						
Takumi Chosokabe	CPO	Otsu 13	S602	Dec/20/44	Balikpapan	
Shiro Kawai	LCDR	64	S308	Dec/24/44	Clark	12th Ku,
Hikotai Ldr.						Tainan Ku,
						201 Ku missing
						after bailed out

Name	Rank	Class	Unit	Date	Location	Notes
Tadashi Yoneda	CPO	Pilot 56	S315	Dec/25/44	Clark	251 Ku
Also CPO Masakazu Kuwabara (Hei 3)						
Yoshio Kinoshita	LT	71	S308	Jan/3/45	Taiwan	
Buntai Ldr.						
Also ENS Shigemasa Nishio (Otsu 6)						
Saburo Mitsuoka	CPO	Otsu 11	341 Ku	Jan/4/45	Malcott	Died by strafing
*Shin-ichi Kanaya	LT	71	201 Ku	Jan/5/45	Philippines	Tokko
Akira Sugiura	LT	Res. 8	901 Ku	Jan/5/45	Hainan Islands	
Buntai Ldr.					(Aoyama)	
Also LT Tsunekata Maki (71) and LTJG Takeshi Inoue (Res.11)						
Masami Iwatsubo	CPO	Hei 3	352 Ku	Jan/6/45	Northern Kyushu	
Also LTJG Ko-ichi Sawada (72)						
*Shigenobu Manabe	CPO	Rei 7	252 Ku	Jan/7/45	Philippines	Tokko
Yoshio Yamazaki	WO	Ko 4	S902	Jan/8/45	Balikpapan	
(night fighters)						
Tokushige Yoshizawa	CPO	Hei 7	105 Air	Jan/9/45	New Britain	201 Ku, 204 Ku
			Base Unit			
*Sachio Endo	LT	Otsu 1	S302	Jan/14/45	Atsumi Peninsula	Intercepted
(night fighters)						B-29s, 251 Ku
Yoshio Orihara	CPO	Ko 8	S314	Jan/15/45	Luzon	*Hiyo*, 601 Ku
Tetsuo Endo	LT	67	S314	Jan/16/45	Luzon	
Hikotai Ldr.						
Hajime Toji	CPO	Otsu 9	901 Ku	Jan/16/45	Sanya	*Junyo*
Also CPO Hisashi Kamata (Otsu 10)						
*Minoru Kawazoe	LT	69	S317	Jan/21/45	off Taiwan	Tokko
Hikotai Ldr. Also LT Sei-ichi Saito (71)						
*Hidenobu Sumino	Res.LTJG	Res.13	201 Ku	Jan/25/45	Lingayen Gulf	Tokko
Toshio Imada	CPO	Otsu 10	S602	Feb/6/45	Balikpapan	Intercepted B-24s
Toshio Araki	LT	67	302 Ku	Feb/16/45	Kanto Area	
Hikotai Ldr.						
Jiro Ishikiwa	CPO	Ko 6	252 Ku	Feb/16/45	Kanto Area	*Hiyo*
Also CPO Shigemi Izumi (Ko 6)						
Kozo Kobayashi	LT	71	Tsukuba Ku	Feb/16/45	Kanto Area	202 Ku
Rokuya Yoneyama	CPO	Hei 6	Tsukuba Ku	Feb/16/45	Kanto Area	
Also LT Itaru Yamashita (71) and CPO Shuji Furu-uchi (Ko 6)						
Noboru Matsu-ura	Res.LTJG	Res.11	Yatabe Ku	Feb/16/45	Kanto Area	
Hachiro Sakai	CPO	Otsu 16	601 Ku	Feb/16/45	Kanto Area	
Masayuki Akai	LTJG	72	302 Ku	Feb/17/45	Kanto Area	
Also PO1c Tadashi Saka (Hei 10)						
Shinkai Fujimori	LT	69	Tsukuba Ku	Feb/17/45	Kanto Area	
Takashi Yamazaki	CPO	Hei 3	Yoko Ku	Feb/17/45	Kanto Area	*Zuikaku*
Sanemasa Nanjo	CPO	Ko 7	S304	Feb/17/45	Kanto Area	
Yoshiei Shinohara	WO	Otsu 8	Yatabe Ku	Feb/25/45	Kanto Area	12th Ku,
						Tainan Ku
Toshihide Kihara	LT	71	Tsukuba Ku	Feb/25/45	Kanto Area	
Katsumi Sugi-e	LTJG	72	601 Ku	Feb/25/45	Kanto Area	
Morito Yamaguchi	LT	71	901 Ku	Feb/25/45	Hainan Island	
Buntai Ldr.						
Masajiro Kawato	PO1c	Hei 12	105 Air	Mar/9/45	New Britain	POW, returned
			Base Unit			
Mitsuru Ohnuma	LT	Res.9	203 Ku	Mar/18/45	Southern Kyushu	
(Waseda Univ.) Also PO1c Mitsunori Kojima (Otsu 16)						
Mankichi Sato	CPO	Otsu 11	221 Ku	Mar/18/45	Southern Kyushu	
Kunio Matsuzaki	LT	71	343 Ku	Mar/19/45	Matsuyama	
Buntai Ldr.						

Name	Rank	Class	Unit	Date	Place	Other
Kozo Shima Buntai Ldr.	LT	71	43 Ku	Mar/19/45	Matsuyama	
Shiro Endo	CPO	Ko 7	343 Ku	Mar/19/45	Matsuyama	
Also CPO Yasuharu Nikko (Ko 10), CPO Kikuichi Ishikawa (Otsu 14), PO1c Mitsuo Nakajima (Otsu 16), Res.LTJG Isaburo Inoue (Res.11), PO1c Haruhiko Takeshima (Ko 11), and LdgSea Yo-ichi Saiki (Spe.Otsu 1)						
*Mutsuo Urushiyama Buntai Ldr.	LT	70	S307	Mar/21/45	off Tosa	Escort of Jinrai force
*Yu-ichi Izawa Buntai Ldr.	LT	71	S306	Mar/21/45	off Tosa	Escort of Jinrai force
*Kojiro Murakami	CPO	Otsu 11	S306	Mar/21/45	off Tosa	Escort of Jinrai force
*Goro Tsuda	CPO	Hei 2	S306	Mar/21/45	off Tosa	Escort of Jinrai force
Also CPO Kazuyoshi Kobayashi (Ko 7) and CPO Shun-ichi Nakano (Hei 3)						
Sunao Sugisaki Hikotai Ldr.	LT	69	352 Ku	Mar/31/45	Kyushu	
*Iwao Fumoto	PO1c	Otsu 17	721 Ku	Apr/1/45	Okinawa	Ohka Tokko
Yozo Tsuboi (Doshisha Univ.) Hikotai Ldr.	LT	Res.9	302 Ku	Apr/l/45	Kanto Area	
Hiroyuki Fujishima	CPO	Ko 8	601 Ku	Apr/3/45	Okinawa	
Also CPO Ryu Nakatani (Ko 10)						
Yu-ichi Kobayashi	CPO	Hei 4	205 Ku	Apr/5/45	Okinawa	
Kunio Kanzaki Hikotai Ldr.	LT	68	S312	Apr/6/45	Okinawa	381 Ku
Hiroshi Matano	LT	71	S312	Apr/6/45	Okinawa	
Also ENS Akira Ozeki (Pilot 25), Res. LTJG Chisato Akiyama, and Res. LTJG Shogo Kobayashi (both Res. 11)						
Noboru Yamakawa	LT	Otsu 1	Omura Ku	Apr/6/45	Okinawa	13th Ku, *Kaga*, 15th Ku, Tainan Ku
Shinzo Tabata	CPO	Otsu 14	Omura Ku	Apr/6/45	Okinawa	
Also CPO Kazuo Tanio (Otsu 16)						
Nobutaka Kurata	CPO	Pilot 54	252 Ku	Apr/6/45	Okinawa	*Zuikaku*, 203 Ku, 352 Ku
Also PO2c Masakazu Koyanagi						
*Yohei Watanabe	PO1c	Hei 11	S313	Apr/7/45	Chiba	Rammed into B-29
Koji Chikama Buntai Ldr. Also LT Manabu Ishimori (71)	LT	Otsu 11	S306	Apr/12/45	Okinawa	
Ei-ichi Kawabata Hikotai Ldr.	LT	69	S804	Apr/12/45	Okinawa	
Sho-ichi Takahashi	LT	70	Tsukuba Ku	Apr/12/45	Okinawa	
Ichigo Katayama	LTJG	72	302 Ku	Apr/12/45	Okinawa	
Tatsutoshi Hashimoto	LTJG	72	343 Ku	Apr/12/45	Okinawa	
Also CPO Yoshio Aoyama (Ko 10)						
*Sukenobu Suzuki	LTJG	73	Yatabe Ku	Apr/14/45	Okinawa	
*Tadashi Ohmoto	LTJG	Res.13	Yatabe Ku	Apr/14/45	Okinawa	Tokko
Also ENS Hachiro Sasaki (Res.14) and PO2c Hideo Sumihiro (Otsu 18)						
*Sho-ichi Sugita	PO1c	Hei 3	S301	Apr/15/45	Kanoya	6th Ku, 343 Ku, 201 Ku
Megumi Yoshioka	CPO	Otsu 16	S302	Apr/15/45	Kanoya	
Also CPO Masao Oikawa (Otsu 15)						
Masa-aki Asakawa Hikotai Ldr.	LT	69	332 Ku	Apr/16/45	Naruo	Accident, 341 Ku
Makoto Oku-umi	LTJG	72	601 Ku	Apr/16/45	Kikaigashima	
Teigo Ishida	CPO	Hei 6	S407	Apr/16/45	Okinawa	202 Ku, 204 Ku
Also LdgSea Hitoshi Kotake (Spe. Otsu 2)						

Name	Rank	Class	Unit	Date	Location	Notes
Yasuo Isobe	CPO	Pilot 53	Genzan Ku	Apr/16/45	Okinawa	
Hiroshi Tajiri	LTJG	72	352 Ku	Apr/16/45	Okinawa	
Also LTJG Kazuyoshi Mori (72)						
Kyoji Takahashi	LT	71	252 Ku	Apr/16/45	Okinawa	
Buntai Ldr. Also CPO Yuzo Komatsu (Hei 6)						
Katsuyoshi Tanaka	CPO	Hei 3	343 Ku	Apr/16/45	Okinawa	
Also CPO Katsue Kato (Ko 9)						
Takao Ohtani	CPO	Ko 8	302 Ku	Apr/16/45	Okinawa	
Hachiro Yanagisawa	LCDR	64	S304	Apr/17/45	Okinawa	
Hikotai Ldr.						
Mutsuo Uemura	CPO	Hei 2	S311	Apr/17/45	Okinawa	
*Nobuo Saito	WO	Pilot 42	205 Ku	Apr/17/45	Okinawa	Tokko
Akira Fukuda	LTJG	72	302 Ku	Apr/19/45	Kanto Area	
Also CPO Shiro Toriyama (Hei 3)						
Yoshishige Hayashi	LT	69	S407	Apr/21/45	Southern Kyushu	Intercepted B-29s, 251 Ku
Hikotai Ldr. Also PO1c Toshinobu Shimizu (Otsu 14)						
So-ichiro Yamada	CPO	Otsu 15	Yoko Ku	Apr/21/45	Southern Kyushu	Intercepted B-29s
Takehiko Kobayashi	LT	Eng 52	Omura Ku	Apr/22/45	Southern Kyushu	
Buntai Ldr.						
Shigenobu Nakada	CPO	Pilot 40	201 Ku	Apr/24/45	Luzon	204 Ku, 201 Ku: died on the ground
Also CPO Seiji Kato (Ko 8)						
Saburo Yoneda	ENS	Ko 2	S901	Apr/28/45	Okinawa	
(night fighters)						
Shiro Hamada	CPO	Ko 10	343 Ku	Apr/28/45	Okinawa	
Seiji Jonoshita	LT	70	S31	Apr/29/45	Homeland	Died by wounds: 253 Ku
Hikotai Ldr.						
Seikichi Izawa	CPO	Hei 7	302 Ku	May/3/45	Atsugi	
Katsujiro Nakano	ENS	Pilot 37	S311	May/4/45	Okinawa	3rd Ku, 202 Ku, 221 Ku
Tatsumi Soga	CPO	Otsu 11	S203	May/4/45	Okinawa	
*Yoshio Oh-ishi	WO	Otsu 9	205 Ku	May/4/45	Okinawa	Tokko; *Zuikaku*
*Tei-ichiro Hayashida	WO	Ko 4	Omura Ku	May/4/45	Okinawa	Tokko, *Shoho*
Hideaki Maeda	CPO	Otsu 11	205 Ku	May/9/45	off Taiwan	
Masayoshi Urano	WO	Ko 5	S312	May/11/45	Okinawa	201 Ku
Also CPO Isamu Ishii (Hei 10)						
Hamashige Yamaguchi	WO	Otsu 9	S312	May/11/45	Okinawa	Tainan Ku, 201 Ku
Shigeo Kimura	ENS	Pilot 41	S313	May/11/45	Okinawa	14th Ku, 22 AF, 331 Ku
Also CPO Akira Saito (Hei 2)						
Yoji Amari	ENS	Ko 2(Recon)	S812	May/13/45	Okinawa	*Tone*
Hiroshi Nemoto	Res.ENS	Res.13	Yatabe Ku	May/13/45	Okinawa	Tokko
Also PO2c Gi-ichi Hoshino (Otsu 18)						
Nobuyoshi Osada	WO	Pilot 35	S303	May/14/45	Okinawa	
Also PO1c Tobei Rikitake (Otsu 17)						
Naosuke Yoshida	LCDR	Res. 4	S312	May/28/45	Okinawa	
(Yokohama Tec.) Hikotai Ldr. Also ENS Atsuki Mezaki (Ko 1) and CPO Sakuji Hayashi (Hei 2)						
Nobuyuki Tanabe	CPO	Ko 11	601 Ku	May/29/45	Yokohama	
Also CPO Jun-ichi Hoshino (Hei 10)						
Takao Kate	CPO	Otsu 11	252 Ku	May/29/45	Yokohama	
Seijo Boji	Res. LTJG	Res. 11	302 Ku	May/29/45	Yokohama	
(Ryukoku Univ.)						
Jiro Funakoshi	CPO	Otsu 14	343 Ku	Jun/2/45	Kyushu	
Also CPO Eiji Mikami (Ko 11)						

Susumu Kawasaki	LTJG	Otsu 3	252 Ku	Jun/3/45	Okinawa	352 Ku, 934 Ku
Sunao Nishikane	WO	Otsu 10	S303	Jun/7/45	Southern Kyushu	253 Ku
Naraichl Murai	CPO	Hei 6	601 Ku	Jun/10/45	Kanto Area	252 Ku
Takio Yoshida	WO	Pilot 39	S402	Jun/10/45	Kanto Area	253 Ku
Hideo Ando	WO	Otsu 10	S90	Jun/13/45	Okinawa	
Keijiro Hayashi	LT	70	S407	Jun/22/45	Kyushu	381 Ku;
Hikotai Ldr. Also CPO Takashi Yanagisawa (Ko 9)						
Junji Yamazaki	CPO	Ko 9	S311	Jun/22/45	Okinawa	
Katsuji Kobayashi	CPO	Ko 8	302 Ku	Jun/23/45	Yokohama	204 Ku
Also LT Norio Ueno (72)						
Shuzo Enomoto	LTJG	73	S308	Jun/23/45	Kanto Area	
Also CPO Katsumi Ni-imoto (Ko 8)						
Hitoshi Hikosaka	CPO	Ko 11	332 Ku	Jun/26/45	Kinki	
Shojiro Ishi-i	PO1c	Hei 10	343 Ku	Jul/2/45	Kyushu	
Also CPO Takashi Sakuma (Ko 10)						
Kazu-aki Kinoshita	Res.LT	Res.10	S701	Jul/5/45	Northern Kyushu	
(Tokyo Univ.) Buntai Ldr. Also CPO Kiyoshi Toyohara (Hei 12)						
Mitsuo Asakura	CPO	Ko 10	Tsukuba Ku	Jul/8/45	Kanto Area	
Takashi Shimada	LTJG	73	332 Ku	Jul/10/45	Akashi	
Akira Sugawara	LT	69	Yoko Ku	Jul/11/45	Homeland	
Buntai Ldr. (night fighters)						
Takashi Oshibuchi	LT	68	S701	Jul/24/45	Bungo Strait	251 Ku, 253 Ku, 203 Ku
Hikotai Ldr. Also CPO Jiro Hatsushima (Ko 9)						
Kaneyoshi Muto	ENS	Pilot 32	S301	Jul/24/45	Bungo Strait	13th Ku, 12th Ku, 3rd Ku, Yoko Ku
Eiji Okuda	CPO	Otsu17	S303	Jul/25/45	Southern Kyushu	
*Naoshi Kanno	LT	70	S301	Aug/1/45	Over Yakushima	343 Ku, 201 Ku
Hikotai Ldr. Also CPO Yasuo Yoshioka (Ko 10)						
Seiji Noma	CPO	Otsu 12	S407	Aug/2/45	Southern Kyushu	
Yoshio Saito	LT	71	S304	Aug/5/45	Boso	332 Ku
Buntai Ldr.						
Sei-ichiro Sako	CPO	Ko 7	203 Ku	Aug/7/45	Southern Kyushu	
Also CPO Bu-ichi Kamo (Otsu 13)						
Goro Kitano	LTJG	73	S303	Aug/7/45	Southern Kyushu	
*Take-aki Shimizu	ENS	Ko 1	S901	Aug/8/45	Okinawa	
(night fighters)						
Mitsuo Ishizuka	ENS	Pilot 22	S407	Aug/8/45	Southern Kyushu	
Masayuki Tadami	CPO	Ko 9	S407	Aug/8/45	Southern Kyushu	
Also CPO Sakae Masumoto (Ko 10)						
Keishichiro Hattori	LT	Ki-52	S701	Aug/8/45	Southern Kyushu	
Also CPO Ka-emon Yokobori (Ko 10)						
Hiroshi Ohara	WO	Pilot 50	S401	Aug/9/45	Died from injuries; *Akagi, Shokaku*	
*Shiro Okajima	LTJG	Res.13	721 Ku	Aug/11/45	Okinawa	Tokko;
Also PO1c Minoru Hoshino (Ko 12)						
Sadao Ohshio	LT	72	343 Ku	Aug/12/45	Kyushu	
Kazumasa Sagara	CPO	Ko 10	302 Ku	Aug/13/45	off Honshu	
Eiji Matsuyama	ENS	Ko 3	601 Ku	Aug/14/45	over Lake Biwa	
Also LTJG Kan-ichi Hyodo (Res. 13)						
Mitsuo Taguchi	LT	Pilot 18	302 Ku	Aug/15/45	Kanto Area	
Buntai Ldr.						
Also LTJG Yoshikane Kuramoto (73) and CPO Ikki Takeda (Hei 6)						
Kaoru Tamura	CPO	Otsu 16	252 Ku	Aug/15/45	Chiba Prefecture	
Also CPO Kohei Sugiyama (Otsu 15)						

APPENDIX C

Naval Fighter Pilots

A set of tables listing the pilots trained under the various bases follows. These seek to identify all naval pilots who have specialized on fighters from the time that naval aviation was recognised in Japan through to the end of the Pacific War in August 1945. Since official records are not always complete, assistance has been sought from various sources. However, sections of the list are not fully complete, and for this reason and space considerations, the following classes have not been included: 40th-42nd Aviation Student Courses; Hei 7th Class; Ko 8th Class; Otsu 14th Class, and classes after the 12th in the Flight Reserve Student Course.

Letters in brackets following names indicate the following:- (H) Received Honours; (K) Killed in Action; (A) Death due to an Accident. It will also be found that a number of names are also surrounded by brackets. These indicate pilots who changed from specializing in one type of aircraft to another. In 1936 when the land attack forces were separated from those intended for carrier service, a large number of personnel changed to a different branch at that time. Pilots who converted from other specializations to that of fighter pilot are grouped together at the end of each class listing. From the mid point of the Pacific War onwards, many personnel converted from float fighters and from two-seat reconnaissance floatplanes. Other then these, pilots of night fighters (such as the Gekko) and float fighters who did not convert, are not included.

Finally, designation of aircraft types are in accordance with practices of the period, e.g. carrier-based aircraft, seaplanes, carrier fighters, etc.

1. Aviation Students
The systematic training of pilots began in July 1912 with the 1st Class, Aviation Technical Research Committee Members (Koku Gi jutsu Kenkyo-iin), and ended with its 6th Class. With the formation of Yokosuka Kokutai in April 1916, the name was changed to Aviation Technical Student (Koku Gi jutsu Gakusei). Subsequently, when Kasumigaura Kokutai was established in November 1922, student training was shifted to this base. Beginning with the 12th Class, students were referred to simply as Aviation Students (Hike Gakusei), and this description continued to be employed until the 42nd Class graduated in February 1945.

Prior to the 18th Class, Aviation Student Course, specialization by type of aircraft was not clearly designated due to the early developmental stage of training at that time. Consequently, the names of all personnel attending these early courses have been listed here, ås it is not certain which of them went on to fly fighters. However, from the 19th Class onwards, it has proved possible to identify and list only those who specialized on fighter aircraft. In those cases where a number in brackets follows a name, this indicates the Naval Academy (Naval Engineering College) which they attended prior to being placed upon the Aviation Students Class with which they subsequently learned to fly. As graduates of the Academy, all such pilots commenced their service as commissioned officers.

First name Family name
Before the lst Class
(5 persons)
Shiro Aihara (29)
Kanehiko Umekita (29)
Yozo Kaneko (30)
Sankichi Kawano (31)
Chuji Yamada (33)

1st Class, Aviation Technical
Research Committee Members
(Oct.1912-May 1913; 4 persons)
Tozaburo Adachi (36) (A)
Fumio Inoue (33) (A)
Shokei Hirose (36)
Masatsune Fujise (36)

2nd Class, Aviation Technical
Research Committee Members
(Jun.1913-; 3 [sic] persons)
Teruo Nanba (37)
Hideho Wada (34)

3rd Class, Aviation Technical
Research Committee Members
(Feb.-Aug.1914; 5 persons)
Teizo Iigura (37)
Yukinobu Ohsaki (36)
Takao Takebe (37) (A)
Kishichi Magoshi (37)
Junpei Yamamoto (38)

4th Class, Aviation Technical
Research Committee Members
(Feb.-Jun. 1915; 6 persons)
Shinji Abe (37) (A)
Masaru Kaitani (38)
Torao Kuwabara (37)
Kosaku Tsuda (39)
Senji Tsuyuki (36)
Nakajiro Mikami (38)

5th Class, Aviation Technical
Research Committee Members
(May1915-Mar.1916; 5 persons)
Misato Asada (39)
Naota Goto (39)
Akitomo Beppu (38)
Toshikazu Yashima (39)
Misao Wada (39)

6th Class, Aviation Technical
Research Committee Members
(Dec.1915-Mar.1916;6 persons)
Tamotsu Araki (40)

Takijiro Ohnishi (40)
Munetaka Sakamoto (40)
Torasaburo Shoji (40)
Tomeo Muroi (39)
Saburo Yamaguchi (39)

1st Class, Aviation Technical
Students
(Jun.1916-Jun.1917, 5 persons)
Osamu Imamura (40)
Masuo Kani (40)
Shin-ichi Sakamoto (40) (A)
Morihiko Miki (40)
Toyoo Yamamura (40) (A)

2nd Class, Aviation
Technical Students
(Dec.1916-Dec.1917; 6 persons)
Seiki Kato (40)
Shun-ichi Kira (40)
Tomo Shirase (38)
Sadatoshi Senda (K)
Masami Niwa (40)
Saburo Yamauchi (40) (A)

3rd Class, Aviation
Technical Students
(Dec.1917-Dec.1918; 12 persons)
Rinosuke Ichimaru (41)
Keizo Ueno (41)
Hisao Kato (41)
Seigo Kadowaki (41)
Sotokichi Kokura (41) (K)
Takeo Komaki (40) (A)
Masaki Sakakibara (41)(A)
Munetaka Sakamaki (41)(H)
Koji Shirai (41)
Tsutomu Tomeoka (41) (A)
Jiro Miyake (41) (A)
Ichitaro Yonezawa (41)

4th Class, Aviation Technical
Students
(Dec.1918-Dec.1919; 12 persons)
Hisakichi Akaishi (42)(A)
Kanjo Akashiba (42)
Makoto Awaya (41)
Tomeo Kaku (42) (K)
Akio Kawazoe (42)
Takasane Furuse (42)
Kiyosuke Shimura (41) (A)
Jiro Shimoyama (41)
Shigeho Tameda (42) (A)
Torao Nara (42) (A)
Sadayoshi Yamada (42) (H)

Michiyuki Yamada (42) (K)

5th Class, Aviation Technical
Students
(Dec.1919-Dec.1920; 7 persons)
Yoshiaki Ito (43)
Yoshio Ueda (44) (A)
Komataro Kawaguchi (44) (A)
Ushie Sugimoto (44) (K)
Michio Sumikawa (45) (H)
Tomiyoshi Maehara (44)
Keikichi Mori(Araki) (45)

6th Class, Aviation Technical
Students
(Dec.1920-Jul.1921; 11 persons)
Sanji Iwabuchi (43) (K)
Kaoru Umetani (46)
Tadao Kato (45)
Yoshio Kamei (46) (K)
Masaharu Suganuma (44) (A)
Prince Takehiko (46)
Aizo Nakamura (44) (H)
Atsuma Baba (46) (A)
Tatsuji Fujimatsu (46)
Hidemi Machida (44)

7th Class, Aviation Technical
Students
(Mar.-Nov.1922; 8 persons)
Fujiro Oh-hashi(46)
Asazo Kikuchi (45)
Shigetoshi Miyazaki (46)(K)
Shinosuke Muneyuki (45)(A)
Chikao Yamamoto (46) (H)
Toshiyuki Yokoi (46)
Saburo Wada (46)

8th Class, Aviation Technical
Students
(Dec.1922-Nov.1923; 9 persons)
Masahisa Saito (47)
Kosuke Sasaki (47) (A)
Chihaya Takahashi (47)
Nobukichi Takahashi (47) (K)
Taro Taguchi (47) (H)
Takanari Maeda (47)
Kenji Matsumura (47) (A)
Kenzi Yamazaki (47) (A)
Sakae Yamamoto (46)

9th Class, Aviation
Technical Students
(Jun.1923-May 1924; 10 persons)
Hideo Koda (48) (A)

Ei-ichiro Jo (47) (K)
Hideo Tsukada (48)
Daizo Nakajima (48)
Den Nakajima (48)
Ayao Nishijima (48) (A)
Katsumi Hayami (47) (A)
Masashi Maruyama (48) (A)
Kanzo Miura (47) (H)
Yoshitoshi Miwa (48) (K)

**10th Class, Aviation
Technical Students**
(Dec.1923-Nov.1924; 10 persons)
Toshihiko Odawara (48) (H, K)
Shigetoyo Shirahama (49) (A)
Shinzo Susumu (48) (A)
Yoshiharu Soga (49)
Shigeru Tanno (49) (A)
Tsukasa Noguchi (48) (A)
Saneyasu Hidaka (49) (A)
Yoshitaro Horikoshi (49) (A)
Iwao Minematsu (48)
Shigeshi Wakamatsu (49)

**11th Class, Aviation
Technical Students**
*(May 1924-Mar.1925;
10 persons)*
Ryo-ichi Asaka (49)
Teruyuki Kakita (48)
Sakae Kamura (Yamashita) (49)
Tokujiro Kikuoka (49)
Yoshito Kobayashi (49)
Eikichi Nakajima (49) (A)
Kiyoma Maeda (49) (A)
Toshio Mizunaga (49) (A)
Ryutaro Yamanaka (49)
Masahiro Watanabe (49) (A)

**12th Class, Aviation Students
[all subsequent classes]**
(Feb.1925-Nov.1925; 8 persons)
Kenji Kimura (49)
Toshiki Kurimoto (48) (K)
Shizuo Tateishi (50) (A)
Takashi Tsue (50) (A)
Misao Terada (50)
Chisato Morita (49)
YoshitaneYanagimura (49) (H,K)
Shigeo Watanabe (50)

13th Class,
(Sep.1925-May 1926; 9 persons)
Sadagoro Uchida (50)
Motoharu Okamura (50)
Naota Sata (50)
Haruo Sato (50)
Sho-ichi Sawabe (50) (H, A)
Kunizo Terai (50)
Jiro Tokui (50) (K)
Taro Fukuda (50)

14th Class
(Mar.-Nov.1926; 9 persons)
Tatsuo Aizawa (51) (K)
Sakae Ichikawa (50) (A)
Yasuna Kozono (51)
Akira Kuroi (51) (A)
Masami Kojima (51)
Chujiro Nakano (51) (H)
Nenosuke Nakamura (51) (K)
Takeo Yasunobu (51) (A)
Tetsujiro Wada (51) (A)

15th Class
*(Sep.1926-May 1927;
10 persons)*
Tsuguo Ikegami (51)
Masatsugu Iwao (51) (K)
Yoneji Uto (51)
Kiyoshi Katsuhata (51)
Chu-ichi Kawashima (51) (A)
Shigeo Kurioka (51) (A)
Hidetoshi Tamura (51) (A)
Masae Handa (51) (A)
Kurio Okehata (51) (H,A)
Goro Hirayama (51) (A)

16th Class
(Mar.-Nov.1927; 14 persons)
Takeshi Aoki (51)
Sukemitsu Ito (51)
Ichiro Imamura (51)
Saburo Katsuta (51) (K)
Yoshinari Kojima (51) (A)
Toshikazu Sugiyama (51) (H)
Yoshijiro Suzuki (51)
Toshimasa Taira (51) (A)
Yoshiteru Take (51)
Hisatsugu Tate (51)
Korokuro Tatemi (51)
Mohachiro Tokoro (51) (K)
Shigeki Negoro (51)
Yoshio Fukumori (51) (A)

17th Class
(Sep.1927-May 1928; 11 persons)
Takahisa Amagai (51)
Yasuo Iwai (51) (K)
Sho-ichi Ogasawara (51)
Yusuke Kakinuma (51) (A)
Shunji Kamide (51)
Tokutaro Kubo (51) (K)
Sadao Koike (51) (H)
Keizo Suda (51)
Shin-ichi Nitta (51) (K)
Tatsukichi Mishiro (51)
Minoru Mori (51) (K)

18th Class
(Mar.-Dec.I928; 16 persons)
Shin-ichi Atsumi (52) (A)
Taisuke Ito (52) (H)

Tsutomu Inoue (52) (A)
Koreroku Oh-uchi (52) (A)
Ikuto Kusumoto (52) (K)
Hiroshi Kogure (52)
Masayoshi Saito (52) (A)
Takeo Shibata (52)
Hiroshi Shimizu (52) (K)
Masakazu Suzuki (52)
Nakazaemon Tokunaga (52)
Yorimi Tsuchihashi (52) (A)
Shinroku Nishizawa (52)
Shusaku Fukuoka (52) (A)
Masami Matsumoto (52)
Hideo Muramatsu (52) (A)

19th Class
(Dec.1928-Nov. 1929; 18 persons)
Nokiji Ikuta (52)
Nobuo Ito (52)
Minoru Genda (52) (H)
Asa-ichi Tamai (52)
Shiro Watanabe (52) (A)

20th Class
(Dec.1929-Nov.1930; 4 persons)
Takeo Ide (54) (A)
Shigeharu Nagano (54) (A)
Tadao Funaki (54) (K)
Katsutoshi Yagi (54)

21st Class
(Dec.1930-Nov.1931; 26 persons)
Norito Ohbayashi (55) (K)
Kiyoji Sakakibara (55)
Mikuma Minowa (55)
Saburo Momozaki (55) (A)
Shigema Yoshitomi (55)

22nd Class
(Dec.1931-Nov.1932; 23 persons)
Chikamasa Igarashi (56)
Juroku Shimizu (56) (A)
Ryosuke Nomura (56) (H)
Mochifumi Nango (55) (K)

23rd Class
(Dec.1932-Jul.1933; 28 persons)
Shigeru Itaya (57) (K)
Ryohei Ushioda (57) (K)
Kiyoto Hanamoto (57) (A)
Shichiro Yamashita (57) (K)

24th Class
(Apr.-Nov.1933; 32 persons)
Naoichi Ishikawa (58) (A)
Mambei Shimokawa (58) (A)
Shigeo Taguma (58) (K)
Hideo Teshima (58) (K)
Tadashi Nakajima (58)
Sei-ichi Maki (57)

25th Class
(Nov.1933-Jul.1934; 33 persons)
Takahide Aioi (59)
Yoshimitsu Oku (58) (H, A)
Takashi Kaneko (59) (K)
Mitsugu Kofukuda (59)
Yoshitami Komatsu (59)
Hideki Shingo (59)

26th Class
(Nov.1934-Jul.1935; 34 persons)
Kiyokazu Ikeuchi (60) (A)
Mitsuo Ishikawa (60) (A)
Toshitaka Ito (60)
Harutoshi Okamoto (60)
Tadashi Kaneko (60) (K)
Saburo Shindo (60)
Minoru Suzuki (60)
Masao Yamashita (60)
Tamotsu Yokoyama (59)

27th Class
(Nov.1935-Nov.1936; 35 persons)
Kai Ikeda (61) (K)
Masaji Suganami (61) (K)
Kenzo Tanaka (61) (A)
Sumio Nono (61) (K)

28th Class
(Dec.1936- Sep.1937; 37 persons)
Fusata Iida (62) (H, K)
Bunji Goto (62) (A)
Yoshio Shiga (Yotsumoto) (62)
Naohisa Shinjo (62) (K)
Motonari Suho (62)
Sho-ichi Takahashi (62) (A)

29th Class
(Oct.1937-May 1938; 21 persons)
Kanetake Okazaki (62) (K)
Kiyokuma Okajima (63)
Takeo Kurosawa (63)
Kenjiro Notomi (62) (H, K)
Takumi Hoashi (63) (A)

30th Class
(Dec.1937-Jul.1938; 27 persons)
Masao Asai (63) (K)
Muneyoshi Aratake (63) (A)
Yoshio Kuragane (63)
Masao Sato (63) (K)
Ichiro Mukai (63) (K)

31st Class
(Aug.1938-Mar.1939; 59 persons)
Isamu Ikeda (64) (H, A)
Kiku-ichi Inano (Takabayashi) (64)
Shiro Kawai (63) (K)
Minoru Kobayashi (64) (K)
Aya-o Shirane (64) (K)
Masuzo Seto (64) (K)
Tadatsune Tokaji (64) (K)

Yasushi Nikaido (64) (A)
Masayoshi Murakami (64) (A)
Shigeru Mori (64) (K)
[Converted]
Hachiro Yanagisawa (64) (K)
Kushichiro Yamada (64)

32nd Class
(Sep.1939-Apr.1940; 64 persons)
Masanobu Ibusuki (65)
Yuzuru Nakano (65) (A)
Takaichi Hasuo (65) (K)
Yukio Maki (65) (A)
Masatoshi Makino (65) (K)
Akira Wakao (65) (K)

33rd Class
(Nov.1939-Jun.1940; 29 persons)
Tomoyuki Sakai (66) (H, K)
Tadashi Hara (66) (A)
Moriyasu Hidaka (66)
Iyozo Fujita (66)
Joji Yamashita (66) (K)

34th Class
(Apr.1940-Apr.1941; 48 persons)
Masao Iizuka (66) (K)
Yasuhiro Shigematsu (66) (K)
Yuzo Tsukamoto (66)
Hisayoshi Miyajima (66)
Shigehisa Yamamoto (66)

35th Class
(Nov.1940-Nov.1941; 158 persons)
Shigeru Araki (67) (K)
Nobuhiro Iwasaki (67) (K)
Toshitada Kawazoe (67) (K)
Katsutoshi Kawamata (67) (K)
Katsumi Kurihara (67) (K)
Jun-ichi Sasai (67) (K)
Kiyoharu Shibuya (67) (K)
Masayoshi Baba (67) (K)
Tadashi Hayashiya (67) (K)
Kaoru Yamaguchi (67) (K)
Sadao Yamaguchi (67) (K)
[Converted]
Toshihito Araki (67) (K)
Toshiharu Ikeda (67) (K)
Tetsuo Endo (67) (K)
Shigeo Juni (67) (K)
Takeo Yokoyama (67)

36th Class
(May 1941-Jun.1942; 125 persons)
Shu-ichi Utsumi (68) (K)
Takeyoshi Ohno (68) (K)
Takashi Oshibuchi (68) (K)
Toshio Kato (68) (A)
Mukumi Kanemitsu (68) (K)
Shigeto Kawahara (68) (K)
Masakazu Kusakari (68)

Yo-ichi Kenmochi (68) (K)
Hohei Kobayashi (68) (K)
Saburo Saito (68) (K)
Tadashi Sato (68) (K)
Toshio Shiozuru (68)
Usaburo Suzuki (68) (K)
Kennoshin Tagami (68) (K)
Shunkichi Tashiro (68) (K)
Kenji Nakagawa (67) (H, K)
Daihachi Nakajima (68)
Yoshikazu Noguchi (68) (K)
Kazuto Kuba (68) (K)
Hisanori Fujimaki (68) (K)
Keigo Fujiwara (68) (K)
Suetane Futagami (68) (K)
Kazumasa Mitsumori (68)
Isao Murata (68) (K)
Kunisuke Yuki (68) (K)
Hiroshi Yoshimura (68) (K)
Torio Watanabe (67) (K)
[Converted]
Shinya Ozaki (68) (K)
Keizo Yamazaki (68)

37th Class
(Nov.1941-Feb.1943; 139 persons)
Nobuo Awa (69) (K)
Shiro Itezono (69) (K)
Ichiro Imamura (69) (K)
Kunio Iwashita (69) (H)
Tetsushi Ueno (69) (K)
Takeshi Umemura (69)
Sue-ichi Ohshima (69) (K)
Yoshio Ohba (69) (K)
Takashi Konoshita (69) (K)
Motoi Kaneko (69) (K)
Tsutomu Kawai (69) (K)
Kakuro Kawamura (69) (K)
Kunio Kanzaki (69) (K)
Kisaku Koshida (69) (K)
Ikuro Sakami (69) (K)
Haruo Sagara (69) (K)
Masao Shimada (69) (K)
Naoshi Sugisaki (69) (K)
Takeo Sekiya (69)
Hiro-o Takao (69) (K)
Kenkichi Takasawa (69) (A)
Dai Nakajima (69) (K)
Mitsuteru Hashimoto (69)(K)
Yoshishige Hayashi (69) (K)
Torajiro Haruta (69) (K)
Ken-ichi Ban (69) (K)
Sumio Fukuda (69) (K)
Naoshi Mae (69) (K)
Hiroshi Maeda (69) (K)
Yasuo Masuyama (69)
Shigeo Mizorogi (69) (K)
Nobuhiko Muto (69) (K)
Hiroshi Mori-i (69) (K)
Tatsuo Hirano (Ki 47) (K)
[Converted]

Shiro Kawakubo (69) (K)
Minoru Kawazoe (69) (K)
Katsumi Koda (69) (K)
Juji Torimoto (69) (K)
Hideo Matsumura (69)
Masamichi Minokata (69) (K)

38th Class
(Jun.1942-September 1943;
130 persons)
Masa-aki Asakawa (69) (K)
Jun Abe (70) (A)
Kishichiro Ayukawa (70) (K)
Nobuyuki Ikura (70) (K)
Mutsuo Urushiyama (70) (K)
Kazuo Oh-hata (70) (K)
Mitsuo Ohfuji (70) (K)
Masato Okasako (70) (K)
Keiji Kataki (70) (K)
Hideo Katori (70)
Korekiyo Kawakita (70)(K)
Naoshi Kanno (70) (K)
Seiji Kinoshita (70) (K)
Kunio Kimura (70) (K)
Tadao Sato (70) (K)
Hiroshi Shiosaka (70) (K)
Nobuo Takano (70) (K)
Masao Takahashi (70) (K)
Jiro Takamure (70) (K)
Mizuho Tanaka (70) (K)
Koji Chikama (70) (K)
Takeo Tsutsumi (70) (K)
Keijiro Hayashi (70) (K)
Shigeo Hayashi (70) (K)
Yoshihiro Hayashi (70)
Kakichi Hirata (70)
Kiyoshi Fukagawa (70) (K)
Toshio Fukushima (70) (K)
Susumu Horio (70) (K)

Ken-ichi Honda (70) (K)
Ryutaro Masuda (70) (K)
Masayoshi Masumoto (70) (K)
Takuo Mitsumoto (70)
Yoshiteru Mine (70) (K)
Yutaka Yagi (70) (K)
Fumio Yamagata (70) (K)
Yoshihiro Yoshikawa (70)(K)
[Converted]
Toyohiko Inuzuka (70) (A)
Akira Satomura (70) (K)
Shinkai Fujimori (70) (K)
Takeshi Murakami (70)
Hiroshi Morioka (70)

39th Class
(Jan.1943-Jan.1944; 165 persons)
Yukio Ayukawa (71) (K)
Manabu Ishimori (71) (K)
Katsuo Imawaka (71) (K)
Tadashi Iwano (71) (K)
Yasuhiko Ukimura (70) (K)
Tsuguo Ohtsubo (71) (K)
Takashi Oh-hira (71) (A)
Hiroshi Ohbuchi (71) (K)
Korekiyo Otsuji (71) (K)
Tamon Kawaguchi (71) (A)
Hisao Kawanishi (70) (K)
Tsutomu Kariya (70) (A)
Ei-ichi Kawashima (70) (K)
Sachitoshi Kikuchi (71)
Osamu Kurata (70)
Hiroshi Kotajima (70) (A)
Jun Saito (70)
Osamu Takahashi (70) (K)
Iwao Takeishi (70) (K)
Akira Takeda (70)
Kosuke Tabuchi (70)
Hiroshi Tanaka (70) (K)

Rikio Tomioka (70) (K)
Hidehiro Nakama (70) (K)
Yasusuke Nakamura (71) (K)
Tomoyoshi Nagatomo (71) (K)
Kisuke Hasegawa (70)
Fujio Hayashi (71)
Susumu Fukuoka (71) (K)
Ei-ichi Furusawa (71) (K)
Hiroshi Matano (71) (K)
Yasushi Matsui (71) (K)
Kunio Matsuzaki (71) (K)
Chikashi Matsuyama (71) (A)
Tomiya Miyazaki (70)
Saneo Miyawaki (70) (K)
Sumihiro Meguro (71)
Isao Yagi (70)
Shin Yamauchi (69) (A)
Natomi Yonemitsu (71) (A)
Taisuke Yonemura (71)
Keizo Wadi (71) (K)
Ichiro Watanabe (K)
Iwao Akiyama (Ki- 51) (K)
Toshikazu Taguchi (Ki-50) (K)
Hisaya Hirusawa (Ki-50) (K)
Taka-aki Yokote (Ki-50) (A)
[Converted]
Yukio Seki (70) (K)
Shizuhiko Chikuma (70)
Yoshita Tada (71) (K)

40th Class
(Jun.1943-Jun.1944; 185 persons)

41st Class
(Sep.1943-Jul.1944; 323 persons)

42nd Class
(Mar.1944-Feb.1945; 466 persons)

2. PILOT TRAINEE STUDENTS AND HEI RESERVE ENLISTED TRAINEE CLASS STUDENTS

Pilot Trainee Students

Initially only officers were considered as appropriate for training as pilots within the IJN. However, in March 1914 non-commissioned officers were employed on an experimental basis in a training test programme. As a result – although considerably later – in May 1928 a Flight Trainee (Hikojutsu Renshu-sei) system was instituted. In June 1930 the name of this system was changed to Pilot Trainee (Sojo RenShu-sei, Soren). Training was initially undertaken at Kasumigaura Kokutai, and later at Yatabe, Tsukuba and elsewhere.

The length of training depended on the class or the year when it was undertaken, ranging from five months to as long as a year. Immediately upon graduation from the course, students were promoted to flying seamen (Koku-hei), although this title was changed in June 1941 to flight seamen (Hiko-hei). Training in the actual service employment and use of aircraft was provided immediately after graduation, at which those successful had been awarded their pilot insignia. This subsequent training, termed advanced training (Encho Kyoiku) was provided at the various kokutais; the use of fighter aircraft was for the most part conducted at the Oita Kokutai. In a departure from normal practice, the 54th-56th Classes were not granted their pilot insignia until after they had completed their training at Oita.

As mentioned earlier, the names of all personnel attending Classes 1-12 are listed in their entirety because it was not always possible at this early stage to ascertain which aircraft the students then went on to specialize in. Beginning with the 13th Class, the names listed are those that specialized on fighters. However, during the period of the 13th to 50th Classes, the courses for pilots who were identified for carrier-based service were not yet separate, so are not identified as such.

Hei Flight Reserve Enlisted Trainee Class Students

From the moment when the Flight Reserve Enlisted Trainee Class system (Yoka Renshu-sei Seido) was set up in June 1930, the training of non-commissioned officer flight personnel was undertaken in parallel; in other words the system covered the flight reserve class system itself, and those personnel recruited from NCO ranks in general (Soren, the Pilot Trainee Course). Commencing in October 1940, the latter group of trainees was used as Hei-type Flight Reserve Enlisted Trainees. Following general training which lasted about two months, the trainees were then sent on to the Flight Trainee Course (Hiko Renshu-sei Kyotei). The Flight Trainee Course was used by both Ko and Otsu category trainees. Carrier fighter training for flight trainees was conducted at the Omura and Tokushima Kokutais, in addition to that being undertaken at the Oita Kokutai. Due to training capacity limitations, flight trainee classes were sometimes held in two or more different locations.

Before Regular Training Courses Started
Shinzaburo Yokochi
Saburo Fuji-i
Kyojiro Ueno (A)
Genji Terada
Kogoro Kizaki (A)
Shikazo Koshimizu
Teruki Sakao
Togo Takahashi
Tadashi Hayakawa
Kanekichi Yokoyama

1st Class
(May 1920-May 1921; 8 persons)
Daijiro Aoki (A)
Tsutomu Abe
Torao Ishikawa (A)
Genzo Ohkubo
Go-ichi Kawabe
Fukuzo Takahashi
Sei-ichiro Takahashi (A)
Tomekichi Hinako (A)

3rd Class
(Dec.1922-Aug.1923; 19 persons)
Yoshihisa Ito
Hideyoshi Iwamoto
Hiko-o Egusa (A)
Kenzaburo Ohwatari

Yoshitaka Kajima
Jokichi Kawakami
Kozo Kokumai
Ryonosuke Suzuki
Eiji Sorabayashi
Masami Takahashi
Torakichi Tanaka
Yoshizo Hanawa
Kinji Haryu
Taiki Furuyama
Mamoru Miyoshi
Sakumatsu Meguro
Yotsuo Yashima
Tsunetaro Yamamoto (A)
Yaichi Watanabe (A)

4th Class
(Jun.1923-May 1924, 21 persons)
Mikio Aoki (A)
Shiro Akisawa
Akio Aso
Tomokazu Ejima
Ichiro Oh-ura
Takao Kaji
Yoshinobu Kawano
Masayuki Shimokawa
Isamu Takahashi
Ichiji Takekawa
Tojiro Takeshita
Iwaji Chikada

Seiko Chikano (A)
Takuma Chiku
Tadashi Nakamura (H, A)
Eisaku Nagashima
Hisamatsu Nori (A)
Chojiro Hayashi (K)
Ryogo Fuji-i (A)
Mitsuo Fujita (A)
Soshiro Matsuno (A)

5th Class
(Dec.1923-Nov.1924; 17 persons)
Takashi Abe
Ryosaku Iyama
Masayuki Iwai (A)
Shunji Ohmura
Saburo Kameda (A)
Kiyoshi Kitayama (A)
Gunji Saito
Haruo Suematsu
Heisuke Suzuki
Kaneo Tanaka (A)
Tamizo Tengu
Shin-ichi Nakanishi
Tsuruo Hashiguchi
Genpachi Fukami
Yoshito Fujisada (H)
Shichinosuke Maeda
Isao Matsumoto

6th Class
(Jun.1924-Mar.1925; 17 persons)
Tadao Iwabe
Sakae Ueno
Shizunori Uchimura
Yoshio Oikawa (K)
Shin-ichi Oh-ishi
Keijiro Kaneko (A)
Shinzo Takamura (K)
Kichinosuke Dochi
Sei-ichi Nakano
Senjiro Nishiyama
Tsuneo Fujimaki
Tomeji Fuse
Kiyoshi Maruyama
Takeji Miura (H, K)
Rokuzo Moriyama (K)
Sho-ichi Yamada
Takashi Yokoyama (A)

7th Class
(Feb.1925-Nov.1925; 11 persons)
Jutsuhei Ashizawa
Kesaju Ishikawa (A)
Taneo Inuzuka (K)
Shomatsu Kobayashi
Jin-ichi Sakurazawa
Tsuneshichi Shibata
Yoshitaro Suga (A)
Saijiro Teramoto (H)
Toshio Nakamura (Yasumoto) (A)
Man-ichi Hama (A)
Norikazu Fujisaki

8th Class
(Sep.1925-May 1926; 15 persons)
Kiyoe Aoki
Yakichi Ando
Shojiro Iwahori
Isao Kubota (A)
Ryoji Komatsu (A)
Teruo Komine (A)
Shosuke Sasa-o (K)
Tadashi Sato (A)
Kozo Susa
Hei-ichiro Mase (H, K)
Yoshio Matsuo (A)
Toyotsugu Miyata
Yasumi Murayama
Isao Morikawa
Shigeru Yamada (K)

9th Class
(Feb.-Nov.1926; carrier and floatplanes, 19 persons)
Atau Aoki
Isamu Ito
Yosuke Ohta
Shinji Oh-hara (A)
Nobuyuki Ogawa (A)
Tadashi Ono (Tsuda) (A)
Asakichi Kurokawa (Yoshida)

Shin-ichi Kawano
Miyoji Kobayashi (Tanaka)
Isamu Koine (Mochizuki) (K)
Hitoshi Sasaki
Tomezo Shiraishi
Kumezo Suzuki
Takeo Tozawa
Kunkichi Moto
Toshio Mori
Yoshimi Mori
Makoto Yanai (H)
Motoyuki Yokota

10th Class
(Jul.1926-May 1927; carrier and floatplanes, 17 persons)
Michiyoshi Ichimaru
Gi-ichi Ogura
Naoshi Kai (A)
Shigeo Katayama
Keijiro Kato
Tsurukichi Kawasaki
Shoji Kikuchi
Sakae Kimura
Mamoru Kubota (A)
Takeo Saito
Bukichi Takano
Tokuji Tatemichi
Koji Tariki
Hiroshi Nakanome
Yutaka Haraguchi
Heitaro Hiragawara
Heiji Yamamoto

11th Class
(Jan.-Nov.1927; carrier and floatplanes, 19 persons)
Bungo Abe
Shinsuke Ishi-i
Chikashi Ito (A)
Shinpei Ohsuga
Tomoshiro Ohmori
Hajime Oshikawa
Tadato Katagiri
Sakae Kato (K)
Takeo Koba
Sanshiro Kudo (A)
Shinkichi Goto
Ryoji Sato
Shigejiro Takahashi
Takeyoshi Taguchi (A)
Michimori Higo (A, H)
Takeji Matsumoto
Yoshio Masuda (A)
Yoshio Marugame
Tei-ichi Watanabe

12th Class
(Jul.1927-May 1928; carrier and floatplanes, 16 persons)
Sumio Anami (A)
Akio Arai

Masao Ando
Yuzo Ishi-i
Koshiro Ebina (A)
Nagao Ohmomo
Munetsugu Sato
Tadayoshi Suzuki
Mitsuo Toyoda
Tsuneo Nakahara (K)
Takao Hirama
Rikimatsu Hirayama
Shigezo Fuji-i (K)
Heitaro Morita
Norimasa Yamaguchi
Yataro Yamamoto

13th Class
(Mar.-Dec.1928; carrier and floatplanes, 22 persons)
Yoshio Inoue (A)
Manzo Iwaki (K)
Masao Ono
Ichio Kaneko (A)
Toshio Kuroiwa (K)
Fumiaki Takimoto
Kazuo Takeo
Hideo Funatsu

14th Class
(Jul.1928-May 1929; carrier and floatplanes, 30 persons)
Akira Eitoku (K)
Shuhei Ohya (K)
Wasuke Otokuni
(Seizaburo Kurosaki)
Hajime Koga
Mineichi Shibata
Mitsugu Mori
Yasukichi Yamagawa (Egawa) (A)

15th Class
(Jun.1929-Apr.1930; carrier and floatplanes, 30 persons)
Tomokichi Arai (H, K)
Noboru Ohzeki
Katsuzo Shima
Jiro Chono (K)
Katsuji Matsushima (A)

16th Class
(May 1930-May 1931; carrier and floatplanes, 25 persons)
Masahiro Ishi-i
(Hiratsuchi Ujiki) (K)
Taizo Okamoto (K)
Sadao Kawano
Kiyoto Koga (A)
Eisaku Shibayama
Sakeo Nishiyama

17th Class
(Jun.1931-Mar.1932; carrier and floatplanes, 32 persons)
Sadaaki Akamatsu
Kichiyoshi Ishikawa (A)
Shojiro Ito
Keiji Kubo
Bunji Shimizu (A)
Kiyoteru Terashita
Sakae Fukuchi (H, A)
Koshiro Yamashita (K)

18th Class
(May-Nov.1932; carrier and floatplanes, 40 persons)
Ki-ichi Oda (Nakamura) (K)
Kyushichi Kobayashi (K)
Chiyoyuki Shibata
Sukeichi Tanaka (A)
Matsukatsu Matsuoka (H, A)
Kazumasa Yokoyama (A)
Yoshio Wajima (K)
[Converted]
Mamoru Shimura (K)
Mitsuo Taguchi (K)

19th Class
(Oct.1932-Mar.1938; carrier and floatplanes, 41 persons)
(Yasuyoshi Akaike) (H, A)
Takeo Inaba (K)
Chitoshi Isozaki
Matsuya Kato (Tsukamoto)
Ippei Goto (K)
Taira Tanaka
Ken-ichi Takahashi (K)
Watari Handa
Motoji Hori-uchi

20th Class
(Feb.-Jul.1933; carrier and floatplanes, 37 persons)
(Tomiji Kawaguchi) (K)
Katsuma Shigemi (K)
Tadashi Torakuma(Kimoto) (A)
(Masao Naito)
Koji Hanagaki
Fusaji Miyata
[Converted]
Shinki Sato (H)

21st Class
(Apr.-Sep.1933; carrier and floatplanes, 30 persons)
Hachiro Aizawa
Masumi Okuyama (A)
Shozo Katayama (K)
Tsutomu Kawakami (A)
Saburo Kitahata (K)
Katsuo Nakamura (A)
Kozaburo Yamanaka

22nd Class
(Jun.-Dec.1933; carrier and floatplanes, 28 persons)
Mitsuo Ishizuka (K)
Tomokazu Ema (K)
Takanobu Obata
(Tadashi Koike)
Katsumi Shirakami
Toraichi Takatsuka (K)
(Rikichi Hiwatari) (A)

23rd Class
(Oct.1933-Apr.1934; carrier and floatplanes, 34 persons)
Saneo Imamura (K)
Takenori Kusumoto (A)
Tetsutaro Kumagaya (K)
Nobuo Kuro-iwa (Ito)
Tsukasa Taguchi (A)
Torajiro Nakatsuchi
(Takeshi Furukawa)
[Converted]
Namio Nishimura

24th Class
(Feb.-Sep.1934; carrier planes, 37 persons)
Kazuo Kagawa (A)
Kan-ichi Kashimura (K)
Noboru Kawase
Taketeru Sento (A)
Masumi Tsutsumi (Yamaguchi) (K)
Haruo Nakagawa (A)
Kijiro Noguchi
Hatsuo Hidaka
Akira Yamamoto (K)

25th Class
(Jun.-Nov.1934; carrier planes, 30 persons)
Kyosaku Aoki
Torakichi Inaha (A)
(Ma-ari Uehara)
Makoto Endo (K)
Jihei Kaneko
Hideyoshi Kume (A)
Toshiyuki Sakai (K)
Saburo Sugai (K)
Torata Takiguchi (K)
(Masaichi Tohata) (A)
Chiharu Naito
(Masao Nakajima) (K)
Katsuhiro Hashimoto
(Kunitada Matsushita) (A)
[Converted]
Akira Ozeki (K)

26th Class
(Sep.1934-Mar.1935; carrier planes, 39 persons)
Teruzo Aihara (A)

Hideo Oh-ishi (K)
Takejiro Onozuka (A)
Masa-aki Kawahara (K)
Kunimori Saeki (A)
Hitoshi Sato (K)
Takeo Sugiyama
Tsuyoshi Sumita (K)
(Toyozo Takehara) (A)
Takeo Tanaka (A)
Hachitaro Hayashi
Yoshio Fukui
Akio Matsuba

27th Class
(Jan.-Jul.1935; carrier planes, 38 persons)
Tetsuo Imabayashi (A)
Ryo-itsu Ohkubo (Saka) (K)
Kikue Otokuni
Masaichi Kondo
Yoshimichi Saeki
(Nobuyoshi Sumi)
Yuzaburo Toguchi
Masato Hino (K)
Isao Hiraishi (H, K)
Yoshiharu Fujimoto (A)
Toyoshige Fukunaga (A)
Hitoshi Fukazawa (K)
Shigeru Makino (K)

28th Class
(Feb.-Aug.1935; carrier planes, 32 persons)
Shiro Ayukawa (A)
(Shizuo Okunishi) (K)
Katsuyoshi Ogasawara (K)
Morinori Kurachi (A)
Einosuke Shitama (A)
Kiyonobu Suzuki (K)
Kamezo Tarui
Matsuo Hagiri
(Kinta Hatakeyama)
Kihei Fujiwara
Koichi Magara (K)
Yoshiharu Matsumoto (K)
Toshio Mori
Tadashi Yokote (K)

29th Class
(May-Nov.1935 land planes, 33 persons)
Shigeshi Imamura (K)
(Keitaro Iwase) (K)
Noboru Okukawa (K)
(Harunobu Oda) (K)
Takashi Kikuchi (A)
(Tomihiko Tanaka) (K)
Mitsunori Nakajima
Koji Hayakawa (K)
Suezo Harada (K)
Kinji Fujiwara (K)

Momoto Matsumura (K)
Koji Miyazaki (K)
Tasuke Mukai (K)
Shotaro Yamashita

30th Class
(May-Nov.1935; carrier planes, 41 persons)
Masazumi Ino (K)
Yoshihiro Inoue
Shinya Kashiwakura (H)
(Tatsugoro Kato) (K)
Toshio Kaneko (K)
(Eikichi Koga) (A)
Maresuke Gokan (A)
Hatsutaro Koshimizu (A)
Sozaburo Takahashi (K)
Kosaku Toyoda
Susumu Nakagaki (A)
Tsuguo Ogiwara (K)
Shunzo Hongo (K)
Namitaro Matsushima
Hideyori Matsumoto
Yoshimi Minami (K)

31st Class
(Sep.1935-Mar.1936; carrier planes, 32 persons)
(Norio Ohtake)
Juzo Okamoto (K)
Hisao Ochi (K)
(Harumi Kawano)
Akira Kikuchi (K)
Jinkichiro Kudo (K)
(Yoshitatsu Kurahara) (K)
Sadao Konno (H)
Kuniyoshi Tanaka
Naoshi Teramatsu (K)
Tadashi Nakamoto (K)
(Tsukasa Muramatsu) (K)

32nd Class
(Jan.-Jul.1936; carrier planes, 33 persons)
Fujio Ayabe (K)
(Hisashi Ino)
Sueji Itsukaichi (K)
Nagaharu Umeda (K)
Yukiharu Ozeki (K)
Shizuo Kojima (K)
(Bunpei Kondo) (A)
Kizo Shizu (K)
Toshiyuki Sueda (K)
(Ichijo Numata) (K)
Kaneyoshi Muto (K)
Isao Murata (K)

33rd Class
(Feb.-Sep.1936; carrier planes, 30 persons)
Shigetaka Ohmori (K)

Hiromori Shimomura (K)
(Minoru Tazaki (Ishii))
Bunkichi Nakajima (K)
Hideo Fukawa
(Mitsugu Maeda) (K)
[Converted]
Masao Suitsu

34th Class
(Apr.-Dec.1936; carrier planes, 27 persons)
Ki-ichi Iwase (K)
Tetsuzo Iwamoto
Kiyomi Kuwabara
Morinosuke Hatanaka
Ineo Hamada (K)
Sadao Yamashita (K)
Kyoichi Yoshii (K)

35th Class
(Jun.1936-Feb.1937; carrier planes, 27 persons)
Tsuneyoshi Iseri
Shine Inoue (A)
(Masayuki Tanimoto)
Toru Haneda (K)
Kaname Harada (H)
Gonji Moriyama (K)

36th Class
(Oct.1936-Jun.1937; carrier planes, 26 persons)
Tsumoru Ohkura (K)
(Osamu Kamikawa)
(Takeo Kawashima) (K)
Wataru Kubota (Tsukada) (A)
Mankichi Sawada
Hiromitsu Takahara (K)

37th Class
(Dec.1936-Jul.1937; carrier planes, 21 persons)
Yoshio Ohki (K)
Yu-ichi Tanaka (A)
Katsujiro Nakano (K)
Kazuki Mikami
Masami Takemoto (K)

38th Class
(Mar.-Nov.1937; carrier planes, 25 persons)
Masayoshi Okazaki
Kenji Okabe
Hiroshi Kawanishi (K)
Saburo Sakai (H)
Masaharu Hiramoto (K)
Iwao Hirayama (K)
(Minoru Maeda) (A)
Yanosuke Nagashima
Inuki Hirose
Yuji Yokoyama (K)

39th Class
(May 1937-Jan.1938; carrier planes, 30 persons)
Hisashi Aoyama (K)
Tetsuo Kikuchi (K)
Juzo Saito
Masanosuke Suzuki (K)
Jiro Tanaka (H, K)
Kasuke Hoshiya (K)
Takio Yoshida (K)

40th Class
(Jul.1937-Feb.1938; carrier planes, 36 persons)
Saburo Ishi-i (K)
Toshikazu Iwata (K)
Hiroshi Oh-iwa (K)
Genkichi Ogawa (K)
Soji Chiba (K)
Michisuke Tokuda (K)
Takashi Nakajima
Shigenobu Nakata (K)
Masaharu Higa (K)
Kozaburo Yasui (K)
Toru Yamamoto (K)
Hatsumasa Yamaya (K)
Kazuo Yokokawa

41st Class
(Nov.1937-Jun.1938; carrier planes, 36 persons)
Iki Arita (K)
Isamu Ochi (A)
Shigeru Ohkura (K)
Mitsugu Ohtani (K)
Shigeo Kimura (K)
Tetsuo Sato (K)
Seinoshin Sano (K)
Jiro Shoji (K)
Jusaku Tanabe (A)
Toshiaki Harada
Fusayoshi Murasaki (K)
Hideo Morinio (H)
(Tsumo-o Yamazaki) (K)
Kaname Yoshimatsu (K)

42nd Class
(Feb.-Sep.1938; carrier planes, 62 persons)
Rokuzo Iwamoto (H,K)
Gon-ichiro Ueda (A)
Takeo Okumura (K)
Masakatsu Obata (A)
Iwaji Kuragami
Saneatsu Kuroki (K)
Tamotsu Kojima (K)
Nobuo Saito (K)
Naokichi Suzuki (K)
Norio Tokushige (K)
Jinsaku Nojima (K)
Yoshiro Hashiguchi (K)

Teisaburo Hida (K)
Hiroshi Fujita (K)
Sanae Matsumoto (K)
Takeshi Yamamoto
   (Takahashi) (K)

43rd Class
*(Apr.-Nov.1938; carrier planes,
56 persons)*
(Gore Abe) (K)
Yasujiro Ohno (K)
Takashi Okamoto
Suminori Kawabata (K)
Yoshinao Kodaira
Goro Sakaida (K)
Ryusuke Sato (K)
Sam-ichiro Tanaka (K)
(Kuraichi Nishida) (K)
(Kosaku Hisamatsu) (K)
Hiromichi Hojo (K)
Teizo Hosomura (A)
Takashi Honda
Sukeji Majima (K)
En-ichi Yokoda (H, A)

44th Class
*(Jun.1938-Jan.1939; carrier planes,
54 persons)*
Osamu Ishi-i (K)
(Kosaku Ito) (A)
Kyo-ichiro Ogino (K)
Torakichi Okazaki (K)
Takeo Kanamaru
Hiroshi Kuratomi (K)
Saburo Saito
Yoshihiro Sakuraba (K)
Takeo Takashima (K)
Toshio Tomiya (A)
Masateru Tomoishi (K)
(Teru Nakai) (K)
Yasunobu Nahara
Ryo-ichi Hanabusa
(Masayoshi Higashima) (A)
Kosaku Minato (K)
Shigeru Yano (K)
Isaburo Yawata (K)
Tadao Yamanaka (H)
Mototsuna Yoshida (K)

45th Class
*(Oct.1938-May 1939; carrier
planes, 48 persons)*
Fumio Ito (K)
Shigetsune Ohgane (K)
Sei-ichi Kitaoka (K)
Takashi Kurauchi (K)
Hisao Komori (K)
Chuji Sakagami (K)
Shin-ichi Suzuki (K)
Isao Tahara (K)
Saburo Nozawa (K)

Nobutoshi Furukawa (K)
Kenzo Nihonmori (K)
Saburo Yoshi-i (K)

46th Class
*(Jan.-Sep.1939; carrier
planes, 49 persons)*
Makoto Inoue (K)
Toshio Ohta (K)
Shigeto Sawano (K)
Kaneo Suzuki (K)
Tetsujiro Suzuki (A)
Yoshizo Tanaka (K)
Tomeyoshi Nagata (A)
Yoshio Hirose (K)
Yoshio Yoshida (K)

47th Class
*(Mar.-Oct.1939; carrier
planes, 49 persons)*
Hiromi Ito (K)
Zenji Ohkubo (A)
Enji Kakimoto (K)
Keiji Kikuchi (K)
Masakichi Sonoyama (K)
Isao Doikawa (K)
Tomita Hara (Atake)

48th Class
*(Jun.1939-Jan.1940; carrier
planes, 82 persons)*
Yuji Ando (K)
Hajime Ishigami (A)
Kiyoharu Ishikawa
Nobuo Ogiya (K)
Kishiro Kobayashi (K)
Minoru Shibamura (K)
Sho-ichi Shoji (K)
Sakae Nakazawa (A)
Haruo Nitta (K)
Yasuji Notani (K)
Makoto Bando
Yoshimi Hidaka (K)
Zenpei Matsumoto (K)

49th Class
*(Nov.1939-Jun.1940; carrier planes,
70 persons)*
Takeichi Kokubu (K)
Katsutaro Kobayashi
Sadamu Komachi
Shinpei Sano (K)
Tetsuo Sento (A)
Katsuaki Nagatomo (K)
Toshiaki Honda (K)
Masao Masuyama
Tomikichi Maruta (A)
Nobuo Yamamoto (A)

50th Class
*(Dec.1939-Jun.1940; carrier
fighters, 11 persons)*
Seiji I-ishi (K)
Shizuo Ishi-i (K)
Hiroshi Oh-hara (K)
Masami Shiga
Seiji Tajiri (K)
Yutaka Chiyoshima (K)
Genzo Nagasawa (K)
Shin-ichi Nagata (K)
Sakae Mori (K)
Ichiro Yamamoto (K)
Tatsu Matsumoto (K)

51st Class
*(Dec.1939-Jul.1940; carrier
fighters, 6 persons)*
Shigeru Kawano
Mitsuyoshi Takasuga (K)
Masao Taniguchi (50 advanced)
Toichiro Hanyu (K)
Shunji Horiguchi (A)
Haruo Miyata (A)
[Converted]
Keiji Sunami (H)
Gozo Teruyama

52nd Class
*(Feb.-Jul.1940; no carrier
fighter pilots)*

53rd Class
*(Feb.-Aug.1940; carrier fighters,
11 persons)*
Yasuo Isobe (K)
Shinko Ito (K)
Yoshio Egawa
Michiyuku Kitaoki
Yukio Kitazato (K)
Masa-aki Shimakawa
Ei-ichi Takahashi (K)
Takashi Nakagami (K)
Takao Banno (H, K)
Haruo Fujibayashi
Yoshio Motoyoshi (K)
[Converted]
Katsuaki Nagamawari

54th Class
*(May 1941; Oita Kokutai
graduates, carrier fighters,
21 persons)*
Genkichi Oh-ishi (A)
Hiroshi Okano
Matsuhimaru Kashibo (K)
Teruo Kawamata (K)
Nobutaka Kurata (K)
Kenji Kotani (A)
Mibuichi Shimada (K)
Hiroshi Suyama (K)

Mitsuo Suizu (K)
Minoru Tanaka (K)
Matsutaro Takaoka (K)
Toshiji Nisugi (K)
Shizuki Nishiyama (K)
Osamu Hatanaka (A)
Ko-ichi Fuji-i (A)
Shichijiro Mae (K)
Shiro Murakami (K)
Ichirobei Yamazaki (K)
Ken-ichiro Yamamoto (K)
Ippei Yoshida
Suekichi Yoshimoto (K)
[Converted]
Kokichi Ohtsuki (A)

55th Class
(Jul.1941; Oita Kokutai graduates;
carrier fighters, 9 persons)
Minoru Ishi-i (K)
Masashi Ishida (K)
Shoji Ishihara (K)
Mitsumasa Ujihara (A)
Kyoji Kobayashi (K)
Eiji Sanada (K)
Kihachi Ninomiya (K)
Shigeru Hayashi (K)
Yasuo Matsumoto

56th Class
(Jul.1941; Oita Kokutai graduate;
carrier fighters, 19 persons)
Yoshio Aoki (K)
Sachio Azuma (K)
Koichi Imamura (K)
Ryoichi Iwakawa (K)
Yoshio Iwabuchi (K)
Tomekichi Ohtsuki (K)
Toru Oda (K)
Haruo Kawanishi (K)
Heikichi Kitao (K)
Takeshi Sakamoto
Tsu Suematsu (K)
Mikio Tanikawa (K)
Mitsuru Tsuruoka (A)
Hajime Toyoshima (K)
Ki-ichi Nagano (K)
Takeo Matsuda (K)
Masayoshi Yonekawa (A)
Tadashi Yoneda (K)
Keisaku Yoshimura (K)
[Converted]
Heisaku Sato (K)
Yoshijiro Shirahama

57th Class
(Oct.1941;Hakata Kokutai
graduates; floatplanes, 33 persons)
[Converted]
Hideo Uemura (Saito)
Sei-ichi Enomoto (K)

Kaneo Oita (K)
Jisaku Kaneko (K)
Tomematsu Matsunaga (K)
Kazuo Yamazaki (A)
Suehiro Yamamoto (K)

Hei 2nd Class (12th Class;
Flight Trainee Course)
(Nov.1941; Oita Kokutai
graduates, 37 persons)
Toshiyuki Ichiki
Kiyoshi Ito
Mari Ito (K)
Ken-ichi Inagaki
Naoshi Ubukata (K)
Kiyoshi Ohsuga (K)
Shigeru Kawazu (K)
Saji Kanda (K)
Magoichi Kosaka (K)
Tamio Kobayashi
Takeshi Kobayashi (K)
Yu-ichi Kobayashi (K)
Akira Saito (K)
Ken-ichi Shinozuka (K)
Miyuki Shimozuru
Yozo Sugawara (K)
Chinta Takagi (K)
Wataru Takeda (K)
Sadamu Tamai (K)
Jafuku Tanji (K)
Goro Tsuda (K)
Yoshio Torishima (K)
Tsuneyoshi Nakazawa (K)
Yoshi-ichi Nakaya
Gi-ichi Nakayama (K)
Chuichiro Hata (K)
Ayunosuke Hattori (A)
Sakuji Hayashi (K)
Takeichiro Hidaka (K)
Yoshimitsu Maeda (K)
Toshimi Matsukata (K)
Shigejiro Murakami (K)
Danji Yatsukura (K)
Kuratoshi Yasuda (K)
Tomezo Yamamoto (A)
Koki Yoneda (K)
Masao Watanabe (K)
[Converted]
Tadashi Sakai (K)

Hei 2nd Class
(Nov.1941; Usa Kokutai graduates;
converted from dive-bombers, 18
persons)
Kunishige Iizuka
Mitsuyoshi Inoue (K)
Satoru Ogawa (K)
Tetsuo Kamata
Yuki Kariya (K)
Shigeji Kawai
Katsuhiko Kawasaki (K)

Hikoji Kawada (K)
Minoru Kuranaga (K)
Shigehiro Sugihara (K)
Katsuji Hijiya (A)
Hiroyuki Hihara (K)
Tsunesaku Hayashi
Sachio Hayama (K)
Tadashi Fujimoto (K)
Hidetaro Hosoya (K)
Isamu Miyazaki
Hideo Watanabe

Hei 2nd Class
(Nov.1941; Usa Kokutai graduates:
converted from torpedo planes,
10 persons)
Mutsu-o Uemura (K)
Kiyoshi Ono (K)
Susumu Ohtsuki (K)
Tsugio Kawagishi (A)
Masayoshi Kojima (A)
Kiyoshi Sekiya (K)
Ten-ichi Tanoue (K)
Yoshihira Hashiguchi (A)
Saburo Hotta (K)
Shigeru Mukumoto (K)

Hei 3rd Class (Flight
Trainee Course 17th Class)
(March 1942; Oita Kokutai
graduates, 80 persons,. Omura
Kokutai graduates, unknown
number of persons)
Hisashi Aoki (K)
Yu-ichiro Ando (K)
Takesaburo Ikeda (K)
Izumi Ishihara (Sanada)
Tomitaro Ito (K)
Shigehiko Ito (K)
Yoshifusa Iwasaka (K)
Jisuke Iwase (K)
Jinzo Ueno (A)
Masahiro Ueno (K)
Noboru Uehara (K)
Yoshio Ohsawa
Toru Ohbayashi (A)
Tamotsu Okabayashi (K)
Kunio Okishige (K)
Mitsuyasu Ozaki (K)
Hiroshi Ochi (K)
Jukichi Ono (K)
Ko-ichiro Kato (K)
Masao Kato (K)
Yasutaka Kanoya (K)
Fusakazu Kaneko (K)
Shigeto Kawakami (K)
Soroku Kawakami (K)
Yukio Kitano (K)
Hide-aki Kiyosawa (A)
Sho-ichi Kuwabara (K)
Ichihei Kobayashi (K)

Toshio Komiya (K)
Hiroshi Koyama (K)
Kazuo Goto (K)
Mitsuo Goto (K)
Iwao Gono (K)
Takao Banno (K)
Takamichi Sasamoto (K)
Genshichi Sato (K)
Toshimi Sato (K)
Kiyotaka Sawazaki (K)
Ichizo Shigematsu (A)
Yoshio Shoji (A)
Katsuo Sugano (K)
Sho-ichi Sugita (K)
Kazuo Sugino
Taiji Suzuki (K)
Takashi Sekiya (K)
Masanobu Tai (K)
Takashi Takayama (K)
Toji Tatebe (K)
Takeo Tanimizu
Kazu Tamura (K)
Hideo Tsukamoto (K)
Shizuo Doi (K)
Kiyoshi Tokunaga (K)
Tatsuo Tokunaga (K)
Kazuo Tobita (K)
Sachio Nakatsukasa (K)
Masa-aki Nakane (K)
Toshikazu Nakano (K)
Tomokazu Nakano (K)
Yasuhiro Nakamura (K)
Kiyoshi Nakamoto (K)
Nobuto Nagao (K)
Shin-ichi Nishisaka (K)
Eiji Nishida (K)
Yoshio Nishida (K)
Ippei Ninomiya (K)
Hisahide Hashimoto (K)
Haruo Hamanaka (K)
Kokichi Hamano (K)
Tokutaro Harami-ishi
  (Suzuki) (K)
Kazuyoshi Harano (K)
Shinjiro Hinobeda (K)
Tetsuo Hidaka (K)
Sadao Hirai
Shigeo Hirano (K)
Shin-ichi Hirabayashi
Imio Fukuda (K)
Tsuguo Fukutome (K)
Sho-ichi Fujisada (K)
Aihiko Fujita (A)
Seiji Hosono (K)
Hidemasa Honda (K)
Tomio Maeda (K)
Masanao Maki (K)
Bangoro Myokei (K)
Hachiro Mitsunaga (K)
Nobuyuki Muraoka (K)
Hiroshi Mure (K)

Fukuyoshi Morino (K)
Motosuke Yato (K)
Ryo-ichi Yasuda (A)
Nobutaka Yanami (K)
Kenji Yanagiya
Yoshimi Yamauchi (K)
Shotoku Yamaguchi (K)
Takashi Yamazaki (K)
Kameji Yamane (K)
Shin-ichi Yamawaki
Ryozo Wakabayashi (K)
Seizaburo Watanabe (K)
Kenji Wada (K)

**Hei 3rd Class (Flight Trainee
Course, 18th Class)**
*(May 1942; Omura Kokutai
graduates; carrier fighters,
30 persons)*
Shigemasa Asami (K)
Yoshimizu Arata (K)
Takematsu Imabayashi (A)
Masami Iwatsubo (K)
Sumio Uehara (A)
Sachio Egawa (K)
Yoneji Kanazawa (K)
Ikuji Kaburagi (K)
Motoyasu Kitamura (K)
Sadao Kubo (K)
Kazuo Komaba (K)
Toshikazu Koyanagi (K)
Hachiro Sato (K)
Itaru Shikano
Katsuyoshi Tanaka (K)
Kisaku Tanaka (K)
Sakuji Tanaka (K)
Shiro Tsukahara (K)
Shiro Toriyama (K)
Ryoji Handa (K)
Sakuichi Fukuda (K)
Kichiro Maekawa
Mono-o Makiyama (K)
Matsukichi Matsui (K)
Setsu Matsuyoshi (K)
Kaoru Miyazaki (A)
Toshio Miyanishi (K)
Takashi Yamashita (A)
Makio Yamato (K)
Takamori Yamanaka (K)

Hei 3rd Class
*(Jul.1942; Omura Kokutai
graduates; carrier fighters,
27 persons)*
Shunji Ito (K)
Sadakazu Iwaki (K)
Seijiro Uemura (K)
Kiyoshi Ogawa (K)
Fumio Kato (K)
Yasuo Kanemitsu (K)
Bunji Kamihira (K)

Atsushi Kubo (A)
Susumu Kubota (K)
Masakichi Kurihara
Sei-ichi Kurosawa (K)
Hikoji Goto (K)
Toshikatsu Sato (K)
Fukumitsu Shimada (K)
Shigeru Tanoue (K)
Isamu Tanaka (K)
Toshio Nakazono (H, K)
Yoshio Nakamura
Shichiro Nagai (K)
Hisao Nishimoto
Skin-ichi Hasegawa
Kiju Hitomi (K)
Fukuda
Hidemichi Matsumura (K)
Makoto Murata (K)
Hisayoshi Mori (K)
Hiroshi Yasumatsu (K)

**Hei 4th Class (Flight
Trainee Course, 21st Class)**
*(July 1942; Oita Kokutai
graduates; carrier fighters,
29 persons)*
Tsurumi Ichiki (K)
Moto-omi Ishibashi (K)
Yoshiharu Izumi (K)
Sanjiro Imai (K)
Kazushiro Umezu
Ryoji Oh-hara
Masanori Oka (K)
Yutaka Kimoto (K)
Tomokazu Kobayashi (K)
Ichijiro Saito
Seikichi Sakae (K)
Shogo Sasaki (K)
Kaoru Sato (K)
Shigekazu Suzuki (K)
Masakazu Nakazawa (K)
Mitsuo Nakahara (A)
Masaichi Nakamura (K)
Wataru Nakamichi
Kanekichi Nemoto (K)
Takeshi Hayashi (K)
Noboru Hayashi (K)
Hirano
Saburo Fujiuma (K)
Toshiaki Maeda (K)
Takayuki Murakami (K)
Shin-ichi Yamawaki (K)
Tokuji Yoshizaki (K)
Furuo Yoshimitsu (K)
[Converted]
Shinkichi Ohshima (A)
Sachio Kambara
Shingo Honda (K)
Tsunesaku Sakai (K)
Eikichi Nakazaki (K)

Hei 6th Class (Flight Trainee Course, 23rd Class)
(Sep.1942; Omura Kokutai graduates, 27persons; Tokushima Kokutai graduates, 30 persons; Oita Kokutai graduates, unknown number)
Hideo Iijima (K)
Tokio I-ishi (K)
Kane-ichi Ishi-i (A)
Teigo Ishida (K)
Bunji Ishida (K)
Ryuzo Isobe (K)
Sue-o Inoue (K)
Yoshiyuki Ueki (K)
Kazuo Endo (K)
Michitoshi Egashira (K)
Matsukichi Ohtomo (K)
Chikara Kitakuchi (K)
Kunio Kidokoro (K)
Kensuke Kurosawa (K)
Masakazu Kobayashi (K)
Shojiro Kobayashi (K)
Yuzo Komatsu (K)
Eigoro Saito
Soji Saitoh (K)

Tomoharu Shi-ina (K)
Kenzo Shiokawa
Katsunobu Shiba (K)
Shigeru Shibukawa
Kazuki Takeda (K)
Tadami Takemoto (K)
Tai-ichi Tashiro (K)
Hideo Takanabe (K)
Tsuyoshi Tanaka (K)
Hiroshi Takebe (K)
Iwao Taneda (K)
Yoshitaka Tamura (K)
Kiyoshi Tamaki (K)
Kazuyoshi Tokuhara (K)
Minoru Tomisono (K)
Akira Nakazawa (K)
Hayao Nakabeppu (K)
Yukuo Nanao (K)
Michibiki Nishi (K)
Noboru Nishio (K)
Takeo Hashimoto (K)
Shujiro Hasegawa (A)
Denjiro Baba (K)
Takayuki Hamasaki (A)
Morimasa Hirai (K)
Taro Fukuhara (K)

Muneichi Fujioka (K)
Kinpei Hoshino (K)
Toshimitsu Matsuzaki (A)
Gen-ichi Matsuda (A)
Masanari Mizusawa (K)
Sueo Mizuno (K)
Mitsuzo Miyagaki
Hisakazu Miyamato (K)
Minoru Miwa (K)
Naraichi Murai (K)
Kaname Mori (K)
Masaru Moriyama (K)
Kei-ichi Kadoma (K)
Ryuji Yagi
Yasukichi Yamakawa (K)
Takeo Yamamoto (K)
Hifumi Yamamoto (K)
Hiroshi Yamamoto (K)
Ken-ichiro Yoshioka (K)
Rokuya Yoneyama (K)
[Converted]
Takeji Ohmori (K)
Tadanobu Okamoto (A)
Yo-ichi Katsumata
Takeshi Kato (K)
Toshizo Hyodo

KO, OTSU FLIGHT RESERVE ENLISTED TRAINEE CLASS STUDENTS

The Flight Reserve Enlisted Trainee (Yoka Renshu-sei) system was inaugurated at the instruction of the Ministry of the Navy in December 1929. In June 1930 students of the 1st Class were selected. The scheme paralleled the Army's Youth Flight Enlisted Trainee (Shonen Hiko-hei) programme, the IJN focusing on youths aged 15-17 who had received an education approximately equivalent to that achieved by graduates from the higher elementary schools (Koto Shogakko). Trainees received training for a period of 30-36 months, primarily centred around general educational subjects; at a later date the programme was shortened to two years. After graduating from these classes, trainees were assigned to Flight Trainee Courses (Hiko Renshu-sei Katei) for basic pilot training. Following completion of this latter course, they were then trained in the operational uses of aircraft. From the Flight Trainee Course level onwards, the training was essentially the same as that given under the Hei Flight Reserve Enlisted Trainee Course.

In order to meet the need for an increase in flight personnel which arose during May 1937, it was decided to employ men with a higher educational background than that required for the Flight Reserve Enlisted Trainee Class system. Consequently, such personnel were placed in the Ko Flight Reserve Enlisted Trainee Class system, the first such course being commenced in September of that year. Qualification requirements matched the high school (chugakko) graduation level. For this type of trainee, the programme was reduced from 24 months to 18 months (and later, to just 12 months). As a result of the establishment of the Ko Flight Trainee Courses, the name of the previous Flight Reserve Enlisted Trainee Class was changed to Otsu Flight Trainee Course.

Initially, Flight Reserve Enlisted Trainee Class training was undertaken within Yokosuka Kokutai, but in March 1939 this was moved to Tsuchiura. In November 1940 the programme was made independent as Tsuchiura Kokutai. Later, due to the increased number of personnel being trained, Mie Kokutai was formed in August 1942. Similar groups continued to be established successively, and by the spring of 1945 there were 18 training kokutais specialising in the Flight Reserve Enlisted Trainee Class progamme.

To see the relationship between the Ko, Otsu and Hei Flight Reserve Enlisted Trainee Class system and the Flight Trainee Course system, the reader should now turn to page 409.

Otsu 1st Class
(Jun.1930-May 1933; graduates of Flight Trainee Course; carrier planes, 20 persons)
Yasujiro Abe
Shozo Okabe
Jirokichi Kusunoki
Yoshimitsu Harada (K)
Morio Miyaguchi (K)
Shigeo Miyamato (K)
Noboru Yamakawa (K)
Hiroyuki Yamaguchi (K)
Hitoshi Watanabe (A)
[Converted]
Susumu Ito

Otsu 2nd Class
(Jun.1931-Apr.1935 graduates of Flight Trainee Course; carrier planes, 48 persons)
Sei-ichi Uemura
Masanori Oh-hashi (A)
Suekichi Osanai (K)
Zenji Ono (K)
Yasuji Kanemitsu (A)
Osamu Kudo (K)
Fujikazu Koizumi (K)
Yoshimi Kodama (K)
Sukesada Senda (A)
Jiro Nasu (A)
Hiro-o Natori (A)
Ichiro Higashiyama (H, K)
Mitsuma Hirai (K)
Yoshijiro Minegishi (K)
Masatsugu Miyauchi (A)
Kazunori Miyabe (K)
Sakutsuchi Yamada (A)
Gi-ichi Yamanouchi (K)
Naoshi Watanabe (K)
[Converted]
Minobu Kaga (K)
Gi-ichi Minami (K)

Otsu 3rd Class
(Jun.1932-Apr.1936; graduates of Flight Trainee Course; carrier planes, 46 persons)
Sachi-o Aiso (K)
Kazumi Aramaki (A)
Mitsuo Kaneko (A)
Seizaburo Sugino
Yoshio Suzuki (A)
Torakichi Tanaka (Ozawa)

Ryo-ichi Nakamizo (K)
(Takuo Noda)
(Keigo Noborimoto)
Tsugio Matsuyama (K)
Sachi-o Miyazato (K)
Yuji Mori (K)
Akiji Yamazaki
[Converted]
Susumu Kawasaki

Otsu 4th Class
(May 1933-May 1937; graduates of Flight Trainee Course; carrier planes, 41 persons)
(Sataro Abe) (K)
Yoshiharu Kusakabe (A)
(Tatsu Kotani) (K)
Mutsuo Sagara
Hayato Noda (K)
Yutaka Matsuzaki (A)
Akira Maruyama (K)
Gitaro Miyazaki (K)
Yoshio Yoshie
[Converted]
Kumaji Tsugane (K)

Otsu 5th Class
(Jun.1934-March 1938; graduates of Flight Trainee Course; carrier planes, 43 persons)
Tomo-o Inoue (K)
(Takeru Oh-hara) (H, K)
Shigeru Okamoto (K)
Ei-ichi Kimura (K)
Yoshihiro Kobayashi (K)
Tsutomu Kobayashi (A)
Yu-ichi Sagara (K)
Shigeo Sugio
Kazuo Tsunoda
Masayuki Nakase (K)
Yukuo Hanzawa (K)
Masayuki Mitsumasu (A)
Sahei Yamashita (K)
Satoshi Yoshino (K)
[Converted]
Kunio Shintani (K)
Masahiko Nakamine
Tsutomu Hamada

Otsu 6th Class
(Jun.1935-August1938; graduates of Flight Trainee Course; Carrier planes, 51 persons)

Sueji Ide (K)
Tsutomu Iwai
Tatsuo Uchimura (A)
Ichizo Ohmori
Satoshi Kano (K)
Masao Kawasaki (K)
Susumu Goto (K)
Tsukasa Kondo (H, K)
Toyo-o Sakai (K)
Yasuhisa Sato (K)
Toshiro San-o
Ko-ichi Takafuji (A)
Yukuo Tanaka (K)
Yoshinao Tokuji (K)
Tatsuo Higashinaka
Masami Fukazawa
(Otokichi Funakawa) (K)
Jiro Mitsumoto (K)
Kazuo Muranaka
Tomio Yoshizawa (H, A)
[Converted]
Shigemasa Nishio (K)

Otsu 7th Class
(Jun.1936-Mar.1939; graduates of Flight Trainee Course; carrier planes, 71 persons)
Tomio Inenaga (K)
Jiro Imura (K)
Fumio Ohsumi (K)
Sukemasa Kato (K)
Toshio Kikuchi (K)
Harukichi Kubota (K)
Yasujiro Kawano (K)
Kenta Komiyama (K)
Teruo Sugiyama (K)
Gunji Suzuki (K)
Kazuo Suzuri (K)
Ryo Takahashi
Akira Takamori (A)
Tokiharu Tezuka (K)
Hiroyoshi Nishizawa (K)
Fujio Hayashi (K)
Eijiro Higaki (K)
Hisashi Hide (K)
Shigeo Fukumoto
[Converted]
Fukuju Kawakami
Tetsuji Koga (K)
Taiji Takayama
Kazutoshi Nagano (A)
Hisao Harada (K)
Toshiaki Maeda (A)

Otsu 8th Class

*(Jun.1937-Mar.1940; graduates of
Flight Trainee Course; carrier
fighters, 11 persons)*
Shigeo Okazaki
Nobuo Osanai (A)
Takumi Kai (K)
Ichiro Kaneko (A)
Ichiro Sakai (K)
Ryokei Shinohara (K)
Shigenobu Takahara (K)
Noboru Todaka (K)
Kunimori Nakakariya
Tatsuzo Hasegawa (K)
Akira Hongo (A)
Hisashi Matsumoto (A)
[Converted]
Susumu Ishida (K)
Kyo-ichi Inuzuka (K)
Fujiki Azuma

Otsu 9th Class

*(Jun.1938-Oct.1941; graduates of
Flight Trainee Course, 10th class;
Oita Kokutai graduates,
23 persons)*
Shigenobu Adachi (K)
Masami Arai (K)
Makoto Ueda (K)
Chikatatsu Kaminuma (K)
Sadao Uehara
Kazushi Uto (K)
So-ichi Ohshoya (A)
Sukeo Kawasaki (K)
Yutaka Kimura (K)
Ken-ichi Kumagaya (K)
Takeo Kume (K)
Ryusuke Goto (K)
Noboru Sato (K)
Tsutomu Shibata (K)
Matsumi Suzuki (K)
Akimizu Seki (K)
Shigeru Nomura (H)
Kiyotake Fukuyama (K)
Yoshio Motegi (K)
Masaru Morita (K)
Toyo-o Moriura (K)
Takashi Yokoyama (K)
Kametsugu Watamura

Otsu 9th Class

*(Oct.1941 graduates of Usa
Kokutai: converted from dive-
bombers, 10 persons)*
Ken-ichi Abe
Daizo Ihara (K)
Masuaki Endo (K)
Yoshio Oh-ishi (K)
Yoshizo Ohnishi (K)
Shin Nakano (K)
Noboru Nishiyama (K)
Sa-ichi Matsumoto

Mamoru Morita (A)
Shigematsu Yamashita

Otsu 9 Class

*(Oct.1941 graduates of Kisarazu
Kokutai; converted from Rikko)*
Gi-ichi Oh-hara (A)
Takashi Konno (K)
Hideo Shimizu (K)
Yosaburo Shinomiya (K)
Hajime Tochi (K)
Yoshinori Noguchi (K)
Katsumasa Matsumoto (K)
Toyo-o Mori-ura (K)
Hamashige Yamaguchi (K)
*[Persons of Otsu 9th who
converted from other kind of
aircraft received fighter training
in Izumi Detachment Unit,
Sasebo Kokutai during Nov.1941-
Feb.1942]*

Otsu 10th Class

*(Nov.1938-Mar.1942; graduated
from Oita Kokutai graduates,
35 persons)*
Minoru Awao (K)
Kazuo Ikumi (A)
Mitsuji Ikeda (K)
Yoshihide Ishizawa (K)
Isao Ito (K)
Jiro Iwai (A)
Naoshi Uematsu (K)
Noboru Ohta
Michitaka Kashiwara (K)
Kishio Kadota (K)
Toyonobu Kuzuhara
Sadao Koike (K)
Shigekazu Sasa (K)
Zensaburo Suzuki (A)
Jun-ichi Takahashi (K)
Toru Takeuchi
Hideya Takesawa (K)
Kisaku Takeda (A)
Shunji Nakasawa (K)
Atsuo Nishikane (K)
Shokichi Nishimoto (K)
Kikuo Nishimori (K)
Yoshiyuki Hirata (K)
Daizo Fukumori (K)
Tamotsu Fujita
Masami Futaki (K)
Katsumi Furumoto (K)
Mitsuo Hori (Mikami)
Shigeo Motegi (K)
Tatsuo Morioka (K)
Toshio Morita (K)
Sakae Yamashita
Takeo Yamashiro (K)
Mitsuo Yamada
Isamu Yoshiwara (K)
[Converted]

Toshio Imada (K)
Hisashi Kamata (K)
Shugo Takahashi

Otsu 11th Class

*(June 1939- 21st Class of Flight
Trainee Course; Jul.1942 graduates
of Oita Kokutai, 14 persons)*
Masanori Arimura (K)
Gen-ichi Uchida (K)
Kumaichi Kato (K)
Shigeru Kimata (K)
Matagoro Kimura (K)
Yorihisa Kobayashi (K)
Koji Shimizu (K)
Takeshige Senuma (K)
Kotaro Takano (K)
Akira Takita (K)
Tamaru Tanigaki (K)
Toshio Nagata (K)
Hiroshi Hirai (K)
Tadao Wakimoto (A)
[Converted]
Kenzo Asatsu (K)
Katsuyoshi Ito (K)
Ryozo Soejima
Teruyuki Naoi
Satoyuki Hayase
Saburo Mitsuoka (K)
*(23rd Class of Flight
Trainee Course; Sep.1942
graduates of Oita Kokutai
48 persons)*
Shin-ichi Ando
Taketoshi Ii-o (K)
Masao Ishi-i (K)
Eikichi Ichimura (A)
Hisashi Ichiyanagi
Motoharu Imazeki (K)
Yoshio Imamura (K)
Hiroshi Ueda (K)
Rizo Ohkubo
Minoru Ohta (A)
Rokei Ohmiya (K)
Takemi Okada (K)
Takao Kato (K)
Mitsuo Kusano
Tashiro Koga (K)
Yoshio Goto (K)
Toshiyuki Koba
Shigeru Sasako (A)
Mankichi Sato (K)
Yoshio Shiode (K)
Rokusaburo Shinohara (K)
Masa Shimada (K)
Hideo Suzuki (K)
Hideo Seki
Tatsumi Soga (K)
Kazuo Tachizumi (K)
Minpo Tanaka
Toshio Tanaka
Makoto Terao (K)

Hiroshi Nishimura (K)
Tokuharu Noda (K)
Namio Hashimasa (K)
Hisashi Hayakawa (K)
Tadashi Hirai (K)
Tsukijiro Fujii (K)
Bun-ichi Fujise (K)
Hideaki Maeda (K)
Ryohei Masajima (A)
Kan-ichi Masuda (K)
Kojiro Murakami (H,K)
Kiyoshi Yamazaki (K)
Tadao Yamashita (K)
Rokusuke Yamashita (K)
Iwao Yamamoto (K)
Yoshio Yoshida (K)
Masuo Yoneyama (K)

Otsu 12th Class
(Nov.1939-Jan.1943 graduates of
25th Class Flight Trainee Course;
carrier fighters, 58 persons)
Eizaburo Asakage (K)
Yoshito Ozuhata (K)
Fusao Ariga (K)
Ko-ichi Ishikawa (K)
Asagoro Ishida (K)
Koshichi Izumida (K)
Noboru Ito (K)
Fumio Ito (K)
Jiro Uchida (K)
Eizo Ohta (K)
Takeo Ohtsu (A)
Mamoru Ohtsuka (K)
Sakae Okamura (K)
Shizumu Ono (K)
Tsunemichi Kogami
Yoshiharu Kagami (K)
Tomonobu Kameyama (K)
Shoyo Kijiya (K)
Takashi Kishi (K)
Minoru Kitazaki (K)
Kinya Kunihiro (K)
Yoshinari Kumada (K)
Kusuo Kuriyama (K)
Noboru Koizumi (K)
Kaname Koide (K)
Masao Kotaki (K)
Yukuhiro Kodama (K)
Kaoru Goto (K)
Bun-ichi Goto (K)
Zentoku Sato (A)
Kazuo Shibayama (K)
Eiji Sekiguchi (K)
Kiyoshi Takei (K)
Tadamori Tajima (K)
Shinsaku Tanaka (K)
Tsuguo Terada (K)
Setsuo Tominaga (K)
Katsumi Nakamura (K)
Hideo Ni-imichi
Satoshi Nishioka

Yo-ichi Nishikura (K)
Hiroshi Nishida (K)
Yoshikazu Nishimoto
Seiji Noma (K)
Shigeo Hayashi
Shigeo Hayashi
Noriyuki Haraguchi (K)
Shun-ichi Hiromori (K)
Yoshio Mazaki (K)
Fukukichi Masago
Tadayoshi Matsui (A)
Seisuke Matsuda (K)
Kimiyoshi Miyamato (K)
Shiro Miwa (K)
Toshi-aki Murata (K)
Kiyota Yanagi (K)
Ko-ichi Yamauchi (K)
Tokio Yokoi (K)
Sadatoshi Yoshino (K)
[Converted]
Yu Inoue
Takeshi Tsuji
Yoshiharu Domoto
Koshi Nosue (K)
Keisuke Yamamura
Takafumi Yokobayashi (K)

Otsu 13th Class
(Jun.1940-Mar.1943 graduates of
26th Class Flight Trainee Course;
carrier fighters, 41 persons)
Shigehisa Aoki (K)
Moriji Akiba (K)
Genbo Adachi (A)
Yoshiaki Ikenaga (A)
Yukuo Ikoma (K)
Seiroku Inoue
Koji Ueda (K)
Shigeru Ueno (K)
Kiyomizu Oh-e (A)
Tasuke Okabe (K)
Teruo Ogawa (K)
Etsuo Okimura (K)
Kiyohei Kaneyama (K)
Mineo Kanzaki (K)
Tomio Kitaoka (K)
Masaru Kubo (K)
Taira Kubota (K)
Chosatsu Koga (K)
Shigeo Saito (K)
Tadashi Sakai (K)
Sekizen Shibayama
Yoshi-ichi Shima (K)
Iso-o Sugiura
Tsuneo Suzuki (K)
Hideo Suzuki (K)
Mitsugu Suzuki
Kazuo Sudo (K)
Kaoru Takaiwa (K)
Ken-ichi Takahashi
Harukuni Tanaka (K)
Tsumio Tanaka (K)

Isao Chosokabe (K)
Masaru Tsukiji (K)
Nobuyoshi Nakamura
Shigeru Nishiyama (K)
Yoshimoto Hattori (K)
Koji Fujishiro
Toshihiro Matsunaga (K)
Ichitaro Muramoto (K)
Keizo Yamaguchi
Yaozo Wada (K)
[Converted]
Suehiro Ikeda (K)
Isao Ito
Bu-ichi Kamo (K)
Soto-o Saito (K)
Hideshi Tanimoto (K)
Shinjiro Nakajima (K)
Takeshi Nishio

Ko 1st Class
(Sep.1937-Jun.1939 graduates of
Flight Trainee Course; carrier
planes, 76 Persons)
Junjiro Ito (K)
Tomotoshi Ishikawa (A)
Takeru Imahashi (K)
Toru Ohshima (K)
Keishu Kamihira
Koreo Kimura
Hiroshi Kurihara (K)
Yoshio Koike
Otojiro Sakaguchi
Yoshikane Sasaki (K)
Yoshihiko Takenaka (K)
Katsumi Tanaka (K)
Takio Dannoue (K)
Terusada Chuman (K)
Kazuyoshi Toyoda
Korenobu Nishide (K)
Yoshiaki Hatakeyama (K)
Naoyuki Hayashi (A)
Takashi Hirano (K)
Hideo Maeda (K)
Toshio Makinota
(Minamoto) (K)
Jiro Matsuda
Hisao Murabayashi
Koroku Yusaki (A)
Shigeru Yoshihashi (K)
[Converted]
Shinji Ishida (K)
Take-ichi Kikuchi (K)
Masaro Nagase (K)
Atsuki Mesaki (K)

Ko 2nd Class
(Apr.1938-Dec.1939 graduates of
Flight Trainee Course; carrier
planes, 81 persons)
Akira Atsumi (K)
Yoshio Iwashiro (K)
Shinji Iwama (K)

Tomio Kamei (K)
Ginji Kiyosue (K)
Yuji Sato (K)
Sei-ichi Tsukuda
Yoshikazu Nagahama (A)
Shigenori Nishikaichi (K)
Mitsuomi Noda (K)
Iwao Mita (K)
[Converted]
Masanobu Ibusuki
Kozo Hashimoto (K)
Masanobu Maehara (K)

Ko 3rd Class
(Oct.1938-Apr.1941 graduates of
1st Class, Flight Trainee Course;
carrier fighters, 21 persons)
Gisuke Arita (K)
Susumu Ishihara
Hideo Izumi (K)
Yutaka Ohtani (K)
Jin-ichiro Kawanishi (K)
Sakyo Kikuchi (K)
Tadayoshi Koga (K)
Tokusuke Konishi (A)
Ichiro Kobayashi (K)
Kazuo Kobayashi
Katsumi Kobayashi (K)
Matsutaro Kobayashi (K)
Hitoshi Sasaki
Tsugio Shikada
Taka-aki Shimotsuka (A)
Tamotsu Nakamura (A)
Yoshio Matsu-ura (A)
Takeo Miyazawa (K)
Yukuo Miyauchi (K)
Tsunehiro Yamagami (K)
Takuro Yoshi-e (K)
[Converted]
Yasunori Ono (K)
Masashi Shibata
Tadao Shiratori (K)
Eiji Matsuyama (K)

Ko 4th Class
(Apr.1939-Sep.1941 graduates of
9th Class, Flight Trainee Course;
carrier fighters, 21 persons)
Masao Aoki (K)
Hisashi Ichinose (K)
Tsutomu Ito (K)
Takeo Inoue (K)
Junzo Okutani (K)
Yozo Kawada (K)
Yasuzo Kimura (A)
Eikichi Kito (A)
Tadahiro Sakai (K)
Yoshimi Sakai (K)
Masao Sasakibara
Tomotsugu Sawada (K)
Toshikazu Tamura (K)
Yukihisa Tan (K)

Kunimatsu Nishi-ura (K)
Tamotsu Nishi-oka (A)
Hideyoshi Nomura (A)
Tei-ichiro Hayashida (K)
Susumu Matsuki (K)
Tatsuo Maruyama (K)
Un-ichi Miya (K)

Ko 5th Class
(Oct.1939;15th Class Flight
Trainee Course; Jan.1942 Oita
Kokutai graduates, 28 persons)
Shiro Ishikawa
Makoto Iwamoto (K)
Kazuo Umeki (K)
Masanori Eguchi (K)
Toshiharu Ohkubo (K)
Takeshi Okui (K)
Bungoro Kawamata (K)
Masa-aki Kanda
   (Ko 4th Advanced) (A)
Hachiro Kuwabara (K)
Itsuzo Shimizu (K)
Yutaka Shimizu (A)
Akira Imoto
Shinpei Sugiura (K)
Hiroshi Suzuki (K)
Mikiya Takada (K)
Shigeru Takahashi
Harumi Tomita (K)
Masashi Tomita (K)
Tadashi Nakamoto (K)
Mitsuo Nakayama
Fumi Nishiyama (K)
Yoshiro Nozu (K)
Yukio Hayasaka (K)
Tasuku Fukuyama (K)
Tadao Fujiwara (K)
Minoru Honda
Ya-ichi Yazawa (A)
Iwao Yamada (K)
Chi-une Yotsumoto (K)

Ko 5th Class
(Feb.1942 graduates of
Usa Kokutai, 12 persons)
Masayoshi Urano (K)
Chuji Sakurai (K)
Katsuzo Shimura (K)
Ryo-ichi Sugiura (K)
Gen-ichi Seki (K)
Joji Taniguchi (K)
Toyomitsu Tsujinoue (K)
Norimasa Narahara (K)
Takehiko Baba
Meiji Hikuma (K)
Nao-ichi Maeda (K)
Katsutoshi Maetsuji (K)

K6 6th Class
(Apr.1940-Jul.1942 graduates of
21st Class, Flight Trainee Course;

carrier fighters, 41 persons)
Yasuto Abe (K)
Jiro Ishikawa (K)
Hirotsugu Ishizaki (K)
Shigemi Izumi (K)
Koji Inamura (A)
Kesaji Iriki (K)
Yasushi Okazaki (K)
Yoneki Ochi (K)
Kaneyuki Kamikatahira (K)
Motomu Kawakami (K)
Ichiro Kawabata (K)
Masatsugu Kawamura (K)
Shigeharu Kido (K)
Haruo Kunihiro (K)
Kiyoshi Kojima (K)
Kurakazu Goto (K)
Isao Kondo (K)
Kunio Sakai (K)
Moriji Sako (H)
Soyo Shibata (K)
Katsumi Shimamoto (K)
Toshihisa Shirakawa (K)
Tomio Shiraki (K)
Ei-ichi Sugiyama (K)
Shinpei Takagaki (K)
Izumi Tanaka (K)
Hachiro Tsuboya (K)
Yoshio Terai (K)
Tokihiro Tokuoka (K)
Haruo Tomita (K)
Takashi Nakamichi (A)
Tsuneishi Nakamura (K)
Tetsuro Nihei (K)
Rokuro Fukuda (K)
Takeshi Fuji-i (K)
Ko-ichi Hoshino (K)
Takeo Matsubayashi (K)
Takeo Maruyama (K)
Toshitsune Misawa (K)
Keizaburo Yamasaki (K)
[Converted]
Kiyoshi Akizuki
Mamoru Irio
Katsushi Kurita (K)
Masakazu Suzuki (K)
Misao Cho (K)
Hideji Furu-uchi (K)

Ko 7th Class
(Oct.1940-Nov.1942 graduates of
24th Class, Flight Trainee Course;
carrier fighters, 41 persons)
Shizuo Aoki (A)
Koshiro Agawa (K)
Yonehachiro Ishikura (K)
Shigeru Ishizuka (K)
Yasuo Iguchi (K)
Ichiji Ikeda (K)
Kikuo Ikeda (K)
Osamu Ichikawa (K)
Hiroshi Iwano (K)

Tomimasa Ohkubo (K)
Fumi-o Ohsumi (K)
Ryozo Okada (K)
Tadashi Okamitsu (K)
Sho-ichi Kai (K)
Fusao Kinoshita (K)
Nobuo Konishi (K)
Shigemitsu Kozuma (K)
Misao Sakanoue (A)
Sanpei Shiono
Munenori Shimizu (K)
Ikuzo Shimizu (K)
Toshitaro Sekiguchi (K)
Miyoshi Tanaka (K)

Michio Takeshita (K)
Gihachiro Taniguchi (A)
Yoshinobu Tsurusaki (K)
Yoshio Nakajima (K)
Hiroshi Nagakura (K)
Hitoshi Nagano (K)
Mitsuru Hama (A)
Katsujiro Matsumoto (K)
Yoshiyuki Miura
Jiro Murata
Kazuharu Morimoto (K)
Shigeaki Morita (A)
Ichiro Yamashita (K)
Ken-ichi Yamamoto (A)

Shigetaro Yamamoto (K)
Takashi Yoshi-i (K)
Saburo Yoshida (K)
Kenzo Yonemoto (A)
Takeru Wada (K)
[Converted]
Shiro Endo (K)
Yu-ichi Okada (K)
Ichizen Kobayashi (K)
Masuo Doi
Sanemasa Nanjo (K)
Masayuki Hanamura (K)
Kenjiro Honma (K)
Hidenori Matsunaga

AVIATION RESERVE STUDENTS

The Aviation Reserve Students (Koku Yobi Gakusei) programme was begun in November 1934, built around the Oceanic Division of the Japanese Student Aviation League (Nihon Gakusei Koku Remmei Kaiyo-bu), which was itself renamed the Student Oceanic Aviation Group (Gakusei Kaiyo Hiko-dan), and then the Naval Reserve Aviation Group (Kaigun Yobi Koku-dan). Trainees were generally selected from amongst graduates of universities and colleges; until 1942, the majority were from the oceanographic discipline. Following selection, Aviation Reserve Students were given about two months of general instruction, followed by ten months of pilot training. On completion of this training, they were commissioned as Ensigns; (after 1942 a proportion of these reserve students were integrated into regular active duty service).

During the first three classes, students specializing in land-based aircraft were given additional training on carrier attack aircraft (torpedo-bombers). Those specializing on floatplanes were trained to fly the observation floatplanes. However, from the 4th Class onwards, training was undertaken for the types of aircraft employed by the speciality involved.

Yo (Reserve) - 1st Class
(Nov.1934-Nov.1935; all types of aircraft, 5 persons)
[Converted]
Isamu Matsubara

Yo 2nd Class
(May 1935-Apr.1936; total of all aircraft types, 14 persons, no fighters)

Yo 3rd Class
(Apr.1936-Mar.1937; total of all aircraft types,17persons; no fighters)

Yo 4th Class
(Apr.1937-Jan.1938; total of all types, 12 persons; fighters 2 persons)
Yoshio Murata (K)
Ko-ichi Yoshida (K)
[Converted]
Naonori Yoshida (K)

Yo 5th Class
(Apr.1938-Mar.1939; total of all types of aircraft, 19 persons; no fighters)

Yo 6th Class
(Apr.1939-Apr.1940; total of all types of aircraft, 26 persons; no fighters)

Yo 7th Class
(Apr.1940-Apr.1941; total of all types of aircraft, 32 persons; fighters, 2 persons)
Masahiro Chikanami (K)
Takeshi Morisaki (K)
  [Converted]
Teiji Kagami (K)

Yo 8th Class
(Apr.1941-Apr.1942; total of all types of aircraft, 43 persons)
Mitsuoki Asano (K)

Masao Kuramoto (K)
Akira Sugiura (K)
Akira Tanaka (K)
Ya-ichiro Fukunishi (K)
Shigeo Morimoto (A)
[Converted]
Takeo Hirose

Yo 9th Class
(Jan.1942-Jan.1943; total of all types of aircraft, 34 persons)
Yoshio Aizawa (K)
Mitsuru Ohnuma (K)
Genshichiro Ohyama (K)
Den Katayama (K)
Kenjo Kusaka (A)
Hiroshi Suzuki (K)
She-ichiro Yamada (K)
[Converted]
Yozo Tsuboi (K)
Chobei Morita (K)

Yo 10th Class
(Jan.1942-Jan.1943; total of all types of aircraft, 48 persons; carrier fighters, 13 persons)
Izumi Ashida
Suzuo Ito (K)
Toshio Ushikubo (K)
Akira Kasahara (A)
Kazuo Kayaki
Kazuaki Kinoshita (K)
Masayuki Goto (K)
Kinshi Shimizu (K)
Sanenori Takamatsu (A)
Minoru Tanaka (K)
Kagemitsu Matsuo (K)
Hidezo Moriyama (K)
Suke-ichi Yamashita (K)
[Converted]
Tei-ichi Kato (K)

Yo 11th Class
(Sep.1942-Nov.1943; total of all types of aircraft, 85 persons, carrier fighters, 37 persons)
Yasuhide Aoki (K)
Yoshihiro Aoki
Chisato Akiyama (K)
Hayao Ishibashi (K)
Isaburo Inoue (K)
Takeshi Inoue (K)
Kazuo Uji-ie (A)
Takao Okakura (K)
Katsuzo Kajikawa (K)
Seijo Boji (K)
Masao Katsuta (K)
Takeo Kawaguchi (K)
Hikomori Kimiyama (A)
Yoshitaka Kuno (K)
Masateru Kurokawa (K)
Hiroshi Kojima (A)
Yataro Koizumi (K)

Shogo Kobayashi (K)
Mitsuro Sakamoto (K)
Noriyasu Sato
Masao Shibuya (K)
Yoshito Shimada (K)
Tatsu Nagato (K)
Teruhisa Hatai
Kikumasa Fujita (K)
Yoshishi Busujima (K)
Yoshio Hotta (K)
Noboru Matsu-ura (K)
Takashi Matsumoto (K)
Naoyasu Matsudaira
Mitsuo Yatomi (A)
Tetsuro Yano (K)
Keizo Yamakawa (K)
Takashi Yamazaki (A)
Tatsuo Yui (A)
Sadakatsu Yuchi (K)
Masatsugu Yoshitomi (K)

RELATIONSHIP BETWEEN THE FLIGHT RESERVE ENLISTED TRAINEE CLASS SYSTEM

Flight Reserve Enlisted Trainee Programme Classes			Flight Trainee Course		Remarks
Class	Graduates	Period	Class	Period	
Ko 3 Pilot	253	Oct 1938-Apr 1940	1	Apr 1940-Apr 1941	
Ko 3 Recon					
			2		Only reconnaissance studied
Pilot 54	83		3	Mar 1940-May 1941	
			4		Only reconnaissance studied
			5	Aug 1940-Jun 1941	Only communications studied
Pilot 55			6	Jun 1940-Jul 1941	
Pilot 56			7	Jun 1940-Jul 1941	
			8	Aug 1940-Jul 1941	
Ko 4 Pilot	258	Apr 1939-Sep 1940	9	Sep 1940-Sep 1941	
Ko 4 Recon					
Otsu 9 Pilot	(200)	Jun 1938-Nov 1940	10	Nov 1940-Oct 1941	
Otsu 9 Recon					
Hei 1 Pilot	31	Aug 1940-Nov 1940	11	Nov 1940-Oct 1941	
Pilot 57					Floatplanes only
Hei 2 Pilot	225	Nov 1940-Jan 1941	12	Jan 1941-Nov 1941	
	117		13	Mar 1941-Jan 1942	Only reconnaissance studied
	73		14	Jan 1941-Dec 1941	Only communications studied
Ko 5 Pilot	252	Oct 1939-Mar 1941	15	Mar 1941-Feb 1942	
Ko 5 Recon					
Otsu 10 Pilot	(233)	Nov 1938-May 1941	16	May 1941-Mar 1942	
Otsu 10 Recon					

Class	Number	Date	No.	Date	Notes
			17	Apr 1941-Mar 1942	
Hei 3 Pilot	317	Feb 1941-Apr 1941	18	Jul 1941-May 1942	
				Jul 1941-Jul 1942	
Hei 3 Recon	157	Feb 1941-Apr 1941	17	Apr 1941-Aug 1941	
	131		19	Jul 1941-Apr 1942	Only communications studied
	127		20	Aug 1941-May 1942	Only communications studied
			18	Jul 1931-May 1942	
Hei 4 Pilot	217	May 1941-Jul 1941	21	Sep 1941-Jul 1942	
			22	Mar 1942-Jul 1942	Larger aircraft only
Hei 4 Recon	198	May 1941-Jul 1941	18	Jul 1941-Oct 1941	
Ko 6 Pilot	262	Apr 1940-Sep 1941	21	Sep 1941-Jul 1942	
Ko 6 Recon			21	Sep 1941-Aug 1942	
Otsu 11 Pilot	386	Jun 1939-Sep 1941	21	Sep 1941-Jul 1942	
			23	Nov 1941-Sep 1942	
Otsu 11 Recon					
Hei 5 Recon	201	Jun 1941-Aug 1941	21	Sep 1941-Dec 1941	Only reconnaissance studied
Hei 6 Pilot	262	Aug 1941-Oct 1941	23	Nov 1941-Sep 1942	
Hei 6 Recon	170		23	Nov 1941-Jul 1942	
Ko 7 Pilot	312	Oct 1940-Jan 1942	24	Jan 1942-Nov 1942	
Ko 7 Recon					
Hei 7 Pilot	259	Oct 1941-Jan 1942	24	Jan 1942-Nov 1942	
Hei 7 Recon	170		24	Jan 1942-Sep 1942	
		Dec 1941-Jan 1942	24	Jan 1942-Nov 1942	
Hei 8 Pilot	117	Dec 1941-Mar 1942	25	Mar 1942-Jan 1943	Only floatplanes
Hei 8 Recon	140	Dec 1941-Mar 1942	25	Apr 1942-Sep 1942	Only reconnaissance studied
Hei 9 Recon	209	Dec 1941-Mar 1942	25	Mar 1942-Sep 1942	Only reconnaissance studied
Otsu 12 Pilot	(365)	Nov 1939-Mar 1942	25	Mar 1942-Jan 1943	
Otsu 12 Recon					
Hei 10 Pilot	188	Feb 1942-Mar 1942	25	Mar 1942-Jan 1943	
			26	May 1942-Mar 1943	
Hei 10 Recon			26	May 1942-Nov 1942	
Otsu 13 Pilot	(294)	Jun 1940-May 1942	26	May 1942-Mar 1943	
Otsu 13 Recon					
Otsu 14 Pilot	(325)	Aug 1940-Jul 1942	27	Jul 1942-May 1943	
Otsu 14 Recon					
Hei 11 Pilot	385	May 1942-Jul 1942	27	Jul 1942-May 1943	
Hei 11 Recon			27	Jul 1942-Jan 1943	
Special Hei 11 Pilot	186	Aug 1942-Sep 1942	28	Sep 1942-Sep 1943	Larger aircraft only
Ko 8 Pilot	(455)	Apr 1941-Sep 1942	28	Sep 1942-Jul 1943	
Ko 8 Recon					
Hei 12 Pilot	291	Aug 1942-Sep 1942	28	Sep 1942-Jul 1943	
Hei 12 Recon	177		28	Sep 1942-Mar 1943	
Hei 13 Pilot	158	Sep 1942-Nov 1942	29	Nov 1942-Sep 1943	
Hei 13 Recon	165		29	Nov 1942-May 1943	
Otsu 15 Pilot	(630)	Dec 1940-Nov 1942	29	Nov 1942-Sep 1943	
Otsu 15 Recon			29	Nov 1942-May 1943	
Ko 9 Pilot	790	Oct 1941-Jan 1943	30	Jan 1943-Nov 1943	
Ko 9 Recon			30	Jan 1943-Sep 1943	

Hei 14 Pilot	238	Nov 1942-Jan 1943	30	Jan 1943-Nov 1943	
Hei 14 Recon	159		30	Jan 1943-Jul 1943	
Special Hei 14 Pilot	144	Dec 1942-Mar 1943	31	Mar 1943-Oct 1943	
Hei 15 Pilot	142	Nov 1942-Mar 1943	31	Mar 1943-Jan 1944	
Hei 15 Recon	140		31	Mar 1943-Sep 1943	
Ko 10 Pilot	(1,097)	Apr 1942-May 1943	32	May 1943-Nov 1943	
Ko 10 Recon			33	Jul 1943-Feb 1944	
Otsu 16 Pilot	(1,237)	May 1941-May 1943	32	May 1943-Mar 1944	
Otsu 16 Recon			33	Jul 1943-Mar 1944	
Hei 16 Pilot	141	Jan 1943-May 1943	32	May 1943-Mar 1944	
	118	Feb 1943-May 1943	32	May 1943-Mar 1944	
Hei 16 Recon	133	Jan 1943-May 1943	32	May 1943-Nov 1943	
Hei 17 Pilot	438	Mar 1943-Jul 1943	33	Jul 1943-Mar 1944	Final Hei flight training course
Special Otsu 1 Pilot	(1,585)	Apr 1943-Sep 1943	34	Sep 1943-Jul 1944	
Special Otsu 1 Recon			34	Sep 1943-Feb 1944	
Special Otsu 2 Pilot	(625)	Jun 1943-Nov 1943	35	Nov 1943-Aug 1944	
Special Otsu 2 Recon			35	Nov 1943-Mar 1944	
Ko 11 Pilot	(1,185)	Oct 1942-Nov 1943	36	Nov 1943-Jul 1944	
Ko 11 Recon		Oct 1942-Jan 1944			
		Apr 1943-Mar 1944			
Ko 12 Pilot			37	Mar 1944-Sep 1944	
	(3,242)	Jun 1943-Mar 1944			
Ko 12 Recon	Aug 1943		37	Mar 1944-Sep 1944	
Otsu 17	(1,209)	Dec 1941-Feb 1944	37	Feb 1944-Sep 1944	
Special Otsu 3 Pilot	(526)	Aug 1943-Mar 1944	37	Mar 1943-Sep 1944	
Special Otsu 3 Recon			37	Mar1944-Sep 1944	
				May 1944-Dec 1944	
Otsu 18	(1,480)	May 1942-Mar 1944	38	Jun 1944-	
		Oct 1943-Jul 1944		May 1944-Dec 1944	
Ko 12	(28,111)		38-42		
	Dec 1943-Sep 1944			Feb 1945-	
Special Otsu 4 Pilot	(787)	Oct 1943-May 1944	38	May 1944-Dec 1944	
Special Otsu 4 Recon			38	May 1944-Feb 1945	
Special Otsu 5 Pilot	(938)	Dec 1943-Dec 1944			
Special Otsu 5 Recon			41	Nov 1944-Feb 1945 (Mar 1945)	
Otsu 19	1,500	Dec 1942-Jan 1945	42	Jan 1945-Mar 1945	Flight training Feb 1945 Course closed down
Ko 14	(52,115)	Apr 1944-Mar 1945 Jun 1944			

Claims and Losses during Major Air Battles

This table seeks to summarize data regarding the major aerial engagements in which fighter units of the Imperial Japanese Navy took part during the China Incident and the Pacific War. The numbers of aircraft involved and the totals of claims listed have been extracted from the IJN official records for each such engagement. In order to expose the differences between the claims and losses recorded by the Japanese units and the corresponding figures (both claims and losses) noted by their opponents, the latter figures have been drawn from US, British, Australian, New Zealand, Dutch and Chinese sources where available. Given the circumstances of the fighting in China and the often chaotic circumstances under which the Western Allies were operating during the opening weeks of the Pacific War, the latter figures must, on occasion, be considered as incomplete, and at times, speculative.

For convenience, the types of aircraft involved in each action are identified by the US Navy abbreviations, viz:-
VT Torpedo-bomber (kanko)
VB Dive-bomber (kanbaku)
VF Fighter (sentoki)
VBF Fighter-bomber (bakusen, senbaku)
VA Twin-engined, land-based torpedo-bomber (Rikko)
VA Twin-engined land-based bomber (Ginga)
VOS Reconnaissance floatplane (suitei)

Parentheses indicate 'Probables'
Dest.grnd indicates aircraft destroyed and/or set on fire on the ground

Air Battle	Date	Participating Forces	IJN Claims	IJN Losses	Opposing Records
Attack on Nanking	19 Sep 37	18 VOS, 17 VB, 12 VF	3(6) 3	3 VB, 1 VOS	11 lost; 1 victory
Attack on Canton	21 Sep 37	24 VF, 6 VT, 24 VB	16(1)	5 VF ditched	11 losses
Attack on Nanking	12 Oct 37	11 VF, 9 VA	5; 2 dest.grnd	3 VF	
Attack on Nanking	2 Dec 37	6 VF, 8 VT	13(3)	None	10 losses
Attack on Nanchang	9 Dec 37	8 VF, 15 VA	2; 12 dest.grnd	1 VF	
Attack on Nanchang	22 Dec 37	12 VF, 11 VA	17(4); 13 dest.grnd	1 VF	
Attack on Hankow	18 Feb 38	11 VF, 15 VA	18(1)	4 VF	5 losses, 14 victories
Attack on Nanchang	5 Feb 38	18 VF, 35 VA	42(13)	2 VF	6 losses, 3 victories
Attack on Canton	13 Apr 38	6 VF, 18 VB	15(3)	3 VF	
Attack on Hankow	29 Apr 38	28 VF, 18 VA	51(11)	2 VF, 2 VA	13 losses, 24 victories

430

Attack on Hankow	31 May 38	35 VF, 18 VA	20(2)	1 VF	
Attack on Nanchang	26 Jun 38	28 VF, 18 VA	19(4); 2 dest grnd	None	
Attack on Nanchang	4 Jul 38	23 VF, 26 VA	45(5); 9 dest,grnd	2 VF	
Attack on Nanchang	16 Jul 38	15 VF, 18 VA	10; 3 dest,grnd	None	
Attack on Nanchang	18 Jul 38	6 VF, 14 VB, 5 VT	9(2); 19 dest.grnd	1 VF	
Attack on Hankow	3 Aug 38	29 VF, 18 VA	32(5); 7 dest,grnd	3 VF	
Attack on Nan-yung	Aug 38	6 VF, 5 VB, 4 VT	17	2 VF	
Attack on Liuchow	30 Dec 39	13 VF	14	1 VF	
Attack on Kweilin (Guilin)	10 Jan 40	26 VF, 27 VA	16; 9 dest.grnd	None	
Attack on Chungking	13 Sep 40	13 VF, 27 VA	27	None	13 losses
Attack on Chengtu	4 Oct 40	8 VF, 27 VA	6; 19 dest.grnd	None	
Attack on Kunming	7 Oct 40	7 VF, 27 VA	14; 4 dest.grnd	None	
Attack on Chengtu	26 Oct 40	8 VF	10	None	
Attack on Chengtu	14 Mar 41	12 VF, 10 VT	27(3); 7 dest.grnd	None	16 losses, 6 victories
Attack on Hawaii	7 Dec 41	78 VF, 129 VB, 143 VT	14; 450 dest.grnd	9 VF, 15 VB, 5 VT	Total 347 losses
Attack on Luzon	8 Dec 41	89 VF, 108 VA	25(2); 80 det.grnd	7 VF	5-23 losses + 72-85 dest.grd
Attack on Luzon	10 Dec 41	56 VF, 81 VA	43; 61dest.grnd	2 VF	13 losses
Attack on Singapore	18 Dec 41	11 VF, 26 VA	15(5)	2 VF	
Attack on Singapore	22 Jan 42	9 VF, 52 VA	12	2 VF	
Attack on Darwin	29 Jan 42	36 VF, 71 V13, 81 VT,54 VA	8	1 VF, 1 VB	4 losses; 14 dest.grnd
Marshall interceptions	1 Feb 42	Total 34 VF	17(3)	None(1 VF)	5 victories, 10 losses
Attack on Surabaya	3 Feb 42	58 VF, 37 VA	48; 36 det.grnd	3 VF	20 losses
Attack on Batavia	9 Feb 42	15 VF, 17 VA	12		
Attack on Surabaya	18 Feb 42	8 VF, 23 VA	9(3)		
Attack on Surabaya	19 Feb 42	23 VF, 18 VA	17(3)		
Attack on Colombo	5 Apr 42	36 VF, 38 VB, 53 VT	55(10)	1 VF, 6 VB	21 victories, 25 losses
Attack on Trincomalee	9 Apr 42	38 VF, 91 VT	42(3); 4 dest.grnd	3 VF, 1 VT	15 victories, 14 losses
Attack on Darwin	27 Apr 42	21 VF		19(6)	7 victories, 4 losses
Battle of Coral Sea (attack on US task force)	8 May 42	18 VF, 33 VB, 18 VT	64(5)	2 VF, 9 VB, 8 VT	33 losses; 36 sunk with ships
Battle of Coral Sea (CAP)	"	19 VF	40(7)		
Attack on Port Moresby	28 May 42	26 VF	13(4)	None	
Attack on Midway	4 Jun 42	36 VF, 36 VB, 36 VT	45	2 VF, 1 VB, 1 VT	17 losses
Battle of Midway	"	Total 120 VF	90	13 VF	76 losses
Attack on Guadalcanal	7 Aug 42	17 VF, 9 VB, 27 VA	48(8)	2 VF, 9 VB, 5 VA	11 losses
Attack on Guadalcanal	24 Aug 42	15 VF, 6 VT		2 VF, 3 VT	16 victories, 3 losses
Second Battle of the Solomons (attack on task force)	"	19 VF, 54 VB, 12 VT	10(1)	3 VF, 21 VB	90 victories
Attack on Guadalcanal	30 Aug 42	18 VF	10(4)	7VF	14 victories, 4 losses
Attack on Guadalcanal	13 Sep 42	21 VF, 26 VA	17(2)	4 VF	11 victories, 5 losses
Patrol over convoy	15 Oct 42	Total 52 VF	32(9)	2 VF	8 victories, 7 losses
Attack on Guadalcanal	25 Oct 42	Total 40 VF	10	6 VF	22 victories
Battle of Santa Cruz (attack on US Fleet)	26 Oct 42	47 VF, 59 VB, Total 49	VT 55	6 VF, 31 VB, 21 VT	74 losses
Battle of Santa Cruz (over own Fleet)	"	Total 51 VF	5 VF		
Attack on Guadalcanal	11 Nov 42	44 VF, 9 VB, 25 VA	25(5)	2 VF, 5 VB, 4 VA	11 victories, 7 losses

Attack on Guadalcanal	12 Nov 42	30 VF, 19 VA	20(7)	12 VA	5 victories, 4 losses
Patrol over convoy	14 Nov 42	20 VF	15(7)	7 VF	9 losses
Attack on Isabel	1 Feb 43	81 VF, 15 VB	28	4 VF, 5 VB	21 victories, 8 losses
Patrol over convoy	4 Feb 43	22 VF (+Army)	17	3 VF	17 victories, 10 losses
Interception over Buin	14 Feb 43	27 VF	19	1 VF	5 victories, 10 losses
Attack over Rabaul	1 Apr 43	57 VF	47(7)	9 VF	18 victories, 6 losses
Attack on Guadalcanal	7 Apr 43	157 VF, 67 VB	41(13)	12 VF, 9 VB	39 victories, 7 losses
Attack on Oro Bay	11 Apr 43	71 VF, 21 VB	21(9)	2 VF, 4 VB	17 victories
Attack on Port Moresby	12 Apr 43	124 VF, 43 VA	28(7)	2 VF, 6 VA	22 victories, 2 losses, 15 dest.grnd
Attack on Milne Bay	14 Apr 43	129 VF, 75 VB, 44 VA	45(9)	1 VF, 3 VB, 3 VA	5 victories, 3 losses
Escort of *Yamamoto*	18 Apr 43	6 VF, 2 VA	6(3)	2 VF	5 victories, 1 loss
Attack on Darwin	2 May 43	27 VF, 18 VA	21(4)	None	6 victories, 13 losses
Drive on Russell	13 May 43	54 VF	38(10)	4 VF	16 victories, 3 losses
Attack on Oro Bay	14 May 43	33 VF, 18 VA	15	4 VA	16 victories, 2 losses
Attack on Russell	7 Jun 43	81 VF	41(8)	9 VF	23 victories, 9 losses
Attack on Russell	12 Jun 43	74 VF	33(8)	6 VF	31 victories, 6 losses
Attack on Guadalcanal	16 Jun 43	70 VF, 234 VB	32(5)	15 VF, 13 VB	66 victories, 6 losses
Attack on Rendova	30 Jun 43	72 VF, 8 VB, 26 VA	49(8)	13 VF, 17 VA	58 victories, 17 losses
Attack on Brocks Creek	"	27 VF, 22 VA	16(3)	None	8 victories, 6 losses
Attack on Brocks Creek	6 Jul 43	27 VF, 22 VA	17(3)	2 VA	7 victories, 8 losses
Attack on Rubiana	15 Jul 43	44 VF, 8 VA	19(12)	5 VF, 5 VA	44 victories, 3 losses
Interception over Buin	17 Jul 43	46 VF	58(13)	9 VF	48 victories, 6 losses
Interception over Buin	12 Aug 43	45 VF	33	1 VF; 24 dest.grnd	
Attack on Vella Lavella	15 Aug 43	149 VF, 25 VB, 11 VT, 24 VA	29(9)	9 VF, 8 VB	44 victories
Attack on Brocks Creek	7 Sep 43	36 VF	18(3)	1 VF	7 victories, 5 losses
Interception over Buin	14 Sep 43	Total 258 VF	60(6) 5 VF; 9 dest.grnd		
Attack on Cape Cretin	22 Sep 43	35 VF, 8 VA	14	8 VF, 6 VA	39 victories, 3 losses
Interception over Wake	6 Oct 43	23 VF	10	14 VF	22 victories, 6 losses
Interception over Rabaul	12 Oct 43	34 VF	9	2 VF; 12 dest.grnd	26 victories; 100 dest. Grd.; 5 losses
Attack on Buna	15 Oct 43	39 VF, 15 VB	14	5 VF, 14 VB	47 victories, 1 loss
Attack on Cape Torokina	1 Nov 43	79 VF, 16 VB, 9 VT	3	23 VT, 5 VB, 3 VT	22 victories, 4 losses
Interception over Rabaul	2 Nov 43	115 VF	119(22)	18 VF	68 victories, 19 losses
Interception over Rabaul	5 Nov 43	59 VF	49(20)	2 VF 13 losses	25 victories,
Interception over Rabaul	7 Nov 43	58 VF	16	?	23 victories, 16 dest. Grnd; 5 losses
Interception over Rabaul	11 Nov 43	68 VF	71	11 VF	135 victories, 7 losses
Attack on Calcutta	5 Dec 43	17 VF, 9 VA	6(2)	None	
Interception over Roi	"	53 VF	24(6)	16 VF; 15 dest.grnd	28 victories, 4 losses
Interception over Rabaul	23 Dec 43	99 VF	24	6 VF	30 victories, 3 losses
Attack on Cape Marcus	26 Dec 43	63 VF, 25 VB	20(5)	4 VF, 13 VB	75 victories, 7 losses

Interception over Rabaul	9 Jan 44	72 VF	33	2 VF	21 victories, 5 losses
Interception over Rabaul	17 Jan 44	79 VF	87	?	17 victories, 12 losses
Interception over Rabaul	13 Feb 44	40 VF	23(7)	2 VF	
Interception over Truk	17 Feb 44	Total 64 VF	31	31 VF, 81 dest.grnd	56 victories, 19 losses
Interception over Rabaul	19 Feb 44	36 VF	37(6)	6 VF	23 victories
Interception over Marianas	23 Feb 44	20 VF		11 VF, 30 dest.grnd	60 victories, 6 losses
Interception over Palau	30 Mar 44	22 VF	19(3)	14 VF	70 victories
Interception over Palau	31 Mar 44	66 VF	40(3)	35 VF	25 losses
Attack on Nanning	5 Apr 44	32 VF	9(2)	9 VF	9 victories, 1 loss
Interception over Truk	30 Apr 44	54 VF	32(2)	28 VF	19 victories, 34 dest. grnd; 35 losses
Interception over Marianas	11 Jun 44	Total 139 VF	9	22 VF	11 victories, 70 dest. grnd
Interception over Iwo Jima	15 Jun 44	37 VF	4	16 VF	20 victories, 7 dest. grnd; 2 losses
Attack off Saipan	"	11 VF, 3 VB, 10 VA, 8 VT	2	1 VF, 1 VB	
Attack off Saipan	18 Jun 44	36 VF, 24 VFB, 2 VB, 8 VA	5	13 VF, 1 VFB, 7 VA	
Battle of Philippine Sea (attack on task force)	19 Jun 44	107 VF, 80 VFB, 89 VB, 50 VT	24(8)	60 VF, 45 VFB, 56 VB, 31 VT	269 victories, 29 losses, 73 sunk
Battle of Philippine Sea (combat air patrol)	20 Jun 44	34 VF	26(3)	12 VF	
Interception over Guam	19 Jun 44	56 VF	26(1)	12 VF	
Interception over Iwo Jima	24 Jun 44	59 VF	37(10)	23 VF	29 victories, 6 losses
Interception over Iwo Jima	3 Jul 44	Total 110 VF	39(3)	31 VF	
Interception over Iwo Jima	4 Jul 44	Total 45 VF	26(1)	12 VF	70 victories
Interception over central Philippines	12 Sep 44	41 VF	23(3)	27 VF, 30 dest.grd	45 victories, 36 dest.grnd; 9 losses
Interception over central Philippines	13 Sep 44	24 VF	7	9 VF	
Interception over Manila	21 Sep 44	42 VF	27	20 VF, 10 dest.grnd	38 victories, 10 dest. grnd; 9 losses
Interception over Formosa	12-13 Oct 44		112	Total 312	500 victories and dest.grnd
Attack off Formosa	12-16 Oct 44	Total 650 aircraft			
Interception over Balikpapan	14 Oct 44		38		43 victories, 7 losses
Attack on US Fleet	24 Oct 44	Total 199(126 VF)	32	Total of 67	
Air battle off Philippines (Cape Engano)	"	30 VF, 19 VFB, 2 VB, 5 VT	8(1)		
Attack on US Fleet	25 Oct 44	18 VF, 28 VB	7	Total of 10	
Interception over Luzon	29 Oct 44	Total 142 VF	40	21 VF, 8 dest.grnd	71 victories, 11 losses
Interception over Luzon	5 Nov 44	Total 82 VF	45	32 VF, 33 dest.grnd	25 losses

Interception over Kanto	16-17 Feb 45		98	30 MiA	332 victories, 177 dest. grnd; 39 losses
Interception over Kanto	25 Feb 45		10		25 victories, 30 dest. grnd; 9 losses
Interception over Matsuyama	19 Mar 45	70 VF	52	16 VF
Attack off *Shikoku*	21 Mar 45	30 VF, 18 VA	7(3)	10 VF, 18 VA
Okinawa Tokko escort	3 Apr 45	40 VF	17(6)	8 VF
Okinawa Tokko escort	12 Apr 45	75 VF	25	20 VF
Okinawa Tokko escort	16 Apr 45	93 VF	23(1)	17 VF
Interception over Bungo Strait	24 Jul 45	31 VF	19	6 VF

Stackpole Military History Series

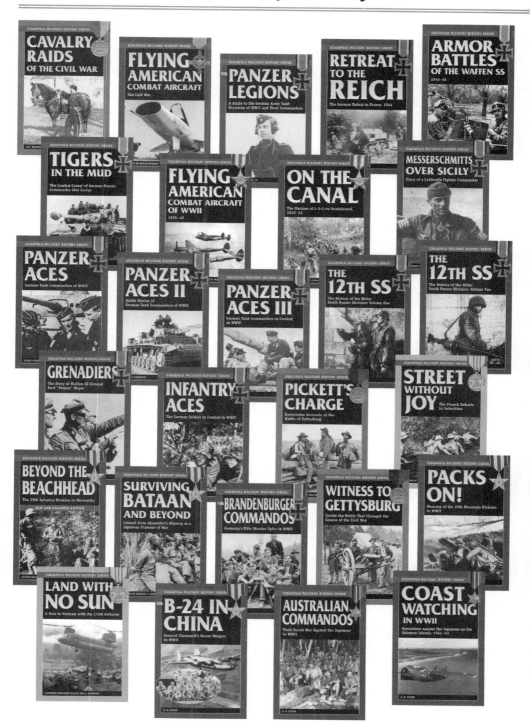

Real battles. Real soldiers. Real stories.

Stackpole Military History Series

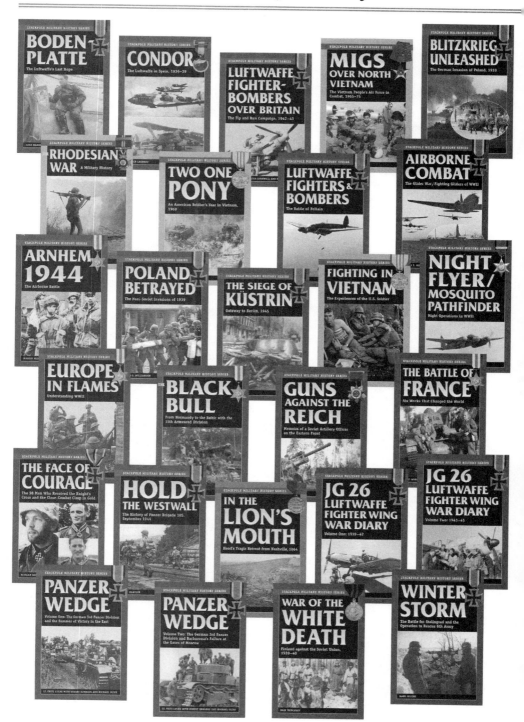

Real battles. Real soldiers. Real stories.

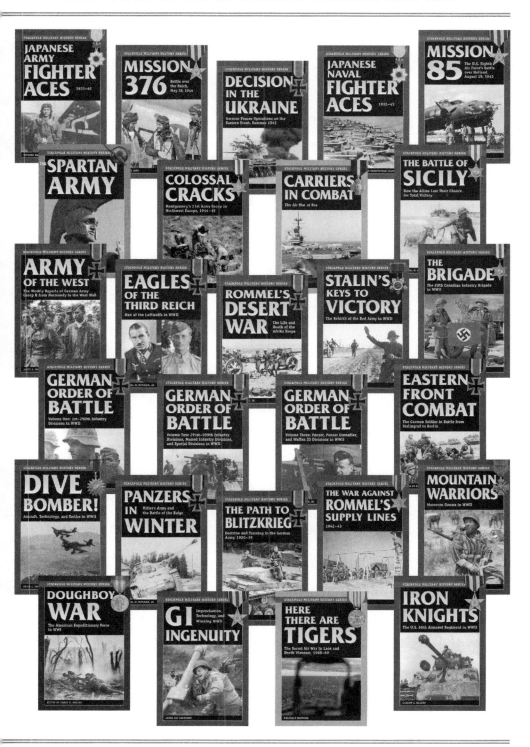

Stackpole Military History Series

Real battles. Real soldiers. Real stories.

Stackpole Military History Series

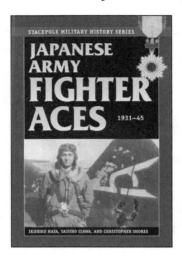

JAPANESE ARMY FIGHTER ACES
1931–45
Ikuhiko Hata, Yasuho Izawa, and Christopher Shores

Beginning in Manchuria in 1931 and China in 1937 and
intensifying after World War II broke out in the Pacific in
1941, the pilots of the Imperial Japanese Army Air Force flew
countless combat sorties for their emperor in Nakajima Ki-43,
Kawasaki Ki-61, and other fighters. Many became aces and
many more lost their lives against the Americans, British,
Chinese, and Soviets in places like Nomonhan, Burma,
Indochina, New Guinea, the Philippines, and Okinawa—and
in desperate missions against B-29 bombers above Japan itself.
This fully illustrated book intertwines the stories of these
fighter aces, their aircraft, and their battles.

Paperback • 6 x 9 • 368 pages • 380 b/w photos

WWW.STACKPOLEBOOKS.COM
1-800-732-3669

Stackpole Military History Series

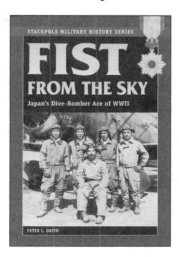

FIST FROM THE SKY
JAPAN'S DIVE-BOMBER ACE OF WORLD WAR II
Peter C. Smith

Known as the "god of dive-bombing," Takashige
Egusa was one of the Imperial Japanese Navy's most
legendary pilots of World War II. In December 1941
he led an eighty-plane attack force against Pearl
Harbor and, two weeks later, assaulted the American
gun batteries on Wake Island. After a series of missions
in the Indian Ocean, Egusa was badly burned at
Midway in 1942. A warrior to the last, he returned
to duty, only to be killed on a desperate raid on
enemy aircraft carriers in the Marianas.

Paperback • 6 x 9 • 192 pages • 70 b/w photos

Stackpole Military History Series

BRINGING THE THUNDER
THE MISSIONS OF A WWII B-29 PILOT IN THE PACIFIC
Gordon Bennett Robertson, Jr.

By March 1945, when Ben Robertson took to the skies above Japan in his B-29 Superfortress, the end of World War II in the Pacific seemed imminent. But although American forces were closing in on its home islands, Japan refused to surrender, and American B-29s were tasked with hammering Japan to its knees with devastating bomb runs. That meant flying low-altitude, nighttime incendiary raids under threat of flak, enemy fighters, mechanical malfunction, and fatigue. It may have been the beginning of the end, but just how soon the end would come—and whether Robertson and his crew would make it home—was far from certain.

Paperback • 6 x 9 • 304 pages • 36 b/w photos, 1 map

WWW.STACKPOLEBOOKS.COM
1-800-732-3669

Stackpole Military History Series

MISSION 376
BATTLE OVER THE REICH, MAY 28, 1944
Ivo de Jong

Some of the U.S. Eighth Air Force's bombing missions of
World War II, such as the raid on the ball-bearing factories at
Schweinfurt, became legendary. Many others did not, but
these more routine missions formed an important part of
Allied strategy. One of them was Mission 376 on May 28, 1944,
when more than 1,200 American B-17s and B-24s took off
from bases in England and headed for targets inside Germany,
where Luftwaffe fighters scrambled to beat them back. With
unprecedented and enthralling detail, this book describes an
"ordinary" bombing mission during World War II.

Paperback • 6 x 9 • 448 pages • 329 b/w photos

WWW.STACKPOLEBOOKS.COM
1-800-732-3669

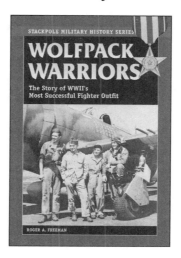

Stackpole Military History Series

THE SAVAGE SKY
LIFE AND DEATH ON A BOMBER OVER GERMANY IN 1944
George Webster

The life expectancy of an American B-17 crew in Europe during World War II was eleven missions, yet crews had to fly twenty-five—and eventually thirty—before they could return home. Against these long odds the bomber crews of the U.S. 8th Air Force, based in England, joined the armada of Allied aircraft that pummeled Germany day after day. Radioman George Webster recounts the terrors they confronted: physical and mental exhaustion, bitter cold at high altitudes, lethal shrapnel from flak, and German fighters darting among bombers like feeding sharks.

Paperback • 6 x 9 • 256 pages • 21 photos

WWW.STACKPOLEBOOKS.COM
1-800-732-3669

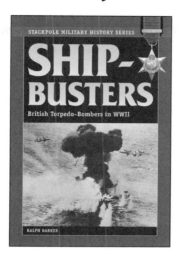

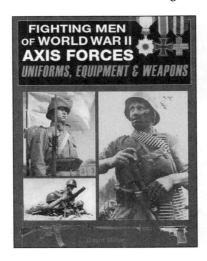